Ocular Drug Delivery Systems

Barriers and Application of Nanoparticulate Systems

Ocular Drug Delivery Systems

Barriers and Application of Nanoparticulate Systems

Edited by
Deepak Thassu and Gerald J. Chader

CRC Press
Taylor & Francis Group
Boca Raton London New York

CRC Press is an imprint of the
Taylor & Francis Group, an **informa** business

First published in paperback 2024

First published 2013
by CRC Press
2385 NW Executive Center Drive, Suite 320, Boca Raton FL 33431

and by CRC Press
4 Park Square, Milton Park, Abingdon, Oxon, OX14 4RN

First issued in hardback 2019

CRC Press is an imprint of Taylor & Francis Group, LLC

© 2013, 2019, 2024 Taylor & Francis Group, LLC

Library of Congress Cataloging-in-Publication Data

Ocular drug delivery systems : barriers and application of nanoparticulate systems / editors, Deepak Thassu and Gerald J. Chader.
   p. ; cm.
  Includes bibliographical references and index.
  ISBN 978-1-4398-4800-5 (hardcover : alk. paper)
  I. Thassu, Deepak. II. Chader, Gerald.
  [DNLM: 1. Eye Diseases--drug therapy. 2. Blood-Retinal Barrier. 3. Drug Delivery Systems--methods. 4. Nanoparticles--therapeutic use. WW 166]

   617.7'061--dc23
   2012015794

ISBN: 978-1-4398-4800-5 (hbk)
ISBN: 978-1-03-291952-2 (pbk)
ISBN: 978-0-429-10549-4 (ebk)

DOI: 10.1201/b12950

Visit the Taylor & Francis Web site at
http://www.taylorandfrancis.com

and the CRC Press Web site at
http://www.crcpress.com

Contents

SECTION I   Basic Considerations

SECTION II   Ocular Barriers

v

SECTION III Ocular Compartment Drug Delivery

SECTION IV Drug Delivery Systems

SECTION V   Technology and Materials Development

Preface

Drug discovery for ocular diseases has made great strides in the last two decades. From cornea to choroid, new drugs have been formulated to address a great variety of ocular diseases. Yet without good drug delivery systems, these drugs are less effective than they could be, possibly even ineffective or able to cause serious side effects. A current example is the use of antineovascular agents in wet age-related macular degeneration, for which the delivery system is simple intravitreal injection, with its potential for infection and complications and a need for frequent reinjection. Even worse, many potentially effective drugs languish on the laboratory shelves of pharmaceutical companies for lack of safe and efficacious delivery vehicles. This problem is compounded in the eye due to the great differences in tissue types that need to be targeted for therapy in different disease situations and the significant uptake barriers posed by ocular structures and biochemical inactivation systems. These barriers include simple but effective scleral/corneal tissue constraints on drug entrance into the eye and the selective permeability of the blood–ocular barriers. Another important aspect of drug delivery is "targeting." To maximize efficacy and safety, drugs need to be directed as best as possible to a specific tissue or cell type once ocular penetration has been achieved. It is usually difficult to have the drug self-target, but a drug delivery system can sometimes be designed to achieve this goal. In all these aspects of ocular therapeutics, the development of nanotechnology applications to drug delivery has opened a new chapter in the treatment of ocular diseases, in particular, the use of nanoparticles.

The term "nanoparticles" encompasses a wide range of solid colloidal particles that can be up to 1000 nm in size but often are <100 nm. A potentially therapeutic drug can be adsorbed, entrapped, or even covalently bound to the matrix of the nanoparticle, which can be constructed to be in particulate, sphere, or capsular form. A nanoparticulate system offers several inherent advantages over other systems, such as naked injection of drug, use of liposomes, and so on. For example, the composition, size, and surface properties of the particle can be controlled to give maximal drug content, penetration, and uptake as well as afford a level of protection against drug toxicity. Because of their size, one of the most interesting and useful properties of nanoparticles is their potential ability to cross the blood–ocular barriers. The nature of nanoparticles also allows for protection of the drug during transport, improving drug lifetime within the eye. Surface modification of the particle (e.g., coating) can also result in a longer half-life of the complex as well as further reducing potential toxicity side effects. The surface-to-mass ratio is high, ensuring, in most cases, an excellent payload delivery of a drug. Similarly, the composition of the particle can be chosen to allow for some specificity in tissue targeting, with drug release that can be either passive or active. There is also a choice in composition with biological (e.g., lipid, dextran, etc.) or nonbiological (e.g., polymers, carbon, and metal) materials and ones of biodegradable or nondegradable nature. To date, most of the materials used have been nondegradable but advances are rapidly being made in the formulation of biodegradable substrata that could further enhance delivery as well as not leave behind a residual footprint. A useful goal here would be the creation of particles with relatively long lives and target specificity, but ones that would ultimately degrade without a permanent effect with regard to both the drug and the nanoparticle delivery system.

There are yet challenges to overcome in maximizing the usefulness of nanoparticles in ocular therapeutics. As pointed out previously, leaving a long-lasting residue in a delicate target tissue such as the retina could result in direct toxicity, inflammatory insult, and so on. The small size and large surface area of the particles could lead to aggregation, making them more difficult to handle prior to use and also more difficult to apply to the eye or, once within the eye, to be transported to a final intraocular target. It is probable though that these problems can be overcome, providing a valuable new tool in the physician's armamentarium for treating eye diseases. By some accounts, the market for such nanotechnology in its full medical application will be over $25 billion by next year, with ocular applications occupying a small but growing fraction of the total.

The present book not only presents the state of the art in the use of nanoparticles in ocular drug delivery systems but also sets the stage for future developments. Covered within are chapters on the basics of nanoparticulate delivery systems and on eye physiology as it relates to drug delivery and action. Also covered are animal models that can be used in evaluating the safety and efficacy of nanoparticulate drug delivery as well as an overview of computer modeling for effective ocular drug delivery. As mentioned previously, there are significant barriers to ocular drug delivery that are not found in most other somatic tissues. These are described in detail, including the specialized blood–retinal barrier. Also covered is the need to target specific tissues and cell types in therapeutic delivery, be it external entities such as the cornea or sclera or internal tissues such as the ciliary body or retina. In this regard, the eye is relatively unique because it has a central "reservoir" in the vitreous body where drugs can be placed for relatively long-term diffusion to anterior as well as posterior segment structures. Somewhat similarly, the sclera and the suprachoroidal space can also be thought of as reservoir systems for ocular drug delivery.

Along with the special biological needs of ocular drug delivery, several chapters of this book address the physical aspects of the delivery systems themselves. A number of currently used and novel ophthalmic drug delivery systems are described, including the use of inserts, nanoliposomes, and stealth-like polymeric nanopolymers in treating eye diseases. Gene-based medicine can also be applied to ocular diseases at the nanolevel. New materials such as photoresponsive polymers and nanosuspensions are powerful new modalities being developed for therapeutic intervention. Finally, targeting of several specific tissues and diseases is covered, along with the design and conduct of preclinical studies and clinical trials. Thus, this book gives both a current evaluation and a future roadmap for developments in ocular drug delivery. The subjects range from biological needs to material challenges and finally to clinical applications for improving drug delivery for conditions where treatments already exist and for unmet needs where effective drugs may be available but yet need a safe, efficient, and efficacious delivery vehicle.

Editors

Deepak Thassu is managing director of Actavis, Owings Mills, Maryland. He is a leading drug delivery expert with more than two decades of pharmaceutical product development experience in research and development with leading pharmaceutical companies in the world. He specializes in the fields of nanotechnology and novel drug delivery systems with emphasis on oral, parenteral and ophthalmic delivery systems for bio/pharmaceuticals. He is an inventor with multiple United States and European Union drug delivery and formulation patents and is the author and/or editor of several books and book chapters. His first book in the area of (pharmaceutical) nanotechnology, *Nanoparticulate Drug Delivery Systems*, was published in 2007 (Taylor & Francis). During his global career he has held several leadership positions. He was chief scientific officer and vice president of pharmaceutical development and lead inventor of NovaSperse technology. He was also chief scientific officer and senior vice president of product development at Holopack International, Columbia, South Carolina. With UCB Inc. and Celltech Americas, he was associate director global product technology development and led several extended release technology and product development projects from concept to clinic. Dr. Thassu holds degrees in pharmaceutical sciences, a PhD in pharmaceutics, and an MBA from Cornell University.

Gerald J. Chader began his research career as an Andelot Fellow in the Department of Biological Chemistry at Harvard Medical School (HMS). Subsequently, he was an assistant professor at HMS in the Howe Laboratory of the Massachusetts Eye & Ear Infirmary and also a tutor of biochemical sciences at Prince House, Harvard College.

He subsequently joined the intramural staff of the National Eye Institute, National Institutes of Health (NIH), where he ultimately served in the Senior Executive Service (Senior Scientific Service) as the chief of the Laboratory of Retinal Cell and Molecular Biology and the director of the Intramural Research Program. At NIH, he helped to set NIH-wide science policy and served on numerous Intramural Research committees, Extramural Study Sections, and the Howard Hughes Scientific Advisory Committee. His research resulted in the discovery and cloning of two proteins critical to the functioning of neural retina. In 1996, he moved to the Foundation Fighting Blindness, a nonprofit foundation, where he served as the chief scientific officer until 2004. Here, he interacted with Congress, meeting with senators and congressmen, and acting as an expert witness before congressional committees as a proponent of increased funding for eye research. He now serves as professor and chief scientific officer of the Doheny Retina Institute, University of Southern California Medical School, Los Angeles, California, where he conducts research on blinding eye diseases and lectures extensively on healthcare priorities.

Dr. Chader's main academic interests are in the areas of medicine and healthcare, specializing in vision research—specifically, in moving basic laboratory research work to clinical trial. His goal is in finding practicable ways to bring preventions, treatments, and cures to patients with untreated or poorly treated eye diseases. At USC, he works with a team that has recently brought an electronic retinal implant (artificial vision) to approval for implantation in patients with inherited retinal degenerative diseases. He also is part of a consortium of investigators (California Project to Cure Blindness) investigating the use of stem cells in treating age-related macular degeneration.

Dr. Chader has over 340 scientific publications and has won several research awards including the ARVO Friedenwald Award and two Alcon Institute Awards. He is a gold fellow of the Association for Research in Vision & Ophthalmology. He holds an honorary doctorate (MD, honoris causa) from the University of Lund, Sweden, and a second honorary doctorate awarded by the University of Pennsylvania. He has served and currently serves on several scientific advisory boards at academic institutions in the United States and Europe and is a past editor of *Investigative Ophthalmology and Visual Science*. He is a founding editor of the review series *Progress in Retinal and Eye Research*, the top ranked journal for vision research.

Contributors

Alexandra Almazan
Allergan, Inc.
Irvine, California

Sheriza Baksh
Office of Pharmaceutical Science
Center for Drug Evaluation and
    Research
US Food and Drug Administration
Silver Spring, Maryland

Sonia Bedi
Parenteral Medication Laboratories
University of Tennessee Health
    Science Center
Memphis, Tennessee

Francine Behar-Cohen
Department of Ophthalmology
Centre de Recherche des Cordeliers
Université Paris Descartes
Paris, France

Damian E. Berezovsky
Emory Eye Center
Emory University
Atlanta, Georgia

Himanshu Bhattacharjee
Department of Pharmaceutical
    Sciences
Parenteral Medication Laboratories
University of Tennessee Health
    Science Center
Memphis, Tennessee

Gerald J. Chader
Doheny Retina Institute
University of Southern California
    School of Medicine
Los Angeles, California

Chi-Chao Chan
Immunopathology Section
Laboratory of Immunology
National Eye Institute
National Institutes of Health
Bethesda, Maryland

Sung Won Cho
Emory Eye Center
Emory University
Atlanta, Georgia

and

Kim's Eye Hospital
Seoul, Korea

Shannon M. Conley
Department of Cell Biology
University of Oklahoma Health
    Sciences Center
Oklahoma City, Oklahoma

José Cunha-Vaz
Department of Ophthalmology
University of Coimbra
and
Association for Innovation and
    Biomedical Research on Light and
    Image
Coimbra, Portugal

Henry F. Edelhauser
Department of Ophthalmology
Emory Eye Center
Emory University
Atlanta, Georgia

Amin Famili
Department of Bioengineering
University of Colorado
Denver, Colorado

Rosa Fernandes
Faculty of Medicine
Institute of Biomedical Research in
    Light and Image
University of Coimbra
Coimbra, Portugal

Bernard F. Godley
Department of Ophthalmology and
    Visual Science
University of Texas Medical Branch
Galveston, Texas

Andreia Gonçalves
Faculty of Medicine
Institute of Biomedical Research in
    Light and Image
University of Coimbra
Coimbra, Portugal

Megumu Higaki
Department of Drug Delivery Institute
Jikei University School of Medicine
Tokyo, Japan

Kris Holt
PharmaNova, Inc.
Victor, New York

Patrick M. Hughes
Formulation and Drug Delivery Sciences
Allergan, Inc.
Irvine, California

Tsutomu Ishihara
Department of Drug Delivery Institute
Jikei University School of Medicine
Tokyo, Japan

Ashwath Jayagopal
Vanderbilt Eye Institute
Vanderbilt University Medical School
Nashville, Tennessee

Uday B. Kompella
Department of Pharmaceutical Sciences
University of Colorado, Denver
Aurora, Colorado

Ed Kraft
Department of Ophthalmology and
    Visual Science
University of Texas Medical Branch
Galveston, Texas

Gabriella Kulp
Department of Ophthalmology and
    Visual Science
University of Texas Medical Branch
Galveston, Texas

James F. Leary
Birck Nanotechnology Center
Purdue University
West Lafayette, Indiana

Susan S. Lee
Allergan, Inc.
Irvine, California

Paul J. Missel
Department of Drug Delivery
Alcon Research, Ltd.
Fort Worth, Texas

Carlo Montemagno
College of Engineering
University of Cincinnati
Cincinnati, Ohio

Muna I. Naash
Department of Cell Biology
University of Oklahoma Health
    Science Center
Oklahoma City, Oklahoma

Timothy W. Olsen
Emory Eye Center
Emory University
Atlanta, Georgia

John S. Penn
Vanderbilt Eye Institute
Vanderbilt University Medical Center
Nashville, Tennessee

Robert Ritch
The Einhorn Clinical Research Center
New York Eye and Ear Infirmary
New York, New York

Kay D. Rittenhouse
Translational Medicine Ophthalmology
Pfizer, Inc.
San Diego, California

Hongwen M. Rivers
Drug Delivery Sciences
Allergan, Inc.
Irvine, California

Michael R. Robinson
Department of Ophthalmology
Allergan, Inc.
Irvine, California

Aron D. Ross
Triton Biomedical, Inc.
Laguna Beach, California

Cheryl L. Rowe-Rendleman
Science-Medical Education
Omar Consulting Group LLC
Princeton, New Jersey

Nakissa Sadrieh
Office of Pharmaceutical Science
Center for Drug Evaluation and
    Research
US Food and Drug Administration
Silver Spring, Maryland

Tsutomu Sakai
Department of Ophthalmology
Jikei University School of Medicine
Tokyo, Japan

Heather Sheardown
Department of Chemical Engineering
McMaster University
Hamilton, Ontario, Canada

Deepak Thassu
Actavis SSL R&D
Owings Mills, Maryland

Joshua R. Trantum
Vanderbilt Eye Institute
Vanderbilt University Medical
    Center
Nashville, Tennessee

Jingsheng Tuo
Immunopathology Section
Laboratory of Immunology
National Eye Institute
National Institutes of Health
Bethesda, Maryland

Vinson M. Wang
Immunopathology Section
Laboratory of Immunology
National Eye Institute
National Institutes of Health
Bethesda, Maryland

Laura A. Wells
McMaster University
Hamilton, Ontario, Canada

Marco A. Zarbin
Institute of Ophthalmology and Visual
    Science
New Jersey Medical School
University of Medicine & Dentistry of
    New Jersey
Newark, New Jersey

Banu S. Zolnik
Office of Pharmaceutical Science
Center for Drug Evaluation and
    Research
US Food and Drug Administration
Silver Spring, Maryland

Section I

Basic Considerations

1 Nanoparticle-Based Therapeutics
An Overview

Deepak Thassu, Kris Holt, and Gerald J. Chader

CONTENTS

1.1 INTRODUCTION

Nanotechnology helps to manipulate particles on both atomic and molecular levels. Researchers began making discoveries using nanotechnology in the late 1950s, although major nanotechnology applications were developed only in the 1980s with the introduction of fullerenes and carbon nanotubes. Only about a decade ago, in the year 2000, the U.S. government funded the National Nanotechnology Initiative, but, since then, nanotechnology has already revolutionized the energy and electronics sectors, and is now set to transform the healthcare sector with the development of treatment and preventive medical care. Importantly, the use of nanotechnology could identify and stop disease/illness even before it starts (Merchant 2010).

Nanoparticles are particles less than 1 μm in diameter. There is a growing trend within the literature to define nanoparticles as being 100 nm or less; a move that supports the development of particles meant to penetrate or pass a barrier but ignores the desirability of larger nanoparticles that can act as a depot for the long-term release of a drug. The term "particle" has similarly been stretched to encompass virtually anything that is in the nanometer range. Originally, this description was intended for discrete solids of essentially pure material such as crystals, semicrystals (materials containing both ordered and amorphous regions), and solid amorphous nanoparticles. The term "nanoparticle" now seems to mean anything that has been manufactured in the nanometer range, regardless of its rigidity, purity, or ability to survive isolated from the matrix or media in which it was formed. The term has been applied to oil droplets in an emulsion, delivery devices, and carriers whose sole shared physical characteristic is that they are less than 1 μm in diameter.

Nanotechnology is one of the growing fields in developing novel therapeutics due to the barriers that nanoparticles can potentially overcome. The eye is an unusual organ in that it is the only organ, besides the skin, that is exposed, allowing for dose administration by systemic, topical, or direct injection routes. Ophthalmic delivery is one of the most interesting and challenging endeavors facing the pharmaceutical scientist. The anatomy, physiology, and biochemistry of the eye render this organ exquisitely impervious to foreign substances. The primitive ophthalmic solution, suspension, and ointment dosage forms are clearly no longer sufficient to combat some present virulent diseases (Hughes and Mitra 1993). In recent years, attempts have been made to design systems that will provide continuous drug delivery to the eye, addressing the side effects and underdosing issues associated with drop installations. Unfortunately, many of the most promising new chemical entities and next-generation therapeutics with greater specificity and/or higher potency are often poorly soluble in water. Ocular administration of these agents in their solution form would be challenging. Therefore, many poorly soluble actives are converted into a salt form for incorporation in aqueous formulations or are developed as dispersed systems. Nanotechnology-based dispersed systems could be an alternate route in developing the novel formulations with accurate dosing, possible reduction of systemic exposure, and better patient compliance (see Table 1.1).

Dissolution, the act of a material dissolving into solution, depends on the intrinsic ability of a compound to dissolve into the surrounding media and the surface area from which dissolution occurs. The math is fairly straightforward: reduction of an amount of particles to one-fourth of their original size gives four times the starting amount of surface area; one-tenth gives ten times the amount of surface area. Most crystalline actives, at the time of manufacture, contain large particles and must be reduced or dissolved before administration. A human

---

TABLE 1.1

Advantages Offered by Nanotechnology-Based Therapeutics

Drug Delivery Issues:
- Improving solubility
- Improving poor bioavailability
- Addressing variability issues observed during fed/fasted conditions
- Pharmacokinetic variability
- Prolonged drug activity

Poorly Water-Soluble Compound Issues:
- Higher dissolution rate
- Smaller effective doses
- Reduced toxicity

Drug Targeting Issues:
- Both active and passive drug targeting can be accomplished
- Stealth nanomaterial can be developed with higher circulation half-life
- Encapsulated nanomaterial can be developed to prolong signaling and delay the degradation

red blood cell is about 5 μm in diameter. A typical micronized active contains particles ranging from 1 to 5 μm. Further reduction into the nanometer range increases the surface area following the general formula as above. One thing to note is that a reduced crystal is still a crystal; the surface area has been increased, but the intrinsic dissolution rate is unaltered. As the particle size decreases in size, a greater proportion of atoms (molecules) are found at the surface compared to those inside. For example, a particle size of 30 nm has 5% of its atoms exposed on the surface, at 10 nm 20%, and at 3 nm 50% of the atoms are exposed on the surface, respectively (The Royal Society 2004, 4). The effective molecular diameter of serum albumin, a common stabilizer used in nanomaterial preparations, is about 7 nm (Boron and Boulpaep 2009); 20–50% of its atoms are exposed on the "surface," but 100% of the molecules are exposed. Single molecules completely exposed (dispersed) to the media in which they reside are dissolved whether they are soluble or not.

A crystalline solid is one in which there is a regularly repeating pattern in the structure, or in other words, there is a long-range order. In fact, describing the single repeat unit completely describes the entire crystal. An amorphous solid is one which does not have a long-range order. In other words, there is no regularly repeating unit. The difference between the intrinsic dissolution rates of crystalline and amorphous materials is highly noticeable due to the additional energy that must be poured into a crystal to disrupt the lattice. Crystals can grow through a process called Ostwald ripening, in which nearby small crystals dissolve and redeposit as part of the crystalline lattice of larger crystals. The driving force is thermodynamics and was alluded to above; a higher percentage of molecules are exposed on the surface of smaller crystals and the relative percentage of surface area among the small particles is much greater than the few large particles. Loss of surface molecules from small crystals means that an even higher percentage of the molecules remaining are exposed at an even greater relative percentage of surface area, so the process drives to completion with the disappearance of small crystals and the formation of a few very large crystals (see Figure 1.1). Only monodispersed (all about the same size—no outliers) crystal suspensions are free from ripening, but the formation of a single seed will upset the balance. Crystal suspensions are generally stabilized against growth, often by the removal of the solvent in which the active is partially soluble. At its most extreme, solvent removal results in lyophilization or freeze-drying to form a dry cake or powder that is resuspended just before administration. Ripening cannot occur in the absence of all solvent, rendering lyophilized products stable against particle size growth for as long as the vacuum seal remains intact. Without solvent removal, crystals must be stabilized through the addition of materials that modify the particle surfaces or the media in which they reside to prevent the dissolution and/or movement of exposed molecules.

Amorphous nanoparticles must also be stabilized against aggregation and growth. Normally, this is accomplished through balancing the zeta potential (see Figure 1.2) (Silver Colloids 2010), born of the electrostatic charges that surround the particles in solution. The ions that carry this charge are arranged in a double layer around the particles. The inner region is called the Stern layer and is where the ions are strongly bound to the particle surface and move without disruption with the particle.

FIGURE 1.1 Ostwald ripening of a nanoparticulate suspension; 2 days on a microscope slide. Large crystals of active are surrounded by a zone completely devoid of particles (starting suspension ~300 nm) in this dark field photomicrograph. Bubble-like structures are smaller crystal seeds that will eventually be consumed by growth of the larger central crystals.

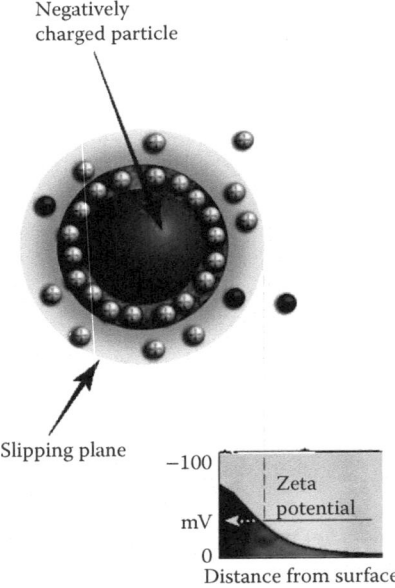

FIGURE 1.2 Separation of charges that define the zeta potential. (Courtesy of Silver-Colloids.com, retrieved September 9, 2011 from http://www.silver-colloids.com/Tutorials/Intro/pcs17.html.)

The outer layer is diffuse in ions and the ions can be transient. Within the diffuse layer and lying very close to the Stern layer is a theoretical boundary called the slipping plane. Within the slipping plane, all of the ions act as a single unit with the particle. The electrostatic potential at this boundary is the zeta potential (Silver Colloids 2011). When the zeta potential is sufficiently high, whether positive or negative, the particles will repel each other, preventing aggregation (see Figure 1.3). Addition of polymeric materials to the solution increases the viscosity and hinders free movement of the particles toward aggregation. Finally, the particle surface can be modified to prevent growth and aggregation. Surface modification also allows one to engineer a variety of interesting properties including mucoadhesiveness, targeting, and stealth to recognition as foreign invaders. Surface modification of nanoparticles will be discussed in much greater detail in Chapter 15. Amorphous nanoparticles can also be stabilized by removal of the solvent for the active through filtration and washing or by lyophilization (see Figure 1.4). The latter may be impossible if the organic solvent employed cannot be removed under vacuum. Dimethyl sulfoxide (DMSO) demonstrates this limitation. Its vapor pressure under vacuum is far below that of water, so lyophilization removes the water but cannot remove the solvent and the system reverts to a drug solution contaminated with surfactants when brought back to room temperature.

Routine generation of particles in the nanometer range requires specialized techniques to accurately differentiate size, shape, and surface morphology. Optical microscopes are of limited use in determining size distributions when the wavelengths of visible light approach the diameter of the particles under observation. This technique is useful in feasibility research, especially when nanoparticles are being

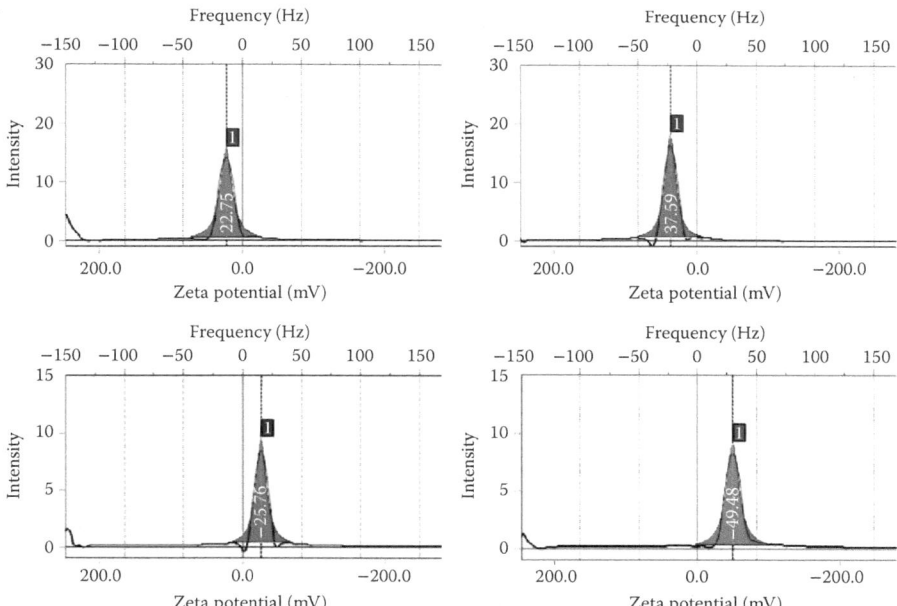

FIGURE 1.3 Four representative zeta potentials of solid amorphous nanoparticulate aqueous suspensions. The samples were in early stages of formulation development and were measured immediately after completing the nanoparticle development process. Examples on the left were below the ±30 mV stability limit and aggregated within hours. Examples on the right were between the desired ±(30–50 mV) range and exhibited limited growth over the next few days. Samples were ultimately stabilized by choosing appropriate polymers and surfactants.

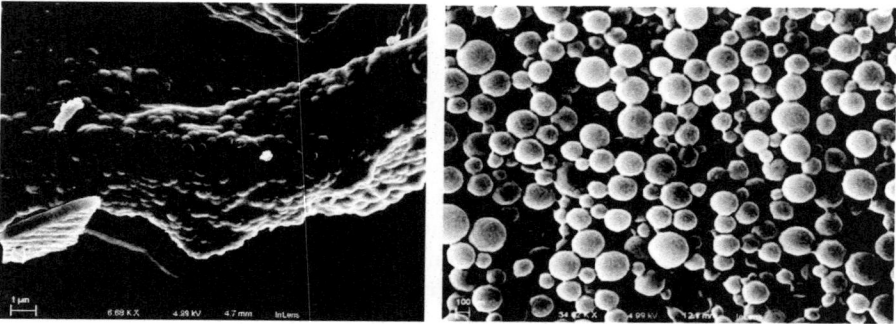

FIGURE 1.4 SEM micrographs of a lyophilized nanosuspension before and after reconstitution.

assembled from individual molecules in solution, such as the solvent exchange reaction. When particles cannot be observed under the microscope, only a general sense of movement, the particle distribution is down to the 250–350 nm range. Different techniques must then be employed to gain meaningful information. One method is to shorten the wavelength of radiation used (e.g., electron beams and x-rays) so that finer detail (the ultimate limit of the best method is around 0.5 nm) is revealed (Herrera and Sakulchaicharoen 2009). Another method is to correlate size to some other attribute. An example of this is photon correlation spectroscopy, in which Doppler shifting of laser light is detected when it is backscattered by a nanoparticulate suspension. The shifting is caused by Brownian movement of the particles while in the beam and the speed of the movement has been correlated to the size of the particles.

1.2 TYPES OF NANOMATERIALS

There are several general types of nanoparticles (Figure 1.5). All tend to be roughly spherical due to the forces at work during assembly, except for milled crystals and nanocrystals, which are formed by critical fluid processes. Each nanoparticle type has its unique chemistry and attributes, and no single type is compatible with the myriad of active pharmaceutical ingredients (APIs) currently under development as new chemical entities or under investigation for repurposing. Nanocrystals and solid amorphous nanoparticles are essentially pure API, while dendrimers, micelles, and liposomes are more appropriately termed nanocarriers or nano delivery devices.

Nanocrystals are probably the simplest form and one of the more difficult to manufacture. Enormous amounts of energy are required to reduce large crystals to nanoparticles by cleaving and recleaving the crystalline lattice. Working against the milling process is the tendency for crystals to "heal" themselves, forming aggregates when two flat crystal faces are within close enough proximity for hydrogen bonding or the appearance of attractive forces. These aggregates are not held together with the same strength as chemical bonds, but the amount of surface area in a nanoparticulate crystalline powder is very large, and therefore so is the sum total of the forces holding together the aggregates. Nanocrystals can also be manufactured by critical fluid processes in which the active is dissolved in a liquefied gas at very high

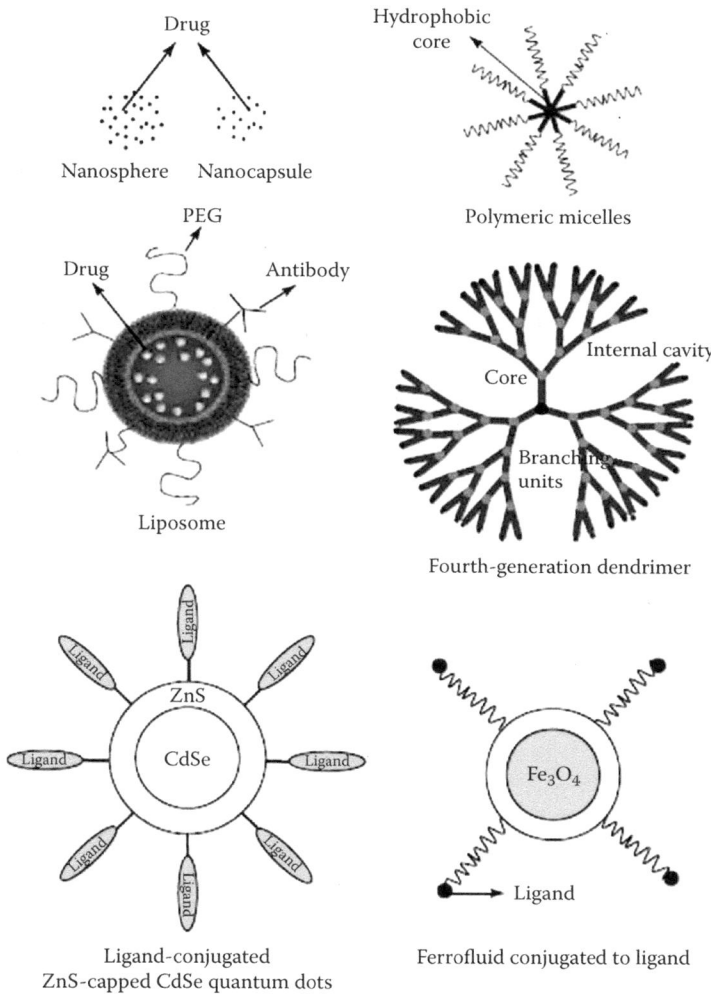

FIGURE 1.5 Several types of nanoparticles and nano delivery devices.

pressure followed by either controlled or explosive decompression. Because of their tendency to aggregate, nanocrystals are rarely presented in pure form. Instead, fillers are added to an extent that they are interposed between the nanocrystals to prevent flat crystal faces from approaching each other. In many cases, these fillers must be added during the milling process to prevent cleaved crystal faces from immediately healing through aggregation. The amount of fillers required to prevent re-aggregation of nanocrystals can easily constitute 90–99% of the final drug product formulation.

Dendrimers are three-dimensional, roughly spherical puffs of highly branched polymeric material attached to a central core. Their surface chemistry is determined by the terminal groups and can be engineered to be hydrophilic, hydrophobic, chiral, and so on. Technically, dendrimers are drug delivery devices in that the active to be administered is held in voids within the puff. In practice, the dendrimers would be manufactured, purified, and then loaded with active as an additional step. Dendrimers must be manufactured from biodegradable polymers if they are intended for parenteral administration. Active loading is low, on the order of 10% or less.

Additional nano delivery devices include ferrofluid-bound and quantum dot-bound ligands. Similar to polymeric micelles, these devices serve as relatively inert porous carriers holding the active around their hydrophobic cores. There are many variations on these themes, but all fall within a category of nanodevices.

Micelles are formed when a surfactant is added to an oil-in-water mixture and stirred. Surfactants have hydrophilic and hydrophobic ends. The hydrophobic ends are immersed within an oil droplet, aligning the hydrophilic ends to the surface of the droplet to stabilize the micelle in the surrounding aqueous media as an emulsion. Micelles are not rigid and can be divided into smaller units with shear energy as long as there is sufficient surfactant to stabilize the increased surface area. Additional materials are added to prevent coalescence of the micelles. The active to be administered is dissolved within the oil portion prior to micelle formation. If the active itself is an oily liquid, fairly high percentages of loading are possible within the micelles; otherwise, it is limited by the concentration achievable when dissolved in the oil and subsequently coated with surfactant.

If the oily core of the micelles is a solid at room temperature, the result is a solid lipid nanoparticle. Solid lipid nanoparticles are manufactured in a hot process. Active is dissolved in melted lipid, usually cholesterol, followed by addition of a hot surfactant solution with high-shear mixing in the form of homogenization or sonication to form micelles. When cooled, the micelles are solid nanoparticles that can be filtered and dried. As with micelles, theoretical loading is about the same but limited by the solubility of the active in hot cholesterol.

Liposomes are formed when phospholipids or cholesterol are added to an aqueous media, self-assembling into spherical vesicles. Similar to cellular membranes, liposome membranes are a bilayer of lipids oriented with their hydrophobic tails comingled together and the hydrophobic heads projecting into the media and into the interior space. Typically, the active is dissolved within the aqueous media so that the liposome encapsulates the drug solution during formation. Although not technically rigid, liposomes behave as if they were solid particles. Generally, they can be filtered or concentrated without coalescence. Theoretical loading is poor because of the limited solubility of the active in aqueous media.

Solid amorphous nanoparticles are manufactured by solvent exchange. The active is dissolved in an organic solvent in which it is fairly soluble and which is miscible with an antisolvent, usually water. The antisolvent is a mixture of surfactants and polymers in solution. When the two are mixed together, the solvent begins to disperse into the antisolvent and the surfactants mediate the controlled precipitation of active. Since the active was originally in solution, the nanoparticles are built up from individual molecules. The strength and composition of the antisolvent determines both the resulting particle size and the selection between amorphous and crystalline growth. Aggregation and flocculation of the resulting nanoparticles are problems encountered during manufacture. Generally, this tendency is countered with the secondary addition of polymers and surfactants to the solution. These additions simultaneously alter the surface properties of the nanoparticles, thus engineering in desired properties that affect their deposition and selectivity. Like nanocrystals, purity approaches 100%, but potency is dependent on the amount of materials in solution required to alter the surface properties and prevent growth (Figures 1.6 and 1.7) (Thassu et al. 2010).

All platforms mentioned in Table 1.2 are the subjects of active research projects and account for billions of dollars in funding. In general, nanomaterials are sought as magic bullets, not panaceas, for very specific disease states or targets of action. As of this writing, nearly all of the types of nanomaterials detailed earlier have at least one entry in the commercial market. An example of a dendrimer system to be added to the list is VivaGel (Starpharma), currently undergoing clinical trials with

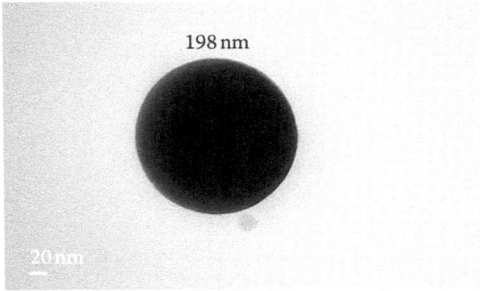

FIGURE 1.6 Transmission electron micrograph of a spherical amorphous nanoparticle. Note the lack of any apparent structure within the particle, demonstrating the amorphous nature. The small dark spot inside at the 2:00 position is a shadow cast by a smaller particle behind the larger and not in the field of focus, similar to the shadow visible outside at the 5:00 position.

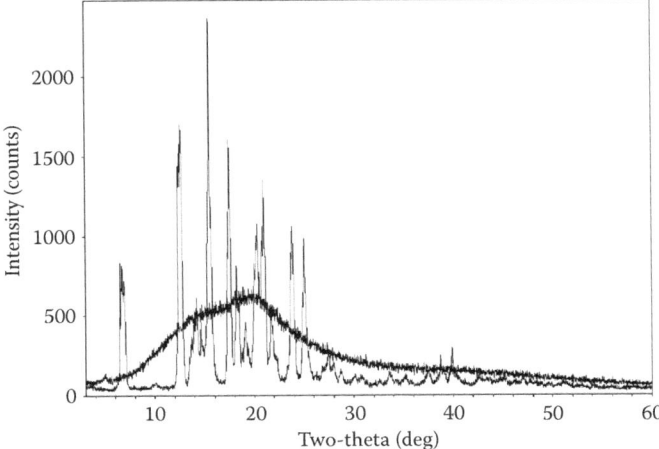

FIGURE 1.7 X-ray diffraction analysis of a pharmaceutical active in powdered form for incorporation into a solid oral dosage and in the form of amorphous solid nanoparticles. The raw active is crystalline, demonstrating internal refraction and reflectance of the beam into narrow channels. The amorphous material is unable to regularize the transmission angle of the beam.

TABLE 1.2

Variety of Commercial Drug Products, All but One of Which Contain Nano-Sized API

Platforms	Drug Name	Indication	Particle Size Range
Liposomal	Doxil	Antineoplastic	75–150 nm
	Abelcet	Antifungal	1.2–8.3 μm
	Daunoxome	Antineoplastic	NA
	Amphotec	Antifungal	85–145 nm
	Ambisome	Antifungal	NA
	Depocyt	Lymphomatous meningitis	7.5–20 nm
	Visudyne	Muscular degeneration	18–104 nm
Nanocrystal	Rapamune	Immunosuppressant	<440 nm
	Emend	Antiemetic	<250 nm
	Tricor	Hypercholesterolemia and hypertriglyceridemia	~200 nm
	Triglide	Hypercholesterolemia and hypertriglyceridemia	NA
	Megace ES	Anorexia, cachexia, or weight loss in AIDS patients	140–150 nm
Other	Feridex	MRI contrast	NA
	Estrasorb	Vasomotor symptoms of menopause	~1 μm
	Abraxane	Metastatic breast cancer	120–150 nm

a possible release date in 2012. Within 5 years of this publication, the number of commercialized nanoscale pharmaceuticals will probably top 100, with hundreds more in corporate pipelines.

1.3 MANUFACTURING PROCESSES FOR NANOMATERIALS

Crystal size reduction is an example of a top-down process of material alteration to enhance solubility through increased surface area. Bottom-up approaches assemble nanoparticles from individual active molecules through absorption, aggregation, precipitation, and/or deposition. The processes employed in bottom-up approaches usually result in noncrystalline forms, thus affecting the intrinsic dissolution rate of the active as well as increasing the surface area. Both top-down and bottom-up approaches are valid and have utility for various actives and needs. There are many nanoparticle and nanodevice production processes that have been patented or are yet under development (Thassu et al. 2007). Both nanoparticle manufacturing processes are high in energy usage, but it is widely established that top-down processes generate more waste, and there are limits to the surface modifications that can be made on the resulting nanoparticles that are required to engineer functionality.

Top-down manufacturing relies on size reduction in order to manufacture nanoparticles and devices. This can be accomplished through crushing, milling, etching, dividing, or other forms of mechanical processing that are usually energy consumptive and/or wasteful of the starting resource. Current and emerging levels in understanding material behavior and properties have decreased the levels of waste and increased throughput, but the energy requirements still remain high. In addition, the majority of the chemical and behavioral properties of the final form are direct descendants of the starting material. Top-down production of nanoparticulate medicines encompasses particle size reduction through milling or grinding, and absorptive or sequestering technologies such as the various forms of micelles and liposomes. In each case, the starting active is divided and redivided to reduce the particle size or placed in solution and encapsulated within the smallest possible vesicle that can be formed for that system, energy input, and processing time.

The bottom-up route to material fabrication requires the building of nanostructures atom by atom or molecule by molecule. The processes employed can be subtle, such as the self-assembly of crystals and liposomes or the formation of loose aggregates and floccs by interactive forces, or can use brute force, such as the use of optical tweezers or electrostatic deposition to physically place individual atoms or molecules at a desired point in space or on a surface. The important consideration for bottom-up fabrication is that properties can be engineered into the particles from the addition of the first unit. Coprocessing, surface modifications, and selection between organized (crystalline) or amorphous are attributes that are planned for, not the result of the manufacturing process by which the nanomaterials arise. It is the possibility of control over growth, stealth, specificity, and stability that has led to enormous expenditures of time and resources in the research and development of bottom-up approaches.

Focusing specifically on the eye, development of functionalized nanoparticles with unique physical, chemical, and biological properties in the range of 1–100 nm enhances the efficacy of ophthalmic delivery systems. Drug delivery from nanoparticles with a variety of surface modifications and size ranges will elucidate the protective mechanisms of the eye against drug absorption. Armed with better understanding of the blood-ocular barrier, topical delivery of macromolecular drugs from functionalized nanoparticles at therapeutically effective levels to the posterior portion of the eye could soon be a possibility.

The rationale for interest in nanoparticulate medicines comes from many directions. Many of the most promising new chemical entities and next-generation therapeutics with greater specificity and/or higher potency are, at best, poorly soluble in water. Since the circulatory system is essentially an aqueous environment, the introduction of insoluble materials in a form not leading to unwanted or adverse events is a challenge. The barriers to certain specific organs such as the brain and eye are selective and restrictive to windows of the size, charge, and lipophilicity of particles and molecules within the bloodstream. Cancerous tissues, especially solid tumors, have their own unique chemistries and typically tend to be warmer and slightly more acidic than the surrounding normal tissues. The explosive growth of cancerous tissue generates an interesting physiologic anomaly in that the epithelial lining of the tumor vasculature has 400 nm gaps between the cells. This feature provides a delivery route for chemotherapeutic agents that are specific to the tissues of the tumor. Much work has been devoted to the manufacture of <200 nm nanoparticles of poorly soluble yet highly cytotoxic compounds that would theoretically circulate systemically until "leaked" into a tumor in concentrations far exceeding any other tissue. Under this theoretical dosing regimen, chemotherapy would be administered based on the size of the tumor to be removed rather than body weight and side effects would be minimal or nonexistent. Surface modification of the nanoparticles engenders stealth capability to prevent opsonization by white blood cells and targets the unique chemistry of the timorous tissue for deposition.

1.4 NONPHARMACEUTICAL NANOMATERIALS IN MEDICAL PRACTICE

The development of nanoparticulate and nanosized materials for biomedical use outside the pharmacy has experienced enormous growth in research expenditures of time and resources, possibly building on the successes of imaging materials. The use of nanoparticulate imaging agents led to high-resolution, low-intensity imaging techniques suitable for real-time monitoring of blood flow, heart function, kidney function, and liver perfusion. Simultaneously, the implant industry was discovering new nano-based techniques for design improvements that yielded devices with improved durability, compatibility, and function. Stents are used to open occluded vasculature but are prone to wild tissue growth. Incorporation of tissue growth inhibitors in or as nanostructured coatings (nanoporous or nanoparticulate, respectively) (see IsoFlux, Inc.) prevents establishment of growth that would eventually reverse the function of the stent. Nanostructured coatings are particularly useful on devices to be implanted, such as artificial joints and replacement teeth (Inframat Corporation). The extremely porous structure of the coating promotes the growth of new bone, imparting a bond that is as strong as, if not stronger than, the material that it replaces. A number of nanoparticulate filling materials have been approved for teeth (Cosmedent, Inc., 3M ESPE, Pentron Labs, Ivoclar Vivadent, Inc) and bone (Angstrom Medica).

1.5 NANODEVICES

This is a growing field that is rapidly moving from the realm of science fiction to reality. The concept is relatively simple: a nanodevice is a detector or instrument that can perform a clearly defined task either in situ or with an extremely small sample. Functions of these devices include detecting individual molecules of biomarkers for a disease state, creation and release of pharmaceutics within the bloodstream from available building materials, detection and removal of undesirable materials such as plaque form arterial walls or cancerous wild cells at the time of their mutation, and the performance of microscopic surgeries such as reversal of aneurisms or the rebuilding of knee cartilage. The device that is closest to reality is the nanodetector. Organizations such as the Lux Bio Group, which grew out of biochemists from the National Institutes of Health (NIH) and a core of chip designers from Intel, are seeking to merge biodetector molecules onto silicon chips capable of detecting single molecules of a biomarker of interest. Some of these dreams have been made real. Offerings currently marketed include the ability to detect circulating tumor cells from breast and colon cancer (Veridex, LLC) and detection of a point mutation in the human Factor V gene (Nanosphere, Inc).

While the technology will likely be disruptive to some areas of the healthcare sector in the future, we expect near-term applications of nanotechnology and nanoparticles primarily to augment the properties of existing products such as drugs, diagnostics, and biomedical implants, rather than creating new markets. Furthermore, we believe that nanotechnology should not be viewed as a single technology but instead as one of several converging sciences that include molecular biology, physics, chemistry, and information technology, which can be greater than the sum of their parts. One of the key challenges for the healthcare industry will be to form collaborative networking groups between these disciplines, which have not historically cooperated on joint research (Credit Suisse Securities (Europe) Limited 2008).

REFERENCES

Boron, W. F., & Boulpaep, E. L. (2009). *Medical Physiology: A Cellular and Molecular Approach*. Philadelphia, PA: Elsevier/Saunders.

Credit Suisse Securities (Europe) Limited. (2008). *The Size of Things to Come: The Rise of Nanotechnology in Healthcare*. London: Credit Suisse Securities (Europe) Limited.

Herrera, J. E., & Sakulchaicharoen, N. (2009). Microscopic and Spectorscopic Characterization of Nanoparticles. In Y. Pathak, & D. Thassu, *Drug Delivery Nanoparticles Formulation and Characterization* (pp. 239–251). New York: Informa Healthcare.

Hughes, P. M., & Mitra, A. K. (1993). *Overview of Cular Drug Delivery and Iatogenic Cytopathologies*. New York: Marcel Dekker.

Merchant, M. (2010). *Nanotechnology in Healthcare*. London: Business Insights, Ltd.

Royal Society. (2004). *Nanoscience and Nanotechnologies: Opportunities and Uncertainties*, p. 4. London: The Royal Society.

Silver Colloids. 2010. The Electric Double Layer. Retrieved September 9, 2011, from http://www.silver-colloids.com/Tutorials/Intro/pcs17.html

Thassu, D., Deleers, M., & Pathak, Y. (2007). *Nanoparticulate Drug Delivery Systems*. New York: Informa Healthcare.

Thassu, D., Holt, K., & Hermans, D. (2010). *Surface Modified Nanoparticles Enhance Delivery of Poorly Soluble Drugs*. Poster at CRS symposium on Poorly Soluble Drugs.

2 Eye Anatomy, Physiology, and Ocular Barriers

Basic Considerations for Drug Delivery

Gerald J. Chader and Deepak Thassu

CONTENTS

2.1 GENERAL EYE ANATOMY

The eye is a small organ whose simple and almost singular purpose is to deliver light signals to the photoreceptor cells of the neural retina and, from there, transmit them to the brain. Though, the eye has evolved and developed many complex structures and biological pathways to ensure that this goal is achieved. Figure 2.1 shows the "globe" character of the eye with the primary tissues and structures identified. The anterior segment consists primarily of the cornea iris, ciliary body (not identified in the figure), and the lens. The main tissues of the posterior segment are the choroid, retinal pigment epithelium (RPE), and neural retina. Specialized structures within the retina include the macula and its central fovea, the small retinal area responsible for central, color, and sharp vision in humans.

The structural and metabolic factors in the anterior segment of the eye that maintain focus of the light image at a particular distance and direct the image to a specific geographical area of the retina (the macula) are a key to good vision. In the posterior segment, the capability of accepting a photic image and converting it into an electrical signal is the basic element of what we call "vision." This includes neuronal processing in both retina and brain, a discussion of the latter being outside the scope of the present chapter. Also included are numerous support functions performed by other cell types and tissues that ensure proper functioning of the neural retina. Nutrition, for example, is suppled both through the retinal vessels for nourishment of the inner retinal neurons and through the choroid for nourishment of the highly metabolically active photoreceptor cells of the outer retina. As a mediator of outer retinal function, RPE cells are on the front line of metabolic transfer of nutrients into the photoreceptors and of waste product removal from the visual cells.

Along with these physiological functions, eye tissues have other qualities that help to protect and sustain the critical internal elements of the visual system: the light-focusing structures of the anterior segment of the eye and the integrity of the neural retina and its signaling process. This starts with the tough, resilient outer surface of the eye, the cornea and sclera, which helps to protect the internal, more delicate

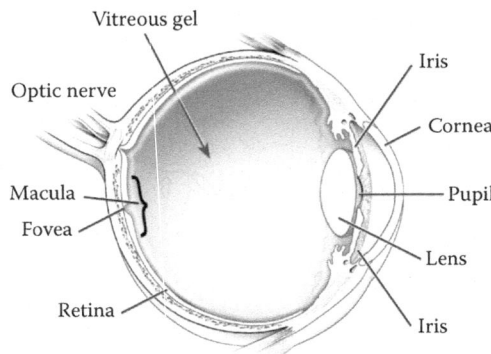

FIGURE 2.1   Major ocular components. (Courtesy of the National Eye Institue.)

tissues from external insult. Similarly, there are mechanisms built into the eye that protect against potential toxins through barrier function: the blood–aqueous barrier and the blood–retina barrier. There are also safeguards against immune attack, essentially making the inside of the eye an immune-privileged area as well as a generally toxin-free site. All these requisites, however, pose both unique problems and interesting opportunities for ocular drug delivery.

The subsequent sections of this chapter give an overview of these phenomena. This includes aspects of both eye anatomy and eye physiology that form the basis of the normal visual process. It is these anatomical and physiological processes that form the ocular barriers that help to protect the visual process from insult but that make drug delivery to the posterior segment—and even to some elements of the anterior segment—more challenging. As such, this chapter acts as an introduction to the specific sections of this book that follow and focuses on current, novel techniques for delivering drugs to target tissues within the eye.

2.2 ANATOMY AND COMPOSITION: EXTERNAL FIBROUS TUNIC

The eye is a functioning organ consisting of diverse tissue and cell types. The basic structure, as shown in Figure 2.1, is essentially a globe consisting of three concentric layers. The outermost layer is called the fibrous tunic and is divided into the cornea and the sclera. Inside this external covering is the vascular tunic or uvea, consisting of the iris and the adjacent ciliary body in the anterior segment and the choroid in the posterior segment of the eye. These structures are responsible for ocular nutrition through the systemic blood supply. Finally, the innermost layer is the nervous tunic, mainly consisting of the neural retina. Within these concentric layers are the fluid or gel-containing reservoirs: the anterior chamber and the smaller posterior chamber, containing the aqueous humor and the larger vitreous cavity filled with vitreous gel.

2.2.1 CORNEA

The external fibrous tunic is tough and resilient, the cornea being the anterior, transparent "window of the eye" accounting for about 15% of the outside ocular surface. It averages only 0.56 mm in thickness (Doughty and Zaman 2000) but provides up to 80% of the refractive power of the eye. The corneal surface is bathed in a thin tear film that helps to protect the eye from irritants and noxious substances and also acts as an effective drug barrier, quickly diluting and sweeping away drops applied to the eye. However, the cornea is the principal route into the eye for topically applied drugs even though it serves as a major barrier to drug absorption, affording both a physical barrier and a metabolic barrier to substances that might enter the eye. The cellular structure of the cornea along with the tear film pose both hydrophilic and lipophilic barriers to substance penetration. Penetration is difficult due to zonula occludens, or tight junctions between the epithelial cells. The hydrophobic nature of the corneal epithelium is probably the greatest impediment to intraocular drug bioavailability. Interestingly, the cornea is rich in nerve fibers but is avascular.

Thus, in spite of the substantial uptake barriers that keep out unwanted substances, the lack of vascularization necessitates the selective uptake of nutrients (including oxygen) through the tear film or the aqueous humor. On the positive side for getting drugs into the eye, the cornea expresses specific transport systems that can facilitate drug entrance into (or out of) the tissue. For example, the multidrug resistance proteins (MRPs) along with the proton-dependent oligopeptide transporter (POT) can be harnessed to better move specifically designed drugs into the eye.

Physically, the cornea is striate in nature, consisting of five discrete layers, two of which are laminar cell layers (Figure 2.2). This essentially makes the cornea a lipid–water–lipid sandwich. On the surface are epithelial cells attached to a layer of basement membrane called Bowman's membrane, a laminate structure mainly composed of collagen fibers. This pairing is one of the first lines of defense of the eye, forming a barrier that is difficult to penetrate. Underneath the epithelium is a thick layer of matrix material called the stroma. It is mainly acellular with only a few cells, such as keratocytes, and is about 500 microns thick, accounting for about 90% of the corneal thickness. It is composed of mostly water (~80%), regularly arrayed collagen fibers (~15%), and mucopolysaccharides (~5%). Finally, on the interior surface is a layer of endothelial cells attached to Descemet's membrane. Descemet's membrane is a latticework of collagen fibers coming together in a nodal pattern. It acts as a barrier to the entrance of potentially toxic organisms into the eye but allows for passage of water and nutrients. The endothelium is composed of simple, mostly hexagonally shaped cells, only a single cell layer in thickness. However, it performs the important task of fluid regulation for the stroma, maintaining a state of deturgescence, compacting the collagen fibers and maintaining tissue transparency. Physiologically, it has no true elastic fibers and thickens with age (birth: 3 microns; adult: 12 microns; aged: up to 30 microns). There is a barrier in the endothelium to penetration by aqueous humor, that is, a tight seal, composed of macula occludens junctions. Endothelial cells are rich in mitochondria to produce the energy needed to pump sodium and water out of the cornea to maintain proper thickness and thus transparency.

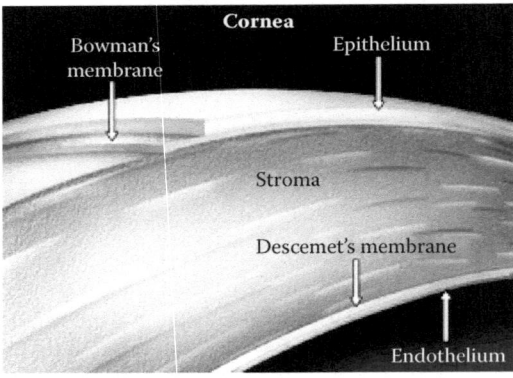

FIGURE 2.2   The five layers of the cornea. (Courtesy of the National Eye Institue.)

The cornea constitutes the classical site of drug entry into the eye, particularly for compounds administered as eye drops. Advantages of topical delivery include easy access and the noninvasive nature of the delivery. Drugs enter through the cornea by both paracellular and transcellular routes. Hydrophilic compounds mainly enter using paracellular routes at the junctional complexes, while lipophilic compounds enter mainly through epithelial cell membranes. Penetration of hydrophilic compounds is low, however, due to small pore size and the fact that many compounds have only limited passage through the corneal stroma. These features account for many of the problems encountered in topical delivery. Along with the need for multiple applications and problems with compliance, it is estimated that only 5% of most drugs actually enter the eye when administered in this manner and most of these are only available to and active in the anterior segment tissues. Compared to the cornea, it is generally considered that sclera and conjunctival absorption of topically applied drugs is greater than corneal penetration. Even here, though, there is very little drug actually made available for action in the posterior segment, where most of the major eye diseases occur. Improvements are being made in delivery formulations that may improve sustained drug delivery to the eye, such as with microadhesive microparticles supplied in a rapidly dissolving tablet form (Choy et al. 2011).

2.2.2 Conjunctiva, Sclera, Tenon's Capsule, and Tear Film

2.2.2.1 Conjunctiva

An area that has critical barrier and uptake functions in the eye but is of increasing importance to drug delivery is the conjunctiva. This is a thin, transparent mucous membrane containing stratified columnar epithelial cells and an underlying basement membrane. It lines the backside of the eyelid (palpebral conjunctiva), looping back to cover the sclera to the edge of the cornea (ocular/bulbar conjunctiva) with the fornix as the transition zone at the junction of the posterior eyelid and the eyeball proper. It has a protective function in keeping foreign substances such as microorganisms out of the eye and contributes to immune surveillance. Ocular lubrication is an important function of the conjunctiva, contributing greatly to the maintenance of the tear film. In the bulbar conjunctiva, goblet cells are present, which secrete mucin that lubricates, protects, and nourishes the cornea.

2.2.2.2 Sclera

The transition zone between the transparent cornea and the opaque sclera is the limbus, an area only 1.5–2.0 mm wide. It is composed of an epithelial cell layer and a stroma but not an endothelial cell layer. The sclera itself is a tough, fibrous layer comprising the external surface of the eye other than the cornea, totaling about 80% of the total surface area. It is a smooth connective tissue that is mainly composed of collagen and elastic fibers. The sclera is avascular, with nutrients coming from anterior vessels in Tenon's capsule/episclera and, in the posterior, the choroid. The main functions of the sclera are to maintain the globe shape of the eye, to resist external force and injury, and to act as a barrier to entrance into the eye for various substances. In line with the globe-like and layered structure mentioned earlier, the sclera itself consists of concentric layers that, externally, even could be considered to include the capsule of Tenon. The sclera proper, though, consists of the episclera, the outermost connective tissue layer, a middle layer of stroma, and the innermost lamina fusca, which is contiguous with the choroid at the back of the eye. Anteriorly, it is continuous with the cornea as mentioned earlier; in the posterior, it connects to the dura mater of the brain through the fibrous sheath of the optic nerve. It exhibits attachment sites for extraocular muscle insertions and is perforated by blood vessels and nerve fibers that penetrate through the posterior sclera foramen at the optic nerve head. Although generally uniform and featureless in nature, there are specific scleral areas with specialized structures and functions. One of these is the trabecular meshwork in the perilimbal area, which allows for aqueous humor drainage into Schlemm's canal and is a major target site of drug action in glaucoma. Another specialized structure is the lamina cribrosa at the optic nerve head, which allows for penetration of the optic nerve through the scleral wall and is also a target site for possible neurotrophic drug action.

The opaque nature of the sclera is mainly due to the irregularity in the placement of the collagen fibers compared to the more regular distances between the fibers and the uniform thickness in the cornea. There is also a higher content of mucopolysaccharide in the cornea that helps to better embed the collagen fibers. Perhaps contrary to what one would think, there are substantial regional differences in scleral chemical composition. Trier et al. (1990), for example, found regional differences in uronic acid, hyaluronic acid, chondroitin sulfate, and dermatan sulfate. Similarly, the thickness of the sclera is not uniform but rather varies widely in different areas. It is about 1 mm in thickness at the posterior pole but only about 0.3 mm around the rectus muscle insertions. Olsen et al. (1998) measured human sclera to be 0.53 mm at the corneoscleral limbus, decreasing to 0.30 at the equator, and then increasing again to approximately 1.0 mm near the optic nerve. This variability allows for some opportunities for drug penetration through drop installation, external drug delivery devices, and so on. Also contrary to what one might expect, the sclera is not fully impermeable to penetration of smaller molecular weight substances. Penetration is dependent on several factors including drug size and degree of polarity. For example, Cruysberg et al. (2005) found that human sclera was permeable to compounds up to about 150 kDa in molecular weight with a moderate influence of transscleral pressure on the process. This allows for some penetration of compounds such as serum albumin, which has a molecular size of 67 kDa, and even of antibodies, a typical IgG molecule, for example, being approximately 150 kDa.

Transverse and lateral diffusion from the site of drug application in the sclera is slow, averaging only 5 and 10 mm from the reservoir site at 4 hours and 3 days, respectively, for a drug such as sulforhodamine (Jiang et al. 2006). The sclera is more permeable to lower molecular weight, water-soluble substances such as antibiotics, supporting the notion that "a local, non-invasive, transscleral drug-delivery method may be reasonable for treating intraocular infections" (Kao et al. 2005). Hydrophilic molecules may be aided by the relatively hydrophilic nature of the resident proteoglycans, while positively charged molecules may be impeded because of interaction with the relatively negatively charged proteoglycans. Importantly, the sclera is permeable to small, single-stranded oligonucleotides. Shuler et al. (2004) were able to deliver a 24-base fluorescein-labeled oligonucleotide across the sclera in *in vitro* experiments, theoretically opening the way for similar gene-based therapies *in vivo*. Thus, a number of substances can penetrate through the scleral wall, but may be subject to clearance through the conjunctival lymphatic and blood vessels as was suggested for steroids by Robinson et al. (2006). A general caution, however, to transscleral drug delivery (and any form of drug delivery for that matter) is that there is a marked effect of eye pigmentation on delivery, particularly to the posterior pole. Lipophilic drugs such as celecoxib and, classically, chloroquine, are readily taken up and bound in pigment granules of the RPE–choroid complex in pigmented rats (Cheruvu et al. 2008). This not only slows uptake into the retina but acts as a drug reservoir, releasing the drug for long periods of time.

Many routes of drug delivery to the interior of the eye are now available, with an increasing number of these utilizing routes of penetration through the sclera. These routes are not simply through topical application and absorption but through sophisticated depots and injections such as in the use of hollow microneedles through which drugs can be infused in a minimally invasive yet efficient manner (Jiang et al. 2009). Several chapters within this book give more extensive coverage of this subject.

2.2.2.3 Tenon's Capsule

Tenon's capsule is another important, discrete structure of the eye. It is a feature of the ocular surface and could be an asset in drug delivery. Anatomically, the capsule is a thin connective tissue membrane that surrounds the eyeball from the optic nerve to the corneal limbus. The capsule separates the eyeball from the surrounding orbital fat and forms a pocket in which the eye can smoothly rotate. A periscleral lymph space separates the capsule from the surface of the sclera. Sub-Tenon's injection of drugs such as corticosteroids in depot form is a clinically tested treatment modality for a number of ocular conditions including inflammation and macular edema (Bonini et al. 2004). However, drug reflux has been reported as a potential problem in some situations involving steroid infusion for diabetic macular edema, resulting in not only poor edema reduction but also a postoperative increase in intraocular pressure (IOP) (Shimura et al. 2009). To counter this problem, methods to enhance delivery have been established, such as specialized cannulae that allow for a safer installation with a lessened chance of reflux or globe perforation.

2.2.2.4 Tear Film and Lacrimal Gland and Meibomian Gland

The tear film coats the anterior surface of the eye and provides both irrigation and flushing to the ocular surface even though it has a thickness of only 3 μm and a volume of only 6–8 μL. There are at least four major functions of tears. First, they lubricate the surface of the eye, allowing for a smooth blinking mechanism. The smooth surface also allows for good and regular refraction of light as it passes into the eye. Tears maintain normal nutrition and oxygenation of the ocular surface cells and structures. Finally, there is a protective or barrier function in sweeping away foreign substances from the ocular surface, be they larger substances such as dirt particles or smaller materials such as drugs. In this regard, tears also exert an antimicrobial protective effect. Tears consist of a complex mixture of substances, some having a structural function and some playing an active role in maintaining surface homeostasis. The largest component, of course, is water, amounting to 98% of the total. Along with this are a number of dissolved salts and macromolecules. There is a specific structure to the tear film that is basic to and necessary for its primary functions. It consists of three layers. First, there is an external lipid layer that delimits evaporation. The second, middle layer is the aqueous or lacrimal layer, the largest of the layers. Finally, there is a mucin layer. Mucins are large glycoproteins that can build up polymeric chains and gels, giving structural stability.

The lacrimal glands juxtaposed to each eye provide the aqueous layer of the tear film and many of its components. For example, they provide growth factors and vitamin A, which are essential for epithelial cell health. The lipid comes from a relatively large number of small meibomian glands positioned close to the rims of the upper and the lower eyelids. The composition of the lipid secretions is complex and has been reviewed by Butovich (2009).

The constant flow of tears along with eye blinks constitute a formidable barrier to externally applied substances. The normal flow of tears is increased with the application of foreign substances such as drugs, quickly washing away much of the material. The blink mechanism also greatly decreases bioavailability when the drop application of a drug is greater than about 30 μL, simply by sweeping away the excess liquid. In general, the eye can accommodate only about a 30 μL volume of applied liquid and not overflow, thus limiting each individual topical application. With the blinking reflex further dispersing the application, the uptake time is limited to a few minutes. The result of all of this is that the total uptake of many applied drugs is generally less than 5% of the initial application, with most of the drug quickly moved into the nasolacrimal duct and then into systemic circulation. However, modifications of drugs used systemically can substantially increase specific ocular uptake and modification of surface drug delivery systems through the use of nanoparticle drug delivery and other such means can substantially increase penetration. "Smart" hydrogels, copolymer vesicles, thermoresponsive polymeric micelles, and so on are being examined as possible carrier systems, along with covalent coupling of drugs to molecules that can be enzymatically recognized and actively transported across cellular membranes.

Numerous journal articles and books are available on the structure of anterior segment tissues, many relating to pathology and drug delivery, such as that of Watson and Young (2004).

2.3 ANATOMY AND COMPOSITION: MIDDLE VASCULAR TUNIC

This layer of tissues within the eye is mainly involved in blood flow, ocular nutrition, and lymphatic drainage and is usually referred to as the uvea. It is composed of several distinct tissues, principally, the iris, the ciliary body, and the choroid. The blood supply to uveal tissues, including some of the nonphotoreceptor layers of the neural retina, comes from arteries that are branches of the ophthalmic artery. Along with the absorption of nutrients originating in the blood vessel system, uveal tissues such as the ciliary body function in the production of fluids, such as the aqueous humor, that bathe the internal surfaces of anterior eye tissues and allow for adequate nutrition of the adjacent tissues, notably the corneal endothelium. Finally, there is a critical barrier function performed by these tissues.

2.3.1 Iris

The iris is a pigmented, muscular ring that is anterior to the lens; it allows for and governs the entrance of light into the interior of the eye. It is attached radially to the cornea and the ciliary body through the pectinate ligaments. It has a contractile diaphragm that controls a central aperture called the pupil. The pupil enlarges or constricts depending on ambient light conditions. Because control of light intensity is an important part of the normal visual process, the iris is often a target of drug delivery for regulation of pupil size. This function is mainly affected by peripheral autonomic action. There is a balance between adrenergic and cholinergic innervation to the iris dilator and sphincter muscles, respectively. The need to regulate pupil size can be due to pathology, treatment regimes, or a temporary need for examination of the posterior pole of the eye. Mydriasis is the excessive dilation of the pupil; sympathetic stimulation of alpha-1 adrenergic receptors causes contraction of the radial muscle, leading to pupil dilation.

Involvement of the iris in drug delivery is mainly passive but the effect can be considerable. Due to relatively heavy pigmentation, the tissue can absorb many drugs, mostly lipophilic in nature, and effectively act as a sink or reservoir for the substance. The effects of this on drug delivery can be twofold. First, the initial dose to target tissues other than the iris can be substantially decreased. Second, although binding to pigment granules is relatively low affinity and nonspecific, the capacity for binding is large, allowing for long-term storage and slow release well past the time expected for normal drug action. This is true for all the pigment-containing cells of the eye, such as RPE cells.

2.3.2 Ciliary Body, Aqueous Humor, and Vitreous Humor

2.3.2.1 Ciliary Body

The ciliary body is a heavily muscled ring of tissue with an extensive capillary bed that, as with the iris, is anterior to the crystalline lens. It consists of two main functional areas: the ciliary muscle and the ciliary processes. It is roughly triangular in the horizontal section and is a forward extension of the choroid. The ciliary processes are arranged in surface folds with the ciliary epithelium covering the fingerlike processes and producing the aqueous humor. The inner layer of the ciliary body faces the vitreous body and is nonpigmented until it reaches the iris; it is pigmented thereafter. The outer layer contains numerous pigment granules and continues posteriorly to be contiguous with the RPE. As with the iris, pigment granules in the ciliary body can act as significant reservoirs for lipophilic drugs. Importantly, the two cell layers are joined together by tight, occluding junctions that form the barrier between the posterior chamber of the eye and the blood vessels that transit the ciliary body.

One of the important functions of the ciliary body is in lens attachment and accommodation, the smooth muscle fibers helping to control the shape of the lens. Attachment to the lens is through a connective tissue named the zonule of Zinn. The ciliary body has sets of smooth muscle fibers: the longitudinal fibers that connect the sclera and the choroid along with the radial and circular fibers. With the assistance of the ciliary muscles, the lens can be rounded or flattened, enhancing focus for near or far objects. Aging results in loss of flexibility of the ciliary body and the lens, resulting in conditions such as presbyopia, a decreased ability to focus on near objects.

Drug diffusion into the ciliary body can be significant since there are no tight junctions in the extensive capillary network to block the entrance of smaller molecules. There are tight junctions, however, in the nonpigmented cells of the epithelium. A ready route of entrance into the aqueous humor is through the ciliary body vessels, diffusing into the iris and then into the aqueous humor. On the other hand, there are extensive metabolic barriers in the ciliary body that pose problems for many drug types that are able to penetrate into the tissue. These are types of drug-metabolizing enzymes that detoxify many classes of drugs and begin their removal from the eye. P450 enzymes and enzymes of conjugation are notable in this regard, with quick clearance of metabolites through the uveal circulation. One could also consider the pigment granules of the ciliary body as part of the ocular detoxification system. In this regard, the granules take up and store the absorbed drug, slowly releasing it for metabolism and removal.

2.3.2.2 Aqueous Humor

Drugs and other molecules that are designed to penetrate the anterior segment of the eye and perhaps target the retina have to contend with three separate liquid milieux: the external tear film, the interior aqueous humor in two separate compartments, and the gel-like vitreous body. The aqueous humor, produced by the ciliary body, is the clear, watery fluid that fills the anterior and posterior chambers between the cornea and the lens. Aqueous humor is continuously secreted by the nonpigmented epithelial cells of the ciliary body and continuously removed at about 5 mL/day. A balance between production and drainage is important in maintaining ocular homeostasis. Mainly, this leads to a steady and normal IOP that helps in keeping normal eye shape and thus focus of the visual image on the proper retinal (macular) position. Abnormally high IOP is a significant risk factor in glaucoma and therefore a major target for drug action. In this regard, the aqueous humor can be a hydrophilic reservoir for drugs and also its turnover can be a path for drug elimination from the eye. After synthesis, aqueous humor is secreted into the small posterior chamber and flows through the pupil into the anterior chamber. Finally, outflow is through drainage channels of the trabecular meshwork and Schlemm's canal into the general circulation. The aqueous humor consists of a clear filtrate of blood serum, lacking both blood cells and larger blood proteins. It also contains specific compounds actively transported across the blood–ocular barrier and deposited in the aqueous humor. Because of the avascular nature of most of the anterior tissues of the eye, the aqueous humor constitutes a major source of nutrition for these structures.

The ciliary body, along with the aqueous humor, is of great interest with regard to drug delivery—as a target tissue itself, as a means of drug delivery, and as a barrier system. The obvious pathology to be targeted would be glaucoma, where there may be an overproduction of aqueous humor as well as problems with drainage from within the eye. There are several classes of glaucoma drugs that target these functions. For example, beta-blockers lower eye pressure by reducing the production of aqueous humor by the ciliary body. Similarly, alpha-agonists and carbonic anhydrase inhibitors lower aqueous humor production. In contrast, miotics and prostaglandin analogs work on outflow, miotics increasing outflow through the trabecular meshwork and prostaglandins increasing the uveoscleral flow.

Since the aqueous humor bathes so many important tissues and cell types, it can be a ready vehicle for drug delivery to these areas of the anterior segment. One important "drug" that is found in very high concentration in the aqueous humor is ascorbate. It is presumed that the vitamin then penetrates the surrounding ocular tissues (Kodama et al. 1985), protecting them from oxidative insult.

2.3.2.3 Vitreous Humor

The vitreous humor is a transparent gel whose function is essentially passive, that is, to remain transparent such that light can cleanly pass through for focus on the macula of the retina. It also helps to maintain ocular turgidity, size, and shape. It is mainly aqueous in composition but contains a network of protein fibrils, polysaccharide moieties, and a few cells interspersed in the matrix. The vitreous humor is approximately 99% water with a specific viscosity of 1.8–2.0 and an index of refraction of $n = 1.33$. Only a tiny amount of collagen fibers and other larger molecules, such as the glycosaminoglycan hyaluronic acid, are present, although their functional significance is great. A number of small solutes can also be found, such as inorganic salts and sugars. As in the aqueous humor, a substance present in relatively high concentration is ascorbic acid, a protective agent that helps to maintain vitreal integrity through its antioxidative properties. Finally, a few cells are present, such as hyalocytes and phagocytes. To promote clear passage of light, the vitreous humor has no penetrating blood vessels. Thus, the strength of the vitreous gel and its mechanical properties is surprising in light of its mainly aqueous composition. The relatively clear optical properties are due to the paucity of macromolecular and cellular interference, but it does have a unique 3-D architecture that is maintained by the hyaluronan gel interposed in a regular collagen meshwork.

The vitreous body constitutes one of the primary targets for ocular drug delivery, affording a large volume depot in the center of the eye and ready access to both anterior tissues and posterior tissues such as the retina. Unlike the aqueous humor, though, which has a distinct flux, movement within the vitreous body is slower and is often thought of as "stagnant"—or sluggish at best. There is, however, a slow but distinct posterior flow due to a pressure differential between the anterior segment and the retinal surface, which is aided by active transport through RPE cells into the choroid. Altogether, it is probable that "mass transport in the vitreous humor is caused by both diffusion and convection," at least in larger animals (Xu et al. 2000). Xu et al. also report, however, that "convection, does not contribute significantly to transport in the mouse eye," indicating that caution must be observed in comparisons of transport mechanisms in eyes of different sizes. Such transport, though, can be dramatically altered with natural aging, where changes in vitreous structure have been long documented (Sebag 1987) or in ocular pathology. With age, the relatively gel-like vitreous body can "liquefy," shrinking and collapsing with perhaps even a posterior vitreous detachment. These factors and the nature of the ultimate target tissue (e.g., ganglion vs. photoreceptor cells of the neural retina) need to be taken into consideration in any drug installation in the vitreous cavity, be it by injection, longer-term inserts, or when drugs are applied to the anterior segment and are simply expected to traverse the vitreous cavity to the posterior pole.

2.3.3 BLOOD–AQUEOUS BARRIER AND IMMUNE PRIVILEGE

2.3.3.1 Blood–Aqueous Barrier

There are two blood–ocular barriers of significance in the eye. The first is the blood–aqueous barrier in the anterior segment, which controls exchange between the blood and the aqueous humor. This barrier helps to regulate the intraocular fluids and metabolism in the anterior chamber tissues such as the cornea and lens. The second is the blood–retina barrier, which regulates flow in the posterior segment of the eye and is a key element in pathologies such as the vascular retinopathies and retinal edema. These features are discussed more extensively in Section 2.4. Both of these barriers are critical physiologically, working together to maintain the eye as a privileged site with a tightly regulated internal ocular milieu that is protected from fluctuations in blood constituents and from toxins and other foreign compounds (including many drugs) in the general circulation. Also, larger molecules that could degrade the visual image if present in the visual axis are filtered out before fluid entrance into the aqueous and vitreous chambers. The blood–aqueous barrier is basically a function of the ciliary body; physically, it is formed by the nonpigmented epithelial cells along with the vascular endothelial cells of the vessels of the iris. Tight junctions between the cells constitute the major physical barrier, although the junctions are somewhat leaky. Permeability is also dependent on IOP, with a significant amount of pressure-dependent diffusion observed. Systemic drugs, however, can enter the posterior and anterior chambers, originating in the ciliary body vasculature and then diffusing into the ciliary stroma and iris and finally into the aqueous humor. A small amount of plasma protein is also found in the aqueous humor—about 1% of that normally found in blood.

Besides physical barriers, metabolic barriers also play a significant role in maintaining strict ocular homeostasis, again with the ciliary body constituting a major repository of drug-metabolizing enzymes within the eye. This barrier constitutes both enzymatic drug detoxification and removal of the primary drug or its detoxified product from the eye through the uveal circulation. Detoxification systems include cytochrome P450, conjugation to form glucuronides or sulfates, acetylation, methylation, and conjugation with glutathione or amino acids. Finally, melanin granules can take up hydrophobic compounds, storing them for subsequent metabolism or future slow release. Thus, the ciliary body poses a formidable physical and metabolic barrier to intraocular drug penetration.

The blood–ocular barrier can be disrupted and temporarily opened using techniques such as the infusion of hyperosmolar agents into the carotid artery.

Prostaglandins and vascular endothelial growth factor (VEGF) can also play a role in barrier breakdown. Temporary disruption can allow for experimental delivery of bolus amounts of drugs to the retina. There are several practical uses for such manipulations, for example in the evaluation of barrier function in pathologies such as the ocular microangiopathies.

Numerous books and journal articles on the blood–ocular barrier are available, such as Cunha-Vaz (1997 and in this book), and on ocular drug toxicity, such as Scroggs and Klintworth (1994).

2.3.3.2 Immune Privilege

Vision depends on the integrity of the visual axis, essentially the path that light takes into the eye, as well as the integrity of the retina that accepts the photic stimulus. Disruption of symmetry in any of these components (tear film, cornea, anterior/posterior chambers, aqueous humor, lens, vitreous body, and finally retina through inflammation and swelling) can distort or effectively eliminate the visual image. Thus, along with the physical and metabolic barriers, the eye has evolved to become an "immune-privileged" site to protect against such disruption (Streilein 1993). In this process, some components of our normal systemic immunity are absent in the eye; this greatly reduces the possibility of inflammation-induced vision loss while still maintaining protection against pathogens. To achieve this, there must be cooperation of both ocular anatomical and molecular specialization with the unique differences in immune expression within the eye. As enumerated by Streilein (2003a), there are five critical features of ocular immune privilege. This starts with the blood–ocular barriers and an absence of lymphatic drainage pathways within the eye. There are also soluble immunomodulatory factors in the aqueous humor, immunomodulatory ligands on the surface of the parenchymal cells, and the presence of indigenous, tolerance-promoting antigen-presenting cells (APCs).

For drug delivery, the presence of ocular immune privilege offers some unique problems as well as opportunities not present systemically in the body. Some of these "therapeutic opportunities" are succinctly reviewed by Streilein (2003b).

2.3.4 CHOROID

There are two major blood pathways in the eye: the retinal vessels that service inner retinal cells and the uveal circulation that services most of the rest of the eye including the outer layer of the retina. The choroid is part of the uveal tract along with the iris and ciliary body but, in fact, forms a functional unit with the neural retina. It is a thin, highly vascularized connective tissue with large lymphatic channels that coats much of the posterior part of the sclera. It is the main source of oxygen and nutrients for the outer (photoreceptor) cells of the neural retina including the pigment epithelium that rests on the choroid. The choroid is also pigmented, aiding in absorption of stray light and reflection. As with the neural retina, it is continuous with the ciliary body but ends at the ora serrata. The choroid is only loosely attached to the sclera with an area of potential separation, the suprachoroidal space that is also present near the ciliary body. This area is important in ocular drug action in that substances can be instilled in the space for intraocular delivery.

Although the function of the choroid is simple—nutrient and oxygen delivery to the outer retina and removal of waste products—the anatomy of the choroid is complex, consisting of three distinct layers. The outermost layer is called Haller's layer and contains vessels of relatively large diameter. Internally, Sattler's layer has vessels of medium diameter, and the final layer, the choriocapillaris, is a layer of smaller-bore capillaries (Figure 2.3). The vessel pattern of the choriocapillaris is relatively unique and varies in different areas. In the posterior pole, there is a marked lobular pattern, while more anteriorly toward the ora serrata, the pattern is less structured with capillaries of larger diameter. Fenestrations in the choriocapillaris that mostly face the RPE allow for quicker transport of nutrients. The choriocapillaris is juxtaposed to a basement membrane, Bruch's membrane (BM) (see Section 2.4.3), which functions as a support for the choroid and the retina and also acts as a partial barrier between the tissues. It is also a critical transit point for nutrients coming into the pigment epithelium/retina complex and for waste products exiting through the pigment epithelium into the choroid. Retinal blood flow is estimated to be about 80 mL/min, amounting to about 5% of the total ocular flow. The choroidal flow is approximately ten-fold this amount—the extremely high blood flow accounting for an arteriovenous oxygen differential of only 3% in the choroid compared to 40% in the retina. It has been estimated that up to 75% of the oxygen and nutrients needed by the retina are supplied through the choroidal circulation. Choroidal blood flow is not always static; it can be affected by autonomic innervation. A significant increase in choroidal flow is observed with increased blood pressure (Polak et al. 2003), while there is a reduced foveal flow with increased severity of age-related macular degeneration (AMD) (Grunwald et al. 2005).

Drug delivery to the retina and other internal ocular tissues is very limited from the choroidal vessels, mostly by factors extrinsic to the choroid, but can be markedly affected by pathologies, especially choroidal neovascularization as seen in

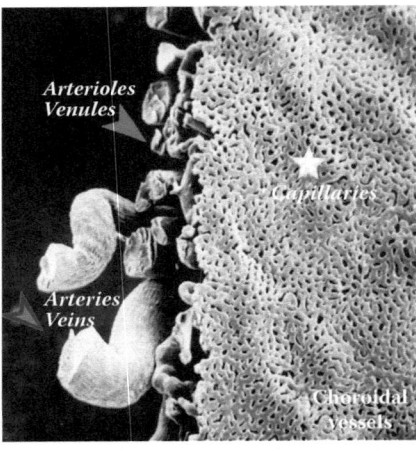

FIGURE 2.3 The three layers of the choroid: larger arteries and veins, smaller arterioles and venules, and the inner capillary bed, the choriocapillaris. (Courtesy of Webvision.)

AMD, myopia, and after ocular trauma where the blood–retinal barrier (BRB) can be breached. A central player in such neovascular conditions is VEGF, which is a prime target in drug therapy for such conditions through use of anti-VEGF agents. Interestingly, though, this is balanced by the finding in an animal model that VEGF is actually essential for maintenance of the normal choriocapillaris (Saint-Geniez et al. 2009). If RPE cell-derived soluble VEGF is not present, changes similar to geographic atrophy are observed in older animals with RPE and photoreceptor cell loss and a decrease in visual acuity.

2.4 ANATOMY AND COMPOSITION: INTERNAL NERVOUS TUNIC

This is the most internal layer of the eye globe and formally is only the neural retina. However, functionally, the retina is intimately apposed to and needs the RPE for proper maintenance, so this cell layer is included in this section along with a short synopsis of the layers of the retina proper.

2.4.1 NEURAL RETINA

The neural retina functions like the photographic film of a camera in capturing an image and thus is the most important tissue of the eye. It is part of the central nervous system (CNS) along with the brain. Its main function is to accept photic stimulation, convert it into an electrical signal, and, with partial processing, transmit the signal to the brain for final synthesis into a visual image. Pathologically, diseases of the retina currently constitute most of the unmet needs in treating blinding ocular conditions. These include glaucoma, diabetic retinopathy, AMD, and the inherited retinal degenerative diseases such as retinitis pigmentosa. Thus, the retina is the major target tissue for drug treatment in the eye but also poses huge problems in delivering active drugs in high enough concentrations to the target retinal cells. The importance of drug delivery to the posterior segment and the routes of delivery has been considered for a number of years (Geroski and Edelhauser 2000; Edelhauser et al. 2010) and is a subject of this book.

The general structure of the neural retina is simple. It is a highly laminate tissue that can functionally be divided into two areas: outer and inner retina (Figure 2.4). The outer retina consists of only the photoreceptor neurons, although functionally the RPE should be included. The role of the photoreceptor cells is to capture photic energy, convert it into an electrophysiological signal, and pass it on to underlying secondary neurons. These other retinal neurons (and glial elements) constitute the inner retina and act more as conventional, brain-like neurons, processing neuronal signals and passing them on to recipient neurons. On the direct path from photoreceptor cells to the brain are bipolar and ganglion cells, but several cell types also are involved in signal processing and in modification, such as amacrine and horizontal cells. Finally, the signal moves down the optic nerve to the brain area, where there is final processing and signal synthesis into a visual image. Photoreceptor cells, both rods and cones, are highly polarized with specialized outer segments containing mainly photopigments such as rhodopsin. These protrude into the interphotoreceptor

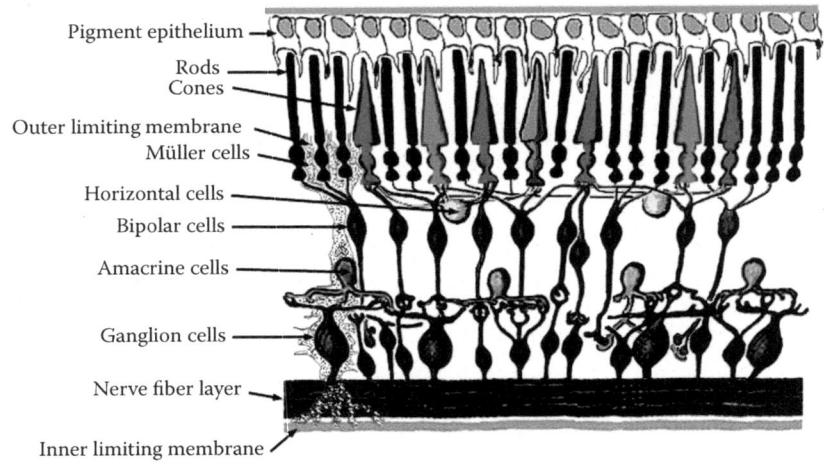

Pigment epithelium
Rods
Cones
Outer limiting membrane
Müller cells
Horizontal cells
Bipolar cells
Amacrine cells
Ganglion cells
Nerve fiber layer
Inner limiting membrane

FIGURE 2.4   A schematic diagram of the structure of the neural retina and the pigment epithelial cell layer. The outer retinal layer consists of the photoreceptor cells; the inner retinal layer consists of all the other neuronal and glial elements. (Courtesy of Webvision.)

matrix (IPM) in fingerlike projections and are enmeshed in a network of pigment epithelial microvillous processes. The other end of the photoreceptor cell is a synaptic terminal that connects to and communicates with the secondary neurons, for example, bipolar cells, of the inner retina.

Glial elements are also present in the retina; Müller cells, for example, are somewhat akin to brain astrocytes and have important structural, metabolic (e.g., detoxification), and even barrier roles in the retina. Müller cells are large and span virtually the whole width of the retina. At the vitreal surface is the inner limiting membrane (ILM) with the Müller cell endfeet contributing in large part to the glia limitans at this inner surface of the retina. The ILM is not a true membrane but rather is formed from fused foot processes of the Müller cells. These have occluding junctions, forming a tight barrier that effectively seals off the retinal cellular elements from the vitreal body. At the outer surface of the neural retina is the outer limiting membrane (OLM). Only the photoreceptor outer segments protrude through the OLM into the IPM. As with the ILM, the OLM is not a true membrane but a series of occluding junctions formed between the Müller cells and the photoreceptor cells. The OLM isolates the outer retina from IPM components in the subretinal space. Along with these barrier functions, Müller cells synthesize factors that induce tight junctions and thus are important in the barrier function of the retinal vasculature. Finally, Müller cells have an important detoxification function in the retina. They express glutamine synthetase, for example, that takes up excess and potentially toxic glutamate and converts it to nontoxic glutamine.

As alluded to earlier and better described in Section 2.4.4, the retinal vascular system plays a key role in maintaining an overall protective barrier function in the eye. In this way it is very similar to the blood–brain barrier (BBB) that protects against unwanted substances that otherwise would enter the brain from the blood circulation. As in the BBB, the most important feature of the BRB is the tight junctions within the endothelial cells of the retinal vasculature. As in many such cases, though, the tight junctions of the BRB do not form a totally impenetrable barrier to the entrance of molecules; rather, they form a selective barrier that keeps out possible toxic agents but allows the entrance of needed substances that help in maintaining neuron homeostasis. Of course, along with potentially toxic substances, most foreign molecules, such as drugs, are excluded from the retinal milieu, necessitating other, more involved, and inventive pathways for drug delivery.

2.4.2 RETINAL PIGMENT EPITHELIUM

The RPE is not formally part of the internal nervous tunic since it is not intrinsically part of the neural retina. It is of neuroepithelial origin and not part of the CNS. However, it forms a functional unit with the outer retinal cells, specifically the photoreceptor cells such that, without proper RPE cell function, photoreceptors soon die.

The RPE consists of a monolayer of pigmented, polygonal neuroepithelial cells that abut the retinal photoreceptor outer segments. RPE cells are highly polarized with distinct apical and basal surfaces. Long microvillous-like processes extend from the apical side of RPE cells and surround the photoreceptor outer segments in the IPM of the subretinal space. These generally are tightly adherent to the outer segments and have support roles in maintaining structural integrity, metabolic function, and waste removal of shed outer segment tips to maintain normal photoreceptor physiology. The basal RPE cell surface is attached to BM, a collagen-rich basement membrane that separates the RPE cell layer from the choroid. Basal infoldings of the RPE cell plasma membrane allow for efficient nutrient uptake across BM from the choroidal circulation and, in the reverse direction, for the removal of waste products from the photoreceptor cells. Laterally, RPE cells are laced together by tight junctions that stop paracellular movement of substances from a somewhat leaky and permissive choroid into the interphotoreceptor space (IPS). These junctions, along with similar junctions in the retinal vascular endothelial cells described earlier, form the basis for the protective barrier function in the posterior segment of the eye.

Within the RPE cells are numerous pigment granules that function in light screening but also play a passive albeit important role in drug delivery as long-term storage depots. RPE cells are biochemically polarized as well, having, for example, a rich supply of Na^+/K^+-ATPase and other enzymes preferentially localized in the apical membranes that control ion fluxes and substance exchange with the IPS. There is also preferential secretion of numerous substances such as neurotrophic factors, for example, pigment epithelium-derived factor (PEDF) into the IPM. Different channels and transport systems (active and facilitated) are present on different surfaces of RPE cells. These allow for selected movement of nutrients such as glucose and amino acids. Asymmetrical transport systems are present on apical and basal surfaces, for example, a chloride–bicarbonate exchange transporter only present on the basal surface and a Na^+-K^+ pump that is present on the apical surface. Such specialization allows for an active water transport from the IPM through the RPE cell and out through the choroid. Organic cation transporter P-gp (P-glycoprotein), glutamate (GAT3) and gamma-amino butyric acid (LAT—L-type amino acid transporter) transporters, monocarboxylate transporter (MRP—multidrug resistance transporter), and many other transport systems are active in RPE cells, allowing for possible piggybacking of specially modified drugs.

Functionally, RPE cells perform many tasks that ensure retinal health and homeostasis. Because of their pigmented nature, they absorb scattered and stray light that otherwise could degrade the visual image. RPE cells control the fluid and nutrient composition that passes through the IPM to the photoreceptors. In the reverse direction, RPE cells control waste product removal from the photoreceptors. Most important here is the daily phagocytosis of shed outer segment tips that helps to maintain proper disc renewal of the outer segments. Because of active posterior fluid flow through the eye and active transport out of the basal portion of RPE cells, retinal adhesion is enhanced. This is also helped by the extensive interdigitation of RPE microvillous processes with the outer segment projections. The RPE contributes to the electrical profile (i.e., electroretinogram) of the retina. RPE cells also partner with photoreceptor cells in maintaining the visual cycle. Retinoids are selectively taken up into RPE cells, stored as retinyl esters, and then moved in proper isomer form to the photoreceptor cells for use with the visual pigments in the visual process. Because of its high metabolic rate and the high surrounding oxygen tension, the retina and, in particular, the photoreceptor cells are subjected to constant oxidative stress, especially with advancing age. Frank et al. (1999) have reported high activity of antioxidant enzymes, particularly in the macular region in eyes with neovascular AMD. This perhaps says something about the etiology of AMD but also underscores the need for delivery of potent antioxidant drugs to this delicate and susceptible area of the eye.

Finally, the RPE cell layer also has a metabolic function in maintaining the BRB. This is through the presence of a number of enzymes that can function in drug metabolism. A prime example is found in an enzyme that is part of the aryl hydrocarbon hydroxylase system, the cytochrome P450 enzyme found in the RPE cytoplasm that is used in detoxification reactions. Interestingly, the aryl hydrocarbon hydroxylase enzyme has long been known to be inducible in RPE cells under stressful conditions (Shichi et al. 1975). All in all, RPE cells contain a relatively large number of such enzymes that, along with drug sequestering in pigment granule depots in choroid and RPE, help to protect retinal cells from toxic agents.

There are a number of excellent articles and books concerning RPE function and barrier protection, including Marmor and Wolfrnsberger (1998).

2.4.3 Bruch's Membrane

As mentioned earlier, BM is a thin basement membrane only 1–3 μm in thickness that separates the choroid from the RPE cells. The structure is complex, consisting of five distinct layers. Abutting the RPE cells is the RPE basement membrane layer (basal lamina). This is followed by an inner collagenous zone, a central band of elastic fibers, and an outer collagenous zone of microfibrils. Finally, there is the basement membrane (basal lamina) of the choriocapillaris. All this leads to a significant but not complete barrier to flow out of the choroidal vessels. However, transport properties change with age and pathology. In general, BM thickens with age, slowing down transport processes in both directions. The pathology of AMD also can slow transport in part because of the inclusion of deposits (such as drusen) around and within BM. Hussain et al., for example, have described the changing macromolecular, diffusional characteristics in the aging human BM with "diffusional transport (in the macular region) in the 9th decade being only 6.5% of that in the 1st decade of life" (Hussain et al. 2010). Thus, for calculations of systemic drug delivery to the retina, along with rates of drug clearance through the posterior segment, changes in BM can have significant effects.

2.4.4 INTERPHOTORECEPTOR SPACE

The IPS (or subretinal space) is important in vision as it constitutes the main route by which nutrients reach the extremely metabolically active photoreceptor neurons and by which waste products are removed from the photoreceptor milieu. The IPS is a unique, well-regulated extracellular compartment. On the retinal side, microvillous processes extending from Müller glial cells form junctional complexes with inner segments of adjacent photoreceptor neurons. As described earlier, the OLM effectively seals this side of the subretinal space. On the other surface of the IPS, tight junctions of the RPE cells effectively block entrance of most substances from access into the IPS and thus to retinal photoreceptors. The IPS is not a discrete open or liquid-containing space but rather is filled with highly structured IPM material and other specialized proteins mainly synthesized by adjoining retinal and RPE cells (Hageman and Johnson 1991). For example, interphotoreceptor retinoid-binding protein (IRBP), active in extracellular retinoid transport, and PEDF, a neurotrophic and antineovascular protein, are found in high concentrations selectively in the IPM.

The IPS has not been a usual target for drug delivery but is recently receiving increased attention as a depository site both for drugs in conditions such as retinal neovascularization and for gene delivery in gene therapy in inherited retinal degenerations such as Leber's congenital amaurosis.

2.4.5 BLOOD–RETINAL BARRIER

The BRB in the posterior segment of the eye works toward the same goals as does the blood–ocular barrier in the anterior segment—that is, a strict regulation of substances that enter the eye and to which the internal tissues are exposed. The BRB has two distinct elements: the first lies within the neural retina itself and the second within the RPE cells. This is logical since the blood–tissue interface comes at the level of both the internal retinal vessels and the choroidal vessels juxtaposed to the RPE. In the retina, the vessels are lined by endothelial cells that are different from those in many peripheral tissues in that they have a strong network of tight junctions (Shakib and Cunha-Vaz 1966). Such junctions are present only in the retina and brain and are composed of belt-like zonula occludens connections. These junctions stop paracellular transport of water-soluble substances, allowing for penetration of only molecules that are transported through the cell itself.

The second element of the BRB is the RPE cell layer. RPE cells are connected by zonula occludens as are the endothelial cells of the retinal vessels, forcing intracellular rather than extracellular uptake of selected substances. Thus, macromolecules and solutes do not passively diffuse into the retina as in most other tissues but rather can enter only through selective processes such as active transport. Recent data indicate that neutrophils could play a role in breakdown of the BRB (Zhou et al. 2010). In particular, they secrete matrix metalloproteinase-9 (MMP-9) that can disrupt the integrity of the RPE cell tight junction, thus increasing permeability. A target for future drug treatment is thus revealed in the possibility of delimiting neutrophil invasion and/or MMP secretion and thus maintaining normal barrier function. Pathological conditions such as uveitis can also affect barrier action. Lin et al. (1991) found that in S-antigen-induced experimental uveoretinitis, there was a "breakdown in the blood-retinal barrier … at the level of the retinal venules." Specifically, the endothelial cells were found to be hypertrophied with a swollen endoplasmic reticulum, allowing for the entrance of a horseradish peroxidase tracer into the retina.

The choroid does not seem to play much of a role in barrier function. It has numerous fenestrations, pinocytotic vesicles, and a lack of tight junctions, thus allowing for relatively easy movement of substances out of the tissue, only to be stopped by the RPE cell tight junctions. Similarly, BM plays only a small role in the barrier function.

2.5 SUMMARY: SURMOUNTING BARRIERS TO DRUG DELIVERY TO THE EYE

Perhaps the simplest and oldest form of drug delivery is the use of eye drops. This is notoriously inefficient, however, because the first barrier the drops meet is the tear film. Less than 5% of most drugs delivered in this way enter into the eye. The remaining 95% is swept away through tear drainage, much entering the general circulation where systemic sequelae can occur. Certainly, drug modification and special compounding can significantly enhance penetration, but for drugs that are expensive, such delivery may not be justified—especially since this route generally does not allow for penetration back to the posterior segment. On the other hand, several techniques such as mixed micellar formulations are showing promise that drug delivery to the posterior segment can, in fact, be both practicable and practical.

Beyond this, there are both hydrophilic and hydrophobic barriers to drug delivery through the cornea. The stroma is a good water barrier, while epithelial and endothelial cells function as effective lipid barriers with only partial penetration of biphasic/nonpolar drugs. Barrier function can change, though, due to age-related changes of the tissues, including corneal endothelium, Descemet's membrane, and so on (Chakravarti et al. 2006). Also, the use of aides such as extraocular, transscleral drug delivery devices and iontophoresis can markedly enhance drug penetration into the eye. A promising technique is the use of hollow microneedles (Patel et al. 2011). Modeling has given needed information on corneal and retinal pharmacokinetics after periocular administration (subconjunctival or posterior subconjunctival) of small, lipophilic drugs such as celecoxib (Amrite et al. 2008). It was found that there was leakage from the periocular space back into the precorneal area and significant drug clearance through the conjunctival/periocular blood and lymphatic systems.

For posterior segment disease, the most common route of delivery presently used is intravitreal injection as in the case of antineovascular agents for AMD. Advantages of this type of direct treatment are that the BRB is circumvented and that a high level of drug is delivered to the retina. Negatives include the possibilities of infection, retinal detachment, the development of cataract or glaucoma, a relatively short time of action, and the lack of patient enthusiasm for such delivery. Implants solve some of these problems, such as for long-term, graded dosing, but present others problems. For example, classical inserts such as Bausch & Lomb Retisert could deliver a drug for 6 months or more. On the other hand, implants need to be replaced, constituting a surgical procedure that can lead to endophthalmitis. Improvements now are focusing on longer-term drug action and refillable reservoirs that limit the need for surgery. The controlled-release implantable devices can be biodegradable or non-biodegradable, with different advantages to both systems. In somewhat the same vein, injectable particles such as microspheres, liposomes, emulsions, and micelles offer specific advantages; some of these, for example, can act as reservoir-like carriers. Ocular inserts such as the Encapsulated Cell Technology (ECT) of Neurotech are good examples of relatively long-term therapy offered by vitreal implants (Tao 2006). The ECT capsule, tethered to the sclera in the vitreous cavity outside of the visual axis, delivers a neurotrophic agent ciliary neurotrophic factor (CNTF) to the retina in pathologies such as retinitis pigmentosa and AMD.

This chapter and several others within this book highlight the difficulty in getting drug penetration within the eye and movement to and action within the retina. This is a compelling problem since posterior segment diseases such as AMD, diabetic retinopathy, and glaucoma constitute most of the current unmet needs in treating blinding ocular conditions. Also, even though intravitreal injection and the use of long-term delivery implants can result in high drug concentrations delivered to the retina, the price to be paid can include infection, cataract induction, and retinal detachment. Systemic drug delivery is an option in some cases but, as with topical drop application, it can result in very little drug accumulating in the posterior segment, while much higher drug concentrations can be a concern in systemic tissues. These problems make alternatives such as a transscleral route of delivery attractive. There are a number of routes that can be used for transscleral drug delivery including subconjunctival, peribulbar, retrobulbar, and posterior juxtascleral pathways. Also, a sub-Tenon's route is possible for drugs destined for retinal use. It has been reported, though, that for drugs such as anecortave acetate that target the posterior segment, even subconjunctival drug administration "resulted in subtherapeutic concentrations in the macular region," but that posterior juxtascleral depot placement allowed for good drug delivery to the macular region for a 6-month period (Kaiser et al. 2007).

Finally, along with various delivery routes, use of special formulations of the drug and specific drug delivery systems can be critical in getting the therapeutic agent to its intended target tissue. Drug targeting, using either passive or active targeting with drug-bound macromolecules, can be an effective delivery method. Binding the small drug moieties to larger macromolecules alters the biodistribution and degradation pattern of the drugs. There can be passive targeting, when the intrinsic nature of the macromolecule controls the drug distribution, or active targeting, when the macromolecule has a particular affinity for a tissue or cell type of choice and thus results in concentration of the agent at or within the target tissue/cell. Nanoparticulate formulations, in particular, can be a deciding factor in successful ocular drug delivery. Advantages of such delivery, particularly for posterior segment diseases, would be numerous, notably longer-term drug release and thus prolonged action along with a longer time between applications. There could be protection of the drug from degradation and also tissue protection from inordinately high drug levels by coupling the nanoparticle release with other systems under development, leading to tissue-specific targeting with lower possibilities of side effects.

Many good articles and reviews are available on ocular drug delivery, such as Yasukawa et al. (2004). An excellent review on drug delivery to tissues of the posterior segment has been published in the ophthalmic literature by Edelhauser et al. (2010).

REFERENCES

Amrite, A. C., H. F. Edelhauser and U. B. Kompella. 2008. Modeling of corneal and retinal pharmacokinetics after perioiular drug administration. *Invest. Ophthalmol. Vis. Sci.* 49:320–32.

Bonini, M. A., R. Jorge, J. C. Barbosa, D. Carlucci, J. A. Caudillo and R. A. Costa. 2004. Intravitreal injections versus sub-tenon's infusion of triamcinolone acetonide for refractory diabetic macular edema. *Invest. Ophthalmol. Vis. Sci.* 46:3845–9.

Butovich, I. A. 2009. The Meibomian puzzle: Combining pieces together. *Prog. Ret. Eye Res.* 20:522–36.

Chakravarti, J., H. Edelhauser and M. Kimos. 2006. Aging changes of mouse corneal endothelium and Descemet's membrane. *Exp. Eye Res.* 83:890–6.

Cheruvu, N. P., A. C. Amrite and U. B. Kompella. 2008. Effect of eye pigmentation on transscleral drug delivery. *Invest. Ophthalmol. Vis. Res.* 49:333–41.

Choy, Y. B., S. R. Patel, J. H. Park, B. E. McCarey, H. F. Edelhauser and M. R. Prausnitz. 2011. Microadhesice microparticles in a rapidly dissolving tablet for sustained drug delivery to the eye. *Invest. Ophthalmol. Vis. Sci.* 52:2627–33.

Cruysberg, L. P., R. M. Nuijts, D. H. Geroski, F. Hendrikse and H. F. Edelhauser. 2005. The influence of intraocular pressure on the transscleral diffusion of high-molecular weight compounds. *Invest. Ophthalmol. Vis. Sci.* 46:3790–4.

Cunha-Vaz, J. G. 1997. The blood-ocular barriers: Past present and future. *Documenta Ophthalmol.* 93:149–57.

Doughty, M. J. and M. L. Zaman. 2000. Human corneal thickness and its impact on intraocular pressure measurements: A review and meta-analysis approach. *Surv. Ophthalmol.* 44:367–408.

Edelhauser, H. F., C. L. Rowe-Rendleman, M. R. Robinson, D. G. Dawson, G. J. Chader, H. E. Grosskiklaus, K. D. Rittenhouse, C. G. Wilson, D. A. Weber, B. D. Kuppermann, K. S. Csaky, T. W. Olsen, U. B. Kompella, V. M. Holers, G. S. Hageman, B. C. Gilger, P. A. Compochiaro, S. M. Witcup and W. T. Wong. 2010. Ophthalmic drug-delivery systems for the treatment of retinal diseases: Basic considerations to clinical applications. *Invest. Ophthalmol. Vis. Sci.* 51:5402–20.

Frank, R. N., R. H. Amin and J. E. Puklin. 1999. Antioxidant enzymes in the macular retinal pigment epithelium of eyes with neovascular age-related macular degeneration. *Am. J. Ophthalmol.* 127:694–709.

Geroski, D. H. and H. F. Edelhauser. 2000. Drug delivery for posterior segment eye diseases. *Invest. Ophthalmol. Vis. Sci.* 41:961–4.

Grunwald, J., T. Metelitsina, J. DuPont, G.-S. Ying and M. Maguire. 2005. Reduced foveolar choroidal blood flow in eyes with increasing AMD severity. *Invest. Opthalmol. Vis. Sci.* 46:1033–8.

Hageman, G. and L. Johnson. 1991. Structure, composition and function of retinal interphoto-receptor matrix. *Prog. Ret. Res.* 10:207–49.

Hussain, A., C. Starits, A. Hodgett and J. Marshall. 2010. Macromolecular diffusion characteristics of aging human Bruch's membrane: Implications for age-related degeneration (AMD). *Exp. Eye Res.* 90:703–10.

Jiang, J., D. H. Geroski, H. F. Edelhauser and M. R. Prausnitz. 2006. Measurement and prediction of lateral diffusion within human sclera. *Invest. Ophthalmol. Vis. Sci.* 47:3011–6.

Jiang J., J. Moore, H. Edelhauser and M. Prausnitz. 2009. Intrascleral drug delivery to the eye using hollow microneedles. *Pharm. Res.* 26, 395–403.

Kaiser, P. K., M. F. Goldberg and A. A. Davis. Anacortave Acetate Clinical Study Group. 2007. Posterior juxtascleral depot administration of anacortave acetate. *Surv. Ophthalmol.* 52:S62–9.

Kao, J. C., D. H. Geroski and H. F. Edelhauser. 2005. Transscleral permeability of fluorescent-labelled antibiotics. *J. Ocul. Pharmacol. Ther.* 21:1–10.

Kodama, T., I. Kabasawa, O. Tamura and V. N. Reddy. 1985. Dynamics of ascorbate in the aqueous humor and tissues surrounding ocular chambers. *Ophthal. Res.* 17:331–7.

Lin, W. L., E. Essner and H. Shichi. 1991. Breakdown of the blood-retinal barrier in S-antigen-induced uveoretinitis in rats. *Graefes Arch. Clin. Exp. Ophthalmol.* 229:457–63.

Marmor, M. F. and T. J. Wolfensberger. 1998. *The Retinal Pigment Epithelium: Function and Disease.* New York: Oxford University Press.

Olsen, T. W., S. Y. Aaberg, D. H. Geroski and H. F. Edelhauser. 1998. Human sclera: Thickness and surface area. *Am. J. Ophthalmol.* 125:137–41.

Patel, S. R., A. S. Lin, H. F. Edelhauser and M. R. Prausnitz. 2011. Suprachoroidal drug delivery to the back of the eye using hollow needles. *Pharm. Res.* 28:166–76.

Polak, K., E. Polska, A. Luksch, G. Dorner, G. Fuchsjager-Mayrl. O. Findl, H. G. Eichler, M. Wolzt and L. Schmetterer. 2003. Choroidal blood flow and arterial blood pressure. *Eye* 17:84–8.

Robinson, M. R., S. S. Lee, H. Kim, S. Kim, R. J. Lutz, C. Galban, P. M. Bungay, P. Yuan, N. S. Wang, J. Kim and K. S. Czaky. 2006. A rabbit model for assessing the ocular barriers to the transscleral delivery of triamcinolone acetonide. *Exp. Eye Res.* 82:479–87.

Saint-Geniez, M., T. Kurihara, E. Sekiyama, A. E. Maldonaldo and P. A. D'Amore. 2009. An essential role for RPE-derived soluble VEGF in the maintenance of the choriocapillaris. *Proc. Natl Acad. Sci. USA* 106:18751–6.

Scroggs, M. W. and G. K. Klintworth. 1994. Drugs and toxins. In *Pathobiology of Ocular Disease*, eds. A. Garner and G. K. Klintworth, 1163–88. New York: Marcel Dekker.

Sebag, J. 1987. Age-related changes in human vitreous structure. *Graefes Arch. Clin. Exp. Ophthalmol.* 225:89–93.

Shakib, M. and J. G. Cunha-Vaz. 1966. Studies on the permeability of the blood-retinal barrier IV. Role of the junctional complexes of the retinal vessels on the permeability of the blood-retinal barrier. *Exp. Eye Res.* 5:229–34.

Shichi, H., S. A. Atlas and D. W. Nebert. 1975. Genetically regulated aryl hydrocarbon hydroxylase induction in the eye: Possible significance of the drug-metabolizing enzyme system for the retinal pigmented epithelium-choroid. *Exp. Eye Res.* 21:557–67.

Shimura, M., K. Yasuda, T. Nakazawa, T. Shiono, T. Sakamoto and K. Nishida. 2009. Drug reflux during posterior subtenon infusion of triamcinolone acetonide in diffuse diabetic macular edema not only brings insufficient reduction but also causes elevation of intraocular pressure. *Graefes Arch. Clin. Exp. Ophthalmol.* 247:907–12.

Shuler, R., P. Dioguardi, C., Henji, J. Nickerson, L. Cruysburg, and H. Edelhauser. 2004. Scleral permeability of a small, single-stranded oligonucleotide. *J. Ocul. Pharmacol. Ther.* 20:159–68.

Streilein, J. W. 1993. Ocular immune privilege and the Faustian dilemma. *Invest. Ophthalmol. Vis. Sci.* 37:1940–50.

Streilein, J. W. 2003a. Ocular immune privilege: The eye takes a dim but practical view of immunity and inflammation. *J. Leukoc. Biol.* 74:179–85.

Streilein, J. W. 2003b. Ocular immune privilege: Therapeutic opportunities from an experiment of nature. *Nat. Rev. Immunol.* 3:879–89.

Tao, W. 2006. Application of encapsulated cell technology for retinal degenerative diseases. *Expert Opin. Biol. Ther.* 6:717–26.

Trier, K., A. B., Olsen and T. Ammitzboll. 1990. Regional glycosaminoglycans composition of the human sclera. *Acta Ophthalmol.* 68:304–6.

Watson, P. G. and R. D. Young. 2004. Scleral structure, organization and disease. A review. *Exp. Eye Res.* 78:609–23.

Xu, J., J. Heys, V. H. Barocas and T. W. Randolph. 2000. Permeability and diffusion in vitreous humor: Implications for drug delivery. *Pharmaceut. Res.* 17:664–9.

Yasukawa, T. Y. Ogura, T. Yasuhiko, H. Kimura, P. Wiedemann and Y. Honda. 2004. Drug delivery systems for vitreoretinal diseases. *Prog. Ret. Eye Res.* 23:253–81.

Zhou, J., S. He, N. Zhang et al. 2010. Neutrophils compromise retinal pigment epithelial barrier integrity. *J. Biomed. Biotech.* 2010:289–360.

3 Animal Models to Evaluate Ocular Nanoparticular Drug Delivery Systems

Vinson M. Wang, Jingsheng Tuo,
and Chi-Chao Chan

CONTENTS

3.1 INTRODUCTION

There has been a great deal of advancement in the field of nanotechnology and its application to ocular diseases. Nanotechnology refers to the use of particles or molecules that are on the nanometer scale, generally with at least one dimension under 100 nm (Nguyen et al. 2010). Due to their small size, nanoparticles have immense potential as vehicles for gene therapy and drug delivery, along with a plethora of other applications in the biomedical field, such as diagnostics and bioimaging (Cai et al. 2008; Foy et al. 2010; Powell et al. 2010). Cutting-edge work with nanotechnology has shown promise for the treatment of multiple diseases, including ocular diseases such as retinal degeneration and glaucoma (Chen et al. 2006; Zarbin et al. 2010).

This chapter will focus on recently developed drug delivery systems that employ the use of nanoparticles and the testing of these systems in animal models. Effective delivery of drugs to ocular tissue is especially challenging due to the presence of the blood–retinal barrier (BRB) and the penetrating permeability (e.g., hydrophilic or hydrophobic) of the cornea and sclera (Kim et al. 2009). Conventional methods of drug delivery into the eye, such as intravenous and intraocular injections, are limited by poor bioavailability, quick clearance, and potential toxicity of drugs at

TABLE 3.1

Advantages and Disadvantages of Nanoparticles

Nanoparticulate Platform	Advantages	Disadvantages
Polymeric nanoparticles	• Consistent formulation • Stable structure	• Difficult to fabricate • Small drug payload
Gold nanoparticles	• Small size • Easy to fabricate • Stable structure • Very monodisperse	• Limited surface modifications • Potential long-term adverse effects • Small drug payload
Lipid nanoparticles	• Larger drug payload • Nontoxic • Versatile surface modifications	• Large size • Low stability • Polydispersion • Low encapsulation efficiency
Nanocrystals	• Very stable	• Low drug retention time
Nanofilters	• Extremely large drug reservoir	• Low drug retention time

the target tissues (Wadhwa et al. 2009b). Due to the less-optimal drug delivery systems currently available, there is a strong need for the development of better drug delivery methods. Even though the field of nanotechnology has its own limitations, it provides many exciting opportunities to improve conventional systems. The advantages and disadvantages of nanoparticles are summarized in Table 3.1. By taking advantage of the unique physical and biochemical properties of nanoparticulate drug delivery systems, we may be able to increase the bioavailability and residence time of drugs in targeted tissues, as well as reduce the toxicity of drugs.

Nanoparticles include monolithic nanoparticles (nanospheres), in which the drug is adsorbed, dissolved, or dispersed throughout the matrix, and nanocapsules, in which the drug is confined to an aqueous or hydrophobic core surrounded by a shell-like wall (Figure 3.1). When nanoparticles enter the body, the drug in the nanoparticles is usually released from the matrix by diffusion, swelling, erosion, or degradation. The main technological advances of nanoparticles as drug carriers are increasing stability, carrier capacity, ability to incorporate both hydrophilic and hydrophobic substances, and routes of administration.

Although numerous drug delivery systems incorporating the use of nanotechnology are available, the systems showing the furthest development and greatest potential involve polymeric, lipid, and gold nanoparticles. Although each system has its own unique characteristics, these systems have all been tested on animal models and shown potential for use as an alternative to conventional drug delivery systems (Kim et al. 2009; Nagarwal et al. 2009; Seyfoddin et al. 2010). Animal models provide the optimal means to analyze the effectiveness and limitations of new nanoparticulate drug delivery systems while reducing the risk to humans. Even though animal and human physiology differs significantly, the basic toxicity, kinetics, and physical nature can be effectively examined. Furthermore, disease models, such as induced

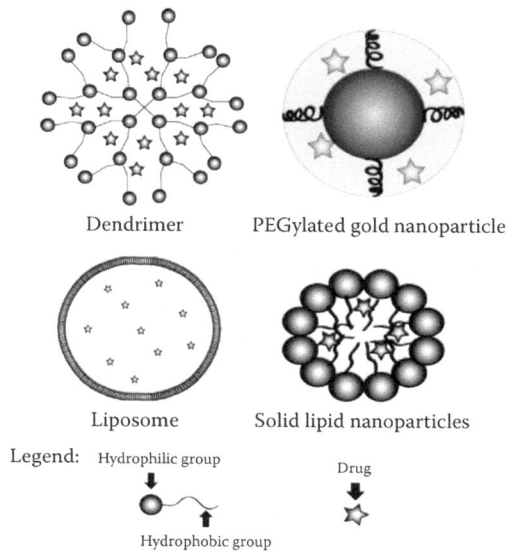

Dendrimer   PEGylated gold nanoparticle

Liposome   Solid lipid nanoparticles

Legend: Hydrophilic group   Drug

Hydrophobic group

FIGURE 3.1   Diagrams of various nanoparticles.

choroidal neovascularization (CNV) lesions associated with age-related macular degeneration (AMD), can be approximated in animals (Grossniklaus et al. 2010; Ideta et al. 2005).

Despite the rapid development of nanoparticulate drug delivery systems, many obstacles still need to be overcome. The full extent of biological interactions associated with the presence of nanoparticulates is still not well understood, thereby presenting a risk for unknown adverse effects in humans. Furthermore, the physical nature of nanoparticles, such as the rate of diffusion, has been largely empirical and needs to be further studied (Orosz et al. 2004). The specific limitations of using nanoparticles will be described in their respective sections. Although limitations associated with the use of nanoparticles exist, nanoparticulate drug delivery systems could make significant impacts on the treatment of many devastating diseases and conditions, such as cancer, cardiovascular disease, Alzheimer's disease, and blindness (Foy et al. 2010; Thomson and Lotery 2009; Vishwakarma et al. 2008; Zarbin et al. 2010).

3.2   POLYMERIC NANOPARTICLES AND DENDRIMERS

Polymeric nanoparticles have been used as ocular drug delivery carriers because of their ability to stably incorporate therapeutic drugs and provide controlled release over prolonged periods of time. These nanoparticles, also called dendrimers, are typically symmetric around the core and often adopt a spherical three-dimensional "tree"-like structure (Figure 3.1). These nanoparticles, prepared from polymers, are usually administered through topical application or injection. Polymeric micelles have been constructed via a wide range of biodegradable and water-soluble compounds, such as chitosan, poly(D,L-lactic acid) (PLA), poly(lactic-*co*-glycolic acid), poly-ε-caprolactone, and poly(ethylene glycol)-*block*-poly(L-lysine) [PEG-*b*-P(Lys)] (Barbu et al. 2006; Ideta et al. 2005; Nemoto et al. 2007).

One of the most promising topically applied polymeric drug delivery systems involves the use of chitosan nanoparticles (de Salamanca et al. 2006). Because chitosan is the second most abundant polysaccharide found in nature, it is relatively inexpensive, nontoxic, and biodegradable (Alonso and Sanchez 2003; Wadhwa et al. 2009a). Furthermore, chitosan is a strong mucoadhesive molecule because of ionic interactions between the positively charged amino groups of chitosan and the negatively charged sialic acid residues of mucin (Ludwig 2005). Hence, chitosan-based micelles have longer retention times, which allow for greater drug penetration through ocular membranes. In addition to chitosan's bioadhesive properties, chitosan also improves drug permeability by opening the tight junctions of epithelial cells (Han et al. 2004). De Campos et al. (2001) showed that cyclosporin A (CsA)-loaded chitosan nanoparticles increased the therapeutic drug level in the ocular tissue of rabbits compared to aqueous solutions of CsA. Through *in vitro* studies, De Campos et al. (2001) also demonstrated that the rate of drug release was fastest during the first hour and was followed by a slow gradual release over the next 24 hours. These experiments illustrated chitosan nanoparticle's ability to increase bioavailability and provide a gradual, controllable release.

In another study, Badawi et al. (2008) confirmed chitosan nanoparticles' capacity to maintain intimate contact with the cornea and gradually release indomethacin to both internal and external ocular tissues. In this study, 20 μL of chitosan nanoemulsion and 20 μL of 0.1% indomethacin were applied to the cul-de-sac of each conjunctivum of 24 rabbits. After 1, 2, 4, and 6 hours, the rabbits were sacrificed and the eyes were enucleated for pharmacokinetic analysis. The concentration of indomethacin was significantly higher in the cornea and aqueous humor of rabbit eyes treated with nanoemulsion. The results reported by De Campos and Badawi illustrated the potential use of chitosan nanoparticles as an ocular drug delivery system.

The application of polyion complex (PIC) ions has also received considerable attention as a viable polymeric nanoparticulate drug delivery method. PIC micelles have shown promise as a dendrimer porphryin (DP) carrier in photodynamic therapy (PDT) (Tamaki 2009). PDT is used to treat CNV in exudative/neovascular AMD (Bressler and Treatment Age Related Macular Degeneration 1999). Ideta et al. (2005) proposed the use of a PEG-*b*-P(Lys)–based PIC micelle that significantly improved current PDT therapies. They demonstrated that PIC micelles could localize DP, a molecule that produces cytotoxic singlet oxygen molecules when exposed to laser irradiation, in laser-induced CNV lesions in rats. Furthermore, free DP disappeared from CNV lesions within 24 hours, while DP-loaded PIC micelles persisted much longer. In addition to reporting a sustained presence of DP in CNV lesions, PIC micelles concentrate in CNV lesions through the enhanced permeability retention (EPR) effect (Kimura et al. 2001). According to the EPR effect, drug carriers aggregate in areas of the tissue where leaky pathological vasculature allow for greater penetration compared to normal epithelial tight junctions (Torchilin 2007). PIC micelles take advantage of the EPR effect and can deliver large quantities of DP to the CNV lesions of AMD patients. Ideta et al. (2005) also demonstrated its effectiveness in PDT treatment. Fifteen minutes following intravenous injection of DP, PDT lasers were applied to the CNV lesions. The fluorescein angiogram from DP-loaded PIC treatment showed a decrease of leakage compared to free DP treatment or PDT treatment alone. Additionally, PIC micelles are able to reduce phototoxicity. Four hours following injection of DP-loaded micelle and Photofrin (a compound currently used in PDT treatment), the rats were exposed to broadband visible light. Interestingly, skin phototoxicity was not observed in DP-loaded micelle treated rats but was detected in Photofrin-treated rats (Ideta et al. 2005).

Other notable animal models that have evaluated the effectiveness of polymeric nanoparticles include experimental autoimmune uveoretinitis (EAU) in rats (Kang et al. 2009; Sakai et al. 2006). Sakai et al. (2006) induced EAU in rats by subcutaneous injection of *Mycobacterium tuberculosis* H37Ra. Thirteen days after EAU induction, the rats received intravenous injections of betamethasone phosphate (BP) solution, poly(lactic acid) nanoparticle–loaded betamethasone phosphate (BP-PLA NP), or saline. Among the three groups, BP-PLA NP–treated rats showed the lowest histology scores 7 days after treatment. These results suggest that polymeric nanoparticles can effectively enhance the effect of BP in controlling intraocular inflammation.

Recently, Kang et al. (2009) induced murine retinoblastoma to evaluate the effectiveness of carboplatin-loaded nanoparticles for the treatment of human retinoblastoma. The transgenic mice carrying the simian virus 40 large T-antigen (TAg) driven by the human luteinizing hormone β–subunit promoter gene (LHβ) was used. These mice, first developed by Windle et al. (1990), exhibit heritable ocular tumors with histological, ultrastructural, and immunohistochemical features identical to those of human retinoblastoma. Forty LHβ-TAg transgenic mice were injected with carboplatin-loaded poly(amidoamine) dendrimer, free carboplatin solution, and phosphate-buffered saline (PBS) (Kang et al. 2009). The eyes receiving carboplatin-loaded nanoparticles showed significantly smaller tumor mass compared with eyes treated with free carboplatin solution and PBS.

Numerous polymeric-based nanocarriers have been evaluated using animal models. This class of nanoparticles has been extensively studied and proven to be effective drug carriers for both topical administration and injection. Polymeric nanocarriers have been developed to successfully increase bioavailability at target regions and to decrease systemic toxicity. Furthermore, these particles have all shown the ability to maintain a sustained release of the drug over longer periods of time, thereby decreasing the frequency of treatment required. The main limitations to the use of polymeric nanoparticles, such as formulation stability, particle size uniformity, and control of drug release rate, have largely been resolved (Nagarwal et al. 2009). Polymeric nanoparticles have shown significant advantages in animal model studies and should be further investigated in clinical applications.

3.3   GOLD NANOPARTICLES

Although there has been limited research on the use of gold nanoparticles (otherwise known as colloidal gold and nanogold) as drug delivery vehicles in the eye, this class of nanoparticles is appealing because of its minute particle size. Several studies have shown that gold nanoparticles can be created with a uniform particle size distribution around 5 nm (Kim et al. 2009; Mukherjee et al. 2005). This is important because particles under 20 nm can pass the BRB, thereby allowing intravenous delivery of drugs (Kim et al. 2009). It is ideal to deliver drugs via intravenous injection because of the risks involved with intraocular injections. Kim et al. (2009) examined the ability of gold nanoparticles to pass through the BRB and ocular toxicity effects in C57BL/6 mice. They reported that 20-nm gold nanoparticles were distributed in all layers of the retina 24 hours after intravenous injections, while 100-nm gold nanoparticles were not found in the retina. Furthermore, intraretinal gold nanoparticles did not affect the cellular viability of retinoblastoma cells, retinal endothelial cells, and astrocytes.

Instead of being a potential delivery system, gold nanoparticles are also therapeutically effective. A study analyzed the antiangiogenic properties of gold nanoparticles in mice (Mukherjee et al. 2005). Mukherjee et al. demonstrated that gold nanoparticles bind to heparin-binding proteins, such as vascular endothelial growth factor (VEGF) and basic fibroblast growth factor, thereby inhibiting subsequent angiogenic pathways. They developed a nude mouse ear model by intradermal injection of adenoviral vector VEGF virus into nude mice ears. These mice exhibited VEGF165-induced angiogenesis. Gold nanoparticles were then injected intravenously to test the efficacy of these nanoparticles in inhibiting angiogenesis. Mukherjee et al. demonstrated that gold nanoparticles interacted with VEGF to decrease its proliferative activity. This animal model study is relevant to ocular therapies since elevated VEGF expression is accepted as one of the major factors that contributes to neovascular/exudative AMD.

In another study, goat immunoglobulin G (IgG) adsorbed to gold nanoparticles was delivered to the subretinal space of rabbit eyes (Hayashi et al. 2009). One week after subretinal injection, goat IgG was found in the retinal pigment epithelium (RPE) and the photoreceptor cells. Although IgG was successfully targeted to photoreceptor cells and the RPE, retinal degeneration was observed after the injection of IgG-adsorbed gold nanoparticles, goat IgG with PBS, or only gold nanoparticles. The retinal degeneration may be due to the anatomical characteristics of the rabbit retina (Berglin et al. 1997) or due to mechanical damage of the RPE cells from the subretinal injection (Hayashi et al. 2009). Interspecies differences in anatomical structure may limit the usefulness of making general conclusions regarding the efficacy and safety of nanoparticle delivery systems.

Current research on gold nanoparticles shows no toxicity effects. However, because of their minute size, gold nanoparticles are also much more quickly cleared from the body compared with larger-sized nanoparticles. Furthermore, the complete long-term effects of gold nanoparticles have not been thoroughly explored. Gold nanoparticles may sometimes aggregate to form larger particles and consequently become trapped in the human body. Therefore, extensive research is still required to improve our understanding of the long-term biological effects of gold nanoparticles. Gold nanoparticles can be stabilized by heterobifunctional polyethylene glycol (PEG). The heterobifunctional PEG ligands contain a dithiol group for stable anchoring onto the gold surface and a terminal carboxyl group for coupling of a particular drug to the outside of the PEG shell (Figure 3.1).

3.4 LIPID NANOPARTICLES

Liposomes and solid lipid nanoparticles (SLNs) are the major lipid-based drug delivery platforms. Lipid nanoparticles are especially attractive because they are mainly formulated from phospholipids or triglycerides, resulting in easily degradable and nontoxic drug delivery vehicles (Mehnert and Mader 2001). Furthermore, lipid nanoparticles have a much larger payload of drugs compared to polymeric or gold nanoparticles.

Liposomes are composed of an aqueous core surrounded by a phospholipid bilayer (Figure 3.1). Their size, which is largely dependent on their method of formulation, typically ranges from 50 to 400 nm in size. Liposomes are formed via self-assembly and can be prepared by many methods, such as ultrasonication and reverse-phase evaporation (Sakai et al. 2008). Liposomes are incredibly useful because they can transport both hydrophilic drugs in the aqueous core and hydrophobic drugs in the lipid bilayer. In addition to the variety of drugs they can carry, liposomes are also amenable to versatile surface modifications, such as conjugation to targeting moieties or attachment to PEG, which can help nanoparticles evade immune recognition (Caliceti and Veronese 2003).

There has been a long history of using a liposomal drug delivery system for the treatment of ocular diseases. From as early as the 1980s, liposome-encapsulated drugs have been tested in animal models and shown to be effective in therapies for glaucoma, proliferative vitreoretinopathy, and pseudomonas keratitis (Alvarado 1989; Assil et al. 1991; Frucht-Perry et al. 1992). However, the liposomes used in these studies were not characterized and their diameters might have been too large to be considered nanoparticles. Extensive studies have been conducted to evaluate liposomal delivery of verteporfin, a benzoporphyrin derivative, for PDT therapy of CNV (Husain et al. 1996; Kramer et al. 1996; Miller et al. 1995). These animal model experiments in monkeys were important for determining the safety, efficacy, and optimal parameters of PDT with verteporfin. Without these animal models, verteporfin therapy would not have advanced to clinical trials, such as the treatment of AMD with PDT investigation, or the verteporfin in PDT trials (Michels and Schmidt-Erfurth 2001).

More recently, Moon et al. (2006) showed that subconjunctival injection of liposome-bound, low-molecular-weight heparin (LMWH) could increase the absorption rate of subconjunctival hemorrhage in rabbits. The liposomes used in this study were formulated from 1,2-dioleoyl-3-trimethylammonium-propane. Moon et al. reported that rabbits injected with liposome-bound LMWH showed significantly faster subconjunctival hemorrhage absorption rates compared to rabbits injected with free LMWH, liposome injection, or no injection. In another study, the final concentration of bevacizumab (Avastin) was much greater in rabbit eyes after 42 days when injected with liposomal-encapsulated bevacizumab compared to free bevacizumab (Abrishami et al. 2009). The lipids used in this study were formulated from egg phosphatidylcholine and 1,2-dipalmitoyl-*sn*-glycero-3-phosphocholine. Both these studies offered more detailed characterization of the liposomes compared to previous studies. The mean sizes of the liposomes used in these studies were approximately 550 and 220 nm, respectively.

Although these studies demonstrated the effectiveness of liposome-based drug delivery, they also pointed out several limitations and challenges associated with this drug delivery platform. Because of the large size of liposomes compared to polymeric or gold nanoparticles, they cannot pass through the BRB (Abrishami et al. 2009). Therefore, liposomal encapsulated drugs cannot take advantage of the lower risks associated with intravenous injections. In addition to their large size, liposomes are not stable and their size can change depending on the solvent, temperature, and pH. This becomes a great concern when liposomes are injected into animal models, where the physical and chemical environment surrounding the liposomes is completely altered compared to the characterization conditions. Unpredicted changes to the size of liposomes are troubling because small changes in nanoparticle size can greatly affect the routes of action and clearance. Although both studies used cholesterol to stabilize the liposomes, this raises issues related to formulation. The proportion of cholesterol used must be optimized such that the liposomes are stable enough to remain intact after injection into the animal, but not too stable to hinder the effective release of encapsulated drugs. Because of the unique physiological environment of each animal, it is difficult to predict optimal formulations for humans from animal studies. Abrishami et al. also reported that encapsulation efficiencies were less than 50%. Low encapsulation efficiencies become a significant concern in cases where the encapsulated drug is expensive.

Another class of lipid nanoparticles that is receiving a considerable amount of attention is SLNs. Like liposomes, SLNs are also composed of naturally occurring lipids. However, instead of liposomes' phospholipid bilayer, SLNs have a solid lipid core with the polar triglyceride heads facing the outside aqueous phase (Figure 3.1) (Sawant and Dodiya 2008; Wadhwa et al. 2009b). SLNs are generally smaller than liposomes and range from as low as 25 nm to several hundred nanometers, depending on the method of formulation (Mukherjee et al. 2009). SLNs can be formulated by a variety of methods, such as ultrasonification (Hou et al. 2003), hot/cold homogenization (Helgason et al. 2009), and emulsification/evaporation (Sjostrom and Bergenstahl 1992). SLNs can be loaded with hydrophobic or hydrophilic drugs and can also be modified to control rates of drug release and SLN targeting. However, because SLN particle sizes are much smaller than liposomes, they have a limited drug-loading capacity.

One of the earliest studies that successfully utilized SLNs as an ocular drug delivery method was developed by Cavalli et al. (2002). In this study, 100-nm-diameter tobramycin-loaded SLNs (hexadecyl phosphate-based formulation) and free tobramycin were topically applied to male New Zealand albino rabbit eyes. They recorded that the concentration of tobramycin in the aqueous humor after 6 hours was significantly higher in the rabbit eyes that were treated with tobramycin-loaded SLNs. Cavalli et al. proposed that the increased bioavailability of tobramycin might be due to their extremely small size and entrapment in the mucin layer covering the corneal epithelium. This experiment confirms its effectiveness as an antibiotic for bacterial keratitis (Frucht-Perry et al. 1992).

Currently, SLNs have been successfully used to deliver CsA, a strong immunosuppressant agent for uveitis (Nussenblatt et al. 1985; Seyfoddin et al. 2010). CsA-loaded SLNs (based on a formulation of mono-, di-, and tri-acylglycerols of behenic acid) were topically administered to the cul-de-sac of male New Zealand albino rabbits (Gokce et al. 2009). SLN characterization showed that the mean particle diameter was 225.9 ± 5.5 nm with a negative surface charge. Gokce et al. reported that the concentration in the aqueous humor of the rabbits was significantly higher in the CsA-loaded SLNs compared to commercially available CsA ophthalmic emulsions. Although this study showed the effectiveness of SLN as a drug delivery vehicle, it also demonstrated the instability of SLN molecules. Following autoclaving the SLN particles for sterility, the mean particle size increased significantly.

Another study also used SLN to deliver CsA to the eye (Basaran et al. 2010). Basaran et al. incorporated CsA into cationic SLNs and this formulation was topically applied to sheep eyes. Basaran reported that the CsA concentrations measured in both the aqueous and vitreous humor (<45 ng/mL) at various time points up until 48 hours were significantly lower than the required concentration (0.05–0.3 µg/mL) to suppress immune activity. It was noted that the CsA concentration did not start to decrease during the 48 hours; in fact, the CsA concentration actually continued to increase between the 24- and 48-hour time points. The inability to successfully obtain the desired CsA concentration might be due to the fact that the SLN formulation was too stable to release CsA at a sufficient rate. Formulation stability is a delicate issue that is challenging to optimize because of changing environmental conditions when translating *in vitro* studies to *in vivo* models.

Animal models are essential for determining the safety and efficacy of drugs. Although a variety of animal species are used to evaluate lipid nanoparticles, rabbits are the most popular choice for ocular studies. However, clear morphological and physiological differences between human and rabbit ocular surfaces are well documented. The ratio of the areas of conjunctiva to cornea is two times greater in humans than in rabbits (Watsky et al. 1988). Since topically applied drugs likely penetrate through the cornea first, a decrease in relative cornea surface area in humans may dramatically influence the pharmacokinetic parameters obtained through rabbit models. Furthermore, rabbits blink at a much slower rate compared to humans, which can impact preocular retention rates (Hornof et al. 2005).

An interesting alternative to the use of animal models for the evaluation of drug delivery is the use of bioengineered ocular tissue. Bioengineered human cornea was constructed to evaluate the release and permeation of timolol from surface-modified SLNs (Attama et al. 2009). Attama et al. constructed the human cornea using immortalized human corneal endothelial cells, stromal fibroblasts, and epithelial cells. Following human cornea construction, timolol hydrogen maleate was loaded onto SLNs that had mean diameters of approximately 40 nm. Attama et al. reported that timolol hydrogen maleate–loaded SLNs showed a more sustained permeation across the cornea compared to free timolol hydrogen maleate solution. Furthermore, they noted that bioengineering human cornea tissue might eliminate species-related issues associated with animal studies. Although bioengineered tissue may be advantageous for investigating drug penetration and retention times, it does not allow the examination of potential systematic side effects related to drug delivery. Therefore, the use of rabbits and other animal models are still essential for developing a more complete understanding of nanoparticulate drug delivery systems.

Although lipid nanoparticles are extremely useful as drug delivery vehicles for ocular diseases, there are numerous challenges associated with formulation, targeting, encapsulation, and release of the drug. Animal models have provided support for the safety and efficacy of many lipid-based drug delivery systems. Although animal models are incredibly useful and necessary, specific differences in different species may limit the accuracy of these studies for the prediction of drug delivery characteristics in humans.

3.5   ADDITIONAL OCULAR DRUG DELIVERY SYSTEM PLATFORMS

Even though polymeric, gold, and lipid nanoparticles represent the most developed nanoparticulate drug delivery platforms, there are also other less developed systems, such as nanocrystals and nanoporous filters. Nanocrystals are crystalline particles with dimensions measured in nanometers. Cubosomes, self-assembled liquid crystalline nanoparticles produced from surfactants, have generated significant excitement as a drug delivery system (Garg et al. 2007). Cubosomes are bicontinuous cubic phase liquid crystals that are produced when surfactants are added to water at high concentrations (Garg et al. 2007). The unique microstructure of these liquid crystals allows for a controlled release of solubilized drugs and proteins. In a recent study, cubosomes were used as an ophthalmic delivery system for dexamethasone (Gan et al. 2010). Gan et al. showed that dexamethasone cubosome particles exhibited longer preocular retention times compared to dexamethasone solutions. Furthermore, they found that dexamethasone cubosomes generated higher concentrations of dexamethasone in the aqueous humor of male New Zealand albino rabbits at various time points (up to 240 minutes) following topical instillation Gan et al. also demonstrated that the corneal structure and tissue integrity were unaffected following the application of dexamethasone cubosomes, thereby illustrating its biocompatibility and short-term safety in rabbits.

Another proposed method of ophthalmic drug delivery is the use of nanoporous filters (Orosz et al. 2004). Although not technically considered a nanoparticulate drug delivery system, nanoporous filters use nanotechnology to control the release of drugs. Orosz et al. proposed the use of aluminum oxide filters with 20-nm pores that would limit the rate of drug release to the eye. The nanofilter-based drug delivery system would allow for a large reservoir of the drug to be implanted in the posterior segment of the eye (Orosz et al. 2004). Orosz et al. treated human retinal vascular endothelial cells with hydrogen peroxide and observed the growth of the cells in the presence (loaded into nanoporous filtered capsules) and in the absence of catalase, which counteracted the cytotoxicity of hydrogen peroxide. They found that catalase effectively diffused across the nanoporous filters and the cells survived to form a monolayer. Furthermore, they demonstrated that the reservoir capsule could hold enough drug content to provide sustained release for 4 months. Orosz et al. concluded that capsules based on nanoporous filters could provide a controllable delivery of a drug. Although their experiments demonstrated the potential of nanofiltration-based drug delivery systems, this platform needs to be developed and evaluated through animal models.

3.6 CLINICAL EVALUATION OF NANOPARTICULATE DRUG DELIVERY SYSTEMS

The ultimate goal of any nanoparticulate drug delivery system is its application in treating human diseases. Although animal models can evaluate the preliminary safety and efficacy of a nanoparticulate system, a drug delivery system's actual effectiveness can only be accurately assessed by clinical trials. After *in vitro* and *in vivo* studies have demonstrated that a nanoparticle can maintain a controlled release of a drug in the targeted tissue without creating any serious toxicity concerns, then the nanoparticulate system can be evaluated on humans (Figure 3.2).

Although many nanoparticulate drug delivery systems are currently being developed for clinical applications, there has only been one nanoparticulate drug delivery system that has progressed to a phase I clinical study in ocular disease (Libutti et al. 2010). In this clinical study, Libutti et al. used CYT-6091, which is made of recombinant human tumor necrosis factor-alpha (rhTNF-alpha) bound to the surface of colloidal gold nanoparticles. These gold nanoparticles were thiolyated with PEG and averaged 27 nm in diameter. The addition of PEG masks the gold and rhTNF from the reticuloendothelial system, thereby avoiding phagocytic clearance (Patel and Moghimi 1998). These colloidal gold nanoparticles can be used to treat advanced-stage cancer patients by delivering rhTNF to the tumor vasculature. Once rhTNF has been delivered to the tumor, it induces hyperpermeability in the tumor, which enhances chemotherapies' antitumor effects.

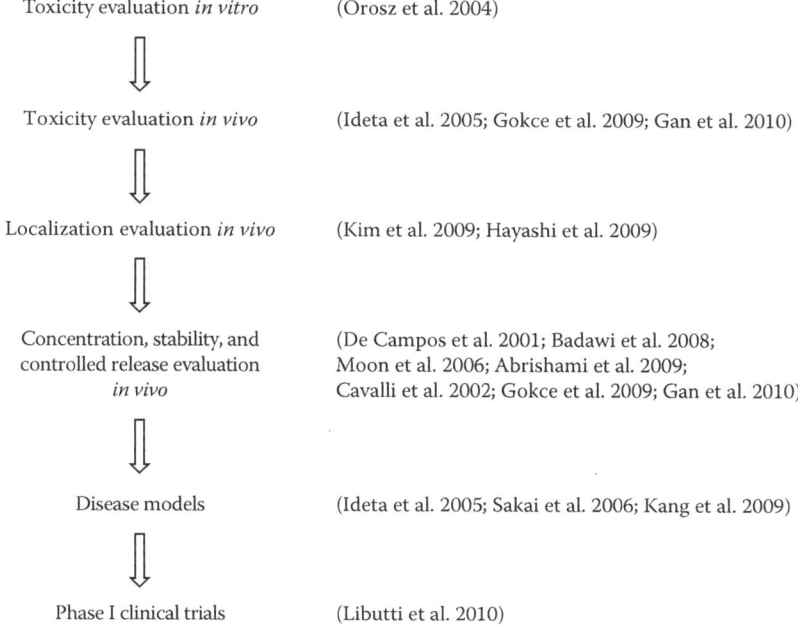

FIGURE 3.2 Developmental process for new drug delivery systems.

Despite TNF's remarkable antitumor effects, numerous clinical trials have revealed their toxic side effects, such as hypotension, hepatotoxicity, and fatigue (Kimura et al. 1987). Hence, it is critical to balance TNF's concentration such that it is both efficacious and nontoxic. PEGylated colloidal gold nanoparticles can sequester TNF within a tumor, thereby reducing the accumulation of the drug in healthy organs.

In the above study by Libutti et al. (2010), 9 of the 29 patients with malignancies had ocular melanoma, with two doses of CYT-6091, and found that the treatment was generally well tolerated. The dose of CYT-6091 was progressively escalated from 50 to 600 $\mu g/m^2$. They found that a predictable, yet controllable fever occurred in the first two patients that were treated with CYT-6091. Subsequently, the rest of the patients were given prophylactic antipyretics and H_2 blockers to counter the fever. Of the 29 patients treated, 18 of them developed at least one episode of hypotension. Furthermore, pharmacokinetic analysis illustrated that the half-life of rhTNF was five times longer when used with the gold nanoparticles. Additional evaluation of safety concerns will be addressed in a planned phase II clinical trial. Moreover, the efficacy of CYT-6091 in combination with chemotherapies will be evaluated to determine if CYT-6091 can effectively enhance antitumor effects.

3.7 CONCLUSION

Nanoparticulate drug delivery systems have received considerable attention in recent years because of their ability to deliver high concentrations of a drug to a targeted location. There has been a strong interest in the utilization of nanotechnology-based drug delivery systems for ocular therapies due to the eye's unique anatomy (Table 3.2). It is particularly challenging to deliver drugs systemically to the eye because of the presence of the BRB and the low permeability of the sclera and cornea. Conventional methods of ocular drug delivery can be limited by poor bioavailability, quick clearance, and potential drug toxicity. Nanoparticulate drug delivery systems have the ability to increase the effectiveness of drugs by increasing the concentration of the drug at the targeted tissue, increasing the retention of the drug through controlled release, and decreasing the toxicity and side effects of therapeutic treatments. Researchers are working to develop a variety of drug delivery platforms using polymeric, gold, or lipid nanoparticles. Each of these systems offers different drug delivery properties because of their unique physical and biochemical characteristics. Hence, each nanoparticulate drug delivery system requires comprehensive characterization and extensive evaluation.

Animal models are usually used after preliminary *in vitro* characterization and before clinical trials. Animals, such as sheep, rabbits, and rodents, have been used to evaluate the toxicity associated with the drugs, retention time of the drug in the eye, and also the effectiveness of the drug-loaded nanoparticle to treat a certain disease. Although animal models are incredibly useful for assessing initial nanocarrier properties, there are many limitations to their use. The anatomy and physiology of each species is different and so the benefits and adverse effects observed in a particular animal model may not translate to humans. Furthermore, the kinetic properties of a drug carrier, such as its permeability rate across ocular membranes and the rate

TABLE 3.2
Ocular Applications of Nanoparticles

Nanoparticulate Carrier	Study	Type of Nanoparticle	Drug or Therapeutic	Animal	Route of Administration	Uses of Nanoparticulate Carrier
Chitosan	DeCampos (2001); Badwai (2008)	Polymeric	CsA, indomethacin	Rabbits	Topical administration	Longer retention time and greater permeability
PIC micelles	Ideta (2005)	Polymeric	Porphryin	Rats	Intravenous injection	Concentration of the drug in CNV lesions, thereby reducing drug toxicity effects
PLA	Sakai (2006)	Polymeric	BP	Rats	Subcutaneous injection	Enhances BP's anti-inflammatory effects
Poly(amidoamine) dendrimer	Kang (2009)	Polymeric	Carboplatin	Mice	Subcutaneous injection	Enhances carboplatin's antitumor effect
Gold	Hayashi (2009)	Gold	IgG	Rabbits	Subcutaneous injection	Assists delivery of IgG to photoreceptor cells and RPE
1,2-Dioleoyl-3-trimethylammonium-propane	Moon (2006)	Lipid	LMWH	Rabbits	Subcutaneous injection	Enhances LMWH's ability to induce faster absorption of subconjunctival hemorrhages
1,2-Dipalmitoyl-sn-glycero-3-phosphocholine	Abrishami (2009)	Lipid	Bevacizumab	Rabbits	Intravitreal injection	Increased residence time of bevacizumab in vitreous
Hexadecyl phosphate	Cavalli (2002)	Lipid	Tobramycin	Rabbits	Topical administration	Increased concentration of tobramycin in the aqueous humor
Glyceryl behenate	Gokce (2009); Basaran (2010)	Lipid	CsA	Rabbits, sheep	Topical administration	Significantly higher concentration of the drug in the aqueous humor
Monoolein and Poloxamer 407	Gan (2010)	Nanocrystals	Dexamethasone	Rabbits	Topical administration	Longer preocular retention time and higher concentration of dexamethasone in the aqueous humor

of controlled release, cannot be accurately predicted in humans due to the different anatomy and physiology between humans and animals.

Ultimately, human trials are required to more precisely evaluate nanoparticulate-based drug delivery systems. Currently, only one nanoparticulate drug delivery system, based on the use of CYT-6091, a gold-based nanoparticle, has progressed to a phase I clinical study in ocular diseases. With the continued evaluation through *in vitro* testing and animal models, we expect to see an increased number of nanoparticulate-based drug delivery systems in the years to come. These new systems can allow physicians to deliver a lower overall concentration of the drug into the system (which would decrease toxicity issues and other adverse effects) while at the same time increase the concentration of the drug at the targeted region. Additionally, patients might require less-frequent drug administration due to the controlled rate of release offered by nanocarriers. Advances in nanotechnology have the potential to significantly change our current standard of care and improve the overall health of patients.

REFERENCES

Abrishami, M., S. Z. Ganavati, D. Soroush, M. Rouhbakhsh, M. R. Jaafari, and B. Malaekeh-Nikouei. 2009. Preparation, characterization, and *in vivo* evaluation of nanoliposomes-encapsulated bevacizumab (avastin) for intravitreal administration. *Retina* 29 (5):699–703.

Alonso, M. J., and A. Sanchez. 2003. The potential of chitosan in ocular drug delivery. *J Pharm Pharmacol* 55 (11):1451–63.

Alvarado, J. A. 1989. The use of a liposome-encapsulated 5-fluoroorotate for glaucoma surgery: I. Animal studies. *Trans Am Ophthalmol Soc* 87:489–514.

Assil, K. K., M. Hartzer, R. N. Weinreb, M. Nehorayan, T. Ward, and M. Blumenkranz. 1991. Liposome suppression of proliferative vitreoretinopathy. Rabbit model using antimetabolite encapsulated liposomes. *Invest Ophthalmol Vis Sci* 32 (11):2891–7.

Attama, A. A., S. Reichl, and C. C. Muller-Goymann. 2009. Sustained release and permeation of timolol from surface-modified solid lipid nanoparticles through bioengineered human cornea. *Curr Eye Res* 34 (8):698–705.

Badawi, A. A., H. M. El-Laithy, R. K. El Qidra, H. El Mofty, and M. El Dally. 2008. Chitosan based nanocarriers for indomethacin ocular delivery. *Arch Pharm Res* 31 (8):1040–9.

Barbu, E., L. Verestiuc, T. G. Nevell, and J. Tsibouklis. 2006. Polymeric materials for ophthalmic drug delivery: Trends and perspectives. *J Mater Chem* 16 (34):3439–43.

Basaran, E., M. Demirel, B. Sirmagul, and Y. Yazan. 2010. Cyclosporine-A incorporated cationic solid lipid nanoparticles for ocular delivery. *J Microencapsul* 27 (1):37–47.

Berglin, L., P. V. Algvere, and S. Seregard. 1997. Photoreceptor decay over time and apoptosis in experimental retinal detachment. *Graefes Arch Clin Exp Ophthalmol* 235 (5):306–12.

Bressler, N. M., and Treatment Age Related Macular Degeneration. 1999. Photodynamic therapy of subfoveal choroidal neovascularization in age-related macular degeneration with verteporfin—One-year results of 2 randomized clinical trials—TAP report 1. *Arch Ophthalmol* 117 (10):1329–45.

Cai, X., S. Conley, and M. Naash. 2008. Nanoparticle applications in ocular gene therapy. *Vision Res* 48 (3):319–24.

Caliceti, P., and F. M. Veronese. 2003. Pharmacokinetic and biodistribution properties of poly(ethylene glycol)-protein conjugates. *Adv Drug Deliv Rev* 55 (10):1261–77.

Cavalli, R., M. R. Gasco, P. Chetoni, S. Burgalassi, and M. F. Saettone. 2002. Solid lipid nanoparticles (SLN) as ocular delivery system for tobramycin. *Int J Pharm* 238 (1–2):241–45.

Chen, J., S. Patil, S. Seal, and J. F. McGinnis. 2006. Rare earth nanoparticles prevent retinal degeneration induced by intracellular peroxides. *Nat Nanotechnol* 1 (2):142–50.

De Campos, A. M., A. Sanchez, and M. J. Alonso. 2001. Chitosan nanoparticles: A new vehicle for the improvement of the delivery of drugs to the ocular surface. Application to cyclosporin A. *Int J Pharm* 224 (1–2):159–68.

de Salamanca, A. E., Y. Diebold, M. Calonge, C. Garcia-Vazquez, S. Callejo, A. Vila, and M. J. Alonso. 2006. Chitosan nanoparticles as a potential drug delivery system for the ocular surface: Toxicity, uptake mechanism and *in vivo* tolerance. *Invest Ophthalmol Vis Sci* 47 (4):1416–25.

Foy, S. P., R. L. Manthe, S. T. Foy, S. Dimitrijevic, N. Krishnamurthy, and V. Labhasetwar. 2010. Optical imaging and magnetic field targeting of magnetic nanoparticles in tumors. *ACS Nano* 4 (9):5217–24.

Frucht-Perry, J., K. K. Assil, E. Ziegler, H. Douglas, S. I. Brown, D. J. Schanzlin, and R. N. Weinreb. 1992. Fibrin-enmeshed tobramycin liposomes: Single application topical therapy of *Pseudomonas keratitis*. *Cornea* 11 (5):393–7.

Gan, L., S. Han, J. Q. Shen, J. B. Zhu, C. L. Zhu, X. X. Zhang, and Y. Gan. 2010. Self-assembled liquid crystalline nanoparticles as a novel ophthalmic delivery system for dexamethasone: Improving preocular retention and ocular bioavailability. *Int J Pharm* 396 (1–2):179–87.

Garg, G., S. Saraf, and S. Saraf. 2007. Cubosomes: An overview. *Biol Pharm Bull* 30 (2):350–3.

Gokce, E. H., G. Sandri, S. Egrilmez, M. C. Bonferoni, T. Guneri, and C. Caramella. 2009. Cyclosporine A-loaded solid lipid nanoparticles: Ocular tolerance and *in vivo* drug release in rabbit eyes. *Curr Eye Res* 34 (11):996–1003.

Grossniklaus, H. E., S. J. Kang, and L. Berglin. 2010. Animal models of choroidal and retinal neovascularization. *Prog Retin Eye Res* 29 (6):500–19.

Han, H. D., D. E. Nam, D. H. Seo, T. W. Kim, B. C. Shin, and H. S. Choi. 2004. Preparation and biodegradation of thermosensitive chitosan hydrogel as a function of pH and temperature. *Macromolecular Res* 12 (5):507–11.

Hayashi, A., A. Naseri, M. E. Pennesi, and E. de Juan, Jr. 2009. Subretinal delivery of immunoglobulin G with gold nanoparticles in the rabbit eye. *Jpn J Ophthalmol* 53 (3):249–56.

Helgason, T., T. S. Awad, K. Kristbergsson, D. J. McClements, and J. Weiss. 2009. Effect of surfactant surface coverage on formation of solid lipid nanoparticles (SLN). *J Colloid Interface Sci* 334 (1):75–81.

Hornof, M., E. Toropainen, and A. Urtti. 2005. Cell culture models of the ocular barriers. *Eur J Pharm Biopharm* 60 (2):207–25.

Hou, D., C. Xie, K. Huang, and C. Zhu. 2003. The production and characteristics of solid lipid nanoparticles (SLNs). *Biomaterials* 24 (10):1781–5.

Husain, D., J. W. Miller, N. Michaud, E. Connolly, T. J. Flotte, and E. S. Gragoudas. 1996. Intravenous infusion of liposomal benzoporphyrin derivative for photodynamic therapy of experimental choroidal neovascularization. *Arch Ophthalmol* 114 (8):978–85.

Ideta, R., F. Tasaka, W. D. Jang, N. Nishiyama, G. D. Zhang, A. Harada, Y. Yanagi, Y. Tamaki, T. Aida, and K. Kataoka. 2005. Nanotechnology-based photodynamic therapy for neovascular disease using a supramolecular nanocarrier loaded with a dendritic photosensitizer. *Nano Letters* 5 (12):2426–31.

Kang, S. J., C. Durairaj, U. B. Kompella, J. M. O'Brien, and H. E. Grossniklaus. 2009. Subconjunctival nanoparticle carboplatin in the treatment of murine retinoblastoma. *Arch Ophthalmol* 127 (8):1043–7.

Kim, J. H., J. H. Kim, K. W. Kim, M. H. Kim, and Y. S. Yu. 2009. Intravenously administered gold nanoparticles pass through the blood-retinal barrier depending on the particle size, and induce no retinal toxicity. *Nanotechnology* 20 (50):505101.

Kimura, H., T. Yasukawa, Y. Tabata, and Y. Ogura. 2001. Drug targeting to choroidal neovascularization. *Adv Drug Deliv Rev* 52 (1):79–91.

Kimura, K., T. Taguchi, I. Urushizaki, R. Ohno, O. Abe, H. Furue, T. Hattori, et al. 1987. Phase I study of recombinant human tumor necrosis factor. *Cancer Chemother Pharmacol* 20 (3):223–9.

Kramer, M., J. W. Miller, N. Michaud, R. S. Moulton, T. Hasan, T. J. Flotte, and E. S. Gragoudas. 1996. Liposomal benzoporphyrin derivative verteporfin photodynamic therapy. Selective treatment of choroidal neovascularization in monkeys. *Ophthalmology* 103 (3):427–38.

Libutti, S. K., G. F. Paciotti, A. A. Byrnes, H. R. Alexander, W. E. Gannon, Jr., M. Walker, G. D. Seidel, N. Yuldasheva, and L. Tamarkin. 2010. Phase I and pharmacokinetic studies of CYT-6091, a novel PEGylated colloidal gold-rhTNF nanomedicine. *Clin Cancer Res* 16 (24):6139–49.

Ludwig, A. 2005. The use of mucoadhesive polymers in ocular drug delivery. *Adv Drug Deliv Rev* 57 (11):1595–639.

Mehnert, W., and K. Mader. 2001. Solid lipid nanoparticles: Production, characterization and applications. *Adv Drug Deliv Rev* 47 (2–3):165–96.

Michels, S., and U. Schmidt-Erfurth. 2001. Photodynamic therapy with verteporfin: A new treatment in ophthalmology. *Semin Ophthalmol* 16 (4):201–6.

Miller, J. W., A. W. Walsh, M. Kramer, T. Hasan, N. Michaud, T. J. Flotte, R. Haimovici, and E. S. Gragoudas. 1995. Photodynamic therapy of experimental choroidal neovascularization using lipoprotein-delivered benzoporphyrin. *Arch Ophthalmol* 113 (6):810–8.

Moon, J. W., Y. K. Song, J. P. Jee, C. K. Kim, H. K. Choung, and J. M. Hwang. 2006. Effect of subconjunctivally injected, liposome-bound, low-molecular-weight heparin on the absorption rate of subconjunctival hemorrhage in rabbits. *Invest Ophthalmol Vis Sci* 47 (9):3968–74.

Mukherjee, P., R. Bhattacharya, P. Wang, L. Wang, S. Basu, J. A. Nagy, A. Atala, D. Mukhopadhyay, and S. Soker. 2005. Antiangiogenic properties of gold nanoparticles. *Clin Cancer Res* 11 (9):3530–4.

Mukherjee, S., S. Ray, and R. S. Thakur. 2009. Solid lipid nanoparticles: A modern formulation approach in drug delivery system. *Indian J Pharm Sci* 71 (4):349–58.

Nagarwal, R. C., S. Kant, P. N. Singh, P. Maiti, and J. K. Pandit. 2009. Polymeric nanoparticulate system: A potential approach for ocular drug delivery. *J Control Release* 136 (1):2–13.

Nemoto, E., H. Ueda, M. Akimoto, H. Natsume, and Y. Morimoto. 2007. Ability of poly-L-arginine to enhance drug absorption into aqueous humor and vitreous body after instillation in rabbits. *Biol Pharm Bull* 30 (9):1768–72.

Nguyen, P., M. Meyyappan, and S. C. Yiu. 2010. Applications of nanobiotechnology in ophthalmology—Part I. *Ophthalmic Res* 44 (1):1–16.

Nussenblatt, R. B., A. G. Palestine, C. C. Chan, M. Mochizuki, and K. Yancey. 1985. Effectiveness of cyclosporin therapy for Behcet's disease. *Arthritis Rheum* 28 (6):671–9.

Orosz, K. E., S. Gupta, M. Hassink, M. Abdel-Rahman, L. Moldovan, F. H. Davidorf, and N. I. Moldovan. 2004. Delivery of antiangiogenic and antioxidant drugs of ophthalmic interest through a nanoporous inorganic filter. *Mol Vis* 10:555–65.

Patel, H. M., and S. M. Moghimi. 1998. Serum-mediated recognition of liposomes by phagocytic cells of the reticuloendothelial system—The concept of tissue specificity. *Adv Drug Deliv Rev* 32 (1–2):45–60.

Powell, A. C., G. F. Paciotti, and S. K. Libutti. 2010. Colloidal gold: A novel nanoparticle for targeted cancer therapeutics. *Methods Mol Biol* 624:375–84.

Sakai, H., T. Gotoh, T. Imura, K. Sakai, K. Otake, and M. Abe. 2008. Preparation and properties of liposomes composed of various phospholipids with different hydrophobic chains using a supercritical reverse phase evaporation method. *J Oleo Sci* 57 (11):613–21.

Sakai, T., H. Kohno, T. Ishihara, M. Higaki, S. Saito, M. Matsushima, Y. Mizushima, and K. Kitahara. 2006. Treatment of experimental autoimmune uveoretinitis with poly(lactic acid) nanoparticles encapsulating betamethasone phosphate. *Exp Eye Res* 82 (4):657–63.

Sawant, K. K., and S. S. Dodiya. 2008. Recent advances and patents on solid lipid nanoparticles. *Recent Pat Drug Deliv Formul* 2 (2):120–35.

Seyfoddin, A., J. Shaw, and R. Al-Kassas. 2010. Solid lipid nanoparticles for ocular drug delivery. *Drug Deliv* 17 (7):467–89.

Sjostrom, B., and B. Bergenstahl. 1992. Preparation of submicron drug particles in lecithin-stabilized O/W emulsions. 1. Model studies of the precipitation cholesteryl acetate. *Int J Pharm* 88 (1–3):53–62.

Tamaki, Y. 2009. Prospects for nanomedicine in treating age-related macular degeneration. *Nanomedicine (Lond)* 4 (3):341–52.

Thomson, H., and A. Lotery. 2009. The promise of nanomedicine for ocular disease. *Nanomedicine (Lond)* 4 (6):599–604.

Torchilin, V. P. 2007. Targeted pharmaceutical nanocarriers for cancer therapy and imaging. *AAPS J* 9 (2):E128–47.

Vishwakarma, K., M. Ramrakhiani, and O. P. Vishwakarma. 2008. Nanotechnology: A boon for medical science. *Int J Nanotechnol Applications* 2 (1):69–73.

Wadhwa, S., R. Paliwal, S. R. Paliwal, and S. P. Vyas. 2009a. Chitosan and its role in ocular therapeutics. *Mini Rev Med Chem* 9 (14):1639–47.

Wadhwa, S., R. Paliwal, S. R. Paliwal, and S. P. Vyas. 2009b. Nanocarriers in ocular drug delivery: An update review. *Curr Pharm Des* 15 (23):2724–50.

Watsky, M. A., M. M. Jablonski, and H. F. Edelhauser. 1988. Comparison of conjunctival and corneal surface areas in rabbit and human. *Curr Eye Res* 7 (5):483–6.

Windle, J. J., D. M. Albert, J. M. O'Brien, D. M. Marcus, C. M. Disteche, R. Bernards, and P. L. Mellon. 1990. Retinoblastoma in transgenic mice. *Nature* 343 (6259):665–9.

Zarbin, M. A., C. Montemagno, J. F. Leary, and R. Ritch. 2010. Nanomedicine in ophthalmology: The new frontier. *Am J Ophthalmol* 150 (2):144–62. e2.

4 Computer Modeling for Ocular Drug Delivery

Paul J. Missel

CONTENTS

4.1 INTRODUCTION

Models for drug delivery seek to understand both the unique dosage form and the physiology of the target for which they are intended. Such understanding is required to make accurate predictions regarding how the dosage form will release and deliver its payload, including the drug's absorption, distribution, and clearance, that is, pharmacokinetics.

Models have utility insofar as they accurately represent the system of interest. But frequently it is difficult to construct an accurate model of a biological system because the anatomy or physiology is poorly defined or poorly understood and key parameters are unknown or are influenced by biologic variability. Thus, by necessity, the theoretician must simplify the model and regard it as an approximation that attempts to describe only a part of the system. The simpler the model, the easier its representation by mathematics, but the greater the risk that what it predicts will not be an accurate reflection of reality. In the limitations of making simplifications, one may be subject to ridicule, as expressed in the old joke about the theoretical physicist who reduced the cow to a simple sphere in an attempt to strip away details unnecessary for the modeling of milk production (Harte 1988).

Sometimes such a radical simplification can help elucidate features and relationships that would not be obvious otherwise. However, with the availability of powerful geometric modeling software, it is becoming possible to more accurately represent the anatomy and to include quantitative aspects of the physiology so that we may move beyond simplistic representations. One might think that the ocular anatomy provides an opportunity for quantitative biology because although it does interact with the body at large, it is relatively self-contained, and it is somewhat accessible to observation and measurement. However, how the ocular physiologic process affects drug absorption, distribution, and clearance are still being worked out (Robinson et al. 2006; Lee and Robinson 2009).

There are different types of numerical modeling that have been applied to the eye. Noncompartmental methods attempt to characterize drug exposure without using detailed models, focusing entirely upon describing the data. Classical compartmental pharmacokinetic modeling constructs a model consisting of compartments, typically assuming first-order transfer of material between compartments. The number of compartments and the equations describing the transfer of material between them is derived empirically by the plasma or tissue concentration versus time data. Physiologic-based pharmacokinetic (PBPK) models attempt to improve upon classical models by associating compartments with specific tissues and attempt to incorporate aspects of anatomy and physiology such as tissue volumes and fluid flow rates. A recent example of a PBPK ocular model and its associated data appears in Figure 4.1 (Amrite et al. 2008). Mixed-effects modeling, which is also called population-based modeling, can use either theoretical compartments (classical pharmacokinetics) or physiological compartments (PBPK). However, the approach is often focused on characterizing population trends through covariates (such as race, gender, etc.), which is enabled by modeling both the typical population behavior and the between-subject variability (Luu et al. 2009).

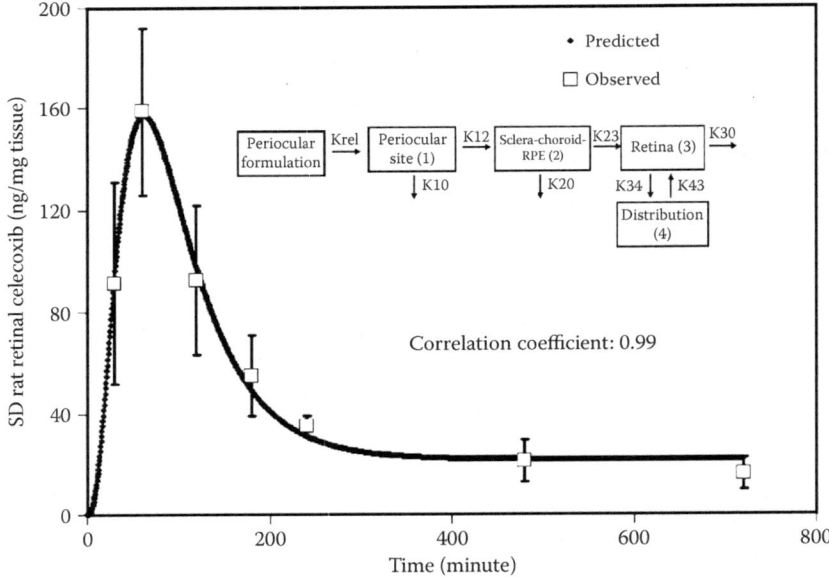

FIGURE 4.1 Model-predicted and model-observed concentrations of celecoxib in the retina after administration of 3 mg celecoxib by periocular injection to Sprague Dawley (nonpigmented) rats. (Adapted from Amrite AC, Edelhauser HF, Kompella UB, *IOVS* 49:320–332, 2008.)

The modeling approach presented in this chapter uses finite element or finite volume methods applied to engineering models, which attempt to accurately represent ocular anatomy and physiology. As such it can be considered as an extension of PBPK compartmental modeling; as the explicit geometric models can associate compartments with particular tissues, physiologic flows can be simulated numerically using flow boundary conditions and some physiologic processes can be represented (Heys 2002a). Geometric models can even accommodate motion of tissues, for example, the motion of the iris during visual accommodation (Heys 2002b). Transfer between adjacent or connected compartments can result from physical processes such as convection and diffusion, as well as biological processes such as active transport.

The ultimate goal is to develop simulation methods that can determine the time course of drug disposition from any mode of administration. First, a very brief description is given of aspects of ocular physiology as it impacts distribution and clearance of drugs. Next, a very brief review will be given of studies that have used anatomical and physiological modeling methods. The utility of these methods will be demonstrated by predicting the distribution and elimination of drugs following intravitreal (IVT) injection and the steady-state distribution of drugs resulting from administration of a juxtascleral device, which delivers drugs from the back of the eye. These modeling exercises will serve as useful tests for the method while simultaneously providing some insight into ocular physiology. Various methods for treating drug release from injected suspensions of particles will be discussed.

4.2 OCULAR PHYSIOLOGY AND DRUG DELIVERY

Drugs administered to the eye will be subject to several physicochemical and physiological processes, most of which will work to clear material from the system and thus constitute an obstacle to drug delivery. Under the process of *diffusion*, a drug will spread out from its initial location by random Brownian motion. *Convection* is the process whereby material is transported along with the flow of fluid, and thus understanding and modeling the ocular fluid flow patterns is paramount. In the anterior chamber, the fluid is nonviscous and is typically treated as a simple fluid. The cells that line the ciliary body secrete fluid that is ultimately derived from plasma at a steady rate (To et al. 2002). This fluid (aqueous humor) flows around the iris and exits the aqueous compartment through a thin band-like structure, the trabecular meshwork (Bron et al. 1997).

All other tissues can be treated as porous media, which are characterized by a hydraulic resistance, a consequence of which is the intraocular pressure. The hydraulic resistance of the trabecular meshwork is perhaps the property that is most variable and which determines the individual intraocular pressure. The tissues can also be characterized by elastic constants, and thus it is possible to simulate the change in pressure that results from an intraocular injection (Kotliar et al. 2007). Intraocular pressure gives rise to a flow of fluid that percolates through the porous media and constitutes another potential fluid flow pattern, which may contribute to convection.

In addition to the process of convective diffusion, various physicochemical factors can cause drugs to interact differently with various tissues, such as chemical *partitioning* or *binding*. Such processes will give rise to discontinuities in concentration at boundaries between tissues. Thus, simulation methods may be required that enable the dependent variable to experience a discontinuity at an interface (Missel 2000; Rim et al. 2005). These processes are particularly important for understanding the *transport* of lipophilic drugs through the cornea or through the multiple layers of the retina, choroid, and sclera.

Convection and diffusion are processes that work globally throughout the ocular anatomy. *Vascular* and *lymphatic* processes work locally to clear drugs from the system. Depending upon the properties of the drug, important locations for vascular clearance can include the conjunctiva (for topically applied drugs), the choroid (Ranta et al. 2010), and the iris/ciliary body (Missel 2012). The episclera and lymphatic system provides important contributions to clearance outside the sclera (Robinson et al. 2006; Chan et al. 2010)

4.3 PREVIOUS PHYSICAL/PHYSIOLOGICAL MODELS

One of the earliest applications of this idea, to represent the ocular compartment using an accurate geometric model, occurred before the advent of computers when Maurice (1959) constructed a model of a slice of the vitreous out of stainless steel. Instead of solving the partial differential equations that govern the diffusion of drugs, he took advantage of the fact that the equations describing heat transfer were analogous to those describing material diffusion. The simple expression derived in that work describing the relationship between the ratio of drug concentrations in the aqueous and vitreous compartments versus the rate of egress from the vitreous informed his later work (Maurice 1976) and will be demonstrated shortly to be very useful as a guide for evaluating the ability of various geometric models to predict the clearance of drugs after IVT injection.

In the late 1980s at Alcon, a similar approach was used based upon an electrical analogy as the differential equations describing voltage in a charge-free medium are identical to those describing diffusion of material in a source-free medium (Keister 1987). A large model of half of the eye was constructed out of stainless steel, using a plastic lens to represent the cornea. The model was filled with a weak salt solution (0.007% sodium nitrate) and connected to a voltage source. A probe was used to measure the change in potential with position and used to map the contours for concentration of drugs (Keister 1988). While this and the Maurice (1959) model were quite cumbersome, they nevertheless provided a means for constructing an accurate geometric model of the eye, admittedly using many more physical resources than is required by modern software.

In the 1990s, at least two different groups began using finite element software to simulate diffusion in geometric models for the eye. One group in Japan used extemporaneous finite element code, incorporating results from used *ex vivo* tissue transport experiments to identify appropriate constants for diffusivity and partitioning (Ohtori and Tojo 1994). The geometric contained only the vitreous compartment in the shape of a simple cylinder. Transfer to other compartments such as the aqueous humor and posterior tissues was incorporated using boundary conditions. The model was improved to use a spherical shape for the eye and to include the aqueous compartment (Tojo 1999, 2004). Another group in Canada used commercially available

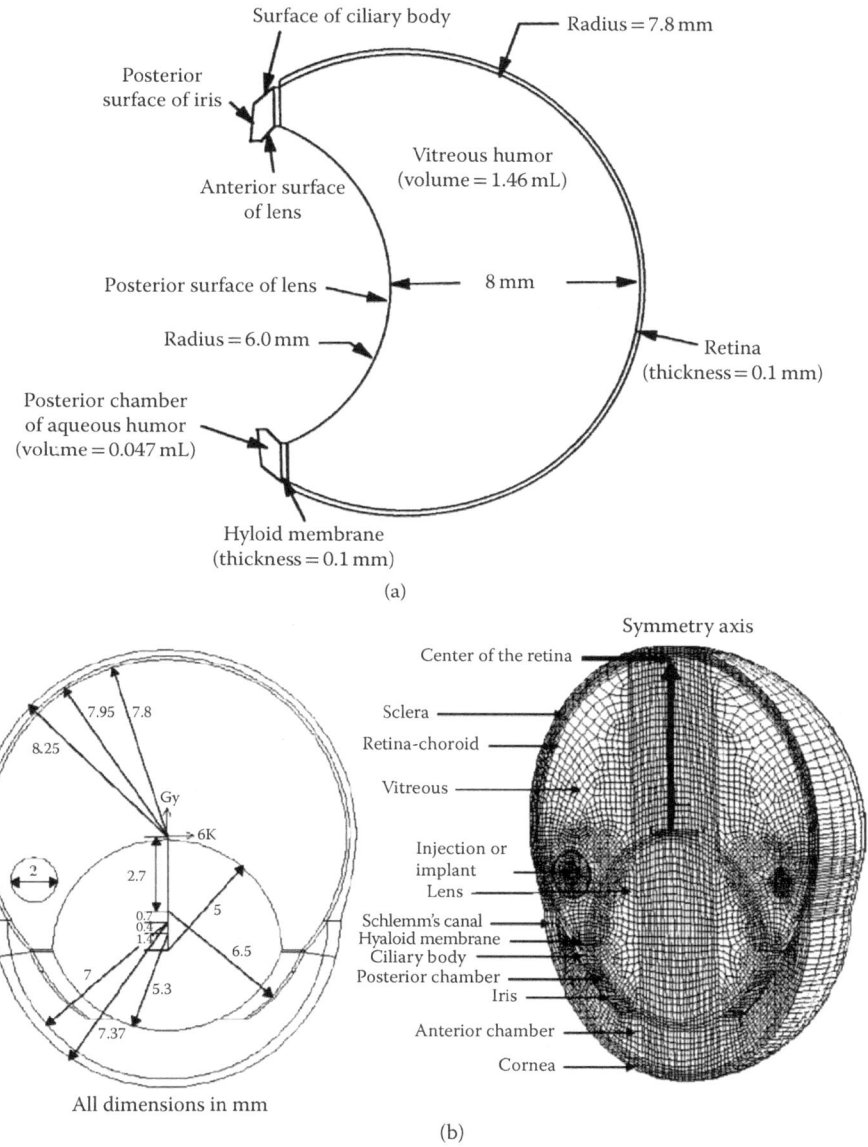

FIGURE 4.2 (a) Geometric model for the posterior portion of the rabbit eye used in finite element analysis study of Friedrich et al. (Adapted from Friedrich S, Cheng Y-L, Saville B., *Ann Biomed Eng*, 25:303–14,1997.) (b) Geometric model and mesh for the entire rabbit eye used in the study of Park et al. (Adapted from Park J, Bungay PM, Lutz RJ et al., *J Controlled Rel* 105:279–95, 2005.)

finite element software to construct models for the vitreous that had an anatomically accurate shape (Friedrich et al. 1997). This model was comprised mainly of the vitreous, containing only the posterior portion of the aqueous humor and a single thin shell outside the vitreous to represent the retina (Figure 4.2a).

In the last decade, at least three other groups have made serious improvements in both anatomical accuracy and incorporating more features of ocular physiology. Models proposed by Barocas et al. incorporated the effects of hydraulic pressure, active transport, choroidal blood flow, and saccadic eye movements (Stay et al. 2003; Balachandran and Barocas 2008, 2011). The geometric model was comprised mainly of the vitreous, but in one study tissue layers for the retina, choroid, and sclera were also present (Balachandran 2008; the choroid and sclera were lumped together in this model). The model proposed by Park et al. (2005) contains all tissues internal to the eye and simulates Navier–Stokes flow in the aqueous humor and hydraulic pressure-driven Darcy law flow through porous media in other tissues (Figure 4.2b). Models developed at Alcon include these same features and have also incorporated the effects of hydraulic pressure, choroidal elimination, partitioning, delivery from IVT and juxtascleral devices, particle dissolution from suspension depots, and differences among animal species (Missel 2000, 2002b; Missel et al. 2006, 2010a,b; Missel 2012) (Figure 4.3; Table 4.1).

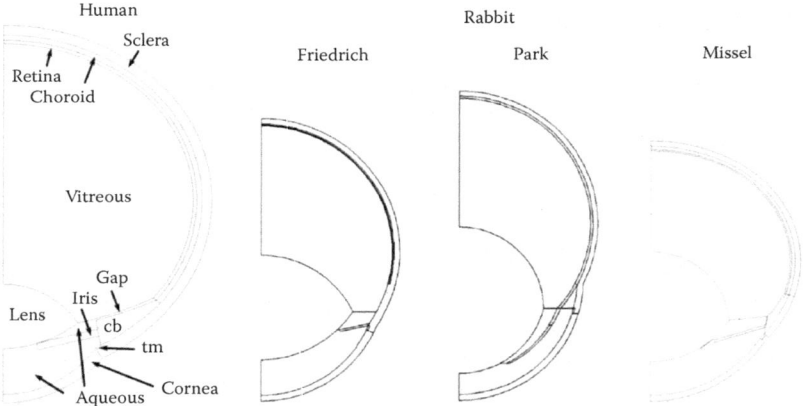

FIGURE 4.3 Scale comparison of various ocular models. The Friedrich model of Figure 4.2 was modified to add the cornea and the remainder of the anterior chamber. The model of Park is the same as in Figure 4.2. The remaining models were taken from Missel. (Adapted from Missel P, *Pharm Res* [in press], 2012.)

TABLE 4.1

Characteristics of the Various Ocular Models in Figure 4.4

| | Human | Rabbit | | |
		Friedrich	Park	Missel
Vitreous volume (mL)	4.85	1.462	1.579	1.52
Aqueous volume (mL)	0.238	0.234	0.245	0.325
Axial length (mm)	23.21	17.33	19.02	16.05
Half-width (mm)	12.42	8.25	8.25	9.00

4.4 COMPLEXITIES OF IVT INJECTION

Before attempting to understand the process of drug release from particles injected into the eye, we must first understand what happens during the injection process. The process of IVT injection is quite complex, as it is the injection of one fluid into another with very different properties (Lin 1997). Moreover, the intraocular pressure is elevated because the tissues of the outer sheath are elastic (Kotliar et al. 2007). Once the needle is withdrawn, a portion of the injected material may flow away from the region of injection by following along the path of the needle (Maurice 1987). Some of it may actually be expelled through the needle hole, but a portion may be drawn into the various layers between the tissues in the outer sheath, that is, the layers between the retina and choroid or the choroid and sclera. If the site of injection is toward the anterior, it is possible that a portion of the bolus may actually be expelled immediately into the aqueous humor. If it is near the outer tissues, then the excess pressure close to the needle opening may be sufficient to drive several percent of the injected bolus directly into the retina, choroid, and sclera (Figure 4.4) (Missel et al. 2003).

Thus, before the first experimental time point occurs, even if it is only a few minutes after injection, usually a fraction (typically 40–80%) of the total injected radioactivity is recovered from all tissues. Drugs which are in intimate contact with the exterior tissues can be cleared very quickly by the choroidal vasculature even if they are not expelled out through the needle hole left behind in the sclera. The initial

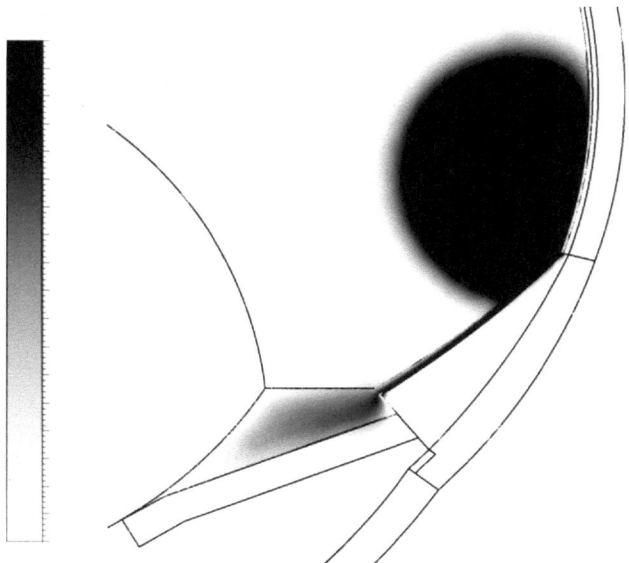

FIGURE 4.4 Simulated distribution of a drug after injection of 20 µL for 20 seconds through a 30-gauge needle inserted a few mm behind the limbus with very shallow penetration into the vitreous. (After Missel PJ, Chastain JE, Kiehlbauch CC, Dahlin DC, *Pharm Res* 27:1530–1546, 2010a.)

shape of the portion of the injection that remains in the vitreous has an irregular shape (Molokhia et al. 2009). The injected fluid does not necessarily mix immediately with the vitreous but can move to another location under the influence of gravity (Johnson and Maurice 1984).

Another factor that adds to the complexity is the variability of the rheological properties of the vitreous fluid among various animal species and changes in the vitreous that occur upon aging (Sebag 1987; Lee et al. 1992, 1994). These differences in vitreous rheology will most likely cause variability in the initial distribution and possibly also the rate of clearance after IVT injection or device implantation. Large differences in clearance of IVT-injected triamcinolone acetonide (TAC) and vascular endothelial growth factor have been observed after vitrectomy (Beer et al. 2003; Lee et al. 2010), but the delivery profile of dexamethasone from an IVT implant was unaffected by vitrectomy (Chang-Lin et al. 2011).

4.5 CLEARANCE AFTER IVT INJECTION

Although all the complications described in Section 4.4 are worthy of separate investigation, eventually the drug diffuses away from its irregular shape and establishes a quasi-steady-state distribution in a drug concentration that decays steadily with time under the influences of the physiologic flows and vascular sinks. Thus, simulations in which the initial condition for the bolus is a simple sphere located in a particular region of the vitreous may not capture all the intricacies of a particular injection site or modality; they are perfectly acceptable for determining the clearance behavior after the quasi-steady-state distribution is established.

In considering the mechanism for clearance of IVT-injected materials, we begin with two small molecules, sucrose and fluorescein. Once injected, these small molecules are transported within the vitreous mainly by diffusion and are relatively unaffected by the convective flow that may be present. For the moment, we will therefore neglect convection and consider only clearance by diffusion. In Section 4.9, the time dependence of a drug concentration in an idealized spherical vitreous following a central bolus injection is derived for the case in which the exterior of the sphere is subject to the following surface condition:

$$-D\frac{\partial C}{\partial r}=\alpha C_{\mathrm{s}} \tag{4.1}$$

where D is the diffusion coefficient, C_{s} is the concentration just inside the sphere, α is a constant, and r is the spatial coordinate. The expression for the average drug concentration in the vitreous as a function of time t following IVT injection is given by

$$C_{\mathrm{v}}(t)=6C_{\mathrm{l}}L\sum_{n=1}^{\infty}\frac{\exp\left(-\beta_{n}^{2}Dt/a^{2}\right)\left[1-\beta_{n}\frac{b}{a}\cot\left(\frac{\beta_{n}b}{a}\right)\right]\sin\left(\frac{\beta_{n}b}{a}\right)}{\beta_{n}^{2}\left[\beta_{n}^{2}+L(L-1)\right]\sin\beta_{n}} \tag{4.2}$$

where a is the radius of the sphere, b is the radius of the bolus, C_1 is the drug concentration of the injected bolus, and the β_ns are the roots of

$$\beta_n \cot \beta_n + L - 1 = 0 \qquad (4.3)$$

where

$$L = a\alpha/D \qquad (4.4)$$

The β_n roots are such that except for very short times, it is only necessary to consider the root for $n = 1$. Thus, in the quasi-steady state, Equation 4.2 may be approximated by Equation 4.5:

$$C_v(t) \approx C_1 \frac{b^3}{a^3} \exp(-k_f t) \qquad (4.5)$$

where the ratio b^3/a^3 arises from the bolus to vitreous volume ratio and k_f is the rate constant for exponential loss given by Equation 4.6:

$$k_f \equiv \beta^2 D/a^2 \qquad (4.6)$$

Thus, the rate of elimination in the terminal phase is proportional to the diffusivity of the injected substance, the larger materials being cleared more slowly. (See Table 4.2 for the diffusivities used in simulations for various materials injected.)

The concentration distribution once the quasi-steady state has been achieved is illustrated in Figure 4.5 for various materials. (The method used to conduct these simulations will be described shortly.) Panels (a) and (c) depict experimental and simulated concentration profiles after central bolus injection with a solution of fluorescein. In this case, the highest concentration occurs behind the lens and the concentration contours radiate out from this point and are parallel to the retina.

TABLE 4.2
Diffusivities ($cm^2 \cdot s^{-1}$) of Materials Applied in Various Tissue Regions

Material	Molecular Weight	Diffusivity in Vitreous, Aqueous	Diffusivity in Cornea, Sclera	Diffusivity in Retina, Choroid, Iris, Ciliary Body
Fluorescein	332	8.5×10^{-6}	6.375×10^{-7}	6.375×10^{-8}
Sucrose	342	7×10^{-6}	5.25×10^{-7}	5.25×10^{-8}
Bovine albumin	68,000	1.04×10^{-6}	7.8×10^{-8}	7.8×10^{-9}
Dextran D10	10,500	1.62×10^{-6}	1.215×10^{-7}	1.215×10^{-8}
Dextran D67	67,000	6.06×10^{-7}	4.545×10^{-8}	4.545×10^{-9}
Dextran D157	157,000	3.86×10^{-7}	2.895×10^{-8}	2.895×10^{-9}

Source: Adapted from Missel, PJ, *Pharm Res*, DOI 10.1007/s11095-012-0721-9.

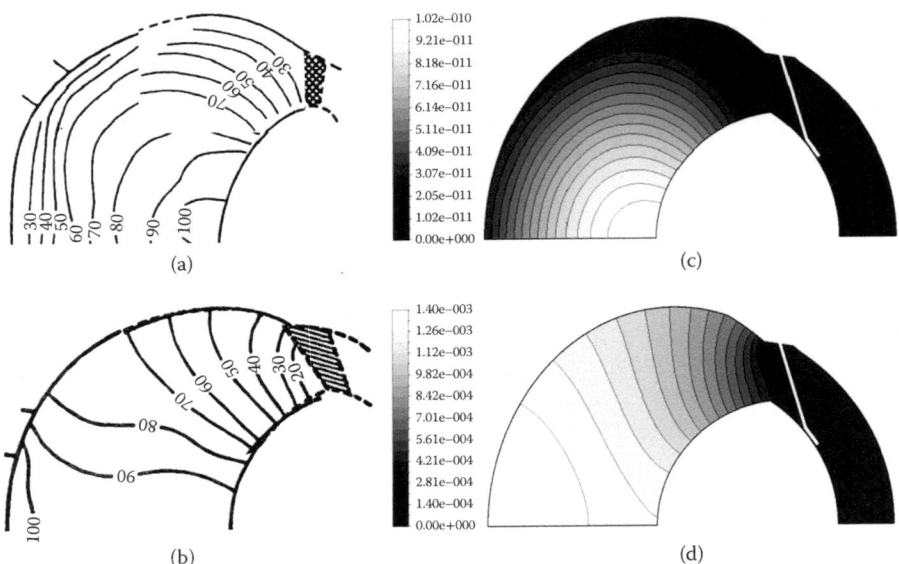

FIGURE 4.5 Spatial distribution in concentration after the quasi-steady state has been established. (a) Composite of fluorescent intensity profiles measured from vitreous slices cut from rabbit eyes frozen immediately after euthanasia 15 hours after central IVT injection of 15 μL injection of 0.2% fluorescein. (b) Similar to (a), but in place of fluorescein a 0.1% solution of 66 kDa fluorescently labeled Dextran had been injected 14 days prior to euthanasia. In (a) and (b), 100 refers to the maximum fluorescence intensity. (c) Concentration profile predicted 48 hours after a simulated central 10 μL bolus injection of fluorescein in the rabbit geometry of Missel from Figure 4.4, vitreous and aqueous compartments only, with permeability sink applied at the outer vitreous surface. (d) Concentration profile predicted 48 hours after simulated central 10 μL bolus injection of sucrose in the same model, but only allowing injected material to be eliminated by the anterior pathway. In (c) and (d), the concentration scale is in units of the concentration of material injected. Note that the grayscale is inverse (lightest shade denotes highest concentration).

The concentration distribution along a path from the point of highest concentration out to the center of the retina approximately follows the r-dependence:

$$C_{ss}(r) = C_0 \frac{a}{r} \frac{\sin\left(\dfrac{\beta r}{a}\right)}{\beta} \tag{4.7}$$

where C_{ss} denotes the quasi-steady state and C_0 denotes the maximum concentration at the geometric center of the distribution. Once the simulated concentration contours achieve the shape shown in panel (c), they maintain this shape while the magnitude decreases exponentially with time.

Panel (b) shows the quasi-steady-state concentration contours for a fluorescently labeled 66 kDa polymer that leaves the vitreous primarily by the anterior pathway. Panel (d) shows the simulated contours established 48 hours after a simulated IVT injection of sucrose, simulated using the same geometric model as in the case of fluorescein in panel (c) but without applying a sink at the exterior vitreous. In both panels (b) and (d), the contours are perpendicular to the retina, with the region of highest concentration remaining in the vitreous residing near the rear retina. The spatial dependence of the contours is distorted from the prediction given from Equation 4.7 because the permeability sink is only applied at the hyaloid membrane. Once the simulated concentration contours achieve the shape shown in panel (d), they maintain this shape while the magnitude decreases exponentially with time, as in the case for fluorescein. However, because material is no longer lost across the boundary behind the vitreous, the rate of decay is much slower than for fluorescein, as shown by the difference in the maximum relative concentration remaining after 48 hours ($1.4 \approx \times 10^{-3}$ in the case of sucrose versus 1×10^{-10} in the case of fluorescein). Although sucrose will freely diffuse into the retina and the exterior tissues, it is not cleared by the choroidal vasculature, and thus it behaves like the fluorescently labeled polymer in panel (b), achieving its quasi-steady state in much less time.

In this modeling method, the rate of transfer between compartments is not determined by a fit to the data. Rather, it is controlled by the geometry and the material properties used in the model. The ability for the simulation to accurately reproduce experimental *in vivo* data is determined by the extent to which the features are reproduced in the model. But exactly which of the following features are required to make the model accurate is not readily apparent:

- Material diffusivity
- Compartment volumes
- Fluid flow rates
- Area of contact available for transfer between compartments
- Shape of compartments

To evaluate the relative importance of these factors, a simple shell model was created as shown in Figure 4.6a (Missel 2012). The vitreous is represented by a semicircle, which when rotated about the symmetry axis becomes a simple sphere of volume 1.52 mL. The aqueous humor compartment is a partial shell, which extends from the symmetry axis to a point halfway along the circular arc of the vitreous. The thickness of the shell is chosen to provide a volume of 0.325 mL for the total aqueous compartment. The interface between vitreous and aqueous compartments extends to an angle of 45° from the symmetry axis. A fluid inlet is placed at the beginning of the point of contact, extending radially outward to the beginning of the iris. The iris follows a curved trajectory at a constant distance from the hyaloid boundary, extends from the fluid inlet toward the symmetry axis, and stops at the pupil opening shortly above the symmetry axis. The iris constitutes a wall boundary to flow of both fluid and drug. Another flow boundary separates the aqueous and vitreous compartments along the remainder of the aqueous shell on the other side of the fluid inlet. The lens is not represented in any way in the model.

Figure 4.6b shows a model comprised only of the vitreous and aqueous regions of one of the rabbit models shown in Figure 4.3. It has a fluid inlet just above the iris and a pressure outlet at the extreme end of the aqueous humor compartment, where the trabecular meshwork would be in the more complete model. The simulation method for these models first solves the flow and pressure equilibrium based on a numerical approximation of the Navier–Stokes equations. The boundary condition on the

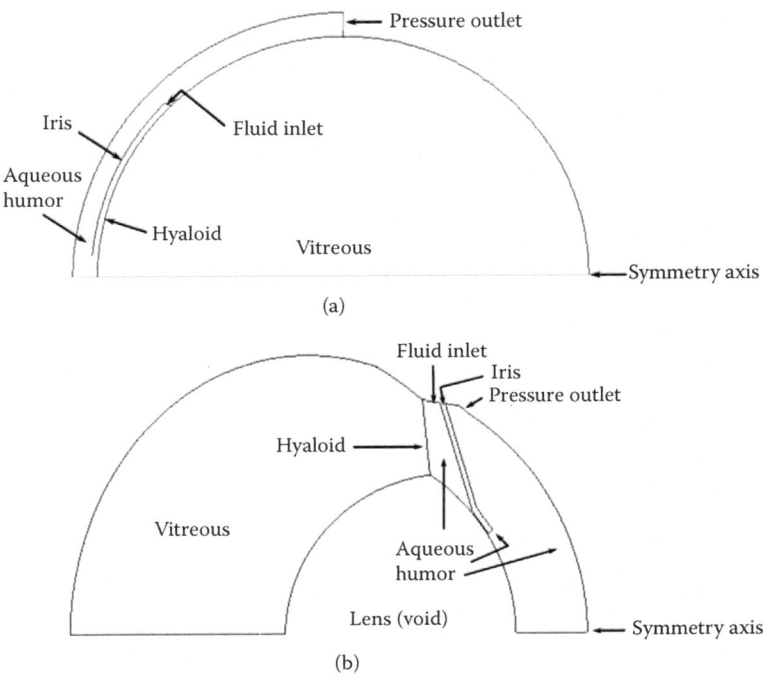

FIGURE 4.6 (a) Simplified ocular model of a spherical vitreous with partial aqueous shell. (b) Rabbit model comprised only of the vitreous and aqueous compartments of the model of Missel shown in Figure 4.3.

fluid inlet sets a mass flow rate of 3 µL/min, which is the rate of production of aqueous humor in the rabbit (Reitsamer et al. 2009). The aqueous humor is treated as a simple fluid with the viscosity of water. All other tissue regions are treated as porous media and assigned an appropriate value for hydraulic resistance (Xu et al. 2000) (see Table 4.3 for the values of hydraulic resistance assigned to each tissue region in the various models).

Figure 4.7 shows the solutions for velocity and pressure in the two models. The fluid flows in an orderly pattern from the fluid inlet to the pressure outlet, achieving its maximum velocity of roughly 10^{-4} M/s near the iris opening in each model. The maximum hydraulic pressure is only about 0.1 mTorr in the simplified shell model and about 4 mTorr in the rabbit model, and although the pressure distribution appears different, each model is essentially at zero hydraulic pressure compared to the physiologic state.

After development of the steady-state flow/pressure solution, the pressure and flow variables are frozen and the advection of the injected bolus is simulated in the flow field for each model. The bolus was represented by a 10 µL spherical region inside which the drug concentration was patched to an initial value of 1. The bolus was placed on the symmetry axis in each model, at the center of the vitreous for the simplified shell model and about halfway between the lens and the rear of the vitreous for the rabbit model. The amount of drug simulated in each compartment is plotted in Figure 4.8, which also contains experimental data of an injection of

TABLE 4.3
Hydraulic Resistance Values Assigned to Various Tissue Regions

Tissue	Hydraulic Resistivity (M^{-2})
Vitreous	1.725×10^{13}
Trabecular meshwork[a]	$\sim 10^{16}$
All other tissues except aqueous humor	9.66×10^{17}

[a] Adjusted to achieve a desired hydraulic pressure. The values in this table are assume the percolating fluid has the viscosity of water at 37°C (6.9×10^{-4} kg/m-s).

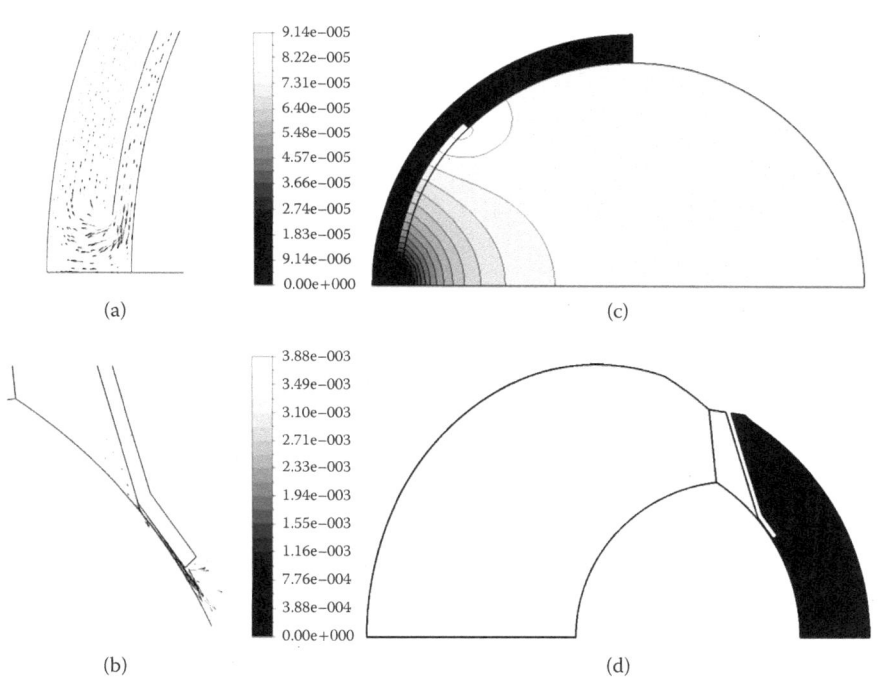

(a)   (b)   (c)   (d)

FIGURE 4.7   Flow and pressure solutions from ocular models as shown in Figure 4.6. (a) Vectors showing direction of fluid flow near the iris opening in the simplified shell model. Vectors are scaled to length according to fluid velocity magnitude, with the longest arrow corresponding to about 7×10^{-5} M/s. (b) Fluid velocity vectors showing flow near iris opening in the rabbit model of Figure 4.7b. The longest vectors correspond to the highest fluid velocity magnitude of about 1.2×10^{-4} M/s. (c) and (d) Pressure distribution colored according to the scales shown, pressure values in Torr. Note that the grayscale is inverse (lightest shade denotes highest pressure).

[14]C-Sucrose (Bito and Salvador 1972). The experimental concentration in the vitreous compartment exhibits an exponential decay over the entire 48 hour period. The data for the aqueous compartment rise steadily for the first 6 hours, then level off and begin declining after 12 hours. This is the beginning of the terminal phase; at this point, the quasi-steady-state concentration profile has been achieved, resembling

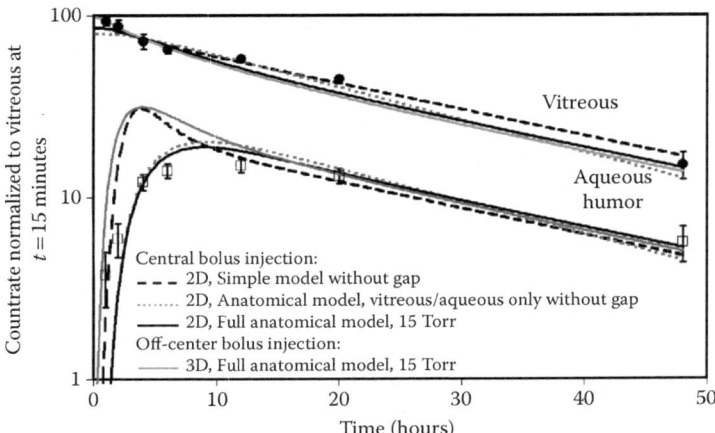

FIGURE 4.8 Simulation of intravitreal injection of ^{14}C-sucrose in various geometric models for the rabbit eye as indicated. All bolus positions were on the symmetry axis except for the three-dimensional model, which was near the pars plana. (Experimental data from Bito LZ, Salvador EV, *Exp Eye Res* 14:233–241, 1972.)

the contours shown in Figure 4.5d. The data in the aqueous compartment exhibit the same exponential rate of decay as the vitreous compartment, maintaining the ratio in the concentration in each compartment approximately constant.

Both models provide an excellent fit to the data in the vitreous compartment. The simplified model overshoots the data for the aqueous compartment at early times, but by 12 hours after injection it reduces to a value that matches the data and continues to match for the remainder of the simulation. The rabbit model does not suffer this overshoot and provides a fairly good description of the entire time course for the data in the aqueous compartment. But during the terminal phase, both models provide an excellent fit to the slope and relative magnitude between compartments. The reason the simplified model experiences the overshoot is because there is no lens in the model to serve as an obstruction around which the drug must diffuse. This is confirmed by the simulation in a full three-dimensional model in which the bolus was injected in the side of the vitreous.

The remaining green curve in Figure 4.8 is from a model with all tissues and conducted at normotensive intraocular pressure. The hydraulic resistances for all interior tissues apart from the aqueous, vitreous, and trabecular meshwork regions were assigned to be that measured for the sclera (Fatt and Hedbys 1970) (see Table 4.3). The flow inlet location and flow rate remained unchanged. The pressure outlet at the outer edge of the aqueous compartment was moved behind the trabecular meshwork. Two additional pressure outlets were applied to the outer sclera and outer cornea. The pressure condition applied to the outer cornea was set to 0 Torr (atmospheric pressure). The pressure condition applied to the outer sclera was set to 10 Torr to approximate the episcleral venous pressure (Funk et al. 1992). The hydraulic resistance of the trabecular meshwork was adjusted until the maximum intraocular pressure achieved the desired value.

Figure 4.9 shows the pressure and velocity solutions for the full model. Most of the pressure drop occurs in the outer shell tissues because the hydraulic resistance in

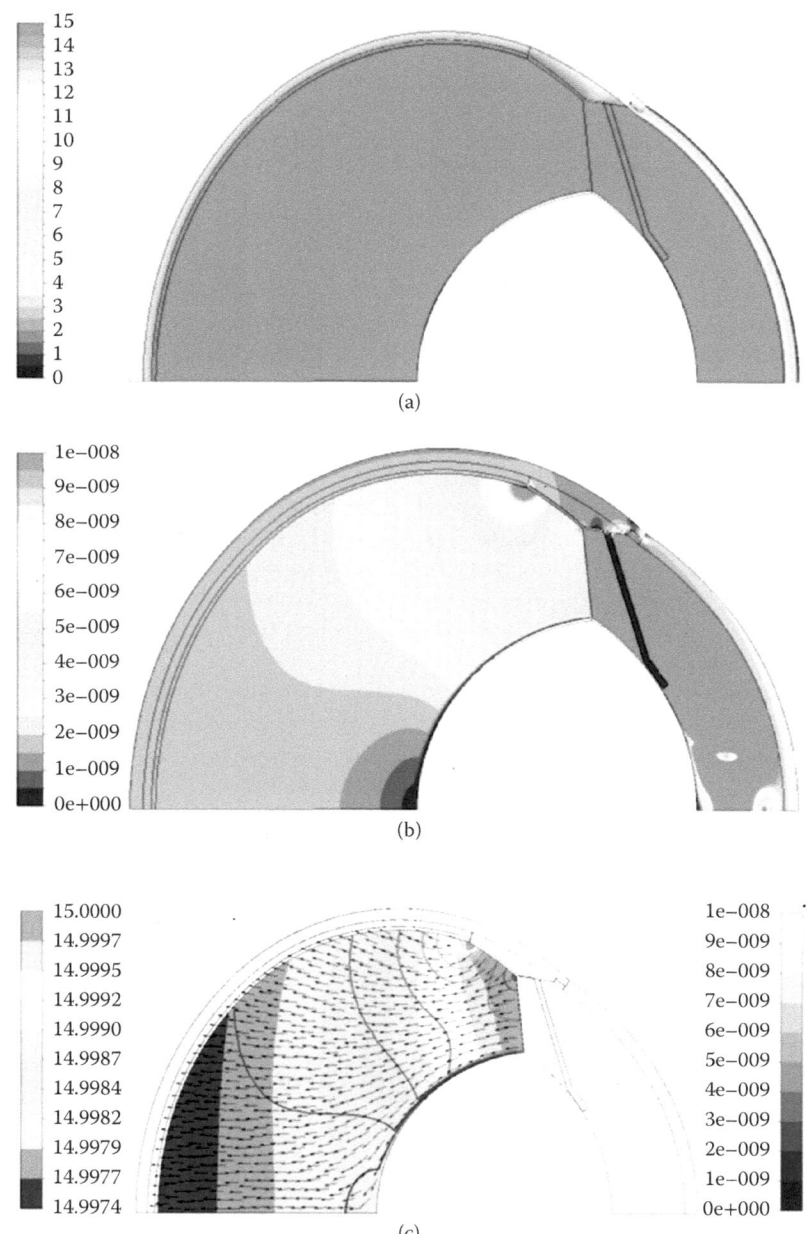

FIGURE 4.9 **(See color insert.)** (a) Pressure distribution in the complete rabbit model with all tissue layers. Pressure boundary conditions were as follows: outer sclera, 10 Torr; outer cornea, 0 Torr. (b) Velocity distribution in M/s, clipped to a maximum velocity of 10^{-8} M/s to show the details of the vitreous. (c) Superposition of velocity and pressure, vitreous region only. Solid colors show the pressure scaled to span the range encountered in the vitreous, line contours show the fluid velocity magnitude, and arrows show the direction of fluid flow.

these tissues is vastly higher than in the vitreous. The velocity solution in the aqueous humor is virtually unchanged from the model without the exterior shell tissues. The velocity solution in the vitreous is quite different in nature than the velocity solution in the aqueous compartment. The velocity of the fluid percolating through the porous medium is derived from the pressure gradient according to the Darcy law equation:

$$\vec{v} = -\frac{K_h}{\eta} \nabla p \qquad (4.8)$$

where K_h is the permeability of the porous medium to fluid flow, η is the fluid viscosity, p is the hydraulic pressure, and $\vec{v}$ is the fluid velocity vector. The weak pressure gradient in the vitreous is plotted in Figure 4.9c, superimposed on the plot of contours of the velocity magnitude, which achieves a maximum value of about 10^{-8} M/s near the front of the vitreous. The direction of flow is shown by the vector arrows.

Assignments must also be made for the diffusivity in the tissue compartments added to the full model. The diffusivity in the cornea and sclera were 1/13th, and the diffusivity in the retina, choroid, iris, and ciliary body were 1/130th, respectively, of the value assigned in the vitreous (see Table 4.2). These ratios of diffusivity matched with those deduced in studies of partitioning and transport of a lipophilic compound (Missel et al. 2010a). The ratio of 1/13 is about a factor of twofold off from the ratio of about 1/6 that was measured for a variety of small (Prausnitz and Noonan 1998) and large (Olsen et al. 1995) molecules.

The simulation results for the full geometry match quite closely with the results of the model that contains only the vitreous and aqueous compartments. From this, we conclude the following:

- The transport of injected ^{14}C-sucrose out of the vitreous is not much affected by the small fluid velocities in the vitreous.
- The slow diffusivity in the exterior tissue layers causes them not to contribute much to the clearance of material from the vitreous.
- All these models provide fairly good predictions for the rate of clearance from the vitreous and for the ratio of concentration in the aqueous and vitreous compartments in the terminal phase.

For the case in which material can only leave the vitreous by diffusion into the aqueous compartment, the following very simple relationship has been derived (Maurice 1959, 1976):

$$\frac{C_a}{C_v} = \frac{k_f V_v}{f} \qquad (4.9)$$

where C_a and C_v are the average concentrations of a drug in the aqueous and vitreous compartments, respectively, and f is the rate of production of aqueous humor. The derivation of Equation 4.9 assumed that the aqueous compartment is well-stirred and that after all the transient effects of the initial distribution have dissipated, the rate of transfer of material out of the vitreous compartment is equal to the rate of transfer of material out of the aqueous compartment mediated by the production and elimination of aqueous humor.

Figure 4.10 shows another method for comparing the clearance behavior of various materials after IVT injection by plotting the concentration ratio C_a/C_v versus k_f on a log–log scale. Compared in the upper panel of Figure 4.10 are simulations and experimental measurements of clearance after IVT injection of sucrose and fluorescein in rabbits. In this manner, the comparison between simulated and experimental assessments of the clearance behavior shown in Figure 4.8 are reduced to individual points on the plot. The points for sucrose fall very close to each other and also fall on the line predicted by Equation 4.9. The experimental point for fluorescein falls quite far from the point for sucrose, having about a ten fold higher rate of clearance from the vitreous and about a ten fold lower aqueous concentration.

In simulating the clearance for fluorescein, an anatomically accurate model for a rabbit similar to that shown in Figure 4.6b was used. An infinite sink (value of concentration fixed to zero) was set on the interfaces of the aqueous humor and the iris and ciliary body, and a partial sink was applied to the outer sclera using the flux condition of Equation 4.1, varying the value of α from 10^{-7} M/s upward. The concentration ratio and clearance rates simulated fall within the range of the data, with an imperfect sink corresponding to a value of α in the range of $1\text{–}2 \times 10^{-7}$ M/s predicting the experimentally observed elimination rate.

From Equation 4.6, k_f is proportional to the diffusivity. Thus, the aqueous/vitreous concentration ratio is also expected to be proportional to the diffusivity. This is more or less borne out by the experimental data for the more slowly diffusing materials, as shown in the lower panel of Figure 4.10, which falls along the line predicted by Equation 4.9. Also compared in this panel are simulations of the clearance behavior of the same materials in several models (see Table 4.2 for the diffusivities used for each material simulated). When the model is comprised only of the vitreous and aqueous compartments but does not include the gap of Petit, simulation results fall directly on the line of Equation 4.9, but the clearance rate for slowly diffusing species is severely underestimated (data points *a*).

The effect of adding the gap is shown by dashed arrow 1; the results fall on a completely different curve and exhibit slight increases in vitreous clearance rate but larger increases in aqueous levels, the deviation increasing as diffusivity diminishes. This deviation cannot be due to an effect of intraocular pressure, which is essentially zero. On adding the exterior shell tissues while still maintaining zero pressure (represented by arrow 2), the simulation results are systematically shifted directly to the left, reducing the elimination rate by about the same proportion independent of diffusivity and without changing the aqueous/vitreous ratio. By increasing the intraocular pressure to the normotensive value (represented by arrow 3), the concentration ratio decreases and the clearance rate increases preferentially for the more slowly diffusing materials, pushing the simulation results (curve *d*) to fall below the curve.

This is because the more slowly diffusing species become more responsive to the convective flows in the vitreous, which though weak reach their maximum values in the anterior. The simulations accurately predict the experimentally observed elimination rate for each injected material, but the aqueous/vitreous concentration ratio is

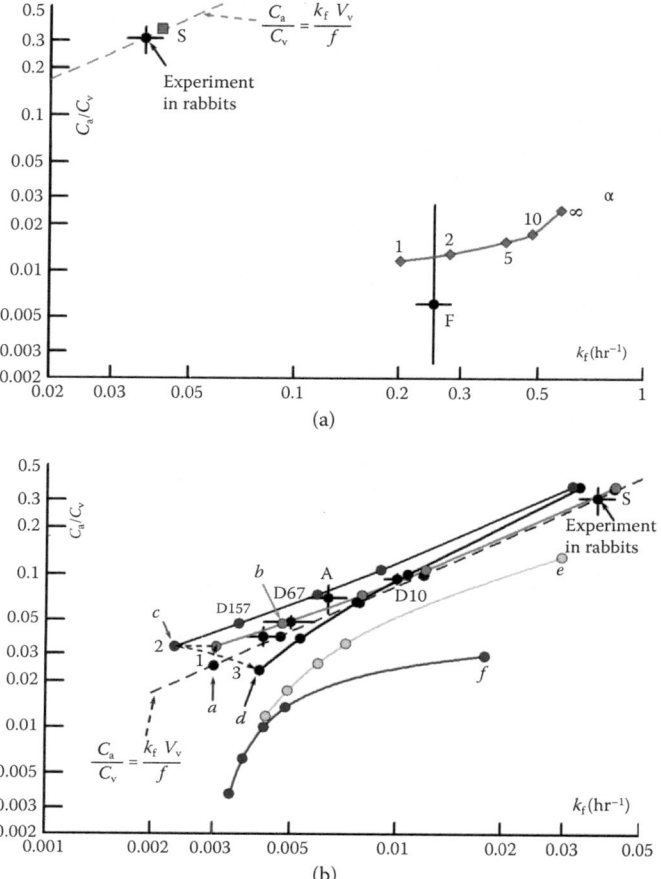

FIGURE 4.10   (a) Log–log plot of aqueous/vitreous concentration ratio in the terminal phase versus vitreous elimination rate. F—fluorescein (data from Cunha-Vaz J, Maurice D, *Doc Ophthalmol* 26:61–72), S—sucrose (data from Bito LZ, Salvador EV, *Exp Eye Res* 14:233–241,1972.) Point plotted as a square represents simulation of clearance of sucrose in the rabbit model (vitreous and aqueous compartments only). Points plotted as diamonds represent simulations of clearance of fluorescein after IVT injection, using the anatomically accurate model for the rabbit eye comprised only of the aqueous and vitreous compartments, similar to Figure 4.6b. Each point is labeled with the value of α used for the strength of the flux condition from Equation 4.1 applied at the outer vitreous boundary in units of 10^{-7} M s^{-1}. (b) Clearance behavior in simulations of intravitreal (IVT) injection of various materials in rabbit ocular models. *a–d*: Simulations based on rabbit model of Missel. (Adapted from Missel P, *Pharm Res* [in press], 2012.) *a*: Vitreous and aqueous compartments only, without gap. *b*: Same model including the gap. *c*: Complete model with gap and all tissues, 10.1 Torr. *d*: Same model, 15 Torr. *e*: Completed model of Friedrich et al., 15 Torr. (Adapted from Friedrich S, Cheng Y-L, Saville B, *Ann Biomed Eng* 25:303–14, 1997, Figure 4.4.) *f*: Model of Park et al., 15 Torr. (Adapted from Park J, Bungay PM, Lutz RJ et al., *J Controlled Rel* 105:279–95, 2005.) Arrows 1 through 3 show how the clearance behavior changes as the features are added to the model of Missel. (Adapted from Missel P, *Pharm Res* [in press], 2012.) Arrow 1: Effect of adding the gap. Arrow 2: Effect of adding the outer shells at zero hydraulic pressure. Arrow 3: Effect of hydraulic pressure.

underestimated for the more slowly diffusing species. A preliminary investigation of the potential influence of the lens (which was omitted from this model) suggests that the elimination rates would not be impacted but the aqueous concentrations would be adjusted slightly upward, bringing the simulations much closer to the experimental measurements (Missel 2012).

Also shown in the figure are predictions of clearance in the rabbit models derived from Friedrich (curve *e*) and Park (curve *f*) shown in Figure 4.3. Note that the simulations for these models included all tissue layers and were conducted at 15 Torr, and thus should be compared with the curve *d* for the model of Missel. Both models systematically underestimate the aqueous concentrations; the Park model also underestimates the clearance rate, presumably because the geometry in these models does not accurately capture the "bottleneck" effect. The lens cavity becomes deeper and narrower in the order Missel → Friedrich → Park, and although the volumes of the vitreous and aqueous compartments are quite similar for all three models (see Table 4.1), the area of contact between the vitreous and aqueous compartments becomes narrower (Figure 4.3). The latter two models also do not contain the gap of Petit. Thus, we find that it is not necessarily the shape of a particular compartment or the exact volume of compartments, but the way in which they are constructed at their interface that controls the rate of transfer between them.

4.6 INJECTION OF PARTICLES IN THE EYE

Thus far we have considered the injection of dissolved materials in the eye (although in some sense one might consider macromolecules as extremely small particles). To model the injection of dispersions of drug-containing particles in the eye, it is necessary to consider the distribution and transport of the particles themselves, and in addition, the release of the drug from the particle. It is beyond the scope of this chapter to consider all the possible mechanisms for drug release, thus attention will be given to only a few examples.

Dextran polymers appear to travel freely within the vitreous and their diffusivity in bovine vitreous decreases compared to water, decreasing more as the molecular weight increases (Peeters et al. 2005). Their diffusivity in the vitreous of other species may not be reduced by the same magnitude because of differences in vitreous composition (Lee et al. 1992, 1994), and rabbit vitreous appeared to have no effect on the diffusivity of intermediate-sized dextrans (Missel 2012). Transport of particles in the vitreous is likely to be influenced by charge. Positively charged particles will form a complex coacervate with negatively charged biopolymers in the vitreous. Peeters et al. (2005) found that negatively charged polystyrene particles bound to collagen fibrils and this binding could be avoided by first coating the particles with polyethylene glycol. Although coating the particles prevented them from binding to the vitreous biopolymer matrix, the diffusivity of coated particles was reduced significantly in bovine vitreous compared to water.

Larger particles tend to be immobilized in the vitreous and form depots, such as has been observed for injection of suspensions of TAC (Beer et al. 2003; Kim et al. 2006).

Injection of large particles in the aqueous humor may lead to an increase in intraocular pressure because particles can provide obstacles to flow in the trabecular matrix. This effect is used in various animal models to create a sustained ocular pressure (Urcola et al. 2006; Sappington et al. 2010).

Drug release from microparticles in the vitreous has been simulated by imposing a drug source on a single particle that varied with the square root of time (Stay et al. 2003). The same principle could be used for multiple particles. However, it is likely that local concentration effects will cause a drug released from one particle to suppress the rate of drug release from particles adjacent to it. Thus, the correlation between *in vivo* and *in vitro* release rates may be affected by the similarities or differences in the flow conditions operational in each situation (Missel et al. 2004). This would also be the case for particles that erode by a dissolution mechanism.

The suppression of dissolution by drug crystals in close proximity leads to an interesting implication, the insensitivity of depot lifetime on particle size. Dissolving crystals can be represented explicitly in the vitreous and positioned in any desired geometric arrangement. Figure 4.11a shows an example of the quasi-steady-state profile of TAC drug concentration in ocular tissue predicted 14 days after IVT injection of a suspension of 276 equally sized TAC drug particles dispersing 16 mg in a 100 μL spherical depot (Missel et al. 2010b). An anatomically accurate model was used for the rabbit eye. An infinite sink for dissolved TAC was applied in the retina. The values of drug diffusivity in the various tissue compartments were quite similar to those shown for sucrose in Table 4.2.

Simulation methods are available that can control the movement of the boundary between solid and dissolved drug in response to the drug flux resulting from the convective diffusion–dissolution process. By successively reducing the particle size while maintaining the total mass of TAC and the depot volume constant, it is possible to estimate the particle size at which the dissolution rate becomes independent of size. Figure 4.11b summarizes simulations for a series of suspension designs in which 4 mg TAC was confined to 100 μL spherical depots. Figure 4.11c shows a similar summary for a more concentrated suspension, 16 mg TAC confined to 100 μL depots. The dissolution rate asymptotically approaches a constant value as the particle size is reduced. The particle size at which the dissolution rate becomes independent of size is surprisingly large. Even in the case when the drug is dispersed throughout the entire vitreous volume, the size-independent limit is achieved for particles having a size of 77 μm diameter or less (Missel et al. 2010b). This is much larger than the particle size used in typical ophthalmic drug formulations. Figure 4.12a demonstrates that the dissolution rate in the size-independent limit provides reasonable predictions of the dissolution rate observed experimentally for suspensions of 4 or 16 mg TAC confined to 100 μL spherical depots (Kim et al. 2006).

The aqueous humor concentration profiles shown in the inset of Figure 4.12a agree approximately with the values observed from aqueous humor paracentesis measurements in humans (Beer et al. 2003). Figure 4.12b shows how the dissolution rate and aqueous humor concentration–time profiles vary with bolus position. The dissolution rate is surprisingly independent of position, suggesting that the anterior

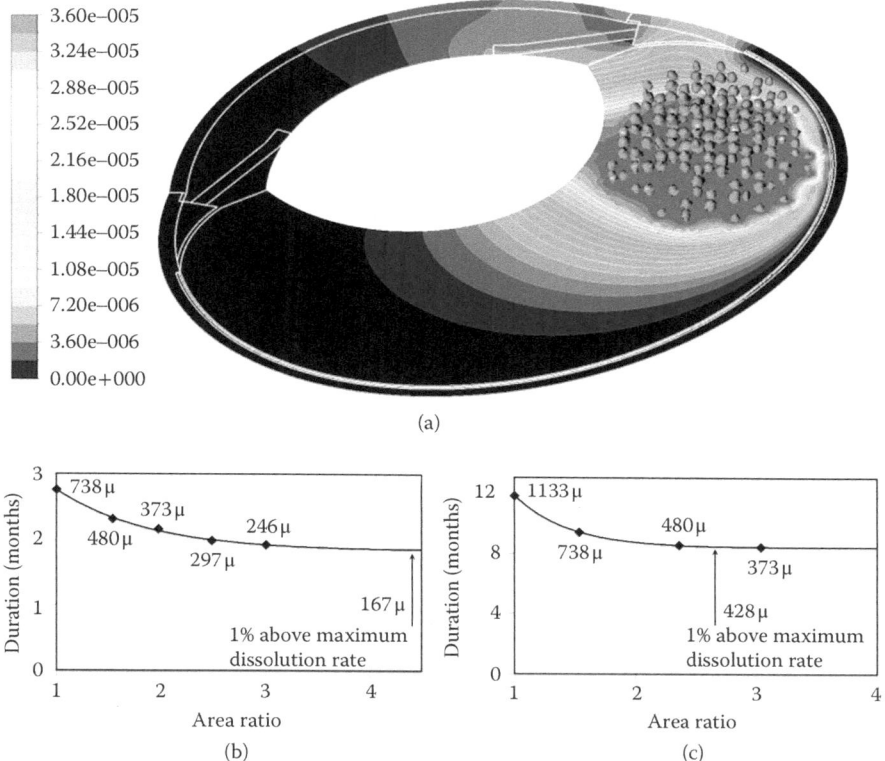

FIGURE 4.11 (See color insert.) (a) Simulated quasi-steady-state triamcinolone acetonide (TAC) concentration profile 14 days after injection of a 100 μL depot containing 16 mg of a drug divided among 276 equally sized particles. (b) Predicted influence of particle size on time to total dissolution of TAC after IVT injection of suspensions comprised of equal-sized particles, 4 mg TAC confined to a 100 μL spherical depot. (c) Same as (b), but with 16 mg TAC confined to a 100 μL spherical depot.

and posterior sinks are balanced in the system. The aqueous humor concentration, however, depends quite significantly on position, with the higher concentrations occurring for the more anteriorly placed depots.

Figure 4.13a shows another type of spherically symmetric model for the eye. This one is simplified even further than the simplified shell model of Figure 4.6a—not only does it have no lens, it does not even have an aqueous compartment. At the center is a spherical drug bead. This simplified model provides an opportunity to derive an exact expression for the relationship between particle radius and time during the process of surface dissolution:

$$t = \frac{\rho}{DS} \left\{ \left[\frac{R_0^2}{2} - \frac{R_0^3}{3R_v} \right] - \left[\frac{R^2}{2} - \frac{R^3}{3R_v} \right] \right\} \qquad (4.10)$$

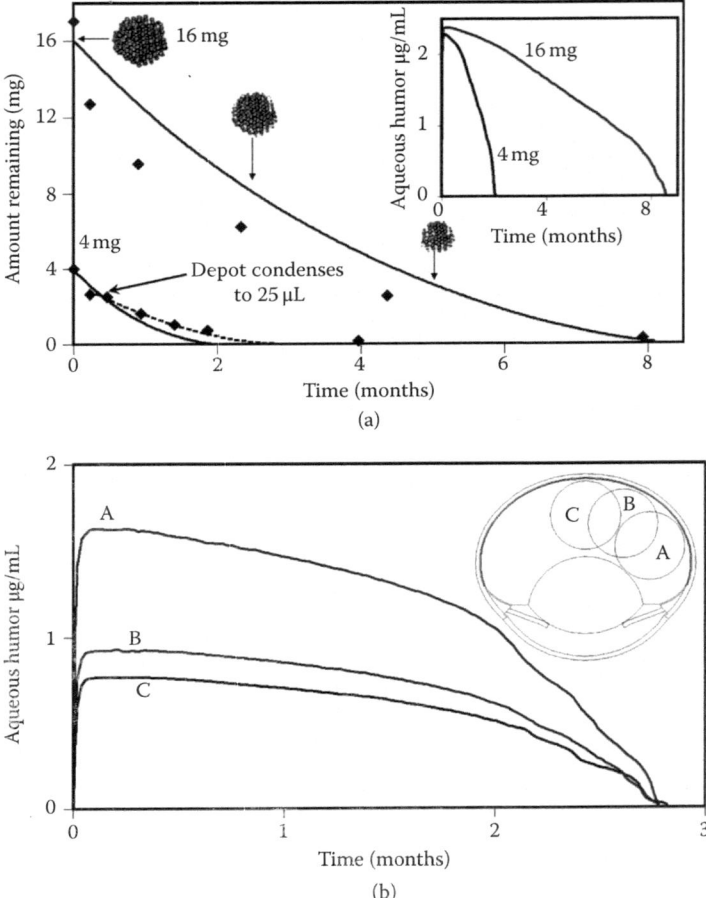

FIGURE 4.12 (a) Dissolution profiles for 4 mg TAC and 16 mg TAC confined to a 100 μL spherical vitreous depot compared with experimental data from Kim et al. (diamonds). (Data from Kim H, Csaky KG, Gravlin L et al., *Retina* 26:523–530, 2006). Curves represent results in the limit of infinitesimally small particles. The dashed curve for 4 mg restarts the simulation by distributing 2.7 mg of drug in a 25 μL depot to approximate the influence of depot condensation observed *in vivo*. The inset shows the mean aqueous humor concentrations predicted. (b) Dissolution profiles for 4 mg TAC divided among 19 equally sized (738 μm) particles confined to a 100 μL spherical vitreous depot located in three different regions of the vitreous as shown in the inset.

where t is time to total particle dissolution, ρ is the density of undissolved drug, D is drug diffusivity in the vitreous, S is drug solubility, R_0 is the initial particle radius, R is the radius at any given time, and R_v is the radius of the vitreous sphere. This relationship can easily be evaluated numerically to obtain a prediction for how the dissolution rate depends on time, as illustrated in Figure 4.13b. In the limit $R_v \to \infty$, it can be shown that R varies linearly with t:

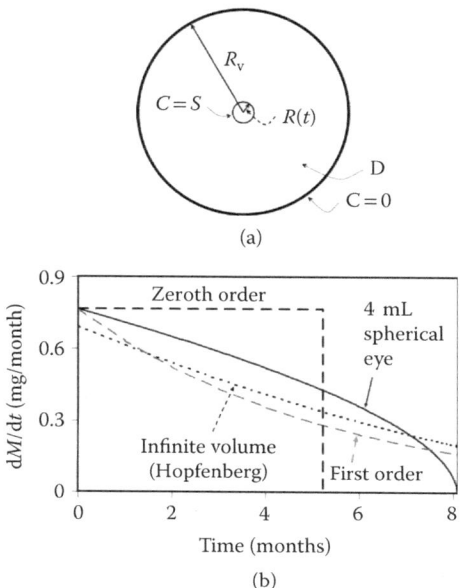

(a)

(b)

FIGURE 4.13   (a) Diagram for a spherical drug bead dissolving in the center of an idealized spherical vitreous domain bounded by an infinite sink. (b) Dissolution rate of a single 4 mg TAC bead in the spherical eye model of (a) (solid curve) compared with the same bead dissolving in an infinite vitreous expanse (dotted curve), perfect first-order release and perfect zeroth-order release (dashed curves).

$$R \underset{\lim R_v \to \infty}{\to} R_0 - \frac{DS}{R_0 \rho} t \qquad (4.11)$$

Hopfenberg derived this same linear time dependence for the radius of spherical particles eroding in an infinite expanse (Hopfenberg 1976). The release rate for the Hopfenberg result is also shown in Figure 4.13b and is similar to the first-order release profile; both the Hopfenberg and first-order release do not end at a decisive point but rather trail out with time. In a finite volume, the erosion process ends abruptly, though not precipitously as does zeroth-order release. However, the release rate is much flatter and higher for most of the release profile compared with first-order or Hopfenberg release. Another useful relationship is the time at which the dissolution process is complete, obtained from Equation 4.10 when R becomes zero:

$$t_{end} = \frac{\rho}{DS}\left[\frac{R_0^2}{2} - \frac{R_0^3}{3R_v}\right] \qquad (4.12)$$

4.7   DELIVERY FROM IMPLANTED DEVICES

Simulation of drug release from implanted devices proceeds along similar lines as for drug particles. In fact, it might be less complicated since fewer releasing surfaces need to be considered. Below are presented simulations for two types of devices releasing a particular lipophilic drug, anecortave acetate. As mentioned in Section 4.5, partition and diffusion coefficients in the various tissues were established independently from *in vitro* partitioning and *ex vivo* tissue transport experiments (Missel et al. 2010a). The partition coefficients for drug in tissues were found to be 4 for retina/choroid and 2.2 for sclera, relative to the water-like vitreous humor. Drug diffusivity in the vitreous was assumed to be 6×10^{-6} cm$^2 \cdot$s^{-1}. Drug diffusivity in the sclera was found to be 1/13th this value, and the diffusivity in the retina/choroid layer was 10 times lower still.

What cannot be estimated from *ex vivo* experiments is the degree to which material is lost by active processes of circulation in the vascular and lymphatic systems. These can only be assessed from *in vivo* experiments in which the circulation is active. The choroidal circulation is a particularly effective sink for removing lipophilic drugs from the vitreous. Instead of applying a boundary condition on the outer vitreous, a volumetric sink condition is applied uniformly throughout the entire choroidal shell. This enables the application of a partial sink that still enables a portion of the drug to reach the sclera. Thus, the model can be used to make region-specific predictions of drug concentration in the sclera. The magnitude of the choroidal sink was estimated by adjusting its value until the clearance rate of anecortave acetate following IVT injection was matched by simulation (Missel et al. 2003).

Figure 4.14a shows a summary of various boundary and volumetric sinks that should be accounted for in the model. In addition to the choroid, a volumetric sink can be applied in the iris and/or ciliary body. The outer sclera is another location for two additional sink effects. The episclera forms another vascular sink, and in addition to this there is the lymphatic system. Recent experiments indicate that these processes are potentially significant barriers to delivery of drugs from the ocular exterior (Robinson et al. 2006; Lee and Robinson 2009; Ranta et al. 2010). Another complicating factor is hydraulic pressure. The model should include a boundary condition to account for the removal of the drug from the system by hydraulic convection. There is also a rather intricate interaction between hydraulic and vascular effects. The mathematics of this has been examined rather closely in one dimension (Missel 2002a) and in three dimensions in yet another spherically symmetric ocular model (Missel 2002b). The relationships in these references serve as useful guides for validating simulation methods to verify that they can reproduce the effect of hydraulic pressure on vascular clearance.

Figure 4.14b shows an example of a simulation of the steady-state drug release profile for a hypothetical spherical device releasing anecortave acetate in the vitreous. The surface of the device was fixed at the solubility limit of the drug. The concentration scale is in units of drug solubility (0.22 µg/mL); higher concentrations exist in the tissues of the retina and choroid since the partition coefficient is higher in these tissues. Since the partition coefficient for the drug is high in the retina and choroid, the maximum concentration in these tissues near the device is higher than the solubility limit by more than a factor of two. The concentration diminishes with distance along a path through the choroid, and thus the highest concentration is in the retina. Also notable is the rather focused nature of the drug distribution in the vitreous, which decreases approximately inversely with distance from the center of the device. The focused nature of delivery from an IVT device has been known for quite some time; Stuart Friedrich predicted similar 1/r-dependent concentration contours for ganciclovir released from the Bausch and Lomb Vitrasert (Friedrich 1996).

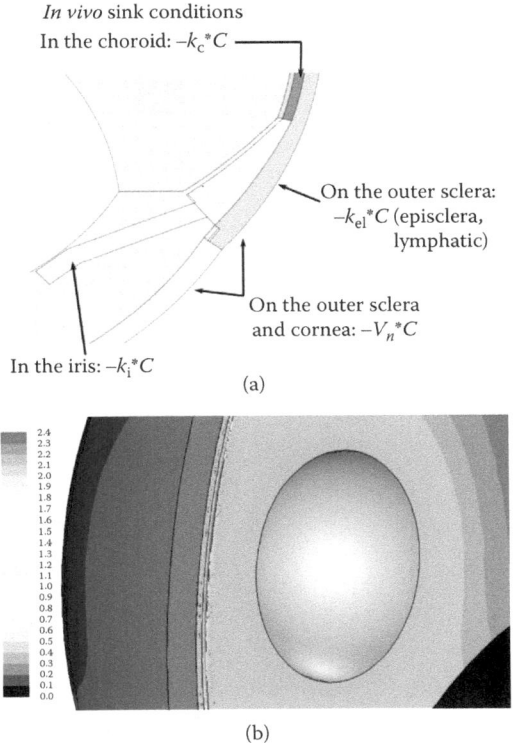

In vivo sink conditions
In the choroid: $-k_c{}^*C$

On the outer sclera:
$-k_{el}{}^*C$ (episclera, lymphatic)

On the outer sclera
and cornea: $-V_n{}^*C$

In the iris: $-k_i{}^*C$

(a)

(b)

FIGURE 4.14 **(See color insert.)** (a) Volumetric and boundary conditions required to properly simulate vascular, lymphatic, and hydraulic sinks that are in operation in the *in vivo* state. (b) Steady-state drug concentration in ocular tissue resulting from a hypothetical spherical IVT device releasing anecortave acetate. The concentration scale is in units of the solubility limit of the drug in buffer. Higher concentrations exist in the retina and choroid because the partition coefficient is high in these tissues. Note that the ocular model does not contain the lens. (Adapted from Missel, PJ et al., *Fluent News*, 15, s8–s10, 2006.)

Figure 4.15b shows a simulation of steady-state concentration contours for the same drug administered using a juxtascleral device, shown in Figure 4.15a (Missel and Yaacobi 2011). The same partition and diffusion coefficients were used for the drug, and the same value was used for the magnitude of the volumetric sink applied in the choroid that was deduced from matching the clearance from IVT injections. In this case, the highest concentrations occur in the sclera but the drug concentration decays rapidly with further distance from the drug payload, which is a simple tablet comprised mainly of the drug. Very little of the drug reaches the vitreous and remains quite focused to just the exterior tissue layers directly underneath the device opening.

Juxtascleral devices containing anecortave acetate implanted in rabbits maintained constant levels of a drug in ocular tissues for 2 years and would have probably continued delivering at the same rate for more than 1 additional year (Missel 2009). Figure 4.15c shows the average drug levels in the retina, choroid, and sclera from a 10 mm circular dissection beneath the drug depot at the 1 year time point.

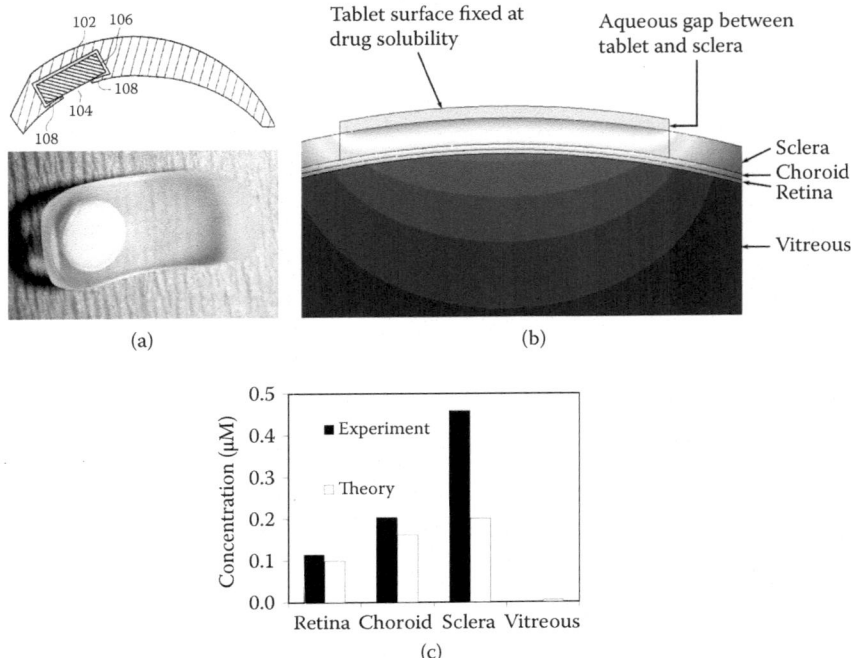

FIGURE 4.15 (See color insert.) (a) Upper panel: Schematic for juxtascleral device: 102—cavity; 106—drug source; 104—exposed opening; 108—circumferential rim to retain the drug source. (Adapted from Missel P, Yaacobi Y, US Patent 7,943,162, 2011.) Lower panel: Photo of prototype device. (b) Simulated steady-state drug concentration in ocular tissue resulting from delivery of anecortave acetate from the device. (c) Comparison between experimental and simulated average concentrations in the measured tissues at 1 year following implantation of juxtascleral devices containing anecortave acetate in rabbits. (Adapted from Missel, PJ, Lecture presented at the Association for Research in Vision and Ophthalmology Summer Eye Research Conference on Ophthalmic Drug Delivery Systems for the Treatment of Retinal Diseases, Basic Research to Clinical Application, Bethesda, MD, 2009.)

The concentration is ranked sclera > choroid > retina >> vitreous. Using no additional adjustable parameters, the simulations predict the same rank order and come close to the values in retina and choroid. The values in the sclera match less well, but less was known about the magnitude of the sinks that were operational outside the sclera. One very important feature of the device is the fact that it provides for unidirectional release of the drug toward the ocular interior and shields the payload from nonproductive loss behind the eye.

4.8 CONCLUSION

For all the power available from using computers, simplified geometric "spherical cow" ocular models continue to provide useful intuitive insight into ocular drug transport and clearance. The simple linear relationship between the rate of egress of material from the vitreous and the mean concentration ratio between the aqueous humor and vitreous compartments, derived more than 60 years ago (Maurice 1959), continues to be a useful starting point for interpretation of the disposition of IVT-injected materials. The key is in determining the circumstances under which the approximation will become impacted by features not contained in the simplified model. For example, a very weak pressure gradient in the vitreous can have a very powerful influence on driving transport for large, slowly diffusing molecules in an intact vitreous. The influence of hydraulic pressure cannot easily be predicted without constructing a full anatomical model.

Much more physiological insight is provided using explicit anatomical/physical modeling compared to parameter fitting methods, which are merely descriptive of the data. The anatomical method offers an alternative means for scaling experimental data from one species to another that may be more appropriate than other simple approaches based entirely upon scaling of compartment volumes. However, this method requires careful construction of the ocular anatomy and assignment of appropriate material properties and boundary conditions. If the model does not accurately reflect key features and flow or clearance processes, its predictive power will be compromised. An example demonstrating this is the unexpected influence exerted by the retrozonular space of Petit on the egress of slowly diffusing molecules, which is mediated by intraocular pressure. It is uncertain how current models are compromised by the features that they currently lack, for example the influence of liquefaction of the vitreous upon aging.

Particle-based drug delivery systems may have utility for ocular delivery, but engineering of such systems needs to take into account the obstacles to delivery constituted by the variety of vascular, hydraulic, and lymphatic clearance mechanisms that are operational in the *in vivo* state. These clearance mechanisms work to focus delivery to a rather confined space, more so for dosage forms placed outside the globe. Without any attempt to shield from these clearance mechanisms (for example, the juxtascleral device), such methods of drug administration are quite likely to fail. Proper characterization of a delivery system needs to account for differences between the *in vivo* and *ex vivo* or *in vitro* conditions. *In vitro* dissolution testing of drug suspensions using standard techniques may be useful as a quality control mechanism, but such methods are not likely to be predictive of the dissolution rate after ocular administration because the hydrodynamic environment of the *in vitro* apparatus is much different than the ocular environment. This is demonstrated by the insensitivity of the dissolution rate for TAC suspensions to differences in particle size.

4.9 APPENDIX

Consider a spherical vitreous region of radius a, the surface of which is subject to the flux boundary condition given in Equation 4.13:

$$-D\frac{\partial C}{\partial r}\bigg|_{r=a} = \alpha\, C|_{r=a} \qquad (4.13)$$

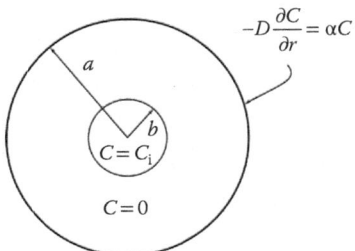

FIGURE 4.16 Diagram defining the problem domain for the time dependence of a drug concentration in a spherical vitreous model following central bolus injection.

At time zero, a spherical bolus of radius b having drug concentration C_i is injected in the center (Figure 4.16). The remainder of the vitreous has zero concentration initially:

$$C(r) = C_i \qquad 0 < r < b \tag{4.14}$$

For constant D, the diffusion equation in spherical radial coordinates is

$$\frac{\partial C}{\partial t} = D\left[\frac{2}{r}\frac{dC}{dr} + \frac{d^2 C}{dr^2}\right] \tag{4.15}$$

The substitution of Equation 4.16 into 4.15 provides the simplified differential Equation 4.17:

$$U(r,t) = r\,C(r,t) \tag{4.16}$$

$$\frac{\partial U}{\partial t} = D\frac{\partial^2 U}{\partial r^2} \tag{4.17}$$

We make the further substitution:

$$U(r,t) = R(r)T(t) \tag{4.18}$$

Equation 4.17 becomes:

$$R\frac{dT}{dt} = DT\frac{d^2 R}{dr^2} \tag{4.19}$$

Dividing both sides of Equation 4.19 by the right-hand side of Equation 4.18, the variables are separated:

$$\frac{1}{T}\frac{dT}{dt} = \frac{D}{R}\frac{d^2 R}{dr^2} \tag{4.20}$$

This equation will be true if each side is set to a constant, which we denote as $-\lambda_n$:

$$\frac{1}{T}\frac{dT}{dt} = -\lambda_n \tag{4.21}$$

Here, the values λ_n are the eigenvalues of the solution. The solution of this equation is the exponential function:

$$T = e^{-\lambda_n t} \tag{4.22}$$

The solution of the spatial equation:

$$\frac{D}{R}\frac{d^2R}{dr^2} = -\lambda_n \tag{4.23}$$

is given by

$$R(r) = \sum_{n=1}^{\infty}\left[A_n \sin\left(\frac{r\sqrt{\lambda_n}}{\sqrt{D}}\right) + B_n \cos\left(\frac{r\sqrt{\lambda_n}}{\sqrt{D}}\right) \right] \tag{4.24}$$

Since $C \sim U/r$, only the sine terms are allowed, otherwise the function would diverge at the origin. Thus, all the coefficients B_n are zero. To find the allowable values of λ_n, we use the boundary condition for Equation 4.13:

$$-D\frac{\partial C}{\partial r}\bigg|_{r=a} = \alpha C\big|_{r=a} \tag{4.25}$$

Note that this condition can be expressed in terms of U using Equation 4.16:

$$-D\left[\frac{1}{r}\frac{dR}{dr} - \frac{R}{r^2}\right]\bigg|_{r=a} = \alpha\frac{R}{r}\bigg|_{r=a} \tag{4.26}$$

where we have also considered only the spatial component $R(r)$ of $U(r,t)$. Substituting the right-hand side of Equation 4.24 for R (recall the coefficients B_n are all zero), the condition of Equation 4.26 for each of the eigenvalues can be expressed as follows:

$$\beta_n \cot\beta_n + L - 1 = 0 \qquad . \tag{4.27}$$

where

$$\beta_n \equiv a\sqrt{\lambda_n}/\sqrt{D} \tag{4.28}$$

and

$$L \equiv \alpha a/D \tag{4.29}$$

Solving Equation 4.A.16 for λ_n,

$$\lambda_n = \beta_n^2 D / a^2 \tag{4.30}$$

The coefficients A_n for each of the eigenfunctions can be determined in the following manner. We begin by multiplying both sides of Equation 4.24 by one of the eigenfunctions and integrating both sides from 0 to a:

$$\int_0^a R(r)\sin\left(\frac{\beta_m r}{a}\right)dr = \sum_{n=1}^\infty A_n \int_0^a \sin\left(\frac{\beta_n r}{a}\right)\sin\left(\frac{\beta_m r}{a}\right)dr \tag{4.31}$$

Note that the identity of Equation 4.28 has been used in the arguments for the sine functions. Since the eigenfunctions are orthogonal over the interval of r from 0 to a, the only term in the series on the right-hand side that is nonzero is the term corresponding to $m = n$:

$$\int_0^a R(r)\sin\left(\frac{\beta_n r}{a}\right)dr = A_n \int_0^a \left[\sin\left(\frac{\beta_n r}{a}\right)\right]^2 dr \tag{4.32}$$

The initial condition from Equation 4.14 must be transformed using the relations from Equations 4.16 and 4.18:

$$R_i(r) = r C_i \qquad 0 < r < a \tag{4.33}$$

Using this initial condition, the expression for A_n obtained from 4.26 becomes:

$$A_n = \frac{C_1 \int_0^b r\sin\left(\frac{\beta_n r}{a}\right)dr}{\int_0^a \left[\sin\left(\frac{\beta_n r}{a}\right)\right]^2 dr} = \frac{2C_1 a\sin\left(\frac{\beta_n b}{a}\right)\left[1 - \beta_n \dfrac{b}{a}\cot\left(\frac{\beta_n b}{a}\right)\right]}{\sin^2\beta_n\left[\beta_n^2 + L(L-1)\right]} \tag{4.34}$$

and thus the full expression for $C(r,t)$ including the time dependences becomes:

$$C(r,t) = \frac{2C_1 a}{r}\sum_{n=1}^\infty \frac{\exp\left(-\beta_n^2 Dt/a^2\right)\left[1 - \beta_n \dfrac{b}{a}\cot\left(\frac{\beta_n b}{a}\right)\right]}{\left[\beta_n^2 + L(L-1)\right]}\frac{\sin\left(\frac{\beta_n b}{a}\right)\sin\left(\frac{\beta_n r}{a}\right)}{\sin^2\beta_n} \tag{4.35}$$

Consider now the case $b = a$. When the bolus fills the entire vitreous, Equation 4.35 becomes identical to Equation 6.40 of Crank (1975, p. 96):

$$C(r,t) = \frac{2C_1 La}{r}\sum_{n=1}^\infty \frac{\exp\left(-\beta_n^2 Dt/a^2\right)}{\left[\beta_n^2 + L(L-1)\right]}\frac{\sin\left(\frac{\beta_n r}{a}\right)}{\sin\beta_n} \tag{4.36}$$

After a long enough time, only the first term contributes significantly. If we ignore the time dependence beyond this point and are interested only in the spatial dependence of the concentration in the established quasi-steady state, we obtain:

$$C_{ss}(r) \propto \frac{\sin\left(\dfrac{\beta_1 r}{a}\right)}{r} \tag{4.37}$$

Araie and Maurice (1991) expressed this spatial dependence normalizing by the concentration in the center, C_0, obtained by taking the limit of Equation 4.37 for r approaching zero:

$$C_{ss}(r) = C_0 \frac{a}{r} \frac{\sin\left(\dfrac{\beta_1 r}{a}\right)}{\beta_1} \tag{4.38}$$

An approximate expression for the average vitreous concentration C_v is provided in Equation 4.5. The exact expression is obtained by integrating Equation 4.35 over the vitreous sphere and dividing by the vitreous volume. The result is

$$C_v(t) = 6C_1 L \sum_{n=1}^{\infty} \frac{\exp\left(-\beta_n^2 Dt/a^2\right)\left[1 - \beta_n \dfrac{b}{a}\cot\left(\dfrac{\beta_n b}{a}\right)\right] \sin\left(\dfrac{\beta_n b}{a}\right)}{\beta_n^2\left[\beta_n^2 + L(L-1)\right]} \frac{}{\sin\beta_n} \tag{4.39}$$

In the special case where $b = a$, the [] in the numerator evaluates to L using Equation 4.27, the sine terms cancel, and the result is

$$C_v(t) = 6C_1 L^2 \sum_{n=1}^{\infty} \frac{\exp\left(-\beta_n^2 Dt/a^2\right)}{\beta_n^2\left[\beta_n^2 + L(L-1)\right]} \tag{4.40}$$

Figure 4.17a shows how the concentration profile changes with time for the situation in which a 20 μL bolus of concentration C_1 is injected in the center of a spherical vitreous having radius 0.78 cm, with diffusivity $D = 6 \times 10^{-6}$ cm^2·s^{-1}, bounded by flux condition 4.13 with a value of α such that the value of L evaluated from Equation 4.29 is 3. These are the same values used in Araie and Maurice (1991) for fluorescein. The step-shape of the bolus begins to smooth rather quickly and eventually begins to take on the shape for the fundamental concentration decay mode ($n = 1$) given in Equation 4.38. Figure 4.17b shows how the various decay modes contribute to the average vitreous concentration over time. After the first hour, only two modes are required, and after 5 hours, the fundamental decay mode is sufficient to describe the subsequent decay in concentration.

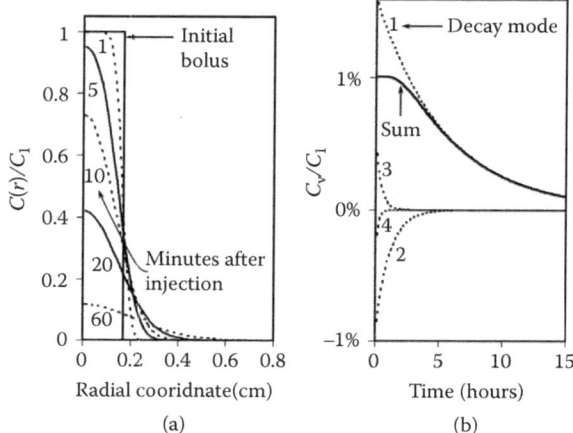

FIGURE 4.17 (a) Plots of decay of concentration from its initial profile immediately after injection of a bolus of radius 0.2 cm having concentration C_1 in a spherical vitreous of radius 0.78 cm bounded by the flux condition of Equation 4.13 for the case in which $L = 3$ and $D = 6 \times 10^{-6}$ cm$^2 \cdot$s^{-1} (calculated using Equation 4.35). (b) Average vitreous concentration normalized by bolus concentration C_1. The amplitude of the first four modes of concentration decay compared with the sum of all modes for the same situation as in (a), calculated using Equation 4.A.28. (Analysis modeled after Acton JR, Squire PT, *Solving Equations with Physical Understanding,* Adam Hilger, Bristol, UK, sections 8.5–8.6,1985.)

REFERENCES

Acton JR, Squire PT. 1985. *Solving Equations with Physical Understanding.* Bristol, UK: Adam Hilger, Sections 8.5–8.6.

Amrite AC, Edelhauser HF, Kompella UB. 2008. Modeling of corneal and retinal pharmacokinetics after periocular drug administration. *IOVS* 49:320–32.

Araie M, Maurice DM. 1991. The loss of fluorescein, fluorescein glucuronide and fluorescein isothiocyanate dextrane from the vitreous by the anterior and retinal pathways. *Exp Eye Res* 52:27–39.

Balachandran RK, Barocas VH. 2008. Computer modeling of drug delivery to the posterior eye: Effect of active transport and loss to choroidal blood flow. *Pharm Res* 25:2685–96.

Balachandran RK, Barocas VH. 2011. Contribution of saccadic motion to intravitreal drug transport: Theoretical analysis. *Pharm Res* 28:1049–64.

Beer PM, Bakri SJ, Singh RJ, Liu W, Peters III GB, Miller M. 2003. Intraocular concentration and pharmacokinetics of triamcinolone acetonide after a single intravitreal injection. *Ophthalmology* 110:681–6.

Bito LZ, Salvador EV. 1972. Intraocular fluid dynamics III. The site and mechanism of prostaglandin transfer across the blood intraocular fluid barriers. *Exp Eye Res* 14:233–41.

Bron AJ, Tripathi RC, Tripathi BJ. 1997. *Wolff's Anatomy of the Eye and Orbit*, 8th ed. New York: Chapman & Hall, Chapter 8.

Chan JE, Pridgen TA, Csaky KG. 2010. Episcleral clearance of sodium fluorescein from a bioerodible sub-tenon's implant in the rat. *Exp Eye Res* 90:501–6.

Chang-Lin J-E, Burke JA, Peng Q, et al. 2011. Pharmacokinetics of a sustained-release dexamethasone intravitreal implant in vitrectomized and nonvitrectomized eyes. *IOVS* 52:4605–9.

Crank J. 1975. *The Mathematics of Diffusion*. London: Oxford University Press, Section 6.3.4, Surface Diffusion.

Cunha-Vaz J, Maurice D. 1969. Fluorescein dynamics in the eye. *Doc Ophthalmol* 26:61–72.

Fatt I, Hedbys B. 1970. Flow of water in the sclera. *Exp Eye Res* 10:243–9.

Friedrich S. 1996. Ocular pharmacokinetic modeling. PhD diss., Toronto, ON: University of Toronto.

Friedrich S, Cheng Y-L, Saville B. 1997. Finite element modeling of drug distribution in the vitreous humor of the rabbit eye. *Ann Biomed Eng* 25:303–14.

Funk RH, Gehr J, Rohen JW. 1992. Short-term hemodynamic changes in episcleral arteriovenous anastomoses correlate with venous pressure and IOP changes in the albino rabbit. *Curr Eye Res* 15:87–93.

Harte J. 1988. Consider a spherical cow: A course in environmental problem solving. Sausalito: University Science Books.

Heys JJ, Barocas VH. 2002a. A Boussinesq model of natural convection in the human eye and the formation of Krukenberg's spindle. *Ann Biomed Eng* 30:392–401.

Heys JJ, Barocas VH. 2002b. Computational evaluation of the role of accommodation in pigmentary glaucoma. *IOVS* 43:700–8.

Hopfenberg HP. 1976. Controlled release from erodible slabs, cylinders, and spheres. In *Controlled Release from Polymeric Formulations*, ed. DR Paul and FW Harris, 26–31. Washington, DC: American Chemical Society Symposium Series 33.

Johnson F, Maurice D. 1984. A simple method of measuring aqueous humor flow with intravitreal fluoresceinated dextrans. *Exp Eye Res* 39:791–805.

Keister JC. 1987. An experimental modeling method for the analysis of diffusion through thin membranes. *J Membrane Sci* 35:73–89.

Keister JC. 1988. Drug transport predicted from measurements of voltage in a water-filled stainless steel model of the eye. Alcon Laboratories, unpublished results.

Kim H, Csaky KG, Gravlin L, et al. 2006. Safety and pharmacokinetics of a preservative-free triamcinolone acetonide formulation for intravitreal administration. *Retina* 26:523–30.

Kotliar K, Maier M, Bauer S, Feucht N, Lohman C, Lanzl I. 2007. Effect of intravitreal injections and volume changes on intraocular pressure: Clinical results and biomechanical model. *Acta Ophthalmol Scand* 85:777–81.

Lee B, Litt M, Buchsbaum G. 1992. Rheology of the vitreous body, part I: Viscoelasticity of the human vitreous. *BioRheol* 29:521–33.

Lee B, Litt M, Buchsbaum G. 1994. Rheology of the vitreous body, part II: Viscoelasticity of bovine and porcine vitreous. *BioRheol* 31:327–38.

Lee SS, Ghosn C, Yu Z, et al. 2010. Vitreous VEGF clearance is increased after vitrectomy. *IOVS* 51:2135–8.

Lee SS, Robinson MR. 2009. Novel drug delivery systems for retinal diseases. *Ophthalmic Res* 41:124–35.

Lin H-H. 1997. Finite element modeling of drug transport processes after an intravitreal injection. Master's diss., Toronto, ON: University of Toronto.

Luu KT, Zhang EY, Prasanna G, et al. 2009. Pharmacokinetic-pharmacodynamic and response sensitization modeling of the intraocular pressure-lowering effect of the EP_4 agonist 5-{3-[(2S)-2-{(3R)-3-hydroxy-4-[3-(trifluoromethyl)phenyl]butyl}-5-oxopyrrolidin-1-yl]propyl}thiophene-2-carboxylate (PF-04475270). *J Ocular Pharm Ther* 331:627–35.

Maurice DM. 1959. Protein dynamics in the eye studied with labelled proteins. *Am J Ophthalmol* 47:361–7.

Maurice DM. 1976. Injection of drugs into the vitreous body. In *Symposium on Ocular Therapy*, ed. T. Leopold and R. Burns, 59–72. London: Wiley.

Maurice DM. 1987. Flow of water between aqueous and vitreous compartments of the rabbit eye. *Am J Physiol* 252:F104–8.

Missel PJ. 2000. Finite element modeling of diffusion and partitioning in biological systems: The infinite composite medium problem. *Ann Biomed Eng* 28:1307–17.

Missel PJ. 2002a. Finite and infinitesimal representations of the vasculature: Ocular drug clearance by vascular and hydraulic effects. *Ann Biomed Eng* 30:1128–39.

Missel PJ. 2002b. Hydraulic flow and vascular clearance influences on intravitreal drug delivery. *Pharm Res* 19:1636–47.

Missel PJ. 2009. Computer modeling of pharmacokinetics from ocular drug delivery. Lecture presented at the Association for Research in Vision and Ophthalmology Summer Eye Research Conference on Ophthalmic Drug Delivery Systems for the Treatment of Retinal Diseases, Basic Research to Clinical Application, Bethesda, MD, July 31–August 1 2009.

Missel PJ. 2012. Simulating intravitreal injections in anatomically accurate models for rabbit, monkey and human eyes. *Pharm Res*. DOI 10.1007/s11095-012-0721-9.

Missel PJ, Chastain JE, Kiehlbauch CC, Dahlin DC. 2003. Predicting ocular drug levels from intravitreal administration of anecortave acetate (AL-3789). Poster presented at the annual meeting of the Biomedical Engineering Society, Nashville, TN, October 1–4 2003.

Missel PJ, Chastain JE, Mitra A, et al. 2010a. *In vitro* transport and partitioning of AL-4940, active metabolite of angiostatic agent anecortave acetate, in ocular tissues of the posterior segment. *J Ocular Pharmacol Ther* 26:137–45.

Missel PJ, Horner M, Muralikrishnan R. 2010b. Dissolution of intravitreal triamcinolone acetonide suspensions in an anatomically accurate rabbit eye model. *Pharm Res* 27:1530–46.

Missel PJ, Stevens L, Chastain J, Yaacobi Y. 2006. A clear vision for drug delivery. *Fluent News* 15:s8–s10.

Missel PJ, Stevens LE, Mauger JW. 2004. Dissolution of anecortave acetate in a cylindrical flow cell: Re-evaluation of convective diffusion/drug dissolution for sparingly soluble drugs. *Pharm Dev Tech* 9:453–9.

Missel PJ, Yaacobi Y. 2011. Drug delivery device, US Patent 7,943,162.

Molokhia SA, Jeong E-K, Higuchi WI, Li SK. 2009. Transscleral iontophoretic and intravitreal delivery of a macromolecule: Study of ocular distribution *in vivo* and postmortem with MRI. *Exp Eye Res* 88:418–25.

Ohtori A, Tojo K. 1994. *In vivo/in vitro* correlation of intravitreal delivery of drugs with the help of computer simulation. *Biol Pharm Bull* 17:283–90.

Olsen TW, Edelhauser HF, Lim JI, Geroski DH. 1995. Human scleral permeability: Effects of age, cryotherapy, trans-scleral diode laser, and surgical thinning. *IOVS* 36:1893–903.

Park J, Bungay PM, Lutz RJ, et al. 2005. Evaluation of coupled convective–diffusive transport of drugs administered by intravitreal injection and controlled release implant. *J Control Release* 105:279–95.

Peeters L, Sanders NN, Braeckmans K, et al. 2005. Vitreous: A barrier to nonviral ocular gene therapy. *IOVS* 46:3553–61.

Prausnitz MR, Noonan JS. 1998. Permeability of cornea, sclera and conjunctiva: A literature analysis for drug delivery to the eye. *J Pharm Sci* 87:1479–88.

Ranta V-K, Mannermaa E, Lummepuro K, et al. 2010. Barrier analysis of periocular drug delivery to the posterior segment. *J Control Release* 148:42–8.

Reitsamer HA, Bogner B, Tockner B, Kiel JW. 2009. Effects of dorzolamide on choroidal blood flow, ciliary blood flow, and aqueous production in rabbits. *IOVS* 50:2301–7.

Rim JE, Pinsky PM, van Osdol WW. 2005. Finite element modeling of coupled diffusion with partitioning in transdermal drug delivery. *Ann Biomed Eng* 33:1422–38.

Robinson MR, Lee SS, Kim H, et al. 2006. A rabbit model for assessing the ocular barriers to the transscleral delivery of triamcinolone acetonide. *Exp Eye Res* 82:479–87.

Sappington RS, Carlson BJ, Crish SD, Calkins DJ. 2010. The microbead occlusion model: A paradigm for induced ocular hypertension in rats and mice. *IOVS* 51:207–16.

Sebag J. 1987. Age-related changes in human vitreous structure. *Graefes Arch Clin Exp Ophthalmol* 225:89–93.

Stay MS, Xu J, Randolph TW, Barocas VH. 2003. Computer simulation of convective and diffusive transport of controlled-release drugs in the vitreous humor. *Pharm Res* 20:96–102.

To C, Kong C, Chan C, Shahidullah M, Do C. 2002. The mechanism of aqueous humour formation. *Clin Exp Optom* 85:335–49.

Tojo K. 2004. A pharmacokinetic model for ocular drug delivery. *Chem Pharm Bull* 52:1290–4.

Tojo K, Nakagawa K, Morita Y, Ohtori A. 1999. A pharmacokinetic model of intravitreal delivery of ganciclovir. *Eur J Pharm Biopharm* 47:99–104.

Urcola JH, Hernandez M, Vecino E. 2006. Three experimental glaucoma models in rats: Comparison of the effects of intraocular pressure elevation on retinal ganglion cell size and death. *Exp Eye Res* 83:429–37.

Xu J, Heys JJ, Barocas VH, Randolph TW. 2000. Permeability and diffusion in vitreous humor: Implications for drug delivery. *Pharm Res* 17:664–9.

5 Considerations for Development of Ophthalmic Nanotechnology-Based Drugs*

Banu S. Zolnik, Sheriza Baksh, and Nakissa Sadrieh

CONTENTS

5.1 INTRODUCTION

Nanotechnology-based drug delivery systems are expected to promise effective therapies and may offer advantages over existing conventional therapies for various routes of administration, including ocular delivery, because they can be designed

---

* Disclaimer: This chapter reflects the current thinking and experience of the authors. However, this is not a policy document and should not be used in lieu of regulations, published FDA guidance documents, or direct discussions with the agency.

to circumvent several natural barriers of the eye [1]. Depending on the type of application, nanotechnology-based drug products may (1) improve the solubility of poorly soluble drugs; (2) protect a therapeutic agent from degradation; (3) change pharmacokinetic and biodistribution profiles, thereby increasing efficacy and reducing toxicity; (4) improve the targeted transport of therapeutics; and (5) improve imaging and diagnostics [2,3]. Nanoparticle-based drug therapy may offer advantages through less-frequent injections, prolonged delivery mechanisms, and adjustable drug delivery kinetics. Readers may refer to recent reviews on the progress of nanotechnology in retinal and eye research and issues related to toxicity of nanoparticles by Diebold et al. and Prow, respectively [4,5].

5.1.1 FDA REGULATORY APPROACHES FOR IDENTIFICATION OF NANOTECHNOLOGY-BASED DRUGS

As of publication of this chapter, the U.S. Food and Drug Administration (FDA) has not adopted any definition for nanomaterial-containing drug products. The Office of the Commissioner at the FDA issued a guidance document in June 2011 entitled "Considering Whether an FDA-Regulated Product Involves the Application of Nanotechnology," in which the agency developed points to consider for nanomaterial-based products. According to the guidance document, an FDA-regulated product will be considered a nanotechnology-based product if it meets the following two criteria:

1. Whether an engineered material or end product has at least one dimension in the nanoscale range (approximately 1 nm to 100 nm)
2. Whether an engineered material or end product exhibits properties or phenomena, including physical or chemical properties or biological effects, that are attributable to its dimension(s), even if these dimensions fall outside the nanoscale range, up to 1 μm [6]

The Office of Pharmaceutical Science in the Center for Drug Evaluation and Research (CDER) at the FDA also issued a manual of policies and procedures (MaPP) entitled *Reporting Format for Nanotechnology-Related Information in CMC Review.* The objective of this MaPP was to identify and track applications for products that are nanomaterial-containing drug products. With the purpose of the MaPP in mind, it defines nanomaterial as "any materials with at least one dimension smaller than 1000 nm" [7].

Using the criteria in the CDER MaPP and the June 2011 FDA guidance document on points to consider, we identified two nanomaterial-containing drug products for ocular delivery approved by the FDA. However, since prospective identification of those formulations containing nanomaterials is difficult and there are no regulatory requirements for nanomaterial-containing drug products to be identified in their labeling, there may be additional FDA-approved formulations in the nanometer size range for ocular delivery. One of the ocular products identified for discussion in this chapter is a liposomal formulation of verteporfin for injection, approved by the FDA in April 2000 for the treatment of patients with predominantly classic subfoveal choroidal neovascularization due to age-related macular degeneration, pathologic myopia, or presumed ocular histoplasmosis. The other drug product is difluprednate ophthalmic emulsion, approved by the FDA in June 2008 for the treatment of inflammation and pain associated with ocular surgery [8].

5.2 REGULATORY PERSPECTIVE ON THE DEVELOPMENT OF NANOMATERIAL-CONTAINING DRUG PRODUCTS FOR OCULAR DELIVERY

In this chapter, our aim is to identify areas one might need to consider when developing nanomaterial-based therapeutics for ocular delivery. It should be noted, however, that currently the regulatory requirements for the development of drugs containing nanoscale materials are no different from those of small-molecule drugs. This chapter is not intended to be a regulatory overview of drug development in general; rather, it focuses more on the considerations for developing drugs containing nanoscale materials intended for ocular delivery. There are unique issues related to combination products (drug–device, drug–biologic, etc.) that are not addressed in this chapter. This chapter has three main sections covering chemistry, manufacturing, and controls (CMC); preclinical and clinical studies.

5.2.1 CHEMISTRY, MANUFACTURING, AND CONTROLS

Under regulations in the United States, the use of a human drug product not previously authorized for marketing in the United States requires the submission of an Investigational New Drug (IND) to the Agency. The FDA's regulations in 21 CFR 312.22 and 312.23, respectively, contain the general principles underlying IND submission and the general requirements for content and format. Section 312.23(a)(7)(i) requires that an IND for each phase of an investigation include sufficient CMC information to ensure the proper identity, strength or potency, quality, and purity of the drug substance and drug product. The type of information submitted will depend on the phase of the investigation, the extent of the human study, the duration of the investigation, the nature and source of the drug substance, and the drug product dosage form [9].

5.2.1.1 Formulation Parameters

Nanomaterial-containing drug products for ocular delivery must have formulations that are inert and not irritating to the eye. Selection of inactive ingredients should be made with consideration of the stability of the drug product, compatibility with the overall formulation, and desired release profile and bioavailability.

The Inactive Ingredients Database provides information on inactive ingredients present in FDA-approved drug products [10]. Once the inactive ingredients are listed in the database, a sponsor may use those inactive ingredients in a similar manner (i.e., same dosage form, potency, and route of administration) in their product with the understanding that the FDA has recognized their safety. However, current regulations do not specify the particle size requirements of inactive ingredients and not controlling for this parameter may impact product safety. If a sponsor chooses to use a nanoscale counterpart of a previously accepted inactive ingredient, then the sponsor may need to investigate the safety of the nanoscale inactive ingredient in case there are questions about the safety, such as a change in absorption profile and indication of irritation potential.

Nanomaterial-containing drug products must meet the same regulatory standards as small-molecule drugs. Other than those typically required by individual review divisions, there are no additional or special CMC requirements for nanomaterial-containing drug products. The previously mentioned *Reporting Format for Nanotechnology-Related Information in CMC Review* MaPP issued by the Office of Pharmaceutical Science identifies several nanotechnology parameters and techniques that FDA chemistry reviewers may look for within applications for products that contain nanomaterials. Physicochemical parameters such as those that are morphology related (size and size distribution of the primary particle, as well as agglomerate/aggregate, molecular weight, structure, shape, and three-dimensional structure), surface related (area, charge, coating composition, coating coverage, reactivity, surface–core interaction, and topology), chemical (chemical composition, purity, stability, solubility, structure, crystallinity, and catalytic activity), and others such as drug loading, potency, functionality, *in vitro* release, and deformability could be assessed by various analytical, spectral, and microscopy techniques [7]. In addition to quality assurance of the product, physicochemical characterization is important because parameters such as particle size, aggregation, agglomeration, and solubility may greatly affect biodistribution [11]. The list of parameters listed in the MaPP is by no means an exhaustive list, rather it serves only to provide examples of techniques which can be utilized to characterize nanotherapeutics, and sponsors are encouraged to propose and justify appropriate methodologies that can adequately measure the specific physicochemical properties of their drugs under investigation.

In vitro studies are an essential part of drug development in that certain features of *in vivo* product performance of drugs may be predicted through results from *in vitro* methods such as drug metabolism for clinical pharmacology and genotoxicity for preclinical safety. Similarly, *in vitro* studies can be very useful in physicochemical characterization. *In vitro* studies for nanoparticles may include release kinetics and uptake of the drug from the delivery system. *In vitro* release studies have been conducted using a modified Franz diffusion cell. Filters have been used to mimic physiological conditions [12]. The use of fresh cornea from rabbits on Franz diffusion cells has also been explored [13]. The limitation of *in vitro* models to predict permeability through the sclera and retinal pigment epithelium should be recognized, as several other barriers, such as clearance (lymphatic/blood), fluid flow, and active transport, exist *in vivo*.

Although the CDER currently has no special requirements for nanotherapeutics, we believe that the existing draft guidance document on the development of liposomal drug products may be applicable to the development of a number of nanotherapeutics. It should be noted that some liposomal products do exceed the nanometer size range or have characteristics unique to liposome products. As stated in *Liposome Drug Products*, there are a number of physicochemical properties that can be subject to characterization [14]. Morphology, liposomal integrity, net charge, volume of entrapment, particle size, phase transition temperature (T_m), *in vitro* release, and osmotic properties are just a few examples. In addition, chemical characterizations of the API, any degradation products, phospholipid content, and antioxidant levels of the liposomal formulation should be characterized if any of these physicochemical properties are attributed to product quality.

As stated in this document, assessment of liposomal integrity upon storage is an important indication of drug product stability. For example, in the case of verteporfin for injection, the spectral properties of the drug product are identified as an important parameter since the efficacy of the treatment depends on the generation of reactive oxygen species using nonthermal light. Additionally, the presence of any structurally related impurities on the spectral properties of the active ingredient should be studied, as this may interfere with the efficacy of the drug. As with any drug, the presence of impurities may impact the manufacturing controls of light labile drugs, and protective measures should be implemented during manufacturing to prevent degradation. Therefore, similar to small-molecule drug development, identification and control of impurities is essential when developing nanotherapeutics. In addition, because liposomal and other complex nanotherapeutics are sensitive to any changes in the manufacturing conditions, it is important to put safety measures in place so that product failure can be tested as early as possible. In general, and in accordance with quality by design (QbD) concepts, critical process controls may help to identify elements which have the most impact on the quality of the product.

5.2.1.2 Sterility and Endotoxin Testing

Sterility testing is required by the regulatory agencies for injectables, inhalation, and ophthalmic drugs. Commonly used sterilization techniques are filtration, gamma irradiation, autoclaving, aseptic production, ethylene oxide sterilization, and high hydrostatic pressure sterilization. The FDA's guidance for industry, entitled *Sterile Drug Products Produced by Aseptic Processing—Current Good Manufacturing Practice*, describes the critical key parameters of aseptic processing [15]. Although various techniques are available, sterilization of nanoparticles may be a challenging task as each technique may affect the product quality and performance of nanoparticles. Sterilization by membrane filtration may be suitable for nanoparticles within a size range that is smaller than the size of the filter; however, if the nanoparticles are larger than the filter size, this method may not be suitable. Methods using elevated temperature, while effective for small molecules, have been shown to affect product performance of polymeric drug delivery systems because they may affect drug release profiles due to changes in morphology and particle size [16]. Another alternative is the utilization of high-energy gamma rays to sterilize pharmaceuticals. One drawback of this method is that free radicals generated upon gamma irradiation may lead to drug instability. For example, it has been shown that poststerilization, gamma irradiation causes some chemical changes on the cyclodextrin nanoparticles; however, these changes were not reported to affect the nanoparticle performance [17]. Assessment of product performance poststerilization will determine whether sterilization affects the product quality. In conclusion, the selection of sterilization method should be decided on a case-by-case basis with consideration for product quality and integrity.

The bacterial endotoxin test is a required test for all human injectable drugs, including biological drugs and ophthalmic and inhalation products. As stated in the United States Pharmacopeia (USP) USP33-NF28 Chapter 85, the Bacterial Endotoxins Test (LAL test) is a test to detect or quantify endotoxins from gram-negative bacteria using amoebocyte lysate from the horseshoe crab (*Limulus polyphemus* or *Tachypleus tridentatus*). There are three techniques for this test: the gel-clot technique, which is based on gel formation; the turbidimetric technique, based on the development of turbidity after cleavage of an endogenous substrate; and the chromogenic technique, based on the development of color after the cleavage of a synthetic peptide–chromogen complex. Presently, the FDA accepts any of the three techniques listed in the USP monograph to test for the presence of endotoxin. However, in the event of equivocal results between the tests, the final decision is made based upon the gel-clot technique unless otherwise indicated in the monograph for the product being tested. There have been reports on issues related to nanoparticle interference with LAL tests [18,19]; however, it is important to recognize that interference issues may stem from a number of factors and may not be inherent to nanoparticles.

5.2.2 PRECLINICAL STUDIES

Preclinical safety assessments for approval of pharmaceuticals regardless of their routes of administration (e.g., topical instillation, parenteral, etc.) include pharmacology, safety pharmacology, pharmacokinetic (ADME), toxicokinetic, acute, subchronic and chronic general toxicology, developmental toxicology, genotoxicity, and for many drugs an assessment of carcinogenic potential.

The FDA guidance *Nonclinical Safety Evaluation of Reformulated Drug Products and Products Intended for Administration by an Alternate Route* describes whether the active ingredient has previously been used by the ocular route; then toxicity studies in two species with complete eye and systemic evaluation for the appropriate duration should be carried out with the new formulation [20]. In certain cases, studies in the most appropriate species may be adequate. If a drug is absorbed systemically following topical installation to the eye, safety concerns must be addressed based on the systemic levels of drug assessed. Pharmacokinetic, toxicokinetic, and additional studies such as drug–drug interaction studies will be required in this situation.

Ocular ADME studies could include the measurement of drug concentration in ocular tissues such as the aqueous humor, conjunctiva, cornea, lens, iris-ciliary body, vitreous humor, retina, and choroid and tear fluid. The evaluation of drug concentration in tear fluid is a challenging task, in general, since high variability is observed within and between subjects. Due to the paucity of data in this area, interference issues are not reported in the literature specifically related to nanomaterial-containing drug products. However, challenges may exist in the area of bioanalytical validation when working with these products.

Interaction with plasma proteins will also affect biodistribution directly. It has been shown *in vitro* that liposomal benzoporphyrin derivative monoacid ring A (BPD-MA) were primarily associated with lipoproteins and subsequently exhibit rapid tissue distribution [21]. For nanotherapeutics encapsulated in a carrier, conducting the PK studies with dual radiolabeling will be informative on the fate of the drug as well as the carrier [22]. Another variable that can affect biodistribution stems from the formulation component of the nanomaterial-containing drug product. For example, the polyethylene glycol (PEG) coat is designed to make the nanotherapeutic stealth and escape uptake by the macrophages, therefore extending circulation half-life. However, it has been reported that accelerated blood clearance of nanoparticles was observed following repeated dosing of the PEG-coated nanotherapeutic [23].

Nanotherapeutics can be designed to have different charge and particle size characteristics; this can present versatility and make formulations containing nanoparticles desirable. However, with the development of any pharmaceutical products, advantages are counterbalanced by some disadvantages which may result in unwanted consequences. In the literature, it has been reported that anaphylactoid-like reactions to liposomal products are seen in a swine model and complement activation by negatively charged phospholipids such as egg phosphatidylglycerol [24]. A similar effect has also been observed in a clinical study, using pegylated phosphatidylethanolamine (PEG-PE) liposomes administered to patients with solid tumors, resulting in immediate hypersensitivity reactions; it has been suggested that this may be due to the presence of negatively charged PEG-PE component in the liposomes [24].

Assessment of particle size is critical and plays an important role in the absorption profile of nanomaterial-containing drug products. Change in particle size may impact the biodistribution and clearance, as shown in a study on retention and ocular distribution of fluorescent polystyrene nanoparticles, where size-dependent disposition kinetics of 20 nm fluorescent polystyrene nanoparticles exhibited rapid clearance from the periocular spaces. In this same study, 200 nm particles were retained at the site of injection for 2 months [25]. Subsequent studies showed that periocular blood and lymphatic circulation was responsible for clearance of 20 nm particles while the 200 nm particles remained at the site of injection [26].

Special ocular toxicity and ocular ADME studies can address toxicity concerns based on the drug levels in the ocular tissues. Examples of ocular toxicity tests can include tests such as ocular irritation tests, aqueous flare, blinking frequency, and corneal epithelial lesions, which should be conducted in different species as well as in albino and pigmented animals. Ocular toxicity tests may require studies with different formulations, such as with aged drug product to investigate the effects of aging and stability. Other studies may be required with appropriate vehicle and saline controls to investigate toxicity from the contribution of impurities in the formulation. The toxicity studies following topical instillation could be conducted in one eye, and the other eye could be utilized as its control.

It is also especially important to support toxicological findings with exposure data, especially when different biodistribution profiles are observed for nanoparticles with different particle sizes. For example, Sakurai et al. have reported that nanoparticles with different sizes may migrate through the retinal layers and tend to accumulate in the retina as well as in the vitreous cavity and trabecular meshwork, following an injection into the vitreous cavity in rabbits [27]. In such cases, nonclinical toxicokinetic data may be used to interpret toxicity findings and their relevance to clinical safety issues.

5.2.3 CLINICAL STUDIES

General principles on the conduct of clinical trials are presented in the International Conference on Harmonisation (ICH) guidance document entitled "E8 General Considerations for Clinical Trials" [28]. Clinical trials can be classified according to the phase of clinical development in which the study is conducted, typically referred to as Phases 1 thorough 4. The results from each phase of drug development are utilized in the planning of subsequent studies. Safety studies include first-time in human studies with a starting dose based on the nonclinical pharmacokinetic, pharmacological, and toxicological studies. Typically, the formulations used in clinical trials should be well characterized and manufactured under GMP conditions. The pivotal nonclinical studies should be done under GMP using the same formulations as those used in the clinical studies. Consequently, if different batches of the nanomaterial-containing drug product are used at different stages of clinical development, adequate testing of each batch is essential. During the development phase, different formulations of the drug products could be used; however, bridging bioequivalence studies (BE) on different formulations should be conducted to link the different clinical studies to each other. Conducting BE studies for a nanomaterial-containing drug product may be challenging as the ADME characteristics of these products may differ significantly from what is typically seen for small-molecule drug development. For example, biodistribution, retention in circulation, and target tissue distribution may be impacted as a result of complex physicochemical properties of nanomaterial-containing drug products. In conducting such BE studies, additional BE parameters (other than AUC and Cmax of the active drug) may be necessary to show that different formulations are bioequivalent. For example, a nanotherapeutic has a targeting moiety and is intended to accumulate in a certain target tissue. Specific parameters such as target tissue concentration of an active and its metabolites may need to be evaluated in order to show that the drug product is bioequivalent to that used in clinical studies since plasma concentration levels of the drug alone may not sufficiently reflect or estimate bioequivalence. For dose selection in cases where nanomaterial-containing drug products are activated with external stimuli such as light, dose range finding studies can be complex since the amount of drug delivered to the target is influenced by external stimuli applied, duration, and intensity in addition to the typical changes in dose of drugs. This may impact the number of patients enrolled in the study or impact the study description since multiple treatments could be conducted, provided patients are subject to washout periods.

Safety assessment in healthy as well as in diseased models may need to be conducted with the appropriate controls (blank nanocarriers, external stimuli alone, etc.) to determine the contribution of nondrug components of the nanomaterial-containing drug product.

5.2.3.1 Postmarketing Safety Reporting

Postmarketing safety reporting is conducted after the approval of a drug. This consists of a combination of Phase IV study commitments, annual reports from the sponsor, and adverse events reported to MedWatch. Phase IV study commitments include special population studies, carcinogenicity studies, and drug interaction studies [29].

These are conditionally agreed upon at the time of approval. The annual reports submitted by the sponsor contain reports to the company of adverse events associated with the drug of interest. The adverse events reported to MedWatch are reports that can be submitted directly to the FDA from healthcare professionals, patients, or patient advocates. These reports are all intended to capture rare, serious, unexpected adverse events associated with the drug through the use of passive surveillance. Although the term "adverse event" covers any undesirable events associated with the use of a drug, a serious adverse event includes, but is not limited to, the following: death, life-threatening, hospitalization, disability, congenital abnormality, or required intervention to prevent permanent damage. Currently, the FDA is embarking on the Sentinel Initiative to proactively gather information on serious and nonserious adverse events. This system will complement the current postmarketing surveillance program.

The FDA Adverse Events Reporting System (AERS) is a surveillance mechanism using spontaneous reporting from consumers, healthcare providers, and drug manufacturers through a program called MedWatch. This is separate from the reports submitted by the applicants to monitor safety signals regarding drug–event combinations. The FDA currently utilizes the Emprica™ Signal data mining program to generate a signal (lower 95% confidence interval limit of the adjusted ratios of the observed events over expected events, also referred to as the EB05) comparing the fraction of all reports of a particular event for a specific drug with the fraction of reports for the same particular event for all drugs. More information on the statistical methods used in Emprica Signal can be found in the literature [30]. Within the context of the FDA's signal detection, further investigation is generally required for drug–event combinations with an EB05 ≥ 2, meaning that the drug–event combination occurred at least two times as often as would be expected [31].

As mentioned previously in this chapter, Visudyne (verteporfin for injection) is a liposomal formulation of poorly soluble verteporfin, a light-activated drug, indicated for the treatment of patients with predominantly classic subfoveal choroidal neovascularization due to age-related macular degeneration, and pathologic myopia. Visudyne therapy is a two-stage process that requires application of nonthermal red light following verteporfin injection [32]. Visudyne was chosen as a model to illustrate the process of postmarketing surveillance through the use of data mining and signal detection with spontaneous reporting in the FDA's AERS system. This drug was chosen because it is an example of a nanotherapeutic with considerable time on the market, making it a suitable candidate to illustrate how the postmarketing surveillance system can be used and the type of data it can generate to help in the evaluation of drug safety, even after approval. However, the postmarketing safety profile of Visudyne is only to be used as a learning tool in this chapter.

5.2.3.2 Data Mining Illustration

Because Visudyne is a nanoscale liposomal formulation, we looked at its adverse event profile in order to illustrate how postmarketing data could be used to evaluate safety, and how the information could be associated with observations with data substituted during the preapproval review process. The following adverse experiences have produced a signal through the AERS database: abnormal angiogram, back pain, fibrosis, renal colic, pain in extremities, and chest pain. However, it cannot be ascertained whether these adverse events are attributable to the liposomal components and characteristics without confirmatory studies.

The adverse event findings mentioned above were also reported by the Photodynamic Therapy (PDT) Users Group in the United Kingdom [33]. They conducted a prospective study of the therapy in 1755 patients being evaluated for improved vision. The study was designed to collect both safety and efficacy information on the verteporfin PDT from 13 UK centers. Each of these centers prospectively collected surveillance data on patients undergoing Visudyne PDTs. The patients were assessed based on improved vision shown by the number of letters they could read in a standard vision test. Side effects experienced by patients in the study included 3.04% with decreased vision, followed by 2.92% of patients with back pain, and 0.58% with a hemorrhage [33].

In another study, The Treatment of Age-Related Macular Degeneration with Photodynamic Therapy (TAP) Study, a randomized control trial on Visudyne PDT found 2% of patients on the treatment experienced back pain while 0% on placebo had back pain [34, 35]. This illustrates that AERS's detecting capability corresponds to current literature on Visudyne PDT's adverse events. However, these adverse events were already identified during development and were listed in Visudyne PDT's label, and therefore these studies did not reveal any new information from what was seen in the clinical trials submitted for approval [32].

It is important to emphasize that the adverse events reported for verteportin for injection are probably not due to the nanocharacteristics of the formulation but should be treated as product specific. By collecting appropriate data during product development, such as information on physicochemical characterization, ADME, and toxicology, one can later follow-up with adverse event profile characteristics in order to determine if certain findings may be linked to nanospecific characteristics.

5.2.3.3 Postmarketing Guidance Documents

When developing nanotechnology-based ophthalmic drugs, one might consider the following FDA guidance documents in regards to postmarketing surveillance and adverse event reporting associated with drugs. This list is by no means comprehensive, nor is it unique to nanotechnology-based ophthalmic drugs; however, it is meant to be a guide for Phase IV planning.

For a general overview of postmarketing surveillance activities, the guidance for industry *Postmarketing Safety Reporting for Human Drug and Biological Products Including Vaccines* published in March 2001 provides a comprehensive explanation of the types of reports that could be submitted as well as special reporting situations [36]. As stated in the guidance, reportable events include "any undesirable event that is associated with the use of a drug or biological product in humans whether or not considered product-related by the applicant." Coding for the adverse events are based on MedDRA terminology.

The July 2003 ICH guidance for industry details the following four types of reports: 15-day reports, periodic reports, follow-up reports, and special reports. The 15-day report is for serious, unexpected adverse events [37]. The applicant has a responsibility to apply due diligence to complete and file the report as soon as possible within 15 days of the event being reported to them.

The periodic reports are quarterly reports submitted to the FDA for the first 3 years of marketing for the drug, and after the first 3 years, they are submitted to the FDA annually. The follow-up reports should contain updated information on the 15-day reports. Follow-up reports should complete the picture and understanding of the adverse event, with all new information highlighted. The follow-up report is intended to capture details missing from the three main components of the 15-day alert and any new adverse experiences associated with the initial adverse event. Lastly, special reporting situations include but are not limited to scientific literature reports, surveillance studies, foreign reports, and lack of effect reports.

5.3 CONCLUSIONS

Nanomaterial-containing drug products constitute a highly specialized area of drug development and therefore pose unique challenges. Moreover, ocular delivery of nanomaterial-containing drug products is an even more specialized area of nanomedicine because of a number of issues related to formulation requirements and the unique biology and physiology of the target organ. Although there is currently no official FDA definition for nanotechnology drugs, there are general criteria that have been identified in order to assess whether FDA-regulated products might have nanotechnology-related characteristics. Nonetheless, there would be great value in a formal definition, as well as in specific guidance documents, which would help sponsors develop drugs that incorporate nanotechnology. We foresee that in the future, many FDA centers, including the CDER, will have nanotechnology-specific guidance documents to help sponsors better develop nanomaterial-containing drug products. With respect to the CDER, a MaPP that was published in 2010 was the first step toward collecting the relevant data on nanospecific properties and subsequently a better understanding of drugs containing nanotechnology. As we collect the relevant information on nanotechnology drugs, the next steps are to evaluate and analyze the collected information and to identify the research and policy gaps. As the field of nanomedicine matures, and as we learn more about drugs containing nanotechnology, the CDER may utilize its postmarketing surveillance methods to retroactively monitor the safety profile of specific drugs with specific nanotechnology-related characteristics. This approach could help identify safety signals that may not be picked up prior to product approval, thus helping to better regulate future products with similar properties. However, associating safety signals with specific physicochemical properties can only be done if those properties are collected during the review process. This is precisely the reason the CDER MaPP formalized the collection of such information in CMC reviews. The regulatory requirements combined with existing postmarketing surveillance of any pharmaceutical drug products, including those containing nanomaterials, is deemed currently adequate; however, as knowledge increases, there may be room for improvement in our regulatory procedures. As regulators, we look to the scientific community to help identify those critical aspects that impact product performance, quality, and safety for all emerging sciences, including nanotechnology.

ACKNOWLEDGMENTS

The authors thank Ms. Taryn Alverson for accessing and collecting FDA reviews for ocular drugs. This project was supported in part by an appointment to the Research Participation Program at the Center for Drug Evaluation and Research administered by the Oak Ridge Institute for Science and Education through an interagency agreement between the U.S. Department of Energy and the U.S. Food and Drug Administration.

REFERENCES

1. H. F. Edelhauser, C. L. Rowe-Rendleman, M. R. Robinson, D. G. Dawson, G. J. Chader, H. E. Grossniklaus, K. D. Rittenhouse et al. Ophthalmic drug delivery systems for the treatment of retinal diseases: Basic research to clinical applications. *Investigative Ophthalmology & Visual Science.* 51:5403–5420 (2010).
2. B. S. Zolnik and N. Sadrieh. Nanotechnology: Past, present and future. *European Biopharmaceutical Review.* October 58–63 (2009).
3. B. S. Zolnik and N. Sadrieh. Regulatory perspective on the importance of ADME assessment of nanoscale material containing drugs. *Advanced Drug Delivery Reviews.* 61:422–427 (2009).
4. Y. Diebold and M. Calonge. Applications of nanoparticles in ophthalmology. *Progress in Retinal and Eye Research.* 29:596–609 (2010).
5. T. W. Prow. Toxicity of nanomaterials to the eye. *Wiley Interdisciplinary Reviews Nanomedicine and Nanobiotechnology.* 2:317–333 (2010).
6. FDA Draft Guidance Document: Considering Whether an FDA-Regulated Product Involves the Application of Nanotechnology Guidance for Industry. http://www.fda.gov/RegulatoryInformation/Guidances/ucm257698.htm (last accessed June 2011).
7. Manual of Policies and Procedures (MaPP) Reporting Format for Nanotechnology-Related Information in CMC Review, MaPP 5015.9. http://www.fda.gov/downloads/AboutFDA/CentersOffices/CDER/ManualofPoliciesProcedures/UCM214304.pdf (last accessed October 2011).
8. Drugs at the FDA. http://www.accessdata.fda.gov/scripts/cder/drugsatfda/ (last accessed October 2011).
9. FDA Guidance for Industry: INDs for Phase 2 and Phase 3 Studies. http://www.fda.gov/downloads/Drugs/GuidanceComplianceRegulatoryInformation/Guidances/ucm070567.pdf (last accessed October 2011).
10. Inactive Ingredients in FDA Approved Drugs. http://www.accessdata.fda.gov/scripts/cder/iig/index.cfm (last accessed October 2011).
11. H. Greim and H. Norppa. Genotoxity testing of nanomaterials—Conclusions. *Nanotoxicology.* 4:421–424 (2010).
12. A. A. Attama, S. Reichl, and C. C. Muller-Goymann. Diclofenac sodium delivery to the eye: *In vitro* evaluation of novel solid lipid nanoparticle formulation using human cornea construct. *International Journal of Pharmaceutics.* 355:307–313 (2008).
13. X. Li, S. F. Nie, J. Kong, N. Li, C. Y. Ju, and W. S. Pan. A controlled-release ocular delivery system for ibuprofen based on nanostructured lipid carriers. *International Journal of Pharmaceutics.* 363:177–182 (2008).
14. FDA Guidance for Industry: Liposome Drug Products. http://www.fda.gov/downloads/Drugs/GuidanceComplianceRegulatoryInformation/Guidances/UCM070570.pdf (last accessed October 2011).

15. FDA Guidance for Industry: Sterile Drug Products Produced by Aseptic Processing Current Good Manufacturing Practice. http://www.fda.gov/downloads/Drugs/Guidance ComplianceRegulatoryInformation/Guidances/ucm070342.pdf (last accessed October 2011).
16. B. S. Zolnik, P. E. Leary, and D. J. Burgess. Elevated temperature accelerated release testing of PLGA microspheres. *Journal of Controlled Release: Official Journal of the Controlled Release Society.* 112:293–300 (2006).
17. E. Memisoglu-Bilensoy and A. A. Hincal. Sterile, injectable cyclodextrin nanoparticles: Effects of gamma irradiation and autoclaving. *International Journal of Pharmaceutics.* 311:203–208 (2006).
18. M. A. Dobrovolskaia, B. W. Neun, J. D. Clogston, H. Ding, J. Ljubimova, and S. E. McNeil. Ambiguities in applying traditional Limulus amebocyte lysate tests to quantify endotoxin in nanoparticle formulations. *Nanomedicine (Lond).* 5:555–562 (2010).
19. M. A. Dobrovolskaia, D. R. Germolec, and J. L. Weaver. Evaluation of nanoparticle immunotoxicity. *Nature Nanotechnology.* 4:411–414 (2009).
20. FDA Guidance for Industry and Review Staff Nonclinical Safety Evaluation of Reformulated Drug Products and Products Intended for Administration by an Alternate Route. http://www.fda.gov/downloads/Drugs/GuidanceComplianceRegulatoryInformation/ Guidances/ucm079245.pdf (last accessed October 2011).
21. A. M. Richter, E. Waterfield, A. K. Jain, A. J. Canaan, B. A. Allison, and J. G. Levy. Liposomal delivery of a photosensitizer, benzoporphyrin derivative monoacid ring A (BPD), to tumor tissue in a mouse tumor model. *Photochemistry and Photobiology.* 57:1000–1006 (1993).
22. B. S. Zolnik, S. T. Stern, J. M. Kaiser, Y. Heakal, J. D. Clogston, M. Kester, and S. E. McNeil. Rapid distribution of liposomal short-chain ceramide *in vitro* and *in vivo*. *Drug Metabolism and Disposition: The Biological Fate of Chemicals.* 36:1709–1715 (2008).
23. T. Ishihara, M. Takeda, H. Sakamoto, A. Kimoto, C. Kobayashi, N. Takasaki, K. Yuki et al. Accelerated blood clearance phenomenon upon repeated injection of PEG-modified PLA-nanoparticles. *Pharmaceutical Research.* 26:2270–2279 (2009).
24. J. Szebeni, L. Baranyi, S. Savay, J. Milosevits, R. Bunger, P. Laverman, J. M. Metselaar et al. Role of complement activation in hypersensitivity reactions to doxil and hynic PEG liposomes: Experimental and clinical studies. *Journal of Liposome Research.* 12:165–172 (2002).
25. A. C. Amrite and U. B. Kompella. Size-dependent disposition of nanoparticles and microparticles following subconjunctival administration. *The Journal of Pharmacy and Pharmacology.* 57:1555–1563 (2005).
26. A. C. Amrite, H. F. Edelhauser, S. R. Singh, and U. B. Kompella. Effect of circulation on the disposition and ocular tissue distribution of 20 nm nanoparticles after periocular administration. *Molecular Vision.* 14:150–160 (2008).
27. E. Sakurai, H. Ozeki, N. Kunou, and Y. Ogura. Effect of particle size of polymeric nano-spheres on intravitreal kinetics. *Ophthalmic Research.* 33:31–36 (2001).
28. ICH Guidance Document: General Considerations for Clinical Trials E8. http://www .ich.org/fileadmin/Public_Web_Site/ICH_Products/Guidelines/Efficacy/E8/Step4/E8_ Guideline.pdf (last accessed October 2011).
29. FDA Data Standards Manual. http://www.fda.gov/Drugs/DevelopmentApprovalProcess/ FormsSubmissionRequirements/ElectronicSubmissions/DataStandardsManualmonographs/ ucm071716.htm (last accessed October 2011).
30. W. DuMouchel. Bayesian data mining in large frequency tables, with an application to the FDA spontaneous reporting System. *The American Statistician.* 53:177–190 (1999).

31. A. Szarfman, S. G. Machado, and R. T. O'Neill. Use of screening algorithms and computer systems to efficiently signal higher-than-expected combinations of drugs and events in the US FDA's spontaneous reports database. *Drug Safety: An International Journal of Medical Toxicology and Drug Experience.* 25:381–392 (2002).
32. Verteporfin for Injection Label. http://dailymed.nlm.nih.gov/dailymed/lookup.cfm?setid =31512723-9ff0-4e18-aa3a-55ab833038c6 (last accessed October 2011).
33. F. D. Ghanchi, J. Fullarton, J. Blake, and S. P. Harding. The introduction of verteporfin photodynamic therapy in the UK: PDT users group (PDTUG) surveillance programme report 1. *Eye (Lond).* 22:671–677 (2008).
34. S. J. Bakri and P. K. Kaiser. Verteporfin ocular photodynamic therapy. *Expert Opinion on Pharmacotherapy.* 5:195–203 (2004).
35. N. M. Bressler and S. B. Bressler. Photodynamic therapy with verteporfin (Visudyne): Impact on ophthalmology and visual sciences. *Investigative Ophthalmology & Visual Science.* 41:624 (2000).
36. FDA Guidance for Industry: Postmarketing Safety Reporting for Human Drug and Biological Products Including Vaccines. http://www.fda.gov/downloads/BiologicsBloodVaccines/ GuidanceComplianceRegulatoryInformation/Guidances/Vaccines/ucm092257.pdf (last accessed October 2011).
37. ICH Guidelines: Post-Approval Safety Data Management: Definitions and Standards for Expedited Reporting. http://www.fda.gov/downloads/RegulatoryInformation/Guidances/ ucm129458.pdf (last accessed October 2011).

Section II

Ocular Barriers

6 Blood–Retinal Barrier
The Fundamentals

*Rosa Fernandes, Andreia Gonçalves, and
José Cunha-Vaz*

CONTENTS

6.1 INTRODUCTION

The concept of the blood–brain barrier (BBB), which first appeared in the literature in 1885, had its origin in the classical experiments of Goldman, who used trypan blue for the first time to demonstrate that there is, at the blood–brain interface, a barrier system that protects the brain. In this experiment, trypan blue injected intravenously induced an intense blue staining of all tissues, with the solitary exception of the brain (Goldman 1913). This was in contrast to the deep blue staining of the brain observed when a small quantity of the same dye was introduced directly into the subarachnoid space. Although no toxic symptoms resulted from intravenous administration, the animals died in a few minutes following subarachnoid administration with convulsions and final paralysis of the central nervous system. Subsequent studies confirmed a distinct set of permeability characteristics at the blood–brain interface.

In 1965 we published our study on the effect of histamine on the permeability of the ocular vessels (Ashton and Cunha-Vaz 1965). Histamine markedly increased the vascular permeability of the various ocular tissues, except for the retina. This behavior of the retinal vessels was similar only to what had been described previously for the cerebral vessels (Cunha-Vaz 1965, 1966; Shakib and Cunha-Vaz 1966).

Following up on our morphological studies and permeability measurements (Cunha-Vaz 1965; Cunha-Vaz and Maurice 1967), I proposed that the blood–retinal barrier (BRB) should be regarded as consisting of two major components, the endothelium of retinal blood vessels (inner BRB) and the retinal pigment epithelium (outer BRB) (Cunha-Vaz et al. 1975). This is a useful oversimplification in order to understand and identify the major barriers that separate the blood from the retina and their alterations in chorioretinal disease. It has proved most useful to understand clinical findings in posterior segment diseases.

Morphological studies, using electron microscopy, were extremely rewarding, demonstrating the presence in the retinal vessels of zonulae occludente between the endothelial cells showing that the retinal endothelial layer has an epithelial-like structure and offering an explanation for the permeability behavior of the retinal vessels (Shakib and Cunha-Vaz 1966). This observation was later confirmed to occur also in the brain vessels, giving support to our present understanding of the important role of the endothelial layers of the retinal and brain vessels in the BRB and BBB, respectively (Reese and Karnovsky 1967).

Our studies showed also for the first time that there was an active transport of an organic anion, fluorescein, out of the vitreous body across the pigment epithelium and across the blood vessels of the retina (Cunha-Vaz 1965; Cunha-Vaz and Maurice 1967).

The term "BRB system" is most useful for clinical purposes and better identifies its major role, that is, regulating the microenvironment of the retina. The BRB system must be viewed as a whole and as regulating both the retina extracellular fluid and the vitreous; it also should include the ciliary processes (Bito 1977). The transport functions of the ciliary epithelia contribute also to regulating fluid movements in the vitreous and to the all-important microenvironment of the retina for adequate visual function.

The tight-junctional endothelium of the retinal vessels appears to be generally analogous to that of the brain vessels and unquestionably represents a critically important and well-documented permeability barrier. In both the brain and retina, the endothelial layer of their vessels is now accepted as the mainstay of these barrier systems.

However, the regulatory roles of the ciliary epithelia in the vitreous fluid and the chorioretinal or retinal pigment epithelium in the retina extra cellular fluid (ECF) have to be taken into consideration and are fundamental components of sophisticated systems of BRB as it maintains the vitreous retina and posterior segment of the eye as "privileged sites" in the body.

The BRB is understood today as playing a fundamental role in retinal function in both health and disease. The major diseases that affect visual function, diabetic macular edema, and "wet" age-related macular degeneration are characterized by a breakdown of the inner and outer BRB, respectively. The BRB is now at the core of our understanding of retinal disease and the development of new therapies such as steroid and anti-vascular endothelial growth factor (anti-VEGF) treatments.

6.2 BLOOD–RETINAL BARRIER

The presence of an intact BRB is essential for the structural and functional integrity of the retina and, in clinical conditions where BRB breakdown occurs, vision may be seriously affected.

The BRB consists of inner and outer components, inner BRB (iBRB) and outer BRB (oBRB), and regulates fluids and molecular movement between the ocular vascular beds and retinal tissues and prevents leakage into the retina of macromolecules and other potentially harmful agents. The iBRB is established by the tight junctions (TJ, zonulae occludentes) between neighboring retinal endothelial cells. These specialized TJ restrict the diffusional permeability of the retinal endothelial layer to values in the order of 0.14×10^{-5} cm/s for sodium fluorescein. The retinal endothelial layer functions as an "epithelium" and, in this way, is directly associated with its differentiation and with the polarization of the BRB function. This continuous endothelial cell layer, which forms the main structure of the iBRB, rests on a basal lamina that is covered by the processes of astrocytes and Muller cells. Pericytes are also present, encased in the basal lamina, in close contact with the endothelial cells, but do not form a continuous layer and, therefore, do not contribute to the diffusional barrier. Astrocytes, Muller cells, and pericytes are considered to influence the activity of the retinal endothelial cells and of the iBRB by transmitting to endothelial cells regulatory signals indicating changes in the microenvironment of the retinal neuronal circuitry.

The oBRB is established by the TJ between neighboring retinal pigment epithelial cells (RPE). The RPE is composed of a single layer of retinal pigment epithelial cells that are joined laterally toward their apices by TJ between adjacent lateral cell walls. The RPE resting upon the underlying Bruch's membrane separates the neural retina from the fenestrated choriocapillaries and plays a fundamental role in regulating access of nutrients from the blood to the photoreceptors as well as eliminating waste products and maintaining retinal adhesion. The metabolic relationship of the RPE apical villi and the photoreceptors is considered to be critical for the maintenance of visual function.

6.2.1 MOLECULAR ARCHITECTURE OF TIGHT JUNCTIONS IN RETINAL BARRIERS

Tight junctions create a paracellular barrier between the vascular lumen and neural layers in the retina, which is important for the maintenance of a confined microenvironment and for proper neuronal function (Cunha-Vaz et al. 1966).

The main functions of the TJ include regulation of ion, water, and nutrient flow between the retina and blood vessels, and protection of the neural retina from inflammatory cells and their toxic products found in the systemic circulation (Gardner et al. 2002; Kaur et al. 2008). It is strongly believed that the disruption of the iBRB (rather than the oBRB) is the primary site of vascular leakage in diabetic retinopathy (Antcliff and Marshall 1999) and retinopathy of prematurity (Zhang et al. 2008; Lepore et al. 2011).

Tight junctions, sealing neighboring endothelial or epithelial cells together, may also serve as regulatory centers to help coordinate several cell processes, such as the regulation of cell morphology, proliferation, and establishment and maintenance of apico-basal polarity (Zahraoui et al. 2000; Yeaman et al. 2004; Matter et al. 2005; Caplan et al. 2008; Cereijido et al. 2008; Mellman and Nelson 2008; Steed et al. 2010).

TJ are dynamic, complex structures in which the extracellular domains of TJ proteins associate with the extracellular domains of proteins on adjacent cells. The branching network of sealing strands of proteins found in retinal endothelial TJ includes a series of transmembrane proteins (junctional adhesion molecules [JAMs], claudins, occludin, and tricellulin) embedded in the plasma membrane (Ikenouchi et al. 2005) and cytoplasmic scaffold proteins (zona occludens [ZO]-1/2/3, MAGI-1, MAGI-3, CASK/LIN-2, MUPP1, AF6, ASIP, PALS1, PATJ and cingulin) (Bauer et al. 2011) (Figure 6.1).

It has also been shown that glial cells and pericytes secrete factors that affect BRB permeability. In fact, the development and maintenance of iBRB may be regulated in part by glial cells (Müller cells and astrocytes) (Janzer and Raff 1987; Tout et al.

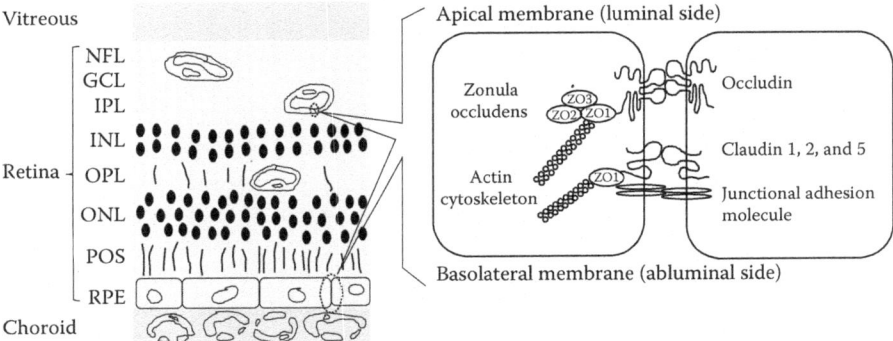

FIGURE 6.1 Representative diagram of the TJ assembly present in endothelial and epithelial cells of the retina, forming the inner and the outer BRB, respectively. TJ are composed of a variety of proteins, namely transmembrane proteins like occludin, claudins, and JAMs, which seal the paracellular space, and scaffolding cytosolic proteins like ZO-1,2,3, which interact with the actin cytoskeleton. NFL, nerve fiber layer; GCL, ganglion cell layer; IPL, inner plexiform layer; INL, inner nuclear layer; OPL, outer plexiform layer; ONL, outer nuclear layer; POS, photoreceptor outer segments; RPE, retinal pigment epithelium.

1993) and pericytes, promoting the integrity of TJ and the nonfenestrated phenotype of retinal endothelial cells (Antonetti et al. 1999).

6.2.1.1 Claudins

The selectivity and permeability of TJ are tissue specific and depend on their protein composition, namely claudins (Furuse et al. 1998a; Morita et al. 1999a; Tsukita and Furuse 1999; Furuse et al. 2001; Van Itallie and Anderson 2006). Cell-culture experiments utilizing siRNA (Hou et al. 2006) or mutational (Colegio et al. 2002) and *in vivo* analysis of transgenic mice have demonstrated that claudins confer barrier properties to the TJ (Morita et al. 1999b).

Claudins are a family of tetraspanning cell–cell adhesion proteins, ranging around 20–27 kDa, containing a short cytoplasmic N-terminus, two extracellular loops (the first loop is larger than the second and is conserved among members of the claudin family) and a C-terminal cytoplasmic domain (Furuse et al. 1998a). Claudin-1, -2, and -5 are the most prominent claudins in inner, outer, and ganglion cell vascular layers of the retina (Barber and Antonetti 2003; Luo et al. 2011). Claudins form TJ strands via interacting with each other between different TJ strands or within individual strands in a homotypic and heterotypic manner (Furuse et al. 1999). Overexpression of claudin-2 in MDCK cells, which normally express claudin-1 and claudin-4, has been shown to decrease transepithelial resistance (TER) values, providing support for the hypothesis that different combinations of claudins determine the barrier properties of individual TJ strands (Furuse et al. 2001). Claudins also mediate calcium-independent cell–cell adhesion, interacting directly with peripheral PDZ-domain-containing proteins, including ZO-1 (Furuse et al. 1998b). Alterations in claudin-5 in retinal vessels have been associated with increased vessel leakage in the early stages of diabetic retinopathy (Leal et al. 2010). Moreover, experiments involving gene deletion of claudin-5 have been shown to result in permeability to molecules under 800 Da and neonatal lethality (Nitta et al. 2003), pointing to the role of this claudin in maintaining proper function and integrity of the blood–tissue barriers.

6.2.1.2 Occludin and Tricellulin

Although claudins confer barrier properties to the TJ, occludin (or MARVELD1) appears to regulate the cell's response to the external signals that control barrier properties. Occludin, expressed ubiquitously in TJ, is a protein of 55.9 kDa, also bearing four transmembrane regions, with two extracellular loops and two intracellular domains (Feldman et al. 2005), and not showing any other similarity to claudins (Furuse et al. 1993, 1998b). At the cytoplasmic surface of cells, the claudins and occludin bind zonula occludens proteins. A special situation emerges at sites where three endothelial or epithelial cells join together. Recently, tricellulin (or MARVELD2), a TJ protein of 63.9 kDa, was found at tricellular contacts (Ikenouchi et al. 2005). The structure of tricellulin resembles that of occludin, with 32% of homology at the C-terminus (Riazuddin et al. 2006). Both occludin and tricellulin have the conserved MARVEL (**MA**L and **r**elated proteins for **ve**sicle trafficking and membrane **l**ink) domain, which has been found in vesicle transport proteins. Recent studies have shown that the tricellulin C-terminus is important for the basolateral translocation of tricellulin, whereas the N-terminal domain is involved in directing the protein to tricellular contacts. The presence of homomeric tricellulin–tricellulin and heteromeric tricellulin–occludin complexes in MDCK C11 cells overexpressing tagged tricellulin constructs led Westphal et al. (2010) to propose that tricellulin and occludin are transported together to the edges of the TJ being occludin and tricellulin complexes dissociated when tricellular contacts are formed. Both tricellulin and occludin were found to interact with ZO-1 via their cytoplasmic C-terminus (Furuse et al. 1994; Riazuddin et al. 2006). Studies using a stable epithelial cell line (MDCK) with occludin expression almost completely reduced through siRNA expression revealed a role for occludin in signal transduction and the regulation of permeability (Yu et al. 2005).

In experimental animal models of streptozotocin-induced diabetes, occludin content and immunoreactivity decrease at cell borders concomitant with increased BRB permeability. In endothelial cells, changes in barrier permeability and decreased occludin content at the cell border were associated with increased occludin phosphorylation (Antonetti et al. 1999). More recently, it has been shown in bovine retinal endothelial cells (BREC) that treatment with VEGF increases occludin phosphorylation and ubiquitination. This leads to a reduced interaction with ZO-1 (Sundstrom et al. 2009), thus contributing to TJ trafficking and subsequent increased vascular permeability (Murakami et al. 2009).

6.2.1.3 Junctional-Associated Adhesion Molecules

JAMs are a family of transmembrane proteins of approximately 40 kDa concentrated at TJ (Martin-Padura et al. 1998; Cunningham et al. 2000; Arrate et al. 2001). In contrast to occludin and claudins, JAMs are single-pass transmembrane proteins, with a transmembrane domain and an extracellular portion folded into two immunoglobulin-like domains (Martin-Padura et al. 1998; Arrate et al. 2001). JAMs are a group of three proteins, named JAM-1, -2, and -3. JAM-1 was shown to be involved in cell–cell adhesion and transmigration of monocytes through endothelial cells (Martin-Padura et al. 1998). JAM-1 is expressed also in epithelial cells, whereas JAM-2 (Aurrand-Lions et al. 2001) and JAM-3 (Palmeri et al. 2000) are expressed in most vascular endothelial cells. JAMs can be found at intercellular contacts, and they play a role in TJ organization and in endothelial polarity, as well as in paracellular permeability (Ebnet et al. 2003; Mandell et al. 2005).

6.2.1.4 Cytoplasmic Accessory Proteins

Other cytoplasmic accessory proteins also are part of the TJ. They interact with each other and link the transmembrane proteins to the actin cytoskeleton. Additionally, they play an important role in the formation and assembly of TJ (Fanning et al. 1998; Itoh et al. 1999; Ikenouchi et al. 2007). ZO-1, ZO-2, and ZO-3 are three isoforms of ZO proteins of approximately 220, 160, and 130 kDa, respectively (Stevenson et al. 1986; Haskins et al. 1998).

ZO-knock-out or ZO-knock-down studies in epithelial cells have revealed that ZO-1 and ZO-2 are important at the initial stages of the polymerization of claudins, determining where claudins should be polymerized, whereas ZO-3 is dispensable *in vivo* in terms of individual viability, epithelial differentiation, and establishment of TJ (Umeda et al. 2004; Adachi et al. 2006). In accordance with the role of ZO-1 and ZO-2 proteins in the formation of the TJ strands, decreased ZO-1 at endothelial cell borders and accumulation in the cytosol of endothelial cells are associated with BRB permeability (Mark and Davis 2002; Leal et al. 2010). Additionally, decreased levels of ZO-1 and ZO-2 induced by VEGF in retinal endothelial cells are associated with increased BRB permeability (Kim et al. 2009).

Besides being indispensable structural components at the junctional site, novel functions have been recently ascribed to ZO proteins, functioning in signal transduction pathways related to gene expression and cell growth (Islas et al. 2002; Bauer et al. 2010).

Many other accessory cytoplasmic proteins have been described (Tsukita et al. 2001). Besides the cytoplasmic scaffold proteins mentioned above, ZO-1-associated acid-binding protein or ZONAB, Rab8, Rab13, Rab3B, 19B1, 7H6, PAR3, PAR6, RhoA, RalA, and Raf-1 are considered signaling proteins of the TJ. These proteins seem to be involved in junction assembly, barrier regulation, gene transcription, and possibly other functions still to be unraveled (Forster 2008).

Disruption of TJ in ocular diseases or induced by drugs can lead to impaired BRB function and compromise vision. Therefore, understanding the mechanisms underlying changes in the expression, subcellular localization, posttranslational modification, and protein–protein interactions of TJ proteins, under physiological and pathological conditions, will allow developing strategies for the modulation of the iBRB to enhance delivery of therapeutics to the retina, an approach for the prevention and treatment of retinal diseases.

6.2.2 REGULATION OF THE MICROENVIRONMENT OF THE RETINA: BRB TRANSPORT SYSTEMS

The presence of TJ in the BRB prevents free diffusion of polar nutrients essential for the metabolism of the retina, and therefore the BRB must contain specific transport proteins that are expressed at the plasma membranes of the retinal epithelial and endothelial cells. Epithelial (oBRB) and endothelial (iBRB) cells exhibit a polarized expression of transport carriers in the apical/luminal and basolateral/abluminal plasma membranes. The orientation of these carriers results in preferential blood-to-retina influx or retina-to-blood efflux transport of substrates or in facilitated transport in either direction depending on the concentration gradient of the solutes across the BRB. Figure 6.2 illustrates the potential routes for facilitated transport across the BRB. One is the blood-to-retina influx transport system that acts as an energy supply system for the retina, since the iBRB supplies metabolic substrates from the circulating blood, such as glucose, amino acids, vitamins, and nucleosides, to the retina. The other is the retina-to-blood efflux transport system that acts to get rid of hydrophobic xenobiotics and neurotransmitter metabolites.

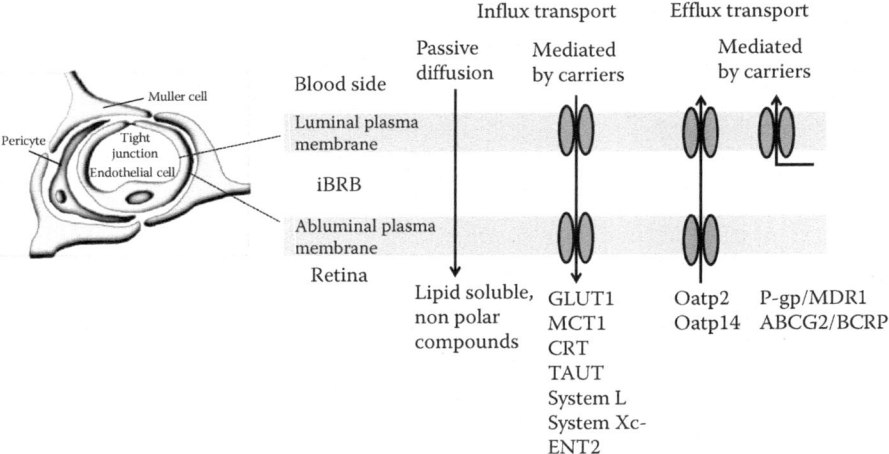

FIGURE 6.2 Several types of transport mechanisms at the iBRB that help maintain retina homeostasis. The iBRB has several transporters that play essential roles in supplying nutrients like glucose, amino acids, and nucleosides to the inner retina and that carry out the efflux transport of drugs, metabolites, and xenobiotics.

6.2.2.1 Energy Transport System

The blood-to-retina influx transporters operating at the iBRB supply hydrophilic substrates to the retina (Table 6.1). Being one of the most metabolically active tissues in the body, the retina uses glucose as its primary energy substrate. Because glucose is hydrophilic, and thus does not freely cross the barrier, its transport is mediated by facilitative glucose transporters, named GLUTs. There has been reported to be an asymmetric distribution of this transporter protein at the human and rat iBRB. This is characteristic of polarized endothelial cells in that intercellular TJ allow for division of the cells into luminal and abluminal surfaces and for the nonuniform distribution of transport proteins within the cell (Kumagai et al. 1996; Fernandes et al. 2003). The higher density of GLUT1 on the abluminal membrane of the retinal endothelial cells suggests that glucose transport is limited at the blood–luminal, rather than the abluminal–interstitial interface. In experimental animal models of type 1 and type 2 diabetes, as well as *in vitro* studies where endothelial cells were exposed to elevated glucose, it has been shown that there is a downregulation of GLUT1 in retinal endothelial cells (Badr et al. 2000; Fernandes et al. 2004).

Besides glucose, the oxidized form of vitamin C, dehydroascorbic acid (DHA), is rapidly transported across the BRB via GLUT1 and accumulates as ascorbic acid in the retina. DHA uptake by GLUT1 is competitively inhibited with D-glucose and its transport from the blood to the retina decreases with increased blood glucose concentrations under diabetic conditions (Minamizono et al. 2006). This can lead to the increased oxidative stress observed in diabetic retinas.

In addition to D-glucose, L-lactic acid appears to be required as an energy source in photoreceptors (Poitry-Yamate et al. 1995), and the transport of this solute

TABLE 6.1

Representative Transporters and Transport Processes at the Inner Blood–Retinal Barrier

Transporter	Substrates	Location	Orientation/ Direction	Expression	References
Energy transport system					
GLUT1	D-Glucose, dehydroascorbic acid	L, A	In	Isolated bovine retinal capillaries, rat retinal vessels	Betz and Goldstein (1980), Takata et al. (1992), Vera et al. (1993)
MCT1	L-Lactate	L, A	In, Ef	Rat retinal endothelium, rat retinal capillary endothelial cell line	Alm and Tornquist (1985), Gerhart et al. (1999), Hosoya et al. (2001a)
CRT	Creatine, metabolites of creatine	L, A	In	Rat retinal endothelium, rat retinal capillary endothelial cell line	Nakashima et al. (2004)
Amino acid transporter system					
TAUT	Taurine, GABA	L, A	In	Primary-cultured human retinal endothelial cells, rat retinal capillary endothelial cell line	Hosoya and Tomi (2005), Tomi et al. (2007), Tomi et al. (2008)
LAT1/4Fhc (system L)	Branched-chain and aromatic amino acids like L-phenylalanine, L-Leucine	L, A	In	Primary-cultured human retinal endothelial cells, rat retinal endothelial cell line	Tomi et al. (2005)
xCT/4F2hc (system Xc−)	Cysteine/glutamate	L, A	In/Ef	Rat retinal endothelial cell line	Tomi et al. (2002)

(Continued)

TABLE 6.1 (Continued)
Representative Transporters and Transport Processes at the Inner Blood–Retinal Barrier

Transporter	Substrates	Location	Orientation/ Direction	Expression	References
Nucleoside transport system					
ENT2	Adenosine, thymidine	L, A	In, Ef	Rat retinal endothelial cell line	Nagase et al. (2006)
Organic anion transport system					
Oatp2	E17βG, organic anions	L, A	Ef	Rat retinal endothelium	Gao et al. (2002), Tomi and Hosoya (2004)
Oatp14	Organic anions	ND	Ef	Rat retinal endothelium	Tomi and Hosoya (2004)
ABC transporters					
P-gp/MDR1	Lipophilic, cationic compounds	L	Ef	Rat retinal endothelial cells	Greenwood (1992), Shen et al. (2003), Hosoya and Tomi (2005)
BCRP/ ABCG2	Xenobiotics, phototoxins	L	Ef	Mouse and rat retinal endothelium, rat retinal capillary endothelial cell line	Asashima et al. (2006)

GLUT, glucose transporter; MCT, monocarboxylate transporter; CRT, creatine transporter; TAUT, taurine transporter; LAT, L-type amino acid transporter; ENT, equilibrative nucleoside transporter; OATP, organic anion transporting polypeptide; P-gp, P-glycoprotein; BCRP, breast cancer resistance protein; GABA, γ-aminobutyric acid; E17βG, estradiol 17-β glucuronide; L, luminal; A, abluminal; ND, not determined; In, influx; Ef, efflux.

between retina and blood seems to be mediated by the monocarboxylate transporter 1 (MCT1). Immunoreactivity for this transporter was found both in luminal and abluminal membranes of rat endothelial cells (Gerhart et al. 1999). It has been shown that the uptake of labeled L-lactic acid is inhibited by protonophores, MCT inhibitors, and a number of other monocarboxylates and monocarboxylic drugs like salicylic and valproic acids (Alm and Tornquist 1985; Hosoya et al. 2001a). These results suggest that these transporters are an attractive route for monocarboxylate drug delivery to the posterior segment of the eye.

Creatine, which plays an essential role in supporting ATP homeostasis in the retina, is transported by creatine transporter (CRT). The storage and transmission of phosphate-bound energy is mediated by the conversion of creatine to phosphocreatine. There has been found to be an asymmetrical distribution of CRT at luminal and abluminal membranes of rat retinal endothelial cells (Nakashima et al. 2004).

In glycolysis, glucose is oxidized to either lactate or pyruvate. Both glycolysis and the tricarboxylic acid cycle generate energy in the form of ATP. Therefore, GLUT1, MCT1, and CRT transporters may act in synergy to maintain energy homeostasis in the retina.

6.2.2.2 Amino Acid Transport System

Taurine is the most abundant free amino acid in the retina. Taurine exerts a number of neuroprotective functions, acting as an osmolyte and antioxidant (Pasantes-Morales et al. 1972). The transport of taurine from the circulating blood to the retina is mediated by the taurine transporter (TAUT). Knockout mice for this transporter exhibit loss of vision due to severe retinal degeneration (Heller-Stilb et al. 2002).

It has been shown that the uptake of labeled taurine and the expression of TAUT mRNA in a rat retinal capillary endothelial cell line are increased under hypertonic conditions (Tomi et al. 2007). Some retinal diseases, such as those relating to ischemia and reperfusion, diabetic retinopathy, macular edema, and neurodegeneration, are associated with fluctuations in cell volume (Pasantes-Morales et al. 1999). Upregulation of taurine transport could markedly affect the achievement of osmotic equilibrium by accumulating taurine in endothelial cells (Tomi et al. 2007). More recently, it has been shown that high-glucose-induced retinal glial cell apoptosis can be inhibited by taurine, and that taurine reverses the diabetes-induced or high-glucose-induced decrease in TAUT expression (Zeng et al. 2010).

L-Type amino acid transporter (LAT1) transports branched-chain and aromatic amino acids like L-phenylalanine and L-leucine, which are essential amino acids as precursors for neurotransmitters and protein synthesis (Kanai et al. 1998; LaNoue et al. 2001). Neutral amino acid carrier (L-system) is composed of LATs and the subunit protein 4F2hc. LAT1 protein was shown to be expressed in rat retinal vessels, primary cultures of human retinal endothelial cells, and TR-iBRB cell line (Tomi et al. 2005). At the BBB, L-dopa, the most widely used drug for Parkinson's disease, is transported by LAT1 (Kageyama et al. 2000). L-Dopa administration was shown to reduce the delay of visual evoked potentials in Parkinson's disease (Bhaskar et al. 1986). In addition to L-dopa, amino acid mimetic-drugs are substrates of LAT1 and are transported to the brain via LAT1 at the BBB (Uchino et al. 2002). At the iBRB,

LAT1 may also play a key role in transporting amino acid mimetic drugs from the circulating blood to the retina.

L-Cystine, one of the precursors of GSH synthesis, is essential to protect the retina against light-induced oxidative stress and to maintain intracellular antioxidants at an appropriate level. The Na^+-independent glutamate/cystine exchange transporter (system X_c^-) is composed of xCT and the subunit protein 4F2hc. This transporter has been shown to be present both in inner and outer BRB (Bridges et al. 2001). When cystine is influxed into the cell, glutamate is effluxed. Therefore, under conditions of oxidative stress in the retina, it is necessary that L-cystine undergoes influx transport from the circulating blood to the retina across the BRB to synthesize the antioxidant glutathione. Several results obtained from *in vivo* and *in vitro* models suggest that L-cystine influx transport at the inner BRB by system X_c^- is induced under oxidative stress by enhanced transcription of the xCT gene with the consequent increase in L-cystine uptake and GSH concentration (Hosoya et al. 2001b; Tomi et al. 2002). Retinal glutathione is preferentially concentrated in Müller cells (Pow and Crook 1995; Schutte and Werner 1998). The identification of the system xc-mediated L-cystine uptake in retinal Müller cells suggests that this transport system is important for L-cystine influx, not only from the circulating blood to the retinal interstitial fluid but also from the interstitial fluid to Müller cells, where it could be used for glutathione synthesis (Tomi et al. 2003). This supply pathway of L-cystine into the retina would be of great value in protecting the retina from oxidative stress.

6.2.2.3 Nucleoside Transport System

Adenosine is an important intercellular signaling molecule that is involved in retinal neurotransmission, blood flow, vascular development, and response to ischemia through cell-surface adenosine receptors (Ghiardi et al. 1999; Lutty and McLeod 2003).

The expression at the mRNA level of several equilibrative nucleoside transporters ENT1, ENT2, CNT1, and CNT2 has been detected in retinal endothelial cells (Nagase et al. 2006). By regulating the concentration of adenosine available to cell surface receptors, these transporters influence the retinal physiological processes mentioned above. ENT2 is also responsible for the uptake of some antiviral and anticancer nucleoside drugs (Yao et al. 2001; Baldwin et al. 2004). This has led to the hypothesis that ENT2 at the iBRB could be a potential route for delivering nucleoside drugs from the circulating blood to the retina.

6.2.2.4 Organic Anion Transport System

In order to maintain a constant milieu in the neural retina, the BRB also carries out the efflux transport of harmful substances like neurotransmitter metabolites, toxins, and xenobiotics (Cunha-Vaz and Maurice 1967).

Members of the family of organic anion transporting polypeptides (OATP) mediate the Na^+-independent transport of a wide range of amphipathic organic compounds, including bile salts, organic dyes, steroid conjugates, thyroid hormones, anionic oligopeptides, numerous drugs, and other xenobiotic substances (Hagenbuch and Meier 2003). Transporters of the family of OATP (OATP2 and OATP14) have been identified in the rat inner and outer BRB (Gao et al. 2002).

6.2.2.5 ABC Transporters

ABC transporters (ATP-binding cassette) are a superfamily of membrane proteins that play a major role in restricting the bioavailability of many drugs in various tissues by pumping agents (with consumption of ATP) from the lipid bilayer or cytoplasm back into the extracellular fluid. In the ocular tissues, the ABC transporters of greatest significance for efflux transport are P-glycoprotein (P-gp), multidrug resistance-associated proteins (MRP), and breast cancer resistance protein (BCRP) (Mannermaa et al. 2006).

The ABC superfamily is subdivided into seven subfamilies based on similarities in domain structure, nucleotide-binding folds, and transmembrane domains (Dean et al. 2001). The general structure of ABC transporters is composed of 12 transmembrane regions, split into two halves, each with a nucleotide-binding domain (NBD) (Altenberg 2004).

P-gp (MDR1) mediates the efflux of a wide range of drugs from the intracellular to the extracellular space (Fojo et al. 1987). The list of its substrates/inhibitors is continually growing and includes anticancer agents, antibiotics, antivirals, calcium channel blockers, and immunosuppressive agents (Dean et al. 2001). P-gp has been shown to be expressed in rat retinal capillaries and cultures of rat retinal endothelial cells (Greenwood 1992; Shen et al. 2003; Hosoya and Tomi 2005). The P-gp-mediated drug efflux pump on the apical plasma membrane of the conjunctiva plays a role in restricting the conjunctival absorption of some lipophilic drugs and xenobiotics.

BCRP (ABCG2) has only one ABC and six putative transmembrane domains, being referred to as a half-ABC transporter, most likely functioning as a homodimer (Krishnamurthy and Schuetz 2006). ABCG2 shows great affinity not only for drugs but also for phototoxic compounds that can cause light-induced damage to the retina (Boulton et al. 2001). BCRP was found to be present in the luminal membrane of mouse capillary endothelial cells by immunolabeling and was shown to be expressed in mouse and rat retinas (Asashima et al. 2006).

These drug efflux pumps at the BRB could act by restricting the distribution of xenobiotics, including drugs and phototoxins, in the retina. Modulation of such efflux mechanisms in conjunction with treatment of ocular tissues in retinal diseases remains a major challenge.

6.3 BRB AND OCULAR IMMUNE PRIVILEGE

The immune response has developed and evolved to protect the organism from invasion and damage by a wide range of pathogens. With time, the immune system has developed destructive responses that are specific for pathogens as well as tissues. Tissue injury might, however, have a devastating effect on the function of an organ such as the eye, which needs to maintain optical stability (Taylor 2011).

The existence of ocular immune privilege is dependent upon multiple factors, such as immunomodulatory factors and ligands, regulation of the complement system within the eye, tolerance-promoting antigen-presenting cells (APCs), unconventional drainage pathways, and with particular relevance, the existence of the blood–ocular barriers.

The blood–ocular barriers provide a relative sequestration of the anterior chamber, vitreous, and neurosensory retina from the immune system, and create the necessary environment for the existence of ocular immune privilege. The evolution of immune privilege as a protective mechanism for preserving the function of vital and delicate organs such as the eye has resulted in a complex system with multiple regulatory safeguards for the control of both innate and adaptative immunity. The consequences of inadvertent bystander tissue destruction by antigen-nonspecific inflammation can be so catastrophic to the organ or host that a finely tuned regulatory system is needed to ensure the integrity of the ocular tissues and maintain optical relationships.

6.4 CLINICAL EVALUATION OF THE BRB

Fluorescein angiography, an examination procedure performed routinely in the ophthalmologist's office, permits a dynamic evaluation of local circulatory disturbances and identifies the sites of BRB breakdown. It is, however, only semiquantitative and its reproducibility depends on the variable quality of the angiograms.

Vitreous fluorometry was developed as a method capable of quantification of both inward and outward movements of fluorescein across the BRB system in the clinical setting. Protocols were devised and tested, and dedicated instrumentation developed (Cunha-Vaz 1997).

With the development of vitreous fluorometry methodologies, a large number of clinical and experimental studies demonstrated well the major role played by alterations of BRB in posterior segment disease. In clinical situations, alterations of the BRB have been measured in aged-related macular degeneration, macular edema, hypertension, diabetes, and so on. Its clinical use, however, has declined because it offers only an overall measurement of the posterior role and because, at the time of its development, there were no drugs available for stabilizing the BRB. Nowadays, vitreous fluorometry is mostly used in experimental research and in drug development.

The major drawback of these methods to evaluate BRB function is the fact that they are invasive, that is, dependent on the injection of a tracer, fluorescein. Using a new method to analyze OCT data, it is now possible to assess differences in the extracellular compartment in eyes with alterations of the BRB (Bernardes et al. 2011). The volume of the extracellular compartment of the neurosensory retina may be measured and comparisons made between areas of normal retina and sites where there is breakdown of the BRB (Figure 6.3).

It is now possible to demonstrate indirectly using OCT the presence of an alteration of the BRB and, furthermore, to identify for the first time whether this alteration involves the inner BRB or the outer BRB.

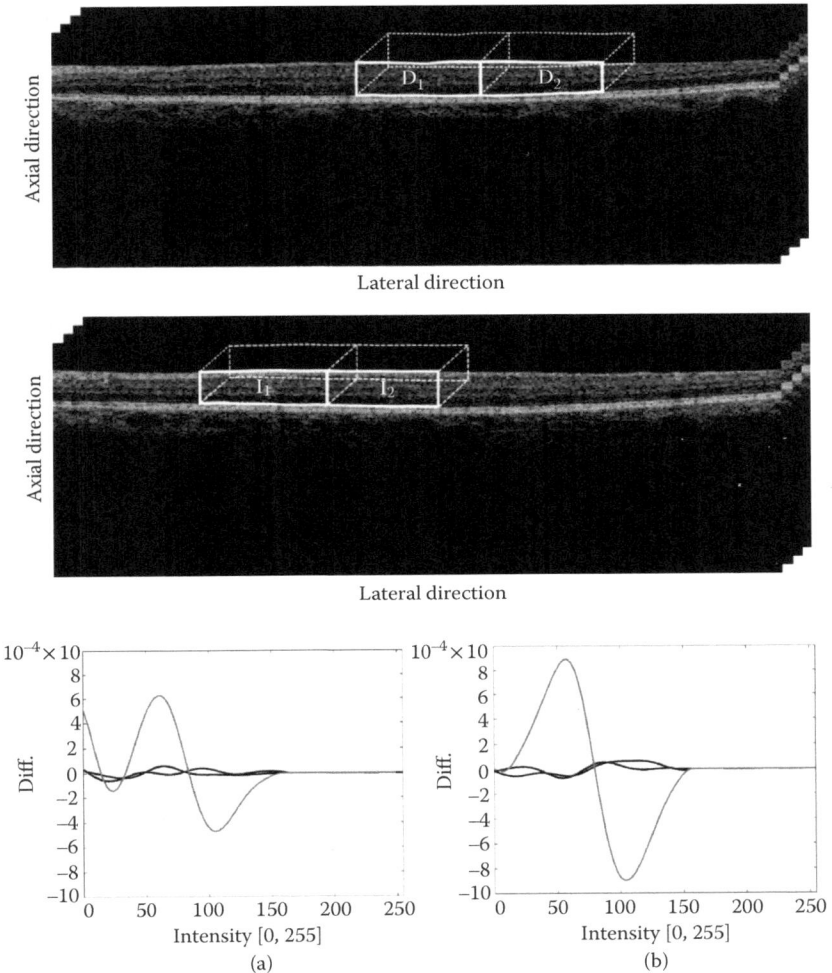

FIGURE 6.3 OCT fundus reference with delimited areas (intact BRB—I and disrupted BRB—D). No apparent difference can be seen in the B-scans passing through disrupted and intact areas. Differences are shown in plots (a) and (b), with black lines representing differences between similar BRB regions (either intact or disrupted BRB) and the gray line representing the differences between dissimilar BRB regions (intact/disrupted BRB regions).

6.5 BRB AND MACULAR EDEMA

Macular edema is the result of an accumulation of fluid in the retinal layers around the fovea, contributing to vision loss by altering the functional cell relationship in the retina and promoting an inflammatory reparative response.

Macular edema is a nonspecific sign of ocular disease, not a specific entity. It should be viewed as a special and clinically relevant type of macular response to an altered retinal environment, generally associated with an alteration of the BRB.

It occurs in a wide variety of ocular situations such as uveitis, trauma, intraocular surgery, vascular retinopathies, hereditary dystrophies, diabetes, age-related macular degeneration, and so on. It is the most frequent cause of vision loss in diabetic retinopathy.

The breakdown of endothelial TJ and breakdown of the BRB seen in diabetic macular edema can occur in either nonproliferative or proliferative diabetic retinopathy at any stage. The prevalence of macular edema after 15 years of known diabetes, according to the Wisconsin epidemiologic study of diabetic retinopathy, is ~20% in patients with type 1 diabetes, 25% in patients with insulin-dependent type 2 diabetes, and 14% in patients with noninsulin-dependent type 2 diabetes (Klein et al. 1984).

An alteration of the outer BRB predominates in "wet" age-related macular degeneration. Choroidal neovascularization invades the retina through an alteration of the retinal pigment epithelium and Bruch's membrane layers. These new vessels do not have barrier properties and retinal edema develops with damage to the photoreceptor layer and visual loss. Furthermore, the development of retinal edema is a reliable indicator of development of neovascularization.

Breakdown of the BRB is, therefore, at the basis of the most frequent causes of vision loss in retinal disease, and its involvement is a sign of sight-threatening evolution of the disease.

6.6 RELEVANCE OF BRB TO THE TREATMENT OF RETINAL DISEASES

When administered systemically, drugs must pass the BRB in order to reach therapeutic levels in the retina. Drug entrance into the retina depends on a number of factors, including the plasma concentration profile of the drug, the volume of its distribution, plasma protein binding, and the relative permeability of the BRB. To obtain therapeutic concentrations within the retina, new strategies must be considered such as delivery of nanoparticles, chemical modification of drugs to enhance BRB transport, coupling of drugs to vectors, and so on.

Better understanding of the transport systems at the BRB will be extremely useful for drug design. Efflux pumps must be effectively circumvented to enhance drug absorption across the retina. Modulating a drug substrate targeting an influx transporter offers great potential. In this strategy, drugs must be designed such that the modified compounds become substrates of nutrient transporters, leading to enhanced absorption across the ocular barriers. In addition, efflux is effectively circumvented due to diminished or no affinity of the drug molecule toward efflux pumps due to structural modification and binding to the influx transporter.

Thus, this rational drug design would lead to higher absorption and bioavailability of certain compounds in the retina.

Eye drops are now being developed for the treatment of posterior segment diseases. Newer formulations that achieve high concentrations of the drug in the posterior segment are expected to have a role in the future. Meanwhile, periocular injections are one modality that has offered mixed results. Finally, the last years have seen a generalized and surprising safe utilization of intravitreal injections, a form of administration that circumvents the BRB. Steroids and a variety of anti-VEGF drugs have been administered through intravitreal injections to a large number of patients without significant side effects and have demonstrated good acceptance by the patients.

Intravitreal injections can achieve high drug concentrations in the vitreous and retina in the presence of a well-preserved BRB.

REFERENCES

Adachi, M., Inoko, A., Hata, M. et al. 2006. Normal establishment of epithelial tight junctions in mice and cultured cells lacking expression of ZO-3, a tight-junction MAGUK protein. *Mol Cell Biol* 26:9003–9015.

Alm, A. and Tornquist, P. 1985. Lactate transport through the blood–retinal and the blood–brain barrier in rats. *Ophthalmic Res* 17:181–184.

Altenberg, G. A. 2004. Structure of multidrug-resistance proteins of the ATP-binding cassette (ABC) superfamily. *Curr Med Chem Anticancer Agents* 4:53–62.

Antcliff, R. J. and Marshall, J. 1999. The pathogenesis of edema in diabetic maculopathy. *Semin Ophthalmol* 14:223–232.

Antonetti, D. A., Barber, A. J., Hollinger, L. A. et al. 1999. Vascular endothelial growth factor induces rapid phosphorylation of tight junction proteins occludin and zonula occluden 1. A potential mechanism for vascular permeability in diabetic retinopathy and tumors. *J Biol Chem* 274:23463–23467.

Arrate, M. P., Rodriguez, J. M., Tran, T. M. et al. 2001. Cloning of human junctional adhesion molecule 3 (JAM3) and its identification as the JAM2 counter-receptor. *J Biol Chem* 276:45826–45832.

Asashima, T., Hori, S., Ohtsuki, S. et al. 2006. ATP-binding cassette transporter G2 mediates the efflux of phototoxins on the luminal membrane of retinal capillary endothelial cells. *Pharm Res* 23:1235–1242.

Ashton, N. and Cunha-Vaz, J. G. 1965. Effect of histamine on the permeability of the ocular vessels. *Arch Ophthalmol* 73:211–223.

Aurrand-Lions, M., Duncan, L., Ballestrem, C. et al. 2001. JAM-2, a novel immunoglobulin superfamily molecule, expressed by endothelial and lymphatic cells. *J Biol Chem* 276:2733–2741.

Badr, G. A., Tang, J., Ismail-Beigi, F. et al. 2000. Diabetes downregulates GLUT1 expression in the retina and its microvessels but not in the cerebral cortex or its microvessels. *Diabetes* 49:1016–1021.

Baldwin, S. A., Beal, P. R., Yao, S. Y. et al. 2004. The equilibrative nucleoside transporter family, SLC29. *Pflugers Arch* 447:735–743.

Barber, A. J. and Antonetti, D. A. 2003. Mapping the blood vessels with paracellular permeability in the retinas of diabetic rats. *Invest Ophthalmol Vis Sci* 44:5410–5416.

Bauer, A. T., Burgers, H. F., Rabie, T. et al. 2010. Matrix metalloproteinase-9 mediates hypoxia-induced vascular leakage in the brain via tight junction rearrangement. *J Cereb Blood Flow Metab* 30:837–848.

Bauer, H. C., Traweger, A., Zweimueller-Mayer, J. et al. 2011. New aspects of the molecular constituents of tissue barriers. *J Neural Transm* 118:7–21.

Bernardes, R., Santos, T., Serranho, P. et al. 2011. Noninvasive evaluation of retinal leakage using optical coherence tomography. *Ophthalmologica* 226:29–36.

Betz, A. L. and Goldstein, G. W. 1980. Transport of hexoses, potassium and neutral amino acids into capillaries isolated from bovine retina. *Exp Eye Res* 30:593–605.

Bhaskar, P. A., Vanchilingam, S., Bhaskar, E. A. et al. 1986. Effect of L-dopa on visual evoked potential in patients with Parkinson's disease. *Neurology* 36:1119–1121.

Bito, L. Z. 1977. The physiology and pathophysiology of intraocular fluids. *Exp Eye Res* 25 (Suppl):273–289.

Boulton, M., Rozanowska, M., and Rozanowski, B. 2001. Retinal photodamage. *J Photochem Photobiol B* 64:144–161.

Bridges, C. C., Kekuda, R., Wang, H. et al. 2001. Structure, function, and regulation of human cystine/glutamate transporter in retinal pigment epithelial cells. *Invest Ophthalmol Vis Sci* 42:47–54.

Caplan, M. J., Seo-Mayer, P., and Zhang, L. 2008. Epithelial junctions and polarity: Complexes and kinases. *Curr Opin Nephrol Hypertens* 17:506–512.

Cereijido, M., Contreras, R. G., Shoshani, L. et al. 2008. Tight junction and polarity interaction in the transporting epithelial phenotype. *Biochim Biophys Acta* 1778:770–793.

Colegio, O. R., Van Itallie, C. M., McCrea, H. J. et al. 2002. Claudins create charge-selective channels in the paracellular pathway between epithelial cells. *Am J Physiol Cell Physiol* 283:C142–147.

Cunha-Vaz, J. G. F. D. 1965. *Permeability of the retinal vessels in health and disease*. London: University of London (Institute of Ophthalmology).

Cunha-Vaz, J. G. 1966. Studies on the permeability of the blood–retinal barrier. 3. Breakdown of the blood–retinal barrier by circulatory disturbances. *Br J Ophthalmol* 50:505–516.

Cunha-Vaz J. G. 1997. Optical sensors for clinical ocular fluorometry. *Prog Ret Eye Res* 16:243–270.

Cunha-Vaz, J., Faria de Abreu, J. R., and Campos, A. J. 1975. Early breakdown of the blood–retinal barrier in diabetes. *Br J Ophthalmol* 59:649–656.

Cunha-Vaz, J. G. and Maurice, D. M. 1967. The active transport of fluorescein by the retinal vessels and the retina. *J Physiol* 191:467–486.

Cunha-Vaz, J. G., Shakib, M., and Ashton, N. 1966. Studies on the permeability of the blood–retinal barrier. I. On the existence, development, and site of a blood–retinal barrier. *Br J Ophthalmol* 50:441–453.

Cunningham, S. A., Arrate, M. P., Rodriguez, J. M. et al. 2000. A novel protein with homology to the junctional adhesion molecule. Characterization of leukocyte interactions. *J Biol Chem* 275:34750–34756.

Dean, M., Rzhetsky, A., and Allikmets, R. 2001. The human ATP-binding cassette (ABC) transporter superfamily. *Genome Res* 11:1156–1166.

Ebnet, K., Aurrand-Lions, M., Kuhn, A. et al. 2003. The junctional adhesion molecule (JAM) family members JAM-2 and JAM-3 associate with the cell polarity protein PAR-3: A possible role for JAMs in endothelial cell polarity. *J Cell Sci* 116:3879–3891.

Fanning, A. S., Jameson, B. J., Jesaitis, L. A. et al. 1998. The tight junction protein ZO-1 establishes a link between the transmembrane protein occludin and the actin cytoskeleton. *J Biol Chem* 273:29745–29753.

Feldman, G. J., Mullin, J. M., and Ryan, M. P. 2005. Occludin: Structure, function and regulation. *Adv Drug Deliv Rev* 57:883–917.

Fernandes, R., Carvalho, A. L., Kumagai, A. et al. 2004. Downregulation of retinal GLUT1 in diabetes by ubiquitinylation. *Mol Vis* 10:618–628.

Fernandes, R., Suzuki, K., and Kumagai, A. K. 2003. Inner blood–retinal barrier GLUT1 in long-term diabetic rats: An immunogold electron microscopic study. *Invest Ophthalmol Vis Sci* 44:3150–3154.

Fojo, A. T., Ueda, K., Slamon, D. J. et al. 1987. Expression of a multidrug-resistance gene in human tumors and tissues. *Proc Natl Acad Sci U S A* 84:265–269.

Forster, C. 2008. Tight junctions and the modulation of barrier function in disease. *Histochem Cell Biol* 130:55–70.

Furuse, M., Fujita, K., Hiiragi, T. et al. 1998a. Claudin-1 and -2: Novel integral membrane proteins localizing at tight junctions with no sequence similarity to occludin. *J Cell Biol* 141:1539–1550.

Furuse, M., Furuse, K., Sasaki, H. et al. 2001. Conversion of zonulae occludentes from tight to leaky strand type by introducing claudin-2 into Madin-Darby canine kidney I cells. *J Cell Biol* 153:263–272.

Furuse, M., Itoh, M., Hirase, T. et al. 1994. Direct association of occludin with ZO-1 and its possible involvement in the localization of occludin at tight junctions. *J Cell Biol* 127:1617–1626.

Furuse, M., Hirase, T., Itoh, M. et al. 1993. Occludin: A novel integral membrane protein localizing at tight junctions. *J Cell Biol* 123:1777–1788.

Furuse, M., Sasaki, H., Fujimoto, K. et al. 1998b. A single gene product, claudin-1 or -2, reconstitutes tight junction strands and recruits occludin in fibroblasts. *J Cell Biol* 143:391–401.

Furuse, M., Sasaki, H., and Tsukita, S. 1999. Manner of interaction of heterogeneous claudin species within and between tight junction strands. *J Cell Biol* 147:891–903.

Gao, B., Wenzel, A., Grimm, C. et al. 2002. Localization of organic anion transport protein 2 in the apical region of rat retinal pigment epithelium. *Invest Ophthalmol Vis Sci* 43:510–514.

Gardner, T. W., Antonetti, D. A., Barber, A. J. et al. 2002. Diabetic retinopathy: More than meets the eye. *Surv Ophthalmol* 47 Suppl 2:S253–262.

Gerhart, D. Z., Leino, R. L., and Drewes, L. R. 1999. Distribution of monocarboxylate transporters MCT1 and MCT2 in rat retina. *Neuroscience* 92:367–375.

Ghiardi, G. J., Gidday, J. M., and Roth, S. 1999. The purine nucleoside adenosine in retinal ischemia-reperfusion injury. *Vision Res* 39:2519–2535.

Goldmann, E. E. 1913. Vitalbarfung am Zentralnervensystem. Abhandl. Konigl. Preuss Akad Wiss 1: 1–60.

Greenwood, J. 1992. Characterization of a rat retinal endothelial cell culture and the expression of P-glycoprotein in brain and retinal endothelium *in vitro*. *J Neuroimmunol* 39:123–132.

Hagenbuch, B. and Meier, P. J. 2003. The superfamily of organic anion transporting polypeptides. *Biochim Biophys Acta* 1609:1–18.

Haskins, J., Gu, L., Wittchen, E. S. et al. 1998. ZO-3, a novel member of the MAGUK protein family found at the tight junction, interacts with ZO-1 and occludin. *J Cell Biol* 141:199–208.

Heller-Stilb, B., van Roeyen, C., Rascher, K. et al. 2002. Disruption of the taurine transporter gene (taut) leads to retinal degeneration in mice. *FASEB J* 16:231–233.

Hosoya, K., Kondo, T., Tomi, M. et al. 2001a. MCT1-mediated transport of L-lactic acid at the inner blood–retinal barrier: A possible route for delivery of monocarboxylic acid drugs to the retina. *Pharm Res* 18:1669–1676.

Hosoya, K., Saeki, S., and Terasaki, T. 2001b. Activation of carrier-mediated transport of L-cystine at the blood–brain and blood–retinal barriers *in vivo*. *Microvasc Res* 62:136–142.

Hosoya, K. and Tomi, M. 2005. Advances in the cell biology of transport via the inner blood–retinal barrier: Establishment of cell lines and transport functions. *Biol Pharm Bull* 28:1–8.

Hou, J., Gomes, A. S., Paul, D. L. et al. 2006. Study of claudin function by RNA interference. *J Biol Chem* 281:36117–36123.

Ikenouchi, J., Furuse, M., Furuse, K. et al. 2005. Tricellulin constitutes a novel barrier at tricellular contacts of epithelial cells. *J Cell Biol* 171:939–945.

Ikenouchi, J., Umeda, K., Tsukita, S. et al. 2007. Requirement of ZO-1 for the formation of belt-like adherens junctions during epithelial cell polarization. *J Cell Biol* 176:779–786.

Islas, S., Vega, J., Ponce, L. et al. 2002. Nuclear localization of the tight junction protein ZO-2 in epithelial cells. *Exp Cell Res* 274:138–148.

Itoh, M., Morita, K., and Tsukita, S. 1999. Characterization of ZO-2 as a MAGUK family member associated with tight as well as adherens junctions with a binding affinity to occludin and alpha catenin. *J Biol Chem* 274:5981–5986.

Janzer, R. C. and Raff, M. C. 1987. Astrocytes induce blood–brain barrier properties in endo-thelial cells. *Nature* 325:253–257.

Kageyama, T., Nakamura, M., Matsuo, A. et al. 2000. The 4F2hc/LAT1 complex transports L-DOPA across the blood–brain barrier. *Brain Res* 879:115–121.

Kanai, Y., Segawa, H., Miyamoto, K. et al. 1998. Expression cloning and characterization of a transporter for large neutral amino acids activated by the heavy chain of 4F2 antigen (CD98). *J Biol Chem* 273:23629–23632.

Kaur, C., Foulds, W. S., and Ling, E. A. 2008. Blood–retinal barrier in hypoxic ischaemic conditions: Basic concepts, clinical features and management. *Prog Retin Eye Res* 27:622–647.

Kim, J. H., Lee, Y. M., Ahn, E. M. et al. 2009. Decursin inhibits VEGF-mediated inner blood–retinal barrier breakdown by suppression of VEGFR-2 activation. *J Cereb Blood Flow Metab* 29:1559–1567.

Klein R., Klein B. E., Moss S. E., et al. 1984. The Wisconsin epidemiologic study of diabetic retinopathy. IV. Diabetic macular edema. *Ophthalmology* 91:1464–1474.

Krishnamurthy, P. and Schuetz, J. D. 2006. Role of ABCG2/BCRP in biology and medicine. *Annu Rev Pharmacol Toxicol* 46:381–410.

Kumagai, A. K., Vinores, S. A., and Pardridge, W. M. 1996. Pathological upregulation of inner blood–retinal barrier Glut1 glucose transporter expression in diabetes mellitus. *Brain Res* 706:313–317.

LaNoue, K. F., Berkich, D. A., Conway, M. et al. 2001. Role of specific aminotransfer-ases in de novo glutamate synthesis and redox shuttling in the retina. *J Neurosci Res* 66:914–922.

Leal, E. C., Martins, J., Voabil, P. et al. 2010. Calcium dobesilate inhibits the alterations in tight junction proteins and leukocyte adhesion to retinal endothelial cells induced by diabetes. *Diabetes* 59:2637–2645.

Lepore, D., Molle, F., Pagliara, M. M. et al. 2011. Atlas of fluorescein angiographic findings in eyes undergoing laser for retinopathy of prematurity. *Ophthalmology* 118:168–175.

Luo, Y., Xiao, W., Zhu, X. et al. 2011. Differential expression of claudins in retinas dur-ing normal development and the angiogenesis of oxygen-induced retinopathy. *Invest Ophthalmol Vis Sci* 52:7556–7564.

Lutty, G. A. and McLeod, D. S. 2003. Retinal vascular development and oxygen-induced retinopathy: A role for adenosine. *Prog Retin Eye Res* 22:95–111.

Mandell, K. J., Babbin, B. A., Nusrat, A. et al. 2005. Junctional adhesion molecule 1 regulates epithelial cell morphology through effects on beta1 integrins and Rap1 activity. *J Biol Chem* 280:11665–11674.

Mannermaa, E., Vellonen, K. S., and Urtti, A. 2006. Drug transport in corneal epithelium and blood–retina barrier: Emerging role of transporters in ocular pharmacokinetics. *Adv Drug Deliv Rev* 58:1136–1163.

Mark, K. S. and Davis, T. P. 2002. Cerebral microvascular changes in permeability and tight junctions induced by hypoxia-reoxygenation. *Am J Physiol Heart Circ Physiol* 282:H1485–1494.

Martin-Padura, I., Lostaglio, S., Schneemann, M. et al. 1998. Junctional adhesion molecule, a novel member of the immunoglobulin superfamily that distributes at intercellular junc-tions and modulates monocyte transmigration. *J Cell Biol* 142:117–127.

Matter, K., Aijaz, S., Tsapara, A. et al. 2005. Mammalian tight junctions in the regulation of epithelial differentiation and proliferation. *Curr Opin Cell Biol* 17:453–458.

Mellman, I. and Nelson, W. J. 2008. Coordinated protein sorting, targeting and distribution in polarized cells. *Nat Rev Mol Cell Biol* 9:833–845.

Minamizono, A., Tomi, M., and Hosoya, K. 2006. Inhibition of dehydroascorbic acid transport across the rat blood–retinal and –brain barriers in experimental diabetes. *Biol Pharm Bull* 29:2148–2150.

Morita, K., Furuse, M., Fujimoto, K. et al. 1999a. Claudin multigene family encoding four-transmembrane domain protein components of tight junction strands. *Proc Natl Acad Sci U S A* 96:511–516.

Morita, K., Sasaki, H., Furuse, M. et al. 1999b. Endothelial claudin: Claudin-5/TMVCF constitutes tight junction strands in endothelial cells. *J Cell Biol* 147:185–194.

Murakami, T., Felinski, E. A., and Antonetti, D. A. 2009. Occludin phosphorylation and ubiquitination regulate tight junction trafficking and vascular endothelial growth factor-induced permeability. *J Biol Chem* 284:21036–21046.

Nagase, K., Tomi, M., Tachikawa, M. et al. 2006. Functional and molecular characterization of adenosine transport at the rat inner blood–retinal barrier. *Biochim Biophys Acta* 1758:13–19.

Nakashima, T., Tomi, M., Katayama, K. et al. 2004. Blood-to-retina transport of creatine via creatine transporter (CRT) at the rat inner blood–retinal barrier. *J Neurochem* 89:1454–1461.

Nitta, T., Hata, M., Gotoh, S. et al. 2003. Size-selective loosening of the blood–brain barrier in claudin-5-deficient mice. *J Cell Biol* 161:653–660.

Palmeri, D., van Zante, A., Huang, C. C. et al. 2000. Vascular endothelial junction-associated molecule, a novel member of the immunoglobulin superfamily, is localized to intercellular boundaries of endothelial cells. *J Biol Chem* 275:19139–19145.

Pasantes-Morales, H., Klethi, J., Ledig, M. et al. 1972. Free amino acids of chicken and rat retina. *Brain Res* 41:494–497.

Pasantes-Morales, H., Ochoa de la Paz, L. D., Sepulveda, J. et al. 1999. Amino acids as osmolytes in the retina. *Neurochem Res* 24:1339–1346.

Poitry-Yamate, C. L., Poitry, S., and Tsacopoulos, M. 1995. Lactate released by Muller glial cells is metabolized by photoreceptors from mammalian retina. *J Neurosci* 15:5179–5191.

Pow, D. V. and Crook, D. K. 1995. Immunocytochemical evidence for the presence of high levels of reduced glutathione in radial glial cells and horizontal cells in the rabbit retina. *Neurosci Lett* 193:25–28.

Reese, T. S. and Karnovsky, M. J. 1967. Fine structural localization of a blood–brain barrier to exogenous peroxidase. *J Cell Biol* 34:207–217.

Riazuddin, S., Ahmed, Z. M., Fanning, A. S. et al. 2006. Tricellulin is a tight-junction protein necessary for hearing. *Am J Hum Genet* 79:1040–1051.

Schutte, M. and Werner, P. 1998. Redistribution of glutathione in the ischemic rat retina. *Neurosci Lett* 246:53–56.

Shakib, M. and Cunha-Vaz, J. G. 1966. Studies on the permeability of the blood–retinal barrier. IV. Junctional complexes of the retinal vessels and their role in the permeability of the blood–retinal barrier. *Exp Eye Res* 5:229–234.

Shen, J., Cross, S. T., Tang-Liu, D. D. et al. 2003. Evaluation of an immortalized retinal endothelial cell line as an *in vitro* model for drug transport studies across the blood–retinal barrier. *Pharm Res* 20:1357–1363.

Steed, E., Balda, M. S., and Matter, K. 2010. Dynamics and functions of tight junctions. *Trends Cell Biol* 20:142–149.

Stevenson, B. R., Siliciano, J. D., Mooseker, M. S. et al. 1986. Identification of ZO-1: A high molecular weight polypeptide associated with the tight junction (zonula occludens) in a variety of epithelia. *J Cell Biol* 103:755–766.

Sundstrom, J. M., Tash, B. R., Murakami, T. et al. 2009. Identification and analysis of occludin phosphosites: A combined mass spectrometry and bioinformatics approach. *J Proteome Res* 8:808–817.

Taylor A.W. 2011. Immunosuppressive and anti-inflammatory molecules that maintain immune privilege of the eye. In *Immunology, Inflammation and Diseases of the Eye*, ed. Darlene A Dartt, Reza Dana, Patricia D'Amore, Jerry Y. Niederkorn, pp. 44–49. Oxford: Elsevier.

Takata, K., Kasahara, T., Kasahara, M. et al. 1992. Ultracytochemical localization of the erythrocyte/HepG2-type glucose transporter (GLUT1) in cells of the blood–retinal barrier in the rat. *Invest Ophthalmol Vis Sci* 33:377–383.

Tomi, M., Funaki, T., Abukawa, H. et al. 2003. Expression and regulation of L-cystine transporter, system xc-, in the newly developed rat retinal Muller cell line (TR-MUL). *Glia* 43:208–217.

Tomi, M. and Hosoya, K. 2004. Application of magnetically isolated rat retinal vascular endothelial cells for the determination of transporter gene expression levels at the inner blood–retinal barrier. *J Neurochem* 91:1244–1248.

Tomi, M., Hosoya, K., Takanaga, H. et al. 2002. Induction of xCT gene expression and L-cystine transport activity by diethyl maleate at the inner blood–retinal barrier. *Invest Ophthalmol Vis Sci* 43:774–779.

Tomi, M., Mori, M., Tachikawa, M. et al. 2005. L-type amino acid transporter 1-mediated L-leucine transport at the inner blood–retinal barrier. *Invest Ophthalmol Vis Sci* 46:2522–2530.

Tomi, M., Tajima, A., Tachikawa, M. et al. 2008. Function of taurine transporter (Slc6a6/TauT) as a GABA transporting protein and its relevance to GABA transport in rat retinal capillary endothelial cells. *Biochim Biophys Acta* 1778:2138–2142.

Tomi, M., Terayama, T., Isobe, T. et al. 2007. Function and regulation of taurine transport at the inner blood–retinal barrier. *Microvasc Res* 73:100–106.

Tout, S., Chan-Ling, T., Hollander, H. et al. 1993. The role of Muller cells in the formation of the blood–retinal barrier. *Neuroscience* 55:291–301.

Tsukita, S. and Furuse, M. 1999. Occludin and claudins in tight-junction strands: Leading or supporting players? *Trends Cell Biol* 9:268–273.

Tsukita, S., Furuse, M., and Itoh, M. 2001. Multifunctional strands in tight junctions. *Nat Rev Mol Cell Biol* 2:285–293.

Uchino, H., Kanai, Y., Kim, D. K. et al. 2002. Transport of amino acid-related compounds mediated by L-type amino acid transporter 1 (LAT1): Insights into the mechanisms of substrate recognition. *Mol Pharmacol* 61:729–737.

Umeda, K., Matsui, T., Nakayama, M. et al. 2004. Establishment and characterization of cultured epithelial cells lacking expression of ZO-1. *J Biol Chem* 279:44785–44794.

Van Itallie, C. M. and Anderson, J. M. 2006. Claudins and epithelial paracellular transport. *Annu Rev Physiol* 68:403–429.

Vera, J. C., Rivas, C. I., Fischbarg, J. et al. 1993. Mammalian facilitative hexose transporters mediate the transport of dehydroascorbic acid. *Nature* 364:79–82.

Westphal, J. K., Dorfel, M. J., Krug, S. M. et al. 2010. Tricellulin forms homomeric and heteromeric tight junctional complexes. *Cell Mol Life Sci* 67:2057–2068.

Yao, S. Y., Ng, A. M., Sundaram, M. et al. 2001. Transport of antiviral 3'-deoxy-nucleoside drugs by recombinant human and rat equilibrative, nitrobenzylthioinosine (NBMPR)-insensitive (ENT2) nucleoside transporter proteins produced in Xenopus oocytes. *Mol Membr Biol* 18:161–167.

Yeaman, C., Grindstaff, K. K., and Nelson, W. J. 2004. Mechanism of recruiting Sec6/8 (exocyst) complex to the apical junctional complex during polarization of epithelial cells. *J Cell Sci* 117:559–570.

Yu, A. S., McCarthy, K. M., Francis, S. A. et al. 2005. Knockdown of occludin expression leads to diverse phenotypic alterations in epithelial cells. *Am J Physiol Cell Physiol* 288:C1231–1241.

Zahraoui, A., Louvard, D., and Galli, T. 2000. Tight junction, a platform for trafficking and signaling protein complexes. *J Cell Biol* 151:F31–36.

Zeng, K., Xu, H., Mi, M. et al. 2010. Effects of taurine on glial cells apoptosis and taurine transporter expression in retina under diabetic conditions. *Neurochem Res* 35:1566–1574.

Zhang, X., Bao, S., Lai, D. et al. 2008. Intravitreal triamcinolone acetonide inhibits breakdown of the blood–retinal barrier through differential regulation of VEGF-A and its receptors in early diabetic rat retinas. *Diabetes* 57:1026–1033.

7 Barriers to Transscleral Drug Delivery to the Retina

Alexandra Almazan, Susan S. Lee, Aron D. Ross, and Michael R. Robinson

CONTENTS

7.1 INTRODUCTION

Diseases of the retina such as age-related macular degeneration, diabetic macular edema, and retinal vein occlusion are leading causes of vision impairment.[1] Methods of drug delivery to the retina include topical, systemic, intravitreal, and periocular administration.[1–3] Topical eye drop administration is limited by the low drug levels achieved in the retina, and the potential for side effects from systemic drug administration makes this route suboptimal, particularly for older patients who are more likely to have comorbid conditions. Due to the lack of efficacy and safety concerns with topical and systemic administration, various methods for local drug delivery to the retina have emerged.[4,5] Intravitreal delivery allows high concentrations of a drug to be achieved in the retina; however, the surgical procedure is associated with a risk of various side effects including cataract,[6,7] retinal detachment,[8,9] and endophthalmitis.[10–12] Transscleral delivery (i.e., subconjunctival, sub-Tenon's, retrobulbar, peribulbar, and intrascleral) has been explored as an alternative to intravitreal injection. The effectiveness of transscleral delivery depends on the ability of a drug to penetrate ocular tissues, such as the sclera, choroid–Bruch's membrane, and retinal pigment epithelium (RPE), to reach the neuroretina.

A variety of transscleral drug delivery methods have been reported in the scientific literature in recent years. Clinical trials of sub-Tenon's infusions of triamcinolone acetonide (TA) have been performed.[13–17] Microparticles administered subconjunctivally have been formulated for sustained release of drugs such as celecoxib[18,19] and budesonide[20] to maintain therapeutic retinal levels. Injectable gels with sustained release of drugs have been administered subconjunctivally.[21,22] Polymeric implants have been constructed as subconjunctival,[23] episcleral,[24] and intrascleral[25,26] discs. Osmotic pumps with the catheter tips placed epi- or intrasclerally have also been utilized to infuse continuous amounts of drug into the eye.[27,28] Lastly, ocular iontophoresis has been utilized to enhance the scleral penetration of drug solutions and nanoparticle-based hydrogels.[29,30]

Although there has been great interest in the potential use of transscleral drug delivery systems for the treatment of retinal disorders, human clinical trials have shown that transscleral delivery of triamcinolone is less effective than intravitreal triamcinolone injections.[14,31] The bioavailability of prednisolone, for example, in the vitreous after subconjunctival injection has been reported to be in the range of 0.01–0.1%.[32] Because drug molecules must cross several layers of tissue before reaching the retina with transscleral delivery, a very steep concentration gradient is established from the sclera to the retina due to the various transport barriers that hinder drug molecules from reaching the retina.

Three barriers to transscleral drug delivery have been identified in the literature to date: static, dynamic, and metabolic. Static barriers include the ocular tissues that pose a physical barrier to drug diffusion; these include the sclera, choroid–Bruch's membrane, and RPE. Dynamic barriers include drug clearance mechanisms through blood and lymphatic vessels, bulk fluid flow due to intraocular drainage, and transporter proteins of the RPE. Lastly, metabolic barriers in the eye reduce drug penetration to the retina by promoting drug degradation. The following review provides an in-depth summary of findings reported in the literature regarding static, dynamic, and metabolic barriers in the eye and closes with a discussion of the future of transscleral drug delivery.

7.2 STATIC BARRIERS

The sclera, choroid, and RPE pose a physical barrier to the diffusion of a drug to the retina following transscleral administration (Figure 7.1). The scleral matrix consists of proteoglycans and collagen fibers, which can hinder drug penetration through the tissue. The choroid and underlying Bruch's membrane, which consists of a complex arrangement of lipids and lipoproteins, can also impede drug movement. The RPE provides a cellular barrier with tight junctions and significantly limits paracellular transport. The extent to which these barriers limit drug diffusion has most often been studied *ex vivo* using a two-chamber Ussing-type apparatus, which measures the passage of solutes across a membrane mounted between two chambers. Permeability, reported as diffusion distance per unit time (centimeters per second), represents the velocity at which drug molecules diffuse through a tissue in response to a concentration gradient.[33] Permeability values are difficult to determine from *in vivo* studies due to the confounding influences of dynamic and metabolic barriers.

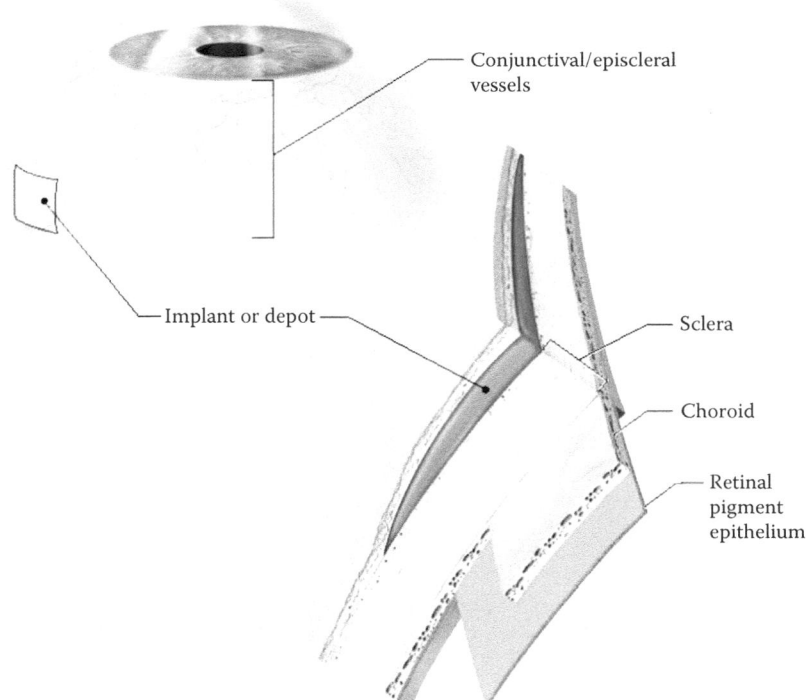

FIGURE 7.1   Static barriers (sclera, choroid, and RPE) limiting transscleral drug delivery.

7.2.1 SCLERA

The permeability of the sclera has been extensively studied using *ex vivo* preparations. Permeability data have been obtained for a variety of molecules including dextrans[34] and PEG molecules[35] of various molecular weights, antibiotics,[36] oligonucleotides,[37] retinal antiangiogenic molecules,[38] and lipophilic compounds.[39,40] Several factors influencing the scleral permeability of molecules have been identified. Scleral permeability has been shown to depend strongly on molecular weight, with permeability being greater for smaller as compared with larger molecules.[34,35,41–43] However, molecular radius appears to be a better predictor of scleral permeability than molecular weight since the permeability of globular proteins has been shown, in rabbit sclera, to be greater than that of linear dextrans of the same molecular weight.[42] Mathematical models[44] and experimental data[42] show that there is a roughly exponential decrease in scleral permeability with increasing molecular radius. Drug movement through the sclera can be largely explained by the structure of the tissue. The sclera consists of a fiber matrix made of collagen and elastin chains, and the diffusion of drugs through the matrix may be determined by the pore diameter and size of the intracellular space. The higher permeability of compounds in the posterior versus anterior sclera[45] may be a result of regional differences in collagen architecture; that is, the collagen fibers are more loosely woven in the posterior sclera.[46] Drug transport may also be affected by the lateral orientation of the scleral fibers as diffusion is hindered in the transverse direction.[47]

Lipophilicity has been shown to be inversely related to permeability in human,[40] rabbit,[48] bovine,[39] and porcine sclera[39]; that is, hydrophilic compounds may diffuse through the aqueous media of proteoglycans in the fiber matrix pores more readily than lipophilic compounds.[40] Negatively charged compounds show higher penetration than negatively charged compounds in bovine and porcine sclera,[39,49] which is likely attributable to the binding of positively charged molecules by negatively charged proteoglycans in the sclera.[50]

Scleral permeability has also been investigated after imposing physical challenges such as hydration in human[51] and rabbit[34] sclera, cryotherapy,[52] transscleral diode laser,[52] surgical thinning,[52] and variations in transscleral pressure.[51,53,54] Increased tissue hydration has been shown to increase solute diffusion,[34,51] while removal of glycosaminoglycans was shown to have no significant effect.[34] Cryotherapy and transscleral diode laser was shown to have no effect on sclera permeability or ultrastructure, whereas scleral permeability was significantly increased by surgical thinning.[52] The effect of transscleral pressure on scleral permeability has been studied in human and rabbit sclera using a Ussing chamber modified to enable manipulation of intraocular pressure (IOP).[54] In both human and rabbit sclera, elevated IOP was found to decrease scleral permeability,[54] despite the tendency for scleral thickness to be decreased as a result of IOP elevation.[51]

7.2.2 BRUCH'S MEMBRANE–CHOROID AND RPE

The permeability of the Bruch's membrane–choroid (BC) complex and the RPE have not been studied as extensively as that of the sclera due to difficulty in separating the two tissue layers; therefore, Bruch's membrane and choroid are always

TABLE 7.1

Solute Physicochemical Properties and Their Effects on *In Vivo* Permeability in Ocular Tissues

Physicochemical Property	Sclera	BC	BC–RPE
Molecular radius	Exponential decrease with increasing molecular radius[27,44]	Decrease with increasing molecular weight[62]	Exponential decrease with increasing molecular radius[57]
Lipophilicity	Decrease with increase in lipophilicity[39,40,48]	Decrease with increase in lipophilicity[39]	Increase with increase in lipophilicity[57]
Charge	Increase with negatively charged solutes[39,49]	Increase with negatively charged solutes[39]	Not studied

studied in tandem.[55] In contrast, the RPE can be easily removed, and its permeability characteristics can be inferred by comparing tissue permeability in the presence and absence of the RPE. Freshly extracted tissues are used in these studies, and the viability of tight junctions in RPE cells can be confirmed by monitoring transepithelial resistance. Cultured monkey RPE cells have been used to measure the flux of fluorescein-conjugated pigment epithelium-derived factor (50 kDa) across the RPE monolayer *in vitro*,[56] and passage of the protein across the RPE monolayer was shown to occur even in the presence of fully formed tight junctions.

The permeability of the BC complex and RPE to hydrophilic carboxyfluorescein and fluorescein isothiocyanate (FITC)-dextrans in bovine tissues decreased exponentially with increasing molecular radius.[57] The BC complex was found to be permeable to proteins with molecular weights up to 150 kDa in dog eyes.[58] Compounds with greater lipophilicity showed less permeability in bovine BC preparations[39] but greater permeability in bovine BC–RPE,[57] suggesting that a direct relationship exists between RPE permeability and lipophilicity. The rate of taurine transport was found to be lower across BC–RPE as compared with BC tissues in both bovine[59] and human[60] samples, indicating that the RPE is the rate-limiting permeability barrier *ex vivo*.

The permeability values for molecules that are passively transported across the RPE are similar in the outward (retina → choroid) and inward (choroid → retina) directions, while molecules that are actively transported show differences in permeability between the two directions. In a study by Pitkänena, the permeability of the RPE to dextran was shown to be similar in both the directions, while the outward permeability to carboxyfluorescein was higher than the inward permeability.[57] The outward transport of carboxyfluorescein may depend in part on carrier-mediated active transport mechanisms[61] (discussed in detail in Section 7.3). The influence of solute physiochemical factors on the *ex vivo* permeability of the sclera, BC, and BC–RPE (summarized in Table 7.1) tends to be similar in all three tissues.

7.2.3 Animal versus Human Studies

In contrast with most studies on scleral permeability, which have been performed using human tissues, studies on the permeability of the choroid, Bruch's membrane, and RPE have most often been conducted using bovine and porcine tissues. Slight intraspecies differences have been reported for scleral diffusion properties among rabbit, human, and bovine tissues.[42] These differences may be attributable to species differences in the scleral ultrastructure or to variations in methods of tissue preparation.

The effects of age on tissue permeability have been studied in human tissue. Age had a significant effect on the permeability of Bruch's membrane and the choroid but was not associated with changes in sclera permeability or ultrastructure.[52] The thickness of Bruch's membrane was shown to increase from 2 μm in the first decade of life to 4.7 μm in the tenth decade.[63] The accumulation of lipid-rich membranous debris and basal laminar deposits with aging may also impede the flux of fluid and solutes.[64] The choroid thins linearly with age by 11 μm per 10 years[63]; however, morphological alteration of Bruch's membrane is the more likely cause of the linear decrease in the permeability of the BC complex to taurine associated with aging.[65] These findings suggest that Bruch's membrane may be a major resistance barrier to the transscleral movement of small solutes.[65] The permeability of the BC complex to serum proteins was also shown to decrease ten-fold from the first to the ninth decade of life,[66] and the diffusion rates for amino acids across the BC complex have been reported to decline linearly with age.[62]

7.3   DYNAMIC BARRIERS

The measurement of tissue permeability *ex vivo* cannot account for the effects of dynamic barriers that are present *in vivo*. *In vivo* studies are necessary to characterize dynamic barriers, which include clearance by lymphatics and blood vessels, bulk fluid flow, and active transport mechanisms involving RPE transporter proteins. Some dynamic barriers cease post-mortem; thus, drug levels in ocular tissues can be overestimated when measurements are obtained *ex vivo*. A summary of experiments investigating the effect of dynamic barriers on transscleral drug delivery *in vivo* is given. Rabbits have typically been used for studies on *in vivo* clearance and the dynamic barriers affecting drugs delivered by the transscleral route. These and other studies on the dynamic barriers affecting transscleral drug delivery are described in Sections 7.3.1 through 7.3.5.

7.3.1   CONJUNCTIVAL/EPISCLERAL CLEARANCE

The conjunctiva is well vascularized in most mammals, and there is considerable variability in episcleral vascularity among species (Figure 7.2).[67–69] *In vivo* studies in different species have demonstrated that a drug present in conjunctival and episcleral tissues can be cleared through the blood and lymphatic vessels.

Ocular lymphatic vessels have been investigated in both humans[70–72] and rabbits.[73] Radioactive tracers injected subconjunctivally are detectable in the cervical lymph nodes,[74] and it has been estimated that ^{131}I-albumin-Evans blue complex migrates to the lymph nodes of the neck within 6 minutes after subconjunctival injection.[75]

The greater retention of subconjunctivally injected microparticles compared to nanoparticles[76] and the longer half-life of albumin compared with [22]Na[77] in subconjunctival tissue suggest that molecular size may influence the conjunctival/episcleral clearance rate. A detailed understanding of the effect of various physicochemical

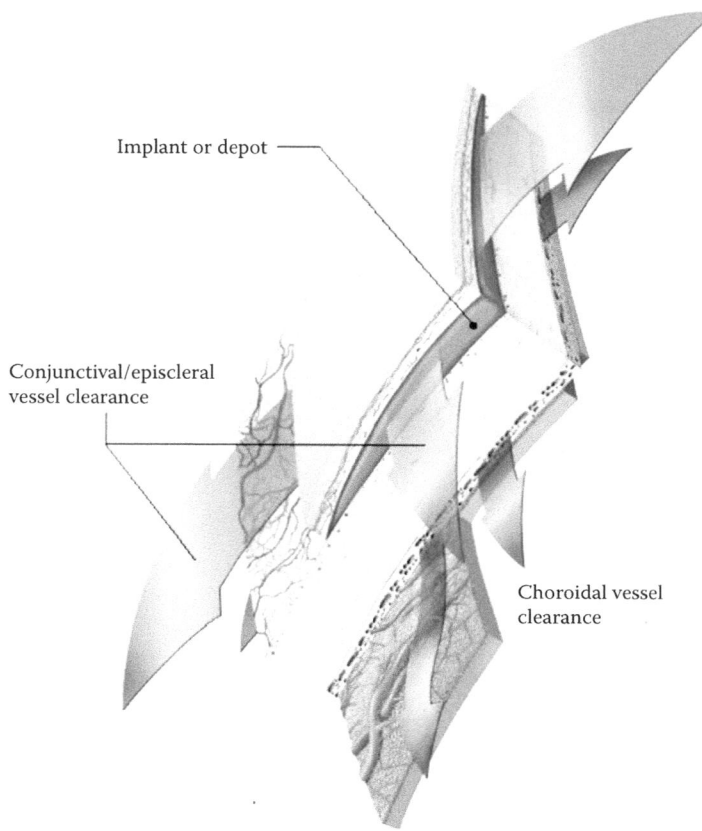

Implant or depot

Conjunctival/episcleral
vessel clearance

Choroidal vessel
clearance

FIGURE 7.2 Elimination pathways in the conjunctiva/episclera and choroid influencing transscleral drug delivery.

factors on drug elimination via the vasculature and lymphatics is lacking, and further studies are needed to clarify these mechanisms.

It has been demonstrated that selective elimination of conjunctival/episcleral clearance mechanisms can enhance intraocular drug penetration. In rabbits, sub-Tenon's injection of triamcinolone acetonide using an incised "conjunctival window" to inhibit local blood and lymphatic clearance in the conjunctiva resulted in higher levels of drug in the vitreous as compared with rabbits that did not have the incision.[78] Studies have also compared the delivery of sodium fluorescein to the retina by periocular injection and by a unidirectional episcleral exoplant that releases dye on the episcleral side but not on the conjunctival side. Sodium fluorescein levels in the retina were higher with unidirectional episcleral exoplant relative to periocular injection. These studies suggest that intraocular drug penetration can be significantly limited by conjunctival/episcleral clearance mechanisms.

The ability of the conjunctival and lymphatic circulation to limit transscleral penetration of hydrophilic compounds has been investigated in live and postmortem rats following administration of a sodium fluorescein–containing transscleral implant and after subconjunctival injection of sodium fluorescein or IgG.[79] The episcleral implant was effective in delivering sodium fluorescein to the subretinal space in post-mortem eyes but was ineffective in doing so when implanted in the eyes of live rats. Subconjunctivally administered sodium fluorescein was shown to reach the retina, despite the maintenance of static barriers, suggesting that static structural barriers play a minor role in limiting drug penetration from the conjunctival space into the retina, while dynamic mechanisms of elimination (e.g., conjunctival blood/lymphatic vessels, episcleral veins, uveoscleral outflow, and the choroidal circulation), which are markedly reduced post-mortem, might be responsible for restricting transscleral drug delivery in live animals. Immunohistochemical evidence indicated that subconjunctivally administered IgG and sodium fluorescein were eliminated via densely distributed blood and lymphatic vessels in the conjunctiva. The lymphatic elimination rate after injection of 75.6 µg of sodium fluorescein into the subconjunctival space was 105 ng/min between 30 and 60 minutes; at least 10% of the dose of sodium fluorescein administered subconjunctivally was eliminated through the conjunctival lymphatic drainage system within the first hour. The authors suggested that small, hydrophilic drugs administered subconjunctivally are eliminated predominantly through the conjunctival blood vessels, while subconjunctivally administered hydrophilic macromolecules are eliminated primarily through the conjunctival lymphatic vessels.

Similar findings were reported in a study conducted in rats to determine the episcleral clearance of sodium fluorescein delivered by a bioerodible sub-Tenon's implant composed of hydroxypropyl methyl cellulose.[80] The kinetics of sodium fluorescein clearance were determined in live animals, following euthanasia, and in animals in which choroidal nonperfusion had been achieved with laser-induced thrombosis of the choroidal vasculature. Sodium fluorescein was efficiently cleared from the episcleral tissue in live rats, with greater than 90% of drug eliminated by 3 hours, while euthanasia led to the blockade of episcleral drug clearance, with drug amounts in the conjunctiva, sclera, and choroid remaining relatively stable over a 7-hour period. Clearance was not affected by elimination of choroidal blood flow. These results suggest that the rapid episcleral clearance of sodium fluorescein following delivery from a bioerodible implant occurs via active physiologic processes rather than by passive diffusion clearance and that the choroid and choroidal circulation do not play a prominent role; thus, the conjunctival blood vessels and/or lymphatics may be responsible for the rapid clearance of transsclerally delivered drugs.

7.3.2 CHOROIDAL CIRCULATION

One of the mechanisms by which drugs can be eliminated during transscleral delivery, and which hinders transport to the retina, involves drug uptake resulting from the rapid blood flow in the choroid. The choroid, due to its high blood perfusion, serves as a sink in which the drug concentration is assumed to be zero in models of ocular drug transport.[81,82]

Detectable drug levels have been reported in the systemic circulation following periorbital drug administration, as in the case of tilisolol, which was shown to be rapidly absorbed systemically when delivered by the periocular injection.[83] However, choroidal clearance cannot be accurately estimated from systemic drug levels in blood samples because drug levels in the systemic circulation reflect clearance from the conjunctival blood vessels as well as the choroidal vasculature. Drugs elimination by the lymphatic system can also add to systemic levels since interstitial fluid is returned to the circulatory system after being filtered through the lymph nodes.

Cryotherapy has been used in experiments to locally eliminate choroidal blood flow. A single freeze/thaw cycle with cryotherapy leads to the formation of a chorioretinal scar but leaves the conjunctiva and sclera intact.[78] The efficacy of sub-Tenon's injection of triamcinolone acetonide has been assessed in rabbits after at least 1 month postcryotherapy to allow time for conjunctival and scleral recovery and for maturation of the chorioretinal scar. The levels of triamcinolone acetonide in the vitreous after sub-Tenon's injection were not higher in rabbits that received cryotherapy as compared with those with intact choroidal blood flow,[78] which suggests that choroidal blood flow does not contribute significantly to drug elimination during transscleral delivery.

7.3.3 Clearance by Bulk Fluid Flow

Drug penetration following transscleral delivery can also be reduced by convective fluid flow in the eye. Drug molecules can move with bulk fluid flow in ocular tissues and be ultimately cleared via the choroidal or conjunctival vasculature.

7.3.3.1 Uveoscleral Outflow

The uveoscleral outflow pathway of aqueous humor drainage, also known as the unconventional route,[84] is characterized by the flow of aqueous humor from the ciliary body into the posterior chamber, through the pupil, and into the anterior chamber, followed by drainage either through the trabecular meshwork into Schlemm's canal and aqueous veins (conventional pathway) or directly through the ciliary body and into the suprachoroidal space (unconventional pathway). From the suprachoroidal space, aqueous humor flows through the scleral channels[85–87] and is ultimately cleared by conjunctival lymphatic vessels (Figure 7.3, uveoscleral flow inset). When ^{131}I-albumin-Evans blue is injected into the suprachoroid space, most of the labeled albumin passes out through the sclera and drains from the conjunctiva through the lymphatic vessels and eventually into the systemic circulation.[88,89]

Because there are no methods for determining uveoscleral outflow directly and in a noninvasive manner, the uveoscleral pathway remains poorly understood.[90] The contribution of uveoscleral drainage to total aqueous humor outflow is known to differ widely among species. The reported rates of uveoscleral drainage range from 3% to 8% in rabbits,[91] up to 80% in mice,[92] and up to 60% in humans. Aging can also influence uveoscleral outflow,[93] and this effect must be taken into account in experimental models. Uveoscleral outflow has previously been regarded as a minor route for aqueous humor drainage; however, the high rate of uveoscleral outflow reported in humans indicates that this pathway may play a more significant role than previously thought.

Drug delivery by the transscleral route can be hindered by the uveoscleral pathway since it generates an outward bulk flow of fluid from the suprachoroidal space.

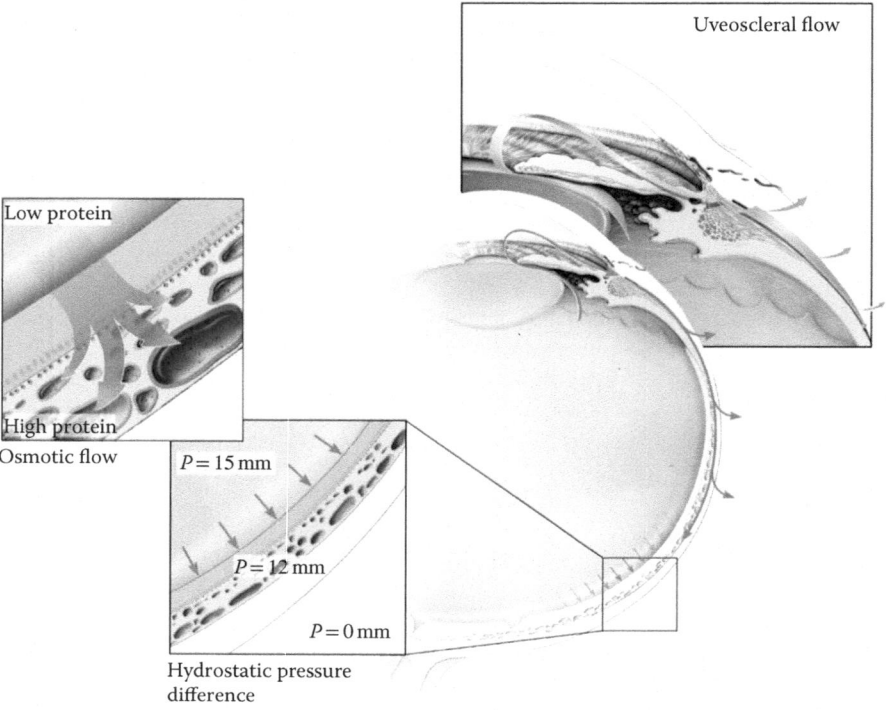

Uveoscleral flow

Low protein

High protein

Osmotic flow

$P = 15\,\text{mm}$

$P = 12\,\text{mm}$

$P = 0\,\text{mm}$

Hydrostatic pressure
difference

FIGURE 7.3   Clearance mechanisms by bulk fluid flow present during transscleral delivery.

Some drug molecules can be carried along with the convective current of the aqueous humor and eventually cleared through the conjunctival vasculature. Fluid and small particles can penetrate directly through the scleral tissue,[94] while larger particles pass through the suprachoroid via the scleral emissary channels that exist along the blood vessels and nerves that traverse the sclera.[85]

7.3.3.2   Hydrostatic and Osmotic Pressure

While most fluid produced by the ciliary body exits the eye via aqueous humor drainage routes, some fluid clearance occurs by passing through the vitreous, retina, and RPE into the choroid. This fluid flow helps to keep the retina attached to the RPE and is created mainly by the difference between osmotic pressures in the choroid and vitreous.[95,96] The choroid is well vascularized, is high in protein content, and has an osmotic pressure in rabbits of 12–14 mmHg.[97] While the osmotic pressure of the choroid is similar to that of blood, the osmotic pressure of the vitreous humor is lower (near 0 mmHg). The osmotic pressure difference between the choroid and the vitreous generates fluid flow in the direction of the choroid (Figure 7.3, osmotic flow inset). The difference in hydrostatic pressure between the suprachoroid and the episcleral tissue also contributes to outward bulk fluid flow.[98] At an IOP of 15 mmHg, the hydrostatic pressure of the posterior suprachoroidal space in cynomolgus monkeys was determined to be 3.7 ± 0.4 mmHg below the IOP.[99] There is a hydrostatic pressure difference of about 12 mmHg between the suprachoroid and the episcleral tissue, which serves as a driving force for outward bulk fluid flow.

7.3.4   RPE TRANSPORTER PROTEINS

The RPE is involved in the regulation of ion and solute transport into the subretinal space. The RPE expresses various transporter proteins, including ion transporters, amino acid transporters, and drug efflux pumps. Drug penetration to the retina can be hindered by transporter proteins that expel solutes toward the choroid. Few studies have been conducted to date on the influence of transporter proteins on ocular drug delivery, and the degree to which they serve as transport barriers is largely unknown. Michaelis–Menten kinetics predicts that transporter proteins will become saturated at high drug concentrations, thereby reducing their effect. However, since drug concentrations are typically low in the RPE following transscleral delivery, it is possible that transporter proteins act as significant barriers to the permeation of various drugs. Many transporter proteins are receptors for specific compounds and do not affect most other solutes. An exhaustive review on transporters present in the blood–retinal barrier (BRB), including the RPE, has been published by Mannermaa et al.[100]

7.3.4.1   Drug Efflux Pumps

Drug efflux pumps act to expel toxic and foreign molecules from cells. They consist of two major efflux transporters: P-glycoproteins (P-gp) and multidrug-resistance-associated proteins (MRPs).[101] Both of these pumps have broad substrate specificity, with P-gp pumps generally expelling large neutral or cationic compounds and MRPs removing large neutral or anionic compounds.[102] Human RPE has been shown to express P-gp and MRP proteins, which direct efflux toward the choroid[103–105] and thereby prevent the entry of drugs into the retina.

7.3.4.2   Organic Ion Transporters

Various ophthalmic drugs are among the potential substrates of organic anion and cation transporters.[100] Active transport of organic anions by the RPE was described initially by Cunha-Vaz and Maurice.[106] Amines and various drugs exist as anions or cations at physiological pH. Organic cation transporters (OCTs) carry endogenous amines such as choline, epinephrine, dopamine, guanidine, and histamine as well as organic cation drugs including anticholinergics, adrenergics, antihistamines, xenobiotics, and vitamins.[100] Amino acid transporters for glutamate,[107,108] taurine,[60] and leucine[109] among others have been identified in human RPE. Many amino acids show a net flux from the retina to the choroid.[110] Solute transport can also be influenced by the relative apical and basolateral concentrations of ions and solute, as in the case of taurine.[59,111] The ocular antihypertensive drug brimonidine, which is used in the treatment of glaucoma, is a substrate for OCTs, and its transport is dependent on extracellular pH, temperature, and concentration.[112] Organic anion-transporting polypeptide-E has been detected in the retinas of rat eyes,[113] but its precise tissue localization remains undetermined. Additional *in vivo* studies are needed to accurately determine the direction of solute transport by transporter proteins.

7.3.5 Drug Binding by Melanin in the Choroid–RPE

Eye pigmentation and ocular melanin content have been shown to influence the transscleral retinal delivery of the lipophilic anti-inflammatory drug celecoxib, as demonstrated by studies in which drug levels in the eyes of Sprague-Dawley (albino) and Brown Norway (pigmented) rats were compared following periocular injection of a conventional celecoxib suspension or celecoxib–polylactic acid (PLA) microparticles.[19] *In vitro,* celecoxib displayed melanin binding with a V_{max} of 0.392 μmol/mg and a k value of 0.08 μM. Melanin was detected at concentrations of 200, 12, and 3 μg/mg tissue, in the choroid–RPE, sclera, and retina, respectively, and was undetectable in the vitreous, lens, and cornea of pigmented rats and in all ocular tissues in albino rats except for the choroid–RPE, in which melanin-like activity was 100-fold that of pigmented rats. Following drug injections, the retinal, vitreal, and choroid–RPE concentrations of celecoxib in pigmented eyes were approximately 1.5-fold higher than those of unpigmented eyes. Administration of celecoxib–PLA microparticles produced even greater differences between rat strains in the drug concentrations achieved in the choroid–RPE, retina, and vitreous on day 8 posttreatment. The significantly lower retinal and vitreal concentrations of celecoxib in pigmented eyes were attributable to binding and retention of celecoxib by melanin in the choroid–RPE. These results have implications for transscleral drug delivery in humans, suggesting not only that melanin can impede drug penetration into the retina and vitreous but also that the effectiveness of drug delivery may differ between individuals with brown versus blue eyes, which differ with respect to melanin content.

7.4 METABOLIC BARRIERS

Due to its exposure to the outside environment, the eye serves as a gateway for the entry of xenobiotics into the body. To protect against the entry of xenobiotics, the eye has evolved metabolizing enzyme systems that can degrade foreign substances that could otherwise injure ocular tissues. Many ocular tissues express drug-metabolizing enzymes; however, the most active sites of ocular xenobiotic metabolism are the ciliary body and RPE.[114] Since these tissues are responsible for detoxifying drugs carried by the systemic circulation, they require higher levels of metabolic enzymes as compared with other tissues.[115,116] Metabolic enzymes present in the RPE can degrade and detoxify drug molecules following transscleral delivery and limit their ability to reach the retina. Prior studies on metabolic enzymes have focused mainly on delineating metabolic pathways, while few studies have examined the effect of metabolic enzymes on ocular pharmacokinetics and drug distribution.[117] Therefore, the impact of metabolic barriers on transscleral drug delivery remains poorly understood.

A brief description of the major metabolic enzymes found in the RPE follows; more extensive reviews have been published by Kumar[114] and Attar et al.[102]

7.4.1 Cytochrome P-450

The cytochrome P-450 (CYP) family of enzymes are the most important of the phase I drug-metabolizing enzymes.[114] Phase I enzymes introduce or unmask polar bodies via oxidation, reduction, or hydrolysis reactions. Metabolites with sufficient polarity are then excreted, while the remaining byproducts are further polarized by phase II enzymes such as acetyltransferase, sulfotransferase, and glutathione *S*-transferase.[114] High levels of CYP enzymes in mouse[118] and bovine RPE have been reported.[115,119] CYP enzymes have broad substrate specificity and metabolize both endogenous compounds and xenobiotics. The CYP3A isozyme, for example, metabolizes more than 50% of all commercial drugs.[120]

7.4.2 LYSOSOMAL ENZYMES

RPE cells play an important role in the formation of the BRB and in regulating the metabolism and transport of solutes. RPE cells are also responsible for digesting shed photoreceptor outer segments, which, if they were otherwise allowed to accumulate, would precipitate photoreceptor cell degeneration and subsequent vision loss. The degradation of photoreceptor outer segments by RPE cells occurs at a very high rate (3000 million disks in 70 years)[121] and involves the action of a variety of lysosomal enzymes. Lysosomal enzymes, also referred to as acid hydrolases, are present in lysosomes and melanosomes[122] and can degrade proteins, polysaccharides, nucleic acids, and lipids in an acidic environment into low-molecular-weight constituents.[123] In bovine eyes, the highest levels of the lysosomal enzymes cathepsin B, cathepsin D, and acid phosphatase are found in the RPE.[123] Other lysosomal enzymes including α-fucosidase, α-mannosidase, and β-*N*-acetylglucosaminidase are also expressed in bovine RPE.[124] Macromolecular drugs as well as xenobiotics are able to cross lysosomal membranes by passive diffusion and can consequently be degraded by lysosomal enzymes during transport across the RPE.[125,126]

7.5 STRATEGIES FOR IMPROVING TREATMENT EFFICACY WITH TRANSSCLERAL DRUG DELIVERY

7.5.1 BIODEGRADABLE IMPLANTS AND MICROSPHERES

It may be possible to overcome barriers to transscleral drug delivery to the retina by using controlled-release implants consisting of a biodegradable polymer-drug matrix.[127,128] Such devices are hydrolyzed over time, which facilitates drug dissolution and diffusion out of the polymer matrix, ideally providing a near constant rate of drug release. Microspheres and minitablets based on the biodegradable polymer PLA have been developed for the transscleral delivery of TA; their release kinetics have been recently tested using *in vitro* and *ex vivo* models.[13] TA-loaded microspheres (average 2-μm diameter) and small, cylindrical TA–PLA minitablets both provided sustained drug release over several days *in vitro*. While the rate of drug release from microspheres was affected by their TA content, the rate of release from PLA–TA minitablets was affected by the PLA:TA ratio. TA–PLA was found to slowly cross the sclera *in vitro* (approximately 21% diffusion through the sclera in 45 days). TA release was essentially linear after a small initial burst, which was attributed to the rapid leakage of TA located on/nearby the surface of the devices. Transscleral implantation of PLA–TA minitablets in rabbit cadaver eyes resulted in the accumulation of TA in the vitreous and aqueous humor. In rabbit sclera, the *in vitro* diffusion profile of TA released from PLA–TA minitablets indicated that the presence of the sclera does not significantly impede the passage of TA, suggesting that the sclera would not pose a significant barrier to TA passage into the vitreous *in vivo*. Calculations based on the *in vitro* release profile of TA–PLA minitablets containing 17-mg TA indicated that the average rate of TA release would be 0.23 mg/day, which exceeds the calculated ocular microcirculation clearance rate of TA after episcleral administration (0.19 mg/day).[129] Thus, assuming a similar release rate *in vivo*, TA–PLA minitablets would be expected to overcome ocular blood barriers and reach the vitreous humor. In addition, implantation of the device into the sclera, either by creating a scleral pocket or by using a scleral patch to cover it, would result in an external fibrous capsule formation, which might act as a barrier to outward diffusion of the drug. Although the amount of TA that entered the interior of the eye was relatively small compared with the TA levels obtained after intravitreal injection,[130] this may be advantageous given that the most common side effects of TA (cataract formation and glaucoma) may be associated with increased aqueous humor TA concentrations.[131]

7.5.2 Unidirectional Episcleral Exoplant

Transscleral drug delivery to the retina and posterior vitreous can be facilitated by the use of drug-eluting episcleral exoplants. Two such devices (3T Ophthalmics, Irvine, CA) have been tested in rabbits using fluorescein as a test agent: (1) a semipermeable rigid polyethylene device loaded with compressed fluorescein pellets and (2) a flexible refillable silicone exoplant containing a fluorescein solution. Both devices were anchored to the sclera using sutures, and their release characteristics were compared with periocular injection of sodium fluorescein.[132] Ocular and systemic drug levels were determined by fluorophotometry at various timepoints after fluorescein administration. The episcleral exoplants were found to facilitate diffusion of fluorescein through the sclera and to produce markedly higher levels in the retina and posterior vitreous (37–44 times higher) as compared with periocular injection of the same amount of fluorescein. Systemic absorption of fluorescein was reduced with the exoplants, as indicated by lower peak plasma levels relative to periocular injection. Episcleral exoplants have also been shown to be effective for transscleral delivery of betamethasone and 6-carboxyfluorescein diacetate to the posterior segment of the eye in rabbits.[133]

7.5.3 Iontophoresis

Ocular iontophoresis is another strategy of interest as a means to overcoming ocular barriers to transscleral drug delivery. Ocular iontophoresis is a noninvasive drug delivery technique that uses a low-intensity electric current to overcome the lens–iris barrier and deliver drugs directly into the vitreous and retina through the choroid or indirectly through the systemic circulation or anterior chamber. With this technique, a donor electrode containing a drug with the same charge as the electrode is placed on the conjunctiva, over the pars-plana area to avoid current damage to the retina, and a return electrode is placed on another body surface; the drug serves as a conductor of the current through the ocular tissues.[29] Iontophoresis enhances drug delivery by three mechanisms: electrophoresis (enhanced movement of ionic species by the applied electric field), electroosmosis (transport of neutral and charged species by an electric field–induced convective solvent flow), and electroporation (alteration of the tissue barrier resulting in increased intrinsic permeability of the membrane).[29] The technique, which is relatively easy to perform, does not produce adverse effects associated with intravitreal and periocular injections such as retinal detachment, endophthalmitis, globe perforation, and ptosis[30] and is associated with greater drug penetration in target tissues. While ocular iontophoresis offers several potential safety advantages over conventional ocular drug administration techniques, the high current intensity used for iontophoresis can conceivably cause ocular tissue damage depending on the site of application, current density, and current duration used.[29]

Two basic device types have been used for ocular iontophoresis: eye-cup solutions and drug-saturated hydrogels. With the former method, a metal electrode connected to a current supply is submerged in a drug solution and an eyecup with an internal diameter of 5–10 mm is placed over the eye using slight suction; the drug solution is then continuously infused into the cup during the iontophoretic treatment. Hydrogels consisting of the drug mixed with hydrophilic polymers have been used in electrophoresis because of their superior behavior, biocompatibility and stability, and modifiable release kinetics.[29,134] Key factors influencing the efficacy of iontophoretic drug delivery to the eye include the current intensity, duration of current application, and the ocular contact area. In addition, the pattern and time course of drug distribution in the vitreous varies considerably among different iontophoretically administered agents due to the inherent physicochemical properties of the drug molecules.[29] A novel approach to ocular iontophoresis involves the administration of charged nanoparticles in hydrogel solutions, which offers several benefits, including ocular drug penetration regardless of a drug's ionic strength and diffusion properties in ocular tissues; controlled release of the drug and prolonged therapeutic activity; and targeting to a specific desired tissue.[135] These properties can be modified by changing the particle size, particle charge, and chemical properties of the nanoparticles or by using different ligands attached to the particle.

Iontophoresis has been extensively studied for the transdermal delivery of various agents such as anesthetics, antibiotics, and pilocarpine, as well as for the delivery of a broad range of ophthalmic drugs, including antibiotics, antiviral and antifungal agents, steroidal and nonsteroidal anti-inflammatory drugs, antimetabolites, and genes.[29,30] Transscleral iontophoresis has shown to be effective for the delivery of dexamethasone and methylprednisolone to control anterior and posterior signs of intraocular inflammation in animal models.[136–139] Clinical studies have demonstrated that ocular iontophoresis can be used to safely and effectively deliver methylprednisolone for the prevention of corneal allograft rejection.[140–143]

Most transscleral iontophoresis studies to date have focused on the delivery of small molecules. The feasibility of delivering macromolecules by ocular iontophoresis has not been thoroughly investigated, and the ocular barriers, penetration, and distribution of macromolecules are not well understood. The internal membrane of the retina has been shown to impede the diffusion of linear molecules larger than 40 kDa and globular molecules larger than 70 kDa.[30] The passive penetration of FITC-labeled bevacizumab (a 149-kDa antibody) through human sclera and the ability of ocular iontophoresis to enhance bevacizumab uptake were recently investigated in an *in vitro* study using a Franz-type diffusion system.[144] Iontophoresis was shown to result in a significant seven-fold enhancement of bevacizumab transport through the sclera, despite the fact that it is an uncharged molecule. Although these results support the feasibility of iontophoresis for the delivery of bevacizumab and other antibodies with a similar charge and molecular weight, the amount of bevacizumab that permeated the sclera was low when compared with intravitreal injection at a typical dose of 1.25 mg monthly. Intravitreal injections, however, are intended to deliver a depot dose for 1 month or more. Given the noninvasive nature and apparent safety of iontophoresis, this technique could be applied more frequently and hence at a lower dose. It is also important to consider that this study in isolated human sclera does not take into account other barriers, both static and dynamic, that limit the ability of drugs administered transsclerally to reach the posterior segment of the eye. Furthermore, the time required to obtain a significant enhancement of drug penetration was quite long (1 hour), indicating that further optimization of the technique is needed.

An *in vivo* study conducted in New Zealand white rabbits used magnetic resonance imaging (MRI) to determine the ocular distribution of iontophoretically administered Galbumin (gadolinium-conjugated bovine albumin; molecular weight 80 kDa), which was used as a surrogate permeant (i.e., similar in molecular weight) for antibody macromolecules such as ranibizumab (149 kDa), bevacizumab (48 kDa), and pegaptanib (50 kDa).[30] Compared with intravitreal injections, iontophoresis enhanced the ocular delivery of Galbumin, which was mainly delivered into the conjunctiva and sclera (in microgram quantities) and then diffused toward the posterior section in the upper hemisphere of the eye. In contrast with the rapid distribution of intravitreally administered Galbumin throughout the anterior chamber and vitreous, Galbumin levels in these regions were below the detection limit following iontophoretic delivery of Galbumin, presumably due to limitations on penetration imposed by the RPE barrier and/or by blood and lymphatic clearance in the aqueous, uvea/ciliary body, or posterior segment. In interpreting these results, it should be noted that the eyes of humans and New Zealand rabbits differ with respect to scleral thickness, the volume and components of the vitreous humor, and the influence of aging and disease states on vitreous humor properties such as viscosity. Thus, extrapolation from rabbits to humans must be done with caution.

7.6   CONCLUSIONS

Studies on transport barriers affecting transscleral drug delivery have so far focused mainly on determining the permeability values of ocular tissues. Such data enable comparisons to be made between the diffusional resistance of tissues and can also be used to calculate the diffusion coefficients of drugs. However, drug transport in the eye cannot be accurately predicted based on permeability values and diffusion coefficients alone since drug distribution can also be affected by dynamic and metabolic barriers.

The influence of dynamic barriers on the penetration of drugs delivered by the transscleral route has been demonstrated in studies using MRI, which provides three-dimensional images noninvasively *in vivo*. Dynamic barriers are active *in vivo* but cease to operate post-mortem, while static barriers, in contrast, are always present. In a study in which MRI scans acquired *in vivo* and postmortem were compared, the penetration of an MRI tracer released from a polymeric episcleral implant into the anterior chamber and vitreous was increased post-mortem but not *in vivo*.[145] Similar results were obtained after subconjunctival injection of manganese ions.[146] The difference in results between *in vivo* and postmortem MRI scans suggests that dynamic barriers, which cease to operate post-mortem, can significantly influence intraocular drug levels. Consequently, transscleral drug penetration cannot be accurately assessed in systems in which only static barriers are operational.

While dynamic barriers, in general, have been shown to affect transscleral drug delivery, the effects of individual dynamic barriers are unknown. Because of the difficulty of isolating individual dynamic barriers in experimental systems, it has been difficult to directly determine the contribution of one dynamic barrier to another. Elucidating the role played by individual dynamic barriers would facilitate the design of new drug delivery methods and devices. For instance, recent studies examining the significance of conjunctival clearance have led to the design of a unidirectional episcleral exoplant[132] that minimizes drug loss from conjunctival and lymphatic vascular clearance. Posterior drug infusion into the suprachoroidal space[147] may also be capable of bypassing conjunctival clearance mechanisms, which would allow for delivery of higher drug doses to the macula. The use of biodegradable PLA-based scleral implants[13] and iontophoresis[29,30] to enhance ocular drug penetration also appear to be promising strategies for the delivery of drugs to the retina.

Studies to date have provided considerable data on the role of static barriers in transscleral drug delivery; however, studies on the role of dynamic and metabolic barriers are lacking. Although tissue permeability is important for drug transport, recent *in vivo* studies have demonstrated that clearance and elimination mechanisms also play a significant role in reducing drug delivery to the retina. Additional research is warranted to determine the effect of dynamic and metabolic barriers on the transport of drugs administered by the transscleral route.

REFERENCES

1. Raghava S, Hammond M, Kompella UB. Periocular routes for retinal drug delivery. *Expert Opin Drug Deliv.* 2004;1(1):99–114.
2. Geroski DH, Edelhauser HF. Drug delivery for posterior segment eye disease. *Invest Ophthalmol Vis Sci.* 2000;41(5):961–964.
3. Hughes PM, Olejnik O, Chang-Lin JE, Wilson CG. Topical and systemic drug delivery to the posterior segments. *Adv Drug Deliv Rev.* 2005;57(14):2010–2032.
4. Ciulla TA, Walker JD, Fong DS, Criswell MH. Corticosteroids in posterior segment disease: An update on new delivery systems and new indications. *Curr Opin Ophthalmol.* 2004;15(3):211–220.
5. Smith CL. Local therapy for cytomegalovirus retinitis. *Ann Pharmacother.* 1998;32(2): 248–255.

6. Kralinger MT, Kieselbach GF, Voigt M et al. Experimental model for proliferative vitreoretinopathy by intravitreal dispase: Limited by zonulolysis and cataract. *Ophthalmologica.* 2006;220(4):211–216.

7. Ozkiris A, Erkilic K. Complications of intravitreal injection of triamcinolone acetonide. *Can J Ophthalmol.* 2005;40(1):63–68.

8. Nicolo M, Ghiglione D, Calabria G. Retinal pigment epithelial tear following intravitreal injection of bevacizumab (Avastin). *Eur J Ophthalmol.* 2006;16(5):770–773.

9. Shah CP, Hsu J, Garg SJ, Fischer DH, Kaiser R. Retinal pigment epithelial tear after intravitreal bevacizumab injection. *Am J Ophthalmol.* 2006;142(6):1070–1072.

10. Sutter FK, Gillies MC. Pseudo-endophthalmitis after intravitreal injection of triamcinolone. *Br J Ophthalmol.* 2003;87(8):972–974.

11. Nelson ML, Tennant MT, Sivalingam A, Regillo CD, Belmont JB, Martidis A. Infectious and presumed noninfectious endophthalmitis after intravitreal triamcinolone acetonide injection. *Retina.* 2003;23(5):686–691.

12. Roth DB, Chieh J, Spirn MJ, Green SN, Yarian DL, Chaudhry NA. Noninfectious endophthalmitis associated with intravitreal triamcinolone injection. *Arch Ophthalmol.* 2003;121(9):1279–1282.

13. Blatsios G, Tzimas AS, Mattheolabakis G, Panagi Z, Avgoustakis K, Gartaganis SP. Development of biodegradable controlled release scleral systems of triamcinolone acetonide. *Curr Eye Res.* 2010;35(10):916–924.

14. Bonini-Filho MA, Jorge R, Barbosa JC, Calucci D, Cardillo JA, Costa RA. Intravitreal injection versus sub-Tenon's infusion of triamcinolone acetonide for refractory diabetic macular edema: A randomized clinical trial. *Invest Ophthalmol Vis Sci.* 2005;46(10):3845–3849.

15. Ohguro N, Okada AA, Tano Y. Trans-Tenon's retrobulbar triamcinolone infusion for diffuse diabetic macular edema. *Graefes Arch Clin Exp Ophthalmol.* 2004;242(5): 444–445.

16. Okada AA, Wakabayashi T, Kojima E, Asano Y, Hida T. Trans-Tenon's retrobulbar triamcinolone infusion for small choroidal neovascularisation. *Br J Ophthalmol.* 2004;88(8):1097–1098.

17. Okada AA, Wakabayashi T, Morimura Y et al. Trans-Tenon's retrobulbar triamcinolone infusion for the treatment of uveitis. *Br J Ophthalmol.* 2003;87(8):968–971.

18. Ayalasomayajula SP, Kompella UB. Subconjunctivally administered celecoxib-PLGA microparticles sustain retinal drug levels and alleviate diabetes-induced oxidative stress in a rat model. *Eur J Pharmacol.* 2005;511(2–3):191–198.

19. Cheruvu NP, Amrite AC, Kompella UB. Effect of eye pigmentation on transscleral drug delivery. *Invest Ophthalmol Vis Sci.* 2008;49(1):333–341.

20. Kompella UB, Bandi N, Ayalasomayajula SP. Subconjunctival nano- and microparticles sustain retinal delivery of budesonide, a corticosteroid capable of inhibiting VEGF expression. *Invest Ophthalmol Vis Sci.* 2003;44(3):1192–1201.

21. Gilbert JA, Simpson AE, Rudnick DE, Geroski DH, Aaberg TM, Jr., Edelhauser HF. Transscleral permeability and intraocular concentrations of cisplatin from a collagen matrix. *J Control Release.* 2003;89(3):409–417.

22. Zignani M, Einmahl S, Baeyens V et al. A poly(ortho ester) designed for combined ocular delivery of dexamethasone sodium phosphate and 5-fluorouracil: Subconjunctival tolerance and *in vitro* release. *Eur J Pharm Biopharm.* 2000;50(2):251–255.

23. Wang G, Tucker IG, Roberts MS, Hirst LW. *In vitro* and *in vivo* evaluation in rabbits of a controlled release 5-fluorouracil subconjunctival implant based on poly(D,L-lactide-co-glycolide). *Pharm Res.* 1996;13(7):1059–1064.

24. Kato A, Kimura H, Okabe K, Okabe J, Kunou N, Ogura Y. Feasibility of drug delivery to the posterior pole of the rabbit eye with an episcleral implant. *Invest Ophthalmol Vis Sci.* 2004;45(1):238–244.

25. Okabe K, Kimura H, Okabe J, Kato A, Kunou N, Ogura Y. Intraocular tissue distribution of betamethasone after intrascleral administration using a non-biodegradable sustained drug delivery device. *Invest Ophthalmol Vis Sci.* 2003;44(6):2702–2707.
26. Okabe J, Kimura H, Kunou N, Okabe K, Kato A, Ogura Y. Biodegradable intra-scleral implant for sustained intraocular delivery of betamethasone phosphate. *Invest Ophthalmol Vis Sci.* 2003;44(2):740–744.
27. Ambati J, Gragoudas ES, Miller JW et al. Transscleral delivery of bioactive protein to the choroid and retina. *Invest Ophthalmol Vis Sci.* 2000;41(5):1186–1191.
28. Okabe K, Kimura H, Okabe J et al. Effect of benzalkonium chloride on transscleral drug delivery. *Invest Ophthalmol Vis Sci.* 2005;46(2):703–708.
29. Eljarrat-Binstock E, Pe'er J, Domb AJ. New techniques for drug delivery to the posterior eye segment. *Pharm Res.* 2010;27(4):530–543.
30. Molokhia SA, Jeong EK, Higuchi WI, Li SK. Transscleral iontophoretic and intravitreal delivery of a macromolecule: Study of ocular distribution *in vivo* and postmortem with MRI. *Exp Eye Res.* 2009;88(3):418–425.
31. Cardillo JA, Melo LA, Jr., Costa RA et al. Comparison of intravitreal versus posterior sub-Tenon's capsule injection of triamcinolone acetonide for diffuse diabetic macular edema. *Ophthalmology.* 2005;112(9):1557–1563.
32. Lee TW, Robinson JR. Drug delivery to the posterior segment of the eye III: The effect of parallel elimination pathway on the vitreous drug level after subconjunctival injection. *J Ocul Pharmacol Ther.* 2004;20(1):55–64.
33. Ambati J, Adamis AP. Transscleral drug delivery to the retina and choroid. *Prog Retin Eye Res.* 2002;21(2):145–151.
34. Boubriak OA, Urban JP, Akhtar S, Meek KM, Bron AJ. The effect of hydration and matrix composition on solute diffusion in rabbit sclera. *Exp Eye Res.* 2000;71(5): 503–514.
35. Hamalainen KM, Kananen K, Auriola S, Kontturi K, Urtti A. Characterization of paracellular and aqueous penetration routes in cornea, conjunctiva, and sclera. *Invest Ophthalmol Vis Sci.* 1997;38(3):627–634.
36. Kao JC, Geroski DH, Edelhauser HF. Transscleral permeability of fluorescent-labeled antibiotics. *J Ocul Pharmacol Ther.* 2005;21(1):1–10.
37. Shuler RK, Jr., Dioguardi PK, Henjy C, Nickerson JM, Cruysberg LP, Edelhauser HF. Scleral permeability of a small, single-stranded oligonucleotide. *J Ocul Pharmacol Ther.* 2004;20(2):159–168.
38. Cruysberg LP, Franklin AJ, Sanders J et al. Effective transscleral delivery of two reti-nal anti-angiogenic molecules: Carboxyamido-triazole (CAI) and 2-methoxyestradiol (2ME2). *Retina.* 2005;25(8):1022–1031.
39. Cheruvu NP, Kompella UB. Bovine and porcine transscleral solute transport: Influence of lipophilicity and the choroid-Bruch's layer. *Invest Ophthalmol Vis Sci.* 2006;47(10):4513–4522.
40. Cruysberg LP, Nuijts RM, Geroski DH, Koole LH, Hendrikse F, Edelhauser HF. *In vitro* human scleral permeability of fluorescein, dexamethasone-fluorescein, meth-otrexate-fluorescein and rhodamine 6G and the use of a coated coil as a new drug delivery system. *J Ocul Pharmacol Ther.* 2002;18(6):559–569.
41. Prausnitz MR, Noonan JS. Permeability of cornea, sclera, and conjunctiva: A literature analysis for drug delivery to the eye. *J Pharm Sci.* 1998;87(12):1479–1488.
42. Ambati J, Canakis CS, Miller JW et al. Diffusion of high molecular weight compounds through sclera. *Invest Ophthalmol Vis Sci.* 2000;41(5):1181–1185.
43. Ahmed I, Gokhale RD, Shah MV, Patton TF. Physicochemical determinants of drug dif-fusion across the conjunctiva, sclera, and cornea. *J Pharm Sci.* 1987;76(8):583–586.
44. Edwards A, Prausnitz MR. Fiber matrix model of sclera and corneal stroma for drug delivery to the eye. *AIChE.* 1998;44(1):214–225.

45. Boubriak OA, Urban JP, Bron AJ. Differential effects of aging on transport properties of anterior and posterior human sclera. *Exp Eye Res.* 2003;76(6):701–713.
46. Curtin BJ. Physiopathologic aspects of scleral stress-strain. *Trans Am Ophthalmol Soc.* 1969;67:417–461.
47. Jiang J, Geroski DH, Edelhauser HF, Prausnitz MR. Measurement and prediction of lateral diffusion within human sclera. *Invest Ophthalmol Vis Sci.* 2006;47(7):3011–3016.
48. Kansara V, Mitra AK. Evaluation of an *ex vivo* model implication for carrier-mediated retinal drug delivery. *Curr Eye Res.* 2006;31(5):415–426.
49. Maurice DM, Polgar J. Diffusion across the sclera. *Exp Eye Res.* 1977;25(6):577–582.
50. Dunlevy JR, Rada JA. Interaction of lumican with aggrecan in the aging human sclera. *Invest Ophthalmol Vis Sci.* 2004;45(11):3849–3856.
51. Lee SB, Geroski DH, Prausnitz MR, Edelhauser HF. Drug delivery through the sclera: Effects of thickness, hydration, and sustained release systems. *Exp Eye Res.* 2004;78(3):599–607.
52. Olsen TW, Edelhauser HF, Lim JI, Geroski DH. Human scleral permeability: Effects of age, cryotherapy, transscleral diode laser, and surgical thinning. *Invest Ophthalmol Vis Sci.* 1995;36(9):1893–1903.
53. Cruysberg LP, Nuijts RM, Geroski DH, Gilbert JA, Hendrikse F, Edelhauser HF. The influence of intraocular pressure on the transscleral diffusion of high-molecular-weight compounds. *Invest Ophthalmol Vis Sci.* 2005;46(10):3790–3794.
54. Rudnick DE, Noonan JS, Geroski DH, Prausnitz MR, Edelhauser HF. The effect of intraocular pressure on human and rabbit scleral permeability. *Invest Ophthalmol Vis Sci.* 1999;40(12):3054–3058.
55. Moore DJ, Hussain AA, Marshall J. Age-related variation in the hydraulic conductivity of Bruch's membrane. *Invest Ophthalmol Vis Sci.* 1995;36(7):1290–1297.
56. Amaral J, Fariss RN, Campos MM et al. Transscleral-RPE permeability of PEDF and ovalbumin proteins: Implications for subconjunctival protein delivery. *Invest Ophthalmol Vis Sci.* 2005;46(12):4383–4392.
57. Pitkänen L, Ranta VP, Moilanen H, Urtti A. Permeability of retinal pigment epithelium: Effects of permeant molecular weight and lipophilicity. *Invest Ophthalmol Vis Sci.* 2005;46(2):641–646.
58. Lyda W, Eriksen N, Krishna N. Studies of Bruch's membrane; flow and permeability studies in a Bruch's membrane-choroid preparation. *Am J Ophthalmol.* 1957;44(5, Part 2): 362–369; discussion 369–370.
59. Hillenkamp J, Hussain AA, Jackson TL, Constable PA, Cunningham JR, Marshall J. Compartmental analysis of taurine transport to the outer retina in the bovine eye. *Invest Ophthalmol Vis Sci.* 2004;45(11):4099–4105.
60. Hillenkamp J, Hussain AA, Jackson TL, Cunningham JR, Marshall J. Taurine uptake by human retinal pigment epithelium: Implications for the transport of small solutes between the choroid and the outer retina. *Invest Ophthalmol Vis Sci.* 2004;45(12): 4529–4534.
61. Kimura M, Araie M, Koyano S. Movement of carboxyfluorescein across retinal pigment epithelium-choroid. *Exp Eye Res.* 1996;63(1):51–56.
62. Hussain AA, Rowe L, Marshall J. Age-related alterations in the diffusional transport of amino acids across the human Bruch's-choroid complex. *J Opt Soc Am A Opt Image Sci Vis.* 2002;19(1):166–172.
63. Ramrattan RS, van der Schaft TL, Mooy CM, de Bruijn WC, Mulder PG, de Jong PT. Morphometric analysis of Bruch's membrane, the choriocapillaris, and the choroid in aging. *Invest Ophthalmol Vis Sci.* 1994;35(6):2857–2864.
64. Bird AC, Marshall J. Retinal pigment epithelial detachments in the elderly. *Trans Ophthalmol Soc UK.* 1986;105(Pt 6):674–682.

65. Hillenkamp J, Hussain AA, Jackson TL, Cunningham JR, Marshall J. The influence of path length and matrix components on ageing characteristics of transport between the choroid and the outer retina. *Invest Ophthalmol Vis Sci.* 2004;45(5):1493–1498.
66. Moore DJ, Clover GM. The effect of age on the macromolecular permeability of human Bruch's membrane. *Invest Ophthalmol Vis Sci.* 2001;42(12):2970–2975.
67. Prince JH, Diesem CD, Eglitis I, Ruskell GL. The Rabbit. In Prince JH, ed. *Anatomy and Histology of the Eye and Orbit in Domestic Animals.* Springfield, IL: Charles C Thomas; 1960:268.
68. Prince JH, Diesem CD, Eglitis I, Ruskell GL. The Dog. In Prince JH, ed. *Anatomy and Histology of the Eye and Orbit in Domestic Animals.* Springfield, IL: Charles C Thomas; 1960:73.
69. Prince JH, Diesem CD, Eglitis I, Ruskell GL. The Eyelids. In Prince JH, ed. *Anatomy and Histology of the Eye and Orbit in Domestic Animals.* Springfield, IL: Charles C Thomas; 1960:46–47.
70. Sugar HS, Riazi A, Schaffner R. The bulbar conjunctival lymphatics and their clinical significance. *Trans Am Acad Ophthalmol Otolaryngol.* 1957;61(2):212–223.
71. Gausas RE, Gonnering RS, Lemke BN, Dortzbach RK, Sherman DD. Identification of human orbital lymphatics. *Ophthal Plast Reconstr Surg.* 1999;15(4):252–259.
72. Singh D. Conjunctival lymphatic system. *J Cataract Refract Surg.* 2003;29(4):632–633.
73. Collin HB. The ultrastructure of conjunctival lymphatic anchoring filaments. *Exp Eye Res.* 1969;8(2):102–105.
74. Gruntzig J, Schicha H, Kiem J, Becker V, Feinendegen LE. Studies on the lymph drainage of the eye 5. Quantitative registration of the lymph drainage from the subconjunctival space with a radioactive tracer (author's transl). *Klin Monatsbl Augenheilkd.* 1978;172(6):872–876.
75. Collin HB. Lymphatic drainage of 131-I-albumin from the vascularized cornea. *Invest Ophthalmol.* 1970;9(2):146–155.
76. Amrite AC, Kompella UB. Size-dependent disposition of nanoparticles and microparticles following subconjunctival administration. *J Pharm Pharmacol.* 2005;57(12): 1555–1563.
77. Maurice DM, Ota Y. The kinetics of subconjunctival injections. *Jpn. J. Ophthalmol.* 1978;22:95–100.
78. Robinson MR, Lee SS, Kim H et al. A rabbit model for assessing the ocular barriers to the transscleral delivery of triamcinolone acetonide. *Exp Eye Res.* 2005;82(3): 479–487.
79. Lee SJ, He W, Robinson SB, Robinson MR, Csaky KG, Kim H. Evaluation of clearance mechanisms with transscleral drug delivery. *Invest Ophthalmol Vis Sci.* 2010;51(10):5205–5212.
80. Chan JE, Pridgen TA, Csaky KG. Episcleral clearance of sodium fluorescein from a bioerodible sub-tenon's implant in the rat. *Exp Eye Res.* 2010;90(4):501–506.
81. Park J, Bungay PM, Lutz RJ et al. Evaluation of coupled convective-diffusive transport of drugs administered by intravitreal injection and controlled release implant. *J Control Release.* 2005;105(3):279–295.
82. Kim H, Lizak MJ, Tansey G et al. Study of ocular transport of drugs released from an intravitreal implant using magnetic resonance imaging. *Ann Biomed Eng.* 2005;33(2): 150–164.
83. Sasaki H, Kashiwagi S, Mukai T et al. Drug absorption behavior after periocular injections. *Biol Pharm Bull.* 1999;22(9):956–960.
84. Bill A. The aqueous humor drainage mechanism in the cynomolgus monkey (Macaca irus) with evidence for unconventional routes. *Invest Ophthalmol.* 1965;4(5):911–919.
85. Inomata H, Bill A. Exit sites of uveoscleral flow of aqueous humor in cynomolgus monkey eyes. *Exp Eye Res.* 1977;25(2):113–118.

86. Krohn J, Bertelsen T. Light microscopy of uveoscleral drainage routes after gelatine injections into the suprachoroidal space. *Acta Ophthalmol Scand.* 1998;76(5):521–527.
87. Krohn J, Bertelsen T. Corrosion casts of the suprachoroidal space and uveoscleral drainage routes in the human eye. *Acta Ophthalmol Scand.* 1997;75(1):32–35.
88. Bill A. The drainage of albumin from the uvea. *Exp Eye Res.* 1964;75:179–187.
89. Bill A. Movement of albumin and dextran through the sclera. *Arch Ophthalmol.* 1965;74:248–252.
90. Fautsch MP, Johnson DH. Aqueous humor outflow: What do we know? Where will it lead us? *Invest Ophthalmol Vis Sci.* 2006;47(10):4181–4187.
91. Bill A. Uveoscleral drainage of aqueous humor: Physiology and pharmacology. *Prog Clin Biol Res.* 1989;312:417–427.
92. Aihara M, Lindsey JD, Weinreb RN. Aqueous humor dynamics in mice. *Invest Ophthalmol Vis Sci.* 2003;44(12):5168–5173.
93. Toris CB, Yablonski ME, Wang YL, Camras CB. Aqueous humor dynamics in the aging human eye. *Am J Ophthalmol.* 1999;127(4):407–412.
94. Inomata H, Bill A, Smelser GK. Unconventional routes of aqueous humor outflow in Cynomolgus monkey (Macaca irus). *Am J Ophthalmol.* 1972;73(6):893–907.
95. Machemer R. The importance of fluid absorption, traction, intraocular currents, and chorioretinal scars in the therapy of rhegmatogenous retinal detachments. XLI Edward Jackson memorial lecture. *Am J Ophthalmol.* 1984;98(6):681–693.
96. Marmor MF. Mechanisms of retinal adhesion. *Prog Retin Res.* 1993;12:179–204.
97. Bill A. A method to determine osmotically effective albumin and gammaglobulin concentrations in tissue fluids, its application to uvea and a note on effects of capillary "leaks" on tissue fluid dynamics. *Acta Physiol Scand.* 1968;73(4):511–522.
98. Bill A. Blood circulation and fluid dynamics in the eye. *Physiol Rev.* 1975;55(3):383–417.
99. Emi K, Pederson JE, Toris CB. Hydrostatic pressure of the suprachoroidal space. *Invest Ophthalmol Vis Sci.* 1989;30(2):233–238.
100. Mannermaa E, Vellonen KS, Urtti A. Drug transport in corneal epithelium and blood-retina barrier: Emerging role of transporters in ocular pharmacokinetics. *Adv Drug Deliv Rev.* 2006;58(11):1136–1163.
101. Dey S, Mitra AK. Transporters and receptors in ocular drug delivery: Opportunities and challenges. *Expert Opin Drug Deliv.* 2005;2(2):201–204.
102. Attar M, Shen J, Ling KH, Tang-Liu D. Ophthalmic drug delivery considerations at the cellular level: Drug-metabolising enzymes and transporters. *Expert Opin Drug Deliv.* 2005;2(5):891–908.
103. Kennedy BG, Mangini NJ. P-glycoprotein expression in human retinal pigment epithelium. *Mol Vis.* 2002;8:422–430.
104. Steuer H, Jaworski A, Elger B et al. Functional characterization and comparison of the outer blood-retina barrier and the blood-brain barrier. *Invest Ophthalmol Vis Sci.* 2005;46(3):1047–1053.
105. Aukunuru JV, Sunkara G, Bandi N, Thoreson WB, Kompella UB. Expression of multidrug resistance-associated protein (MRP) in human retinal pigment epithelial cells and its interaction with BAPSG, a novel aldose reductase inhibitor. *Pharm Res.* 2001;18(5):565–572.
106. Cunha-Vaz JG, Maurice DM. The active transport of fluorescein by the retinal vessels and the retina. *J Physiol.* 1967;191(3):467–486.
107. Maenpaa H, Gegelashvili G, Tahti H. Expression of glutamate transporter subtypes in cultured retinal pigment epithelial and retinoblastoma cells. *Curr Eye Res.* 2004;28(3):159–165.
108. Miyamoto Y, Del Monte MA. Na(+)-dependent glutamate transporter in human retinal pigment epithelial cells. *Invest Ophthalmol Vis Sci.* 1994;35(10):3589–3598.

109. Sellner PA. The blood-retinal barrier: Leucine transport by the retinal pigment epithelium. *J Neurosci.* 1986;6(10):2823–2828.

110. Sellner PA. The movement of organic solutes between the retina and pigment epithelium. *Exp Eye Res.* 1986;43(4):631–639.

111. Kundaiker S, Hussain AA, Marshall J. Component characteristics of the vectorial transport system for taurine in isolated bovine retinal pigment epithelium. *J Physiol.* 1996;492 (Pt 2):505–516.

112. Zhang N, Kannan R, Okamoto CT, Ryan SJ, Lee VH, Hinton DR. Characterization of brimonidine transport in retinal pigment epithelium. *Invest Ophthalmol Vis Sci.* 2006;47(1):287–294.

113. Ito A, Yamaguchi K, Tomita H et al. Distribution of rat organic anion transporting polypeptide-E (oatp-E) in the rat eye. *Invest Ophthalmol Vis Sci.* 2003;44(11):4877–4884.

114. Kumar G. Drug metabolizing enzyme systems in the eye. In Reddy I, ed. *Ocular Therapeutics and Drug Delivery: A Multi-Disciplinary Approach.* Lancaster, PA: Technomic Publishing Company, Inc.; 1996:149–167.

115. Schwartzman ML, Masferrer J, Dunn MW, McGiff JC, Abraham NG. Cytochrome P450, drug metabolizing enzymes and arachidonic acid metabolism in bovine ocular tissues. *Curr Eye Res.* 1987;6(4):623–630.

116. Shichi H, Nebert DW. Drug metabolism in ocular tissues. In Gram TE, ed. *Extrahepatic Metabolism of Drugs and Other Foreign Compounds.* Lancaster, UK: MTP Press Limited; 1980:333–363.

117. Duvvuri S, Majumdar S, Mitra AK. Role of metabolism in ocular drug delivery. *Curr Drug Metab.* 2004;5(6):507–515.

118. Shichi H, Atlas SA, Nebert DW. Genetically regulated aryl hydrocarbon hydroxylase induction in the eye: Possible significance of the drug-metabolizing enzyme system for the retinal pigmented epithelium-choroid. *Exp Eye Res.* 1975;21(6):557–567.

119. Shichi H. Microsomal electron transfer system of bovine retinal pigment epithelium. *Exp Eye Res.* 1969;8(1):60–68.

120. Wrighton SA, Thummel KE. CYP3A. In Levy RH, Thummel KE, Trager WF, Hansten PD, Eichelbaum M, eds. *Metabolic Drug Interactions.* Philadelphia, PA: Lippincott Williams & Wilkins; 2000:115–133.

121. Marshall J. The ageing retina: Physiology or pathology. *Eye.* 1987;1(Pt 2):282–295.

122. Diment S, Eidelman M, Rodriguez GM, Orlow SJ. Lysosomal hydrolases are present in melanosomes and are elevated in melanizing cells. *J Biol Chem.* 1995;270(9):4213–4215.

123. Hayasaka S. Lysosomal enzymes in ocular tissues and diseases. *Surv Ophthalmol.* 1983;27(4):245–258.

124. Hayasaka S, Shiono T. alpha-Fucosidase, alpha-mannosidase and beta-N-acetylglucosaminidase of the bovine retinal pigment epithelium. *Exp Eye Res.* 1982;34(4):565–569.

125. Lloyd JB. Lysosome membrane permeability: Implications for drug delivery. *Adv Drug Deliv Rev.* 2000;41(2):189–200.

126. Lloyd JB. Studies on the permeability of rat liver lysosomes to carbohydrates. *Biochem J.* 1969;115(4):703–707.

127. Lee SS, Hughes P, Ross AD, Robinson MR. Biodegradable implants for sustained drug release in the eye. *Pharm Res.* 2010;27(10):2043–2053.

128. Lee SS, Hughes P, Ross AD, Robinson MR. Advances in biodegradable ocular drug delivery systems. In Kompella U, Edelhauser H, eds. *Drug Product Development for the Back of the Eye.* New York, NY: Springer/American Association of Pharmaceutical Scientists; 2011:185–230.

129. Robinson MR, Lee SS, Kim H et al. A rabbit model for assessing the ocular barriers to the transscleral delivery of triamcinolone acetonide. *Exp Eye Res.* 2006;82(3):479–487.

130. Jermak CM, Dellacroce JT, Heffez J, Peyman GA. Triamcinolone acetonide in ocular therapeutics. *Surv Ophthalmol.* 2007;52:503–522.

131. Gasiorowski JZ, Russell P. Biological properties of trabecular meshwork cells. *Exp Eye Res.* 2009;88:671–675.

132. Pontes de Carvalho RA, Krausse ML, Murphree AL, Schmitt EE, Campochiaro PA, Maumenee IH. Delivery from episcleral exoplants. *Invest Ophthalmol Vis Sci.* 2006;47(10):4532–4539.

133. Kato A, Kimura H, Okabe K, Okabe J, Kunou N, Ogura Y. Feasibility of drug delivery to the posterior pole of the rabbit eye with an episcleral implant. *Invest Ophthalmol Vis Sci.* 2004;45:238–244.

134. Alvarez-Figueroa MJ, Blanco-Mendez J. Transdermal delivery of methotrexate: Iontophoretic delivery from hydrogels and passive delivery from microemulsions. *Int J Pharm.* 2001;215:57–65.

135. Eljarrat-Binstock E, Orucov F, Aldouby Y, Frucht-Pery J, Domb AJ. Charged nanoparticles delivery to the eye using hydrogel iontophoresis. *J Control Release.* 2008;126:156–161.

136. Behar-Cohen FF, Parel JM, Pouliquen Y et al. Iontophoresis of dexamethasone in the treatment of endotoxin-induced-uveitis in rats. *Exp Eye Res.* 1997;65:533–545.

137. Lam TT, Edward DP, Zhu XA, Tso M.O. Transscleral iontophoresis of dexamethasone. *Arch Ophthalmol.* 1989;107:1368–1371.

138. Eljarrat-Binstock E, Raiskup F, Frucht-Pery J, Domb AJ. Transcorneal and transscleral iontophoresis of dexamethasone phosphate using drug loaded hydrogel. *J Control Release.* 2005;106:386–390.

139. Hastings MS, Li SK, Miller DJ, Bernstein PS, Mufson D. Visulex: Advancing iontophoresis for effective noninvasive backto-the-eye therapeutics. *Drug Delivery Tech.* 2004;4:53–57.

140. Behar-Cohen FF, Halhal M, Benezra D, Chauvaud D, Renard G. Reversal of corneal graft rejection by iontophoresis of methylprednisolone. *Invest Ophthalmol Vis Sci.* 2002;43:U504

141. Chauvaud D, Behar-Cohen FF, Parel JM, Renard G. Transscleral iontophoresis of cortcoosteroids: Phase II clinical trial. *Invest Ophthalmol Vis Sci.* 2000;41:S79

142. Halhal M, Renard G, Bejjani RA, Behar-Cohen F. Corneal graft rejection and corticoid iontophoresis: Three case reports. *J Fr Ophthalmol.* 2003;26:391–395.

143. Halhal M, Renard G, Courtois Y, Benezra D, Behar-Cohen F. Iontophoresis: From the lab to the bed side. *Exp Eye Res.* 2004;78:751–757.

144. Pescina S, Ferrari G, Govoni P et al. *In-vitro* permeation of bevacizumab through human sclera: Effect of iontophoresis application. *J Pharm Pharmacol.* 2010;62(9):1189–1194.

145. Kim H, Robinson MR, Lizak MJ et al. Controlled drug release from an ocular implant: An evaluation using dynamic three-dimensional magnetic resonance imaging. *Invest Ophthalmol Vis Sci.* 2004;45(8):2722–2731.

146. Li SK, Molokhia SA, Jeong EK. Assessment of subconjunctival delivery with model ionic permeants and magnetic resonance imaging. *Pharm Res.* 2004;21(12):2175–2184.

147. Olsen TW, Feng X, Wabner K et al. Cannulation of the suprachoroidal space: A novel drug delivery methodology to the posterior segment. *Am J Ophthalmol.* 2006;142(5):777–787.

Section III

Ocular Compartment
Drug Delivery

8 Drug Delivery to the Vitreous Humor

Francine Behar-Cohen

CONTENTS

8.1 INTRODUCTION

To evaluate routes and drug delivery systems designed to target the posterior segment of the eye, pharmacokinetic studies often dose drugs into the vitreous humor. However, few therapeutic targets are located in the vitreous, but rather are located within the retina, retinal pigment epithelium, or choroid. Depending on the route of administration and delivery systems, efficient concentrations can be achieved and maintained retina without being reached in the vitreous (Behar-Cohen et al. 2002). On the other hand, drug concentrations in the vitreous do not always parallel those achieved in the outer retina or in the choroid. Indeed, the vitreous humor is one among other ocular exchange compartments in the complex and not fully understood pharmacokinetic dynamic of the posterior segment of the eye.

However, the vitreous, spherical transparent structure of about 4–4.5 mL, representing about 80% of the volume of the globe, easily accessible and observable, in direct contact with the retina, has appeared as an ideal site to deliver and release drugs, in solutions, suspensions, formulated in solid or semisolid polymers, in lipids or in nonbiodegradable implants and formulated in particulate drug delivery systems.

8.1.1   VITREOUS HUMOR: ITS STRUCTURE AND RELATIONS WITH THE RETINA

The vitreous humor is made of more than 90% of water containing collagen fibers and glycosaminoglycanes (GAG), mostly hyluronans. It is the collagen, though in a low concentration (about 300 µg/mL and less than 0.1% of the vitreous humor), that is responsible for the gel structure of the vitreous. Mostly heterotypic fibrils of collagen II (60–75%), IX (25%), and V/XI (10–25%) constitute the vitreous gel. Those fibrils interact with opticin, a leucin-rich glycoprotein, particularly at the inner limiting membrane (ILM) interface (Hindson et al. 2005). The vitreous gel is attached anteriorly to the ora serrata, ciliary epithelium, and zonular fibers, and posteriorly to all of the retina unless there is a posterior vitreous detachment (PVD). The vitreous cortex is made of densification of the collagen fibrils that directly insert in the ILM at the periphery but that are mostly parallel to the ILM at the posterior pole. In the posterior retina, it is through interactions of heparan sulfates contained in the ILM and opticin that the gel most probably adheres to the ILM (Bishop 2009). In human eyes, the vitreo retinal interface therefore varies in different regions of the retina and particularly at the posterior pole (Figure 8.1). Hyaluronans (90% of the GAG in the vitreous gel) are highly hydrated polyanions forming very long nonsulfated chains

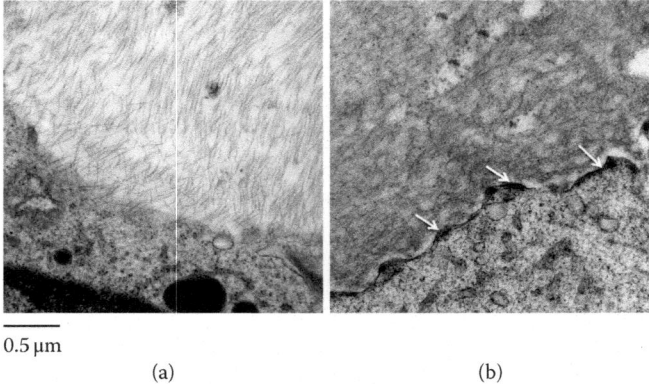

0.5 µm

(a)   (b)

FIGURE 8.1   Transmission electron microscopy images of the vitreo retinal interface in the anterior and posterior retina of the human eye. (a) The vitreous cortex fibrils in the peripheral anterior retina are inserted perpendicularly to the retinal surface and are directly connected to the cells. (b) In the posterior retina, the fibrils are parallel to the retinal surface and densifications (arrows) strengthen the vitreous cortex attachment.

(MW: 2–4 M). Other GAG are chondroitines sulfates, mostly collagen IX and versican (Bishop 2000). In humans, the posterior vitreous cortex is absent at the optic head and extremely thinned in front of the macula, due to the rarefaction of collagen fibrils, where it forms a premacular bursa as shown by Sebag and Balazs (1989). The diffusion and distribution of drugs from the vitreous toward the macula might therefore be different than in other part of the retina. Moreover, drugs may accumulate and concentrate in the premacular bursa.

With aging, the vitreous humor undergoes progressive syneresis and synchisis resulting from aggregation of collagen fibers and segregation of hyaluronans. Liquefaction is followed by collapse of the gel-phase, due in part to decreased hydration of the hyaluronans. The incidence of PVD increases with age, generating several compartments. These structural changes obviously influence the diffusion of drugs and reduce the concentration gradient across the medium. Thus, drug distribution and clearance in an eye with PVD are very different since vitreous barriers are altered (Laude et al. 2010).

8.1.2 Drug Delivery to the Vitreous Humor: From Animal Models to Clinical Applications

Laboratory animals including mice, rats, and rabbits are frequently used as models in pharmacokinetic studies. Much less frequent is the use of primates. Important interspecies differences make the direct translations of data to the human eye unreliable.

First, the vitreous volume varies significantly between the rat (50–55 µL), the rabbit (1.5 mL, about one-third the vitreous volume in man), the monkey (1.9 mL), and the human eye (4 mL). In addition, a difference in vitreous diffusional path length affects drug distribution. For example, the vitreous diffusional path length in the rat and rabbit is estimated to be 4.4 and 9 mm, respectively, compared to the distance in the human of 22 mm. The shorter vitreous diffusional path length in rabbits decreases the half-life by 1.7 times for posterior clearance and is twice as fast for anterior clearance as compared to humans (Maurice and Mishima 1984; reviewed by Laude et al. 2010). Another important interspecies difference is the relation of the vitreous with the posterior retina and the absence of macula in most animal models used for pharmacokinetic studies. There is no simple translation of the drug distribution pattern from the vitreous toward the peripheral retina and to the macula.

Animals used in pharmacokinetic studies are usually young animals with different vitreous structures as compared to the aging eye, which will increase potential differences with drug pharmacokinetics in human aging eyes. Active transretinal transport might be increased by the larger retinal surface and may be different in the macula as compared to other parts of the retina. Finally, many patients are vitrectomized and/or pseudophakic. Vitrectomy reduces the resistance to flow, increasing the role of connective flow on drug movements. The reduced viscosity of the medium in a vitrectomized eye may also increase the diffusion of drugs and the alterations of ocular barriers may increase its clearance, particularly through the anterior route. Vitrectomy, however, might not alter significantly retinal permeability, but this remains to be clarified.

In addition, pathologic states may affect active mechanisms such as efflux proteins or transcellular transport (i.e., diabetes) that intervene in the clearance and uptake of drugs from and toward the retina. Pharmacokinetic studies should therefore also be performed in disease animal models. Therefore, many individual factors in humans can be responsible for variations in drug diffusion and clearance after intravitreous administration that cannot be predicted from animal models.

Another difficulty in developing drug delivery systems into the vitreous is to select the optimal method of sampling the ocular media and tissue without contamination of tissues with the vitreous and allowing dynamic vitreous distribution analysis. The recently developed ocular magnetic resonance imaging analysis of *in vivo* pharmacokinetics are probably optimal and will provide further understanding of drug diffusion and elimination in different animal models (Eter 2010; Liu et al. 2010). Such understanding, particularly in primates, may help reduce the lack of prediction of preclinical pharmacokinetic animal studies.

8.1.3 Advantages of Drug Delivery to the Vitreous Humor

When trying to administer active compounds to target the retina, the simplest and most direct route is injection into the vitreous cavity. Because ocular barriers control exchanges between the eye and systemic circulation and because the total amount of active compound delivered into the vitreous is low compared with the whole body surface, the systemic exposure of intravitrally injected drug is limited. Intravitreous administration can thus be considered as a true "local administration," contrary to periocular injections that do not protect from systemic exposure. This is one of the major advantages of delivering drugs into the vitreous. The second main advantage is the "relative safety" of intravitreous injections (Jager 2004). Similarly, recent advances in 23G and 25 gauges vitreo retinal surgery have demonstrated the feasibility and safety of minimally invasive intravitreous administration procedures.

The eye being a confined organ and the vitreous a gel structure, one could have expected that the vitreous could act as a "reservoir" for drugs, particularly those of a higher molecular weight, since the ILM was thought to be a barrier to molecular higher than 40 KD. As shown by David Maurice using fluorimetry, there are two major routes of elimination from the vitreous (Maurice 1976): the active transretinal pathway (posterior route) and the aqueous humor pathway (anterior route). Interestingly, contrarily to old dogma, molecules as large as antibodies are eliminated through the posterior route and seem to "pass" the ILM. Indeed, it is now accepted that macromolecules and even nanoparticulate systems are able to cross the ILM (Figure 8.2) and be eliminated from the vitreous through the retinal Müller glial (RMG) cells. The rate of drug elimination from the vitreous mostly depends on diffusivity through the vitreous and on retinal permeability. Obviously, the structural state of the vitreous will influence the diffusion of drugs and the ocular barrier integrity will influence the retinal permeability. Vitreous half-lives can be as long as 4–5 days for the larger molecules (such as antibodies). High initial vitreous concentrations allow maintaining a therapeutic concentration up to 4–6 weeks, but initial toxic thresholds must be avoided.

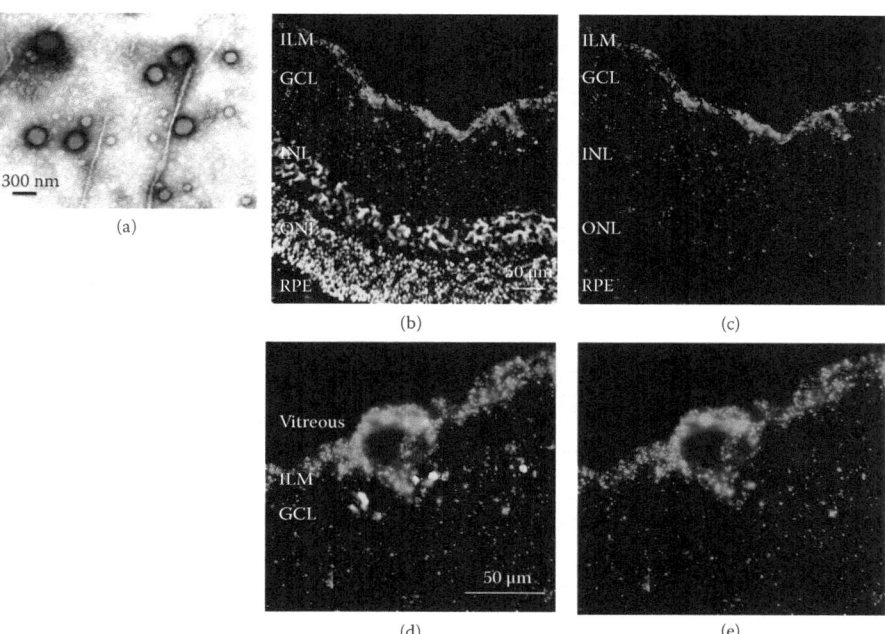

FIGURE 8.2 **(See color insert.)** Intraretinal penetration of a particulate system after intra-vitreous injection. Nanosized (100–300 nm) HMW PEI (polyethylenimine; 25 KD) and green fluorescent oligonucleotides complexes were injected in the vitreous of adult rats (a). Twenty-four hours after injection, the nanoparticles were dispersed homogenously on the retinal surface (b, c) and were concentrated in the ILM and around the vessels (d, e). Nanoparticles were identified in the retinal layers up to retinal pigment epithelial (RPE) cells (b, c). Oligonucleotides remain in the cytoplasm of cells as nanosized complexes as this time point and are not released in the nucleus.

Most attractive is the fact that the human vitreous represents about 80% of the eye volume (around 4 mL), offering a wide liquid empty space for the delivery of implants or devices with limited interference with vision. The clinical development of such polymeric implants has recently confirmed this potential (reviewed by London et al. 2011).

8.1.4 LIMITS OF DRUG DELIVERY TO THE VITREOUS HUMOR

Depending on the physicochemical properties of the drug and on the liquefaction state of the vitreous, the drug might not diffuse homogenously and high, even toxic concentrations could be reached focally. Once in the vitreous, active compounds may not diffuse passively into the retina and some retinal cells may not be targeted efficiently through intravitreous injections. For example, when injected into the vitreous, oligonucleotides do not reach photoreceptor cells (Figure 8.3).

Being anionic, polycationic formulations may form aggregates and induce an inflammatory reaction. Other high-molecular-weight positively charged molecules may be unstable in the vitreous, and protection against degradation may be required. Retinal inflammation and glial cells activation may result from any intervention into

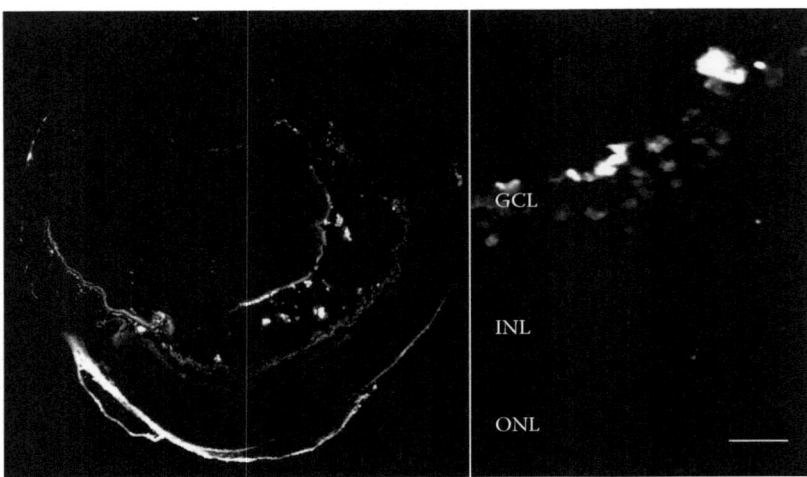

FIGURE 8.3 Penetration of naked 22MER fluorescent-labeled oligonucleotides after intra-vitreous injection. Twenty-four hours after intravitreous injection, oligonucleotides penetrate into ganglion cells and a few cells in the inner nuclear layer, but they do not penetrate into the outer nuclear layer.

the vitreous or with polymers or their bioproducts, which interfere with the drug activity per se. Finally, rarely, tractions induced by implants or injection procedures themselves may create retinal traction, tears, and retinal detachment.

8.2 DRUGS DELIVERY SYSTEMS INTO THE VITREOUS

8.2.1 GENERAL PRINCIPLES

When choosing the vitreous as the site of drug delivery, several questions must be asked:

- What can be administered safely into to the vitreous? What should be forbidden?
- How will the drug and/or the excipients or polymers react with the vitreous?
- What is the targeted tissue, cell, and even molecule?
- Which formulation should be used for which duration of release?
- Where to inject or implant? Does this influence drug concentration at the targeted site?
- What are the risks of retinal lesions? Visual disturbance?

8.2.1.1 General Rules for Intravitreous Drug Release

Any drug, formulation, implant, or device injected or inserted into the vitreous humor should try to limit changes in its chemistry. In physiologic conditions, the human vitreous osmolality is 290 ± 5 mOsm/kg, sodium is 136–145 mmol/L, potassium is 9.5–12 mmol/L, chloride is 105–120 mmol/L, sodium 135 mmol/L, and soluble protein concentration in the vitreous is around 10 mg/mL, which represent almost 15% of the plasmatic protein concentration. The vitreous humor is therefore low in protein and high in potassium as compared to plasma. Any changes in proteins or in ion composition in the vitreous may interfere with hydroionic and proteins active transports in the retina. Two consequences: try and formulate any drugs in a physiologic vitreous milieu and when evaluating the effect of a drug administered or released in the vitreous, select the appropriate control. For example, proteins should be compared to denatured proteins and not to the vehicle alone, because oncotic pressure modification can influence pharmacological responses.

Of course, any toxic agents should be avoided in the vitreous, and nonbiodegradable compounds should be used with caution. To solubilize hydrophobic compounds, benzyl alcohol, ethanol, cremophor (ethylene oxide and castor oil), polysorbate 80, or other nonionic surfactants are commonly used in the pharmaceutical industry and for injectable preparations. These excipients must be used with extreme caution since they may be toxic, induce side effects, and have long-term consequences. Particularly, polyethylene glycol (PEG), which is a polyether compound, is often used to prolong the half-life of a systemically injected drug. Peggylation is often used to modify the surface of particles to avoid opsonization and recognition by the host phagocytes (Tabata YandIkada 1988; de Kozak et al. 2004). Such PEGs should be used with caution when injected in the vitreous because they are not biodegradable in a meaningful timescale. When injected systemically, PEG can be eliminated by the kidney if its molecular weight is less than 10,000 D, but after intravitreous injection, it may accumulate in retinal pigment epithelial (RPE) cells and induce toxicity. The fate of any polymer or its bioproduct should be followed to ensure no potential harmful accumulation in RPE or RMG cells.

Upon degradation, PLA and PLGA release lactic acid that may lower the environmental pH. Rates of degradation and pH changes must be controlled in the vitreous cavity to avoid inflammatory and toxic reactions.

Once in the vitreous, drugs and bioproducts used as therapeutic agents (particularly therapeutic proteins), particulate systems, and emulsions may react with the vitreous gel. Aggregation may modify the diffusivity of the molecules and can induce macrophagic reactions. Such an inflammatory cell reaction is observed with cationic lipids, polymers, or any other cationic molecules. Anionic charged compounds should be preferred to minimize interactions with the vitreous and inflammatory cell reactions. Importantly, aggregation may change the biodisponibility of the compound, playing the role of a slow release system or reducing/suppressing biologic effects.

8.2.1.2 Tissue and Cell Targets: the Potential of Particulate Systems

When designing a drug delivery system, it is important to consider what is the target tissue (neuroretina/RPE/choroid/ciliary body) and the target cell (phtoreceptors/ganglion cells/microglia) and what is the molecular target, particularly whether it is extracellular or intracellular, and if intracellular, whether it is intranuclear. The size and physicochemical properties of a molecule or a drug delivery system change its distribution and targeting. Indeed, a molecule's path of distribution and elimination in the vitreous largely depends on its physicochemical properties and substrate affinity via the retina. Lipophilic compounds such as corticosteroids or particulate systems less than 300 nm and high-molecular-weight proteins that are transported through retinal Müller cells tend to migrate and be eliminated mostly through a transretinal pathway. In particular, the retinal penetration of anti-VEGF antibodies has been studied on animal models, which showed that anti-VEGF full antibodies penetrate throughout the retina and up to the choroid through RMG cell transport and that PVD favored the speed of retinal penetration (Shahar et al. 2006; Dib et al. 2008; Goldenberg et al. 2011). On the other hand, more hydrophilic molecules with low retinal permeability migrate primarily through the anterior hyaloids toward the aqueous humor and will not target retinal cells efficiently.

Particulate drug delivery systems include nano- and microparticles, nano- and microspheres, and nano- and microcapsules. Their size and polymer composition influence markedly their biological behavior *in vivo*. Microparticles act like reservoirs after intravitreous injection and poorly diffuse in the vitreous gel (Ogura and Kimura 1995; Veloso et al. 1997) (Figure 8.4). PLA microspheres were found in the vitreous for 1.5 months in normal rabbit eyes and for 2 weeks in eyes after vitrectomy (Moritera et al. 1992). Nanoparticles, on the other hand, can diffuse in the vitreous and are internalized in the ocular tissues and cells of the anterior and posterior segment (Sakurai et al. 2001; Bourges et al. 2006; Bejjani et al. 2005) (Figure 8.4). The nature of the nanoparticles also influences their affinity for specific cells. We found that nanoparticles made of cationic peptides had increased penetration in RMG cells as compared to PLA/PLGA particles. Specific coating still remains difficult to modelize for this

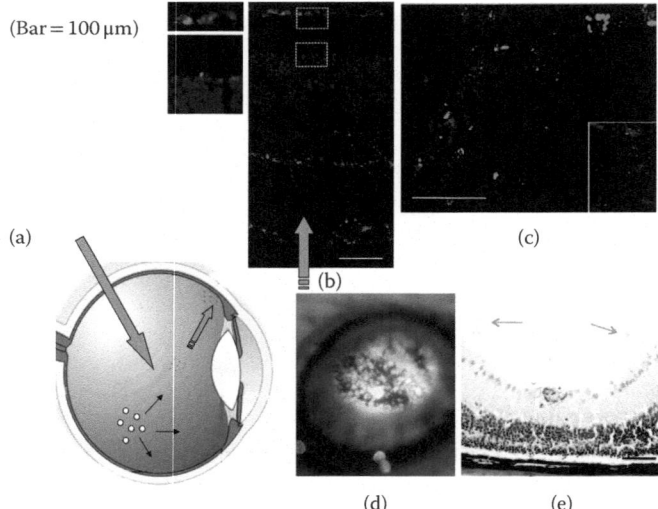

(Bar = 100 μm)

(a)   (b)   (c)   (d)   (e)

FIGURE 8.4 (See color insert.) The fate of micro- and nanoparticles after intravitreous injection in the rat eye. (a) After intravitreous injection, microparticles remain in the vitreous cavity and act as drug reservoir. (b, c) Twenty-four hours after intravitreous injection, 100–300 nm red fluorescent nanoparticles penetrate in astrocytes and RMG cells (b) and in RPE cells (b, inset). Nanoparticles are released from RMG cells at the OLM from apices (b, inset). Nanoparticles are also internalized in the epithelial cells of the ciliary body (c). (d, e) Twenty-four hours after intravitreous injection, polymeric PLGA microparticles remain in the vitreous (d, f arrows). Bar: 100 μm.

purpose because the exact mechanisms of transcellular passage of molecules from the vitreous through RMG cells and release at the RMG apices are poorly understood. Specific and active mechanisms may be involved that remain to be elucidated.

Therapeutic targets for drugs can be located extra- or intracellularly. NPs and liposomes enable the loaded bioactive molecule to cross cell membranes and epithelial barriers by using different internalization pathways. This is one of the most interesting aspects of this technology. The physicochemical characteristics of nanocarriers, such as size, shape, surface charge, surface coating, and surface functionalization with targeting ligands, along with the target cell type, determine the internalization pathway and the intracellular fate. NPs can be targeted to interact with cell receptors of interest to facilitate internalization by receptor-mediated endocytic mechanisms. PLA/PLGA nanoparticles, depending on their size, were found to target RMG and RPE cell (Figure 8.5). RPE cells have the capacity to take up different kinds of NPs (Bourges et al. 2003; Bejjani et al. 2005; Normand et al. 2005). *In vitro*, a high concentration of PLA/PLGA nanoparticles did not affect cell viability in the short term. After intravitreous injection, we found that nanoparticles (<300 nm and slightly anionic) diffuse slowly into the rat vitreous and spread at the retinal surface. After 24 hours, fluorochrome-loaded nanoparticles followed a transretinal pathway through RMG cells and were internalized in RPE cells, where they accumulated and released the fluorochrome for several months (Bourges et al. 2003). This raises the question of the potential toxicity of poorly degraded polymers in RPE cells and the risk of altering the normal phagocytic activity of those cells in the long-term. This remain to be studied.

Liposomes are vesicular lipid systems of a diameter ranging between 50 nm to a few mm. They allow the encapsulation of a wide variety of drug molecules such as proteins, nucleotides, and even plasmids (Ebrahim et al. 2005). Liposomes are biocompatible and biodegradable and are composed of lipids similar to those present in biological membranes. Their membranes are stable and may undergo severe deformation without disruption. Using liposomes, hydrophilic, lipophilic, or amphiphilic

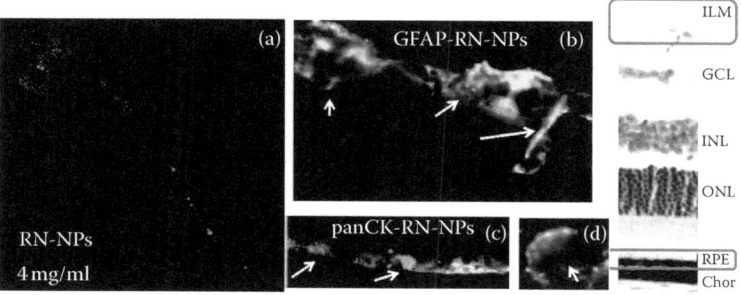

FIGURE 8.5   (See color insert.) RPE and RMG targeting. PLA Nile red nanoparticles (NR-NPs, 200 ± 30 nm, zeta potential −5 ± 0.4 mV) were evaluated *in vitro* on RPE and RMG cells and *in vivo* in the rat eye. (a) *In vitro*, NR-NPs are internalized in RPE and RMG cells and are non-toxic in the short term even at high concentration (4 mg/mL). *In vivo*, 24 hours after intravitreous injection, NR-NPs penetrate into glial GFAP-labeled cells (b, arrow) and in RPE cells (c, arrows). Flat mounting confirmed internalization of NPs in RPE cells (d, arrow).

drugs can be encapsulated, providing a convenient way of obtaining slow drug release from a relatively inert depot without changing the intrinsic characteristics of the encapsulated agents. Depending on the liposomes' size, they can remain in the vitreous and act as a reservoir of drugs for a short period of time or be used to target glial Müller cells.

Another interesting application of nanoparticles and liposomes, injected in the vitreous, is their internalization by activated microglia and resident macrophages and by infiltrating cells, allowing efficient targeting of inflammatory cells (Figure 8.6). De Kozak's group recently demonstrated that the vasoactive intestinal peptide (VIP, 3.3 kDa) was efficient to reduce ocular inflammation and retinal lesions in two experimental models of uveitis: endotoxin-induced uveitis (Lajavardi et al. 2007) and experimental autoimmune uveoretinitis (Camelo et al. 2009). This effect of VIP was detected only when VIP was formulated in liposomes (300–600 nm). VIP, a 28-aa immunosuppressive neuropeptide, is unstable in ocular media and cannot exert its immunosuppressive effect but when formulated in liposomes, it not only remains longer in the eye but also more specifically targets inflammatory cells that naturally engulf liposomes. It has been shown that 24 hours following IVT injection in a normal Lewis rat, liposomes loaded or not with VIP are detected mainly in the posterior segment of the eye, especially in RMG cells and in ocular tissue resident macrophages (Camelo et al. 2007). Particularly in the inflamed eye, liposomes leak from ocular tissues to regional and inguinal lymph nodes and spleen (Camelo

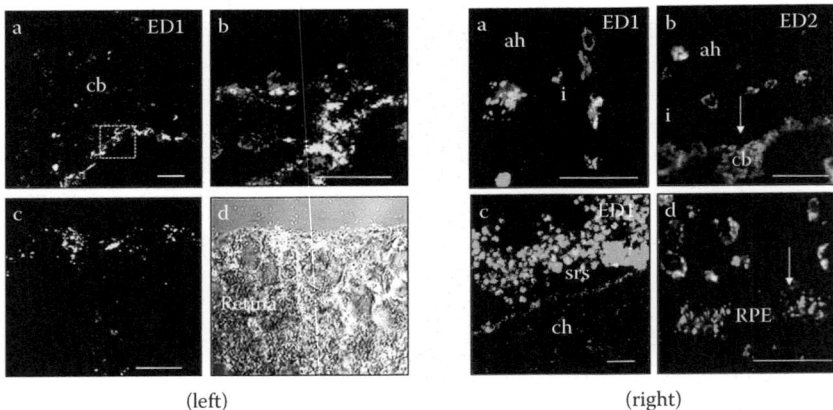

(left)   (right)

FIGURE 8.6   (See color insert.) The fate of nanoparticles in inflamed eyes. The ocular distribution of fluorescent nanoparticles (150 nm, poly(ethylene glycol)-grafted to polyhexadecylcyanoacrylate (PEG-PHDCA)) was followed during EAU in rats. Particles injected before the clinical onset of uveitis (left) are internalized in the epithelial cells of the ciliary body (a, cb) and in few ED1 positive cells within the ciliary body (b, high magnification). They penetrate in RMG cells in the retina (c and phase contrast image in d). When particles were injected at the time of clinical uveitis onset (right), they are internalized mostly in ED1 positive cells in the aqueous humor (ah) and in epithelial cells of the ciliary body (b, cb). In the retina (c), they accumulate massively in the subretinal space (srs) but not in the choroids (ch) and are internalized in ED1 positive cells and in RPE cells (d). Bar = 50 μm. (Images provided by Dr. Yvonne de Kozak.)

et al. 2007, 2008, 2009; Lajavardi et al. 2007). At the systemic level, the treatment diminished the specific delayed-type hypersensibility and the activation of lymphocytes in the lymph nodes, draining the site of immunization. Thus, VIP-loaded liposomes intravitreal treatment modulated both macrophage and T-cell mediated responses in the eye as well as at the systemic level. In the case of inflammation and particularly if the particles are engulfed by inflammatory cells, one must expect the administered compounds to have not only a local but also a locoregional and eventually a systemic effect.

The potential of intracellular delivery has encouraged the use of nanoparticulate systems for nonviral gene delivery, protecting nucleic acids from degradation in the vitreous and enhancing their intracellular delivery. We found that nanoparticles encapsulating a plasmid encoding red nuclear fluorescent protein were localized in the RPE cells 24 hours after intravitreal injection in rats. Effective plasmid expression was achieved after 45 days of injection and expression-associated red fluorescence remained detectable in RPE cells during the following 3 weeks, with no apparent tissue damage or toxicity (Bejjani et al. 2005; Andrieu-Soler et al. 2006). In our experiments, the DNA loading capacity of the nanoparticles was quite low, explaining the limited potential of transfection. Indeed, because only a small fraction of plasmid DNA contained in the nanoparticles efficiently reaches the nucleus without being degraded, high DNA loading without increasing the particle size is required. In this regard, another group has made nanoparticles with plasmid DNA condensed with polycationic polymers compacted in PEG-substituted lysine peptides (Farjo et al. 2006). Efficient expression of a fluorescent reporter gene was achieved in choroid, RPE, and other retinal cells at 24 hours after injection of such particles in mice vitreous. Further improvements of DNA compaction permitted transduction of not only RPE cells but also photoreceptor cells without a toxic or inflammatory reaction, demonstrating the potential for application of nanoparticle-based gene replacement therapy for treatment of human retinal degenerations (Cai et al. 2010). This is obviously one of the most interesting applications of nanoparticles, since even viral vectors do not target photoreceptors efficiently when injected into the vitreous. The question of stability of expression using plasmid DNA remains an open question in photoreceptor cells. However, reinjection in the vitreous can be envisaged two or three time a year without major damage or burden.

Other pharmacologic molecules specifically designed to interfere with photoreceptors' metabolism or functions could be vectorized by specifically designed nanoparticulate systems.

To further control the site and timing of the release of genetic material in retinal cells, we have used nanoparticles made with VP22, a cationic peptide which can be complexed with labeled ODNs, leading to the formation of "vectosomes" (Zavaglia et al. 2003). When injected in the vitreous of rat eyes, vectosomes follow a transretinal migration and accumulate at the external limiting membrane and in the cytoplasm of RPE cells (Normand et al. 2005) (Figure 8.7a). After their internalization by the RPE cells, the vectosomes remain stable within the cytoplasm. Interestingly, VP22 vectosomes demonstrated light-dependent sensitivity. On illumination (white light or laser beam), the internalized vectosomes are destabilized and dissociation between the VP22 and the ODN occurs. The released ODN then migrates from the cytoplasm to the cell nucleus, expressing its genetic load. Fluorescent ODN distribution after

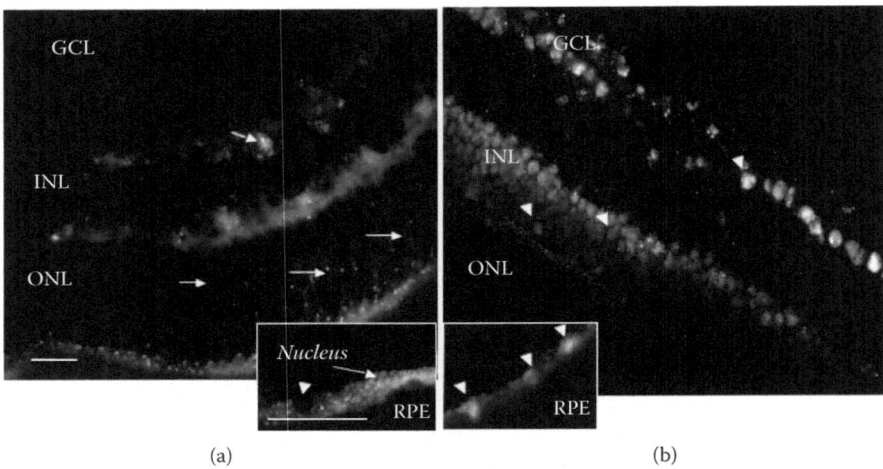

(a)   (b)

FIGURE 8.7 Photosensitive VP22 vectosomes. Vectosomes were injected in the vitreous of a rat. After injection (a), they accumulate in RMG and RPE cells and in a few cells in the inner nuclear layer, where they remain stable in the cytoplasma for at least 3 weeks. After illumination (b), vectosomes dissociate, releasing fluorescent oligonucleotides that are internalized in ganglion cells, cells of the inner nuclear layer, and in RPE cells. Bar = 50 μm.

illumination was mainly found in the nuclei of RGCs, the inner nuclear layer, and in the nuclei of RPE (Figure 8.7b). This strategy is appealing for the potential treatment of eye diseases as it can allow for a controlled and localized release of ODNs using a well-focused laser beam (reviewed by Andrieu-Soler et al. 2006).

Other cationic peptides conjugated with PEG and forming nanoparticles have been used to enhance plasmid DNA delivery to RPE cells and applied to expression of GDNF in light-induced retinal degeneration mice models, but transfection of RPE cells was shown to be effective after subretinal injection and not after intravitreous. Efficiency after intravitreous injection remains to be evaluated (Read et al. 2010).

8.2.1.3 Modalities and Kinetics or Intravitreous Administration of Particulate Systems

One of the obvious advantages of particulate systems is that they are injectable through 23–30G needles, limiting the injection trauma. Where and how to administer particulate systems in the vitreous or vitreous cavity has not been fully explored. The retinal distribution and the vitreous diffusion of micro- and nanoparticles may be influenced by their site of administration. Specific injection systems should therefore be developed to control the depth and orientation of the needle in order to place the particle in a reproducible site within the vitreous. Injection in the temporal inferior quadrant should be preferred to limit visual disturbance and favor macular targeting.

As previously mentioned, anionic particles diffuse more easily through the three-dimensional vitreal network of collagen fibrils than through a cationic one (Kim et al. 2009), but other parameters such as the composition of the particles also influence their diffusion. Vitroctomy, aging, and pathologic conditions are also important factors to consider.

One drawback of particulate systems is their limited half-life in the vitreous. Microparticles of PLA/PLGA last about 1.5–2 months (Moritera et al. 1992), and PLGA nanoparticles efficiently release drugs for about 1 month. Recently, PLGA nanoparticles loaded with dexamethasone were injected in rabbit vitreous. They maintained a sustained release of DEX for about 50 days in the vitreous and provided relatively constant DEX levels for more than 30 days with a mean concentration of 3.85 mg/L^{-1} (Zhang et al. 2009).

In order to prolong the half-life of microparticles, their size can be increased; in this case they could be injected through 23G vitroctomy guides at the end of a surgery or with specific injectors. When using PLA/PLGA particles, release of the drug mainly occurs through diffusion and empty polymers remain in the vitreous cavity for various periods of time. Such polymers degrade in acid lactic that may change the pH at the retinal surface and create a glial reaction. Other types of polymers such as poly(orthoesters) or PCL should be preferred for their longer half-life and the erosion process that occurs for the release of drugs.

In any case, in animal studies evaluating the effect of drugs released from polymeric particulate systems, the control animals must receive blank particles that always have biologic effects per se.

To prolong the half-life of nanoparticles or liposomes in the vitreous and prolong the duration of drug action, combined systems can be used to incorporate liposomes within hyaluronic acid (HA) gel (Lajavardi et al. 2009). We showed recently that gel–VIP–liposomes reduced LPS-induced ocular inflammation even when intravitreally injected 7 days before uveitis induction. The high viscosity and interactions between liposomes and HA gel allowed a sustained release of VIP in ocular tissues and in regional lymph nodes. The incorporation of liposomes in hydrogels delayed the release of VIP, increased the time of presence of liposomes in the eye, and could possibly increase the local versus systemic effect of the treatment.

Alternatively, nanosized particles used for cell targeting and gene delivery can be incorporated in larger microspheres. Nanosized complexes of antisense phosphorothioate oligonucleotides with polyethylenimine (PEI) were encapsulated into poly(lactide-*co*-glycolide) microspheres. Playing with the microspheres' composition controls the porosity, the faster, and the release of nanosized complexes (Gomes dos Santos et al. 2006) (Figure 8.8).

The last but not least important parameters to take into account and to control are the stability of the active compound within the particulate system, particularly proteic compounds and the maintenance of their full activity during the preparation and sterilization processes. The latter sometimes limits progress toward clinical applications.

8.2.2 Applications of Particulate Drug Delivery Systems to Eye Diseases

8.2.2.1 Neurotrophic Factors for Retinal Degenerative Diseases

The neurotrophic effect of recombinant human GDNF (rhGDNF) encapsulated in PLGA microspheres was evaluated in a *rd1/rd1* mouse model (Andrieu-Soler et al. 2005). *In vivo*, rhGDNF microspheres of 27 μm radius, releasing 10 ng/day for at least 3 months, were compared to blank microspheres or microspheres loaded with

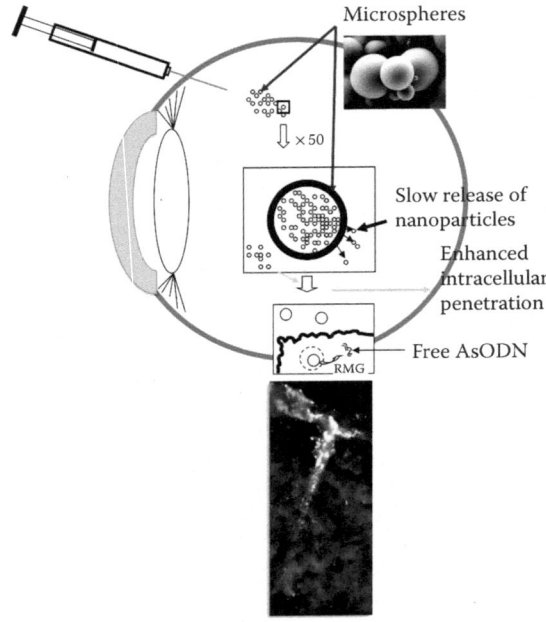

FIGURE 8.8 Microparticles releasing nanoparticles. A slow release of polymeric nanoparticles designed to target RMG and/or RPE cells can be induced by encapsulation in larger polymeric microparticles introduced into the vitreous.

inactivated rhGDNF. Particles were injected once in the vitreous of PN11 mice. At PN28, the eyes receiving the rhGDNF microspheres showed significant rhodopsin positive signals, a higher number of cell nuclei within the ONL, and a significant b-wave increase. No significant intraocular inflammatory reaction was observed after the intravitreous injection of the various microspheres. However, the animals injected with blank PLGA microspheres also showed a significant photoreceptor rescue and glial reaction (Andrieu-Soler et al. 2005). The neuroprotective effect of GDNF-loaded biodegradable microspheres has been demonstrated in animal models of glaucoma (Jiang et al. 2007). Long-term RGC survival in DBA/2J mice, a spontaneous glaucoma model, was also significantly increased with GDNF-loaded biodegradable microspheres. These delivery systems allowed a cumulative GDNF release of 35.4 ng/mg over 71 days, with a nonlinear kinetics, suggesting that burst release may play a role in this rescue (Ward et al. 2007).

A small N-terminal PEDF fragment, $PEDF_{82-121}$, mimicking the PEDF effect on cell survival, protected the retina from ischemic injury when delivered in PLGA nanospheres. Seven days after ischemia reperfusion insult, PLGA- $PEDF_{82-121}$ was more effective at preserving cells in the RGC layer and the IPL thickness than the free peptide. The lower efficacy of the free peptide was probably due to more rapid degradation and faster clearance from the vitreous (Li et al. 2006).

FGF-2 encapsulated in nanoparticles has also been evaluated in RCS rats using NPs (Sakai et al. 2007). Until 8 weeks after administration, NPs were present in the INL, ONL, outer segment debris zone, and RPE, and in the numerous resident microglial/macrophages of these layers. Compared with eyes treated with a single injection of recombinant FGF-2, the treatment significantly increased FGF-2 levels in the retina and the number of protected nuclei in the ONL for at least 6 weeks.

These experiments tend to show that a sustained release of therapeutic proteins from particulate systems into the vitreous may have superior effects than repeated bolus injections.

A novel approach for scavenging reactive oxygen species prominent in retinal degenerative diseases was presented by Chen et al. (2006, 2008) and reviewed by Edelhauser et al. (2008). Cerium oxide nanoparticles (CeO_2, nanoceria particles), which are nontoxic, nonimmunogenic, and protective at a very low dosage, provided protection *in vivo* using a light-damage animal model. In this case, these rare earth particulates are not carriers of a specific drug but are the therapeutic agent itself.

8.2.2.2 Treatment of Infectious Retinitis

At the time when CMV retinitis was one of the major complication of HIV infection, many attempts were made to reduce the need for intravitreous frequent injections of ganciclovir. In a CMV rabbit model, PLGA microspheres loaded with 10 mg ganciclovir were still detected 8 weeks after a single injection (Veloso et al. 1997). Negatively charged albumin nanoparticles loaded with ganciclovir injected in normal eyes showed that 40% of the loaded drug was released within 1 hour, with a slower release rate during the following 10 days (Merodio et al. 2002).

In humans, liposome-encapsulated ganciclovir was injected in one eye and the other eye served as a control, receiving intravitreal-free ganciclovir. The liposome formulations spread diffusely within the vitreous cavity and caused cloudiness, interfering with the patient's visual acuity and the ability of the ophthalmologist to examine the fundus until complete resorption of the formulation occurred 14–21 days after administration. Despite this drawback, week examination showed neither progression of the CMV retinitis nor new lesions in the eye treated with liposomes, whereas the eye injected with the solution of ganciclovir showed reactivation of old CMV retinitis. Liposome-encapsulated ganciclovir also reduced the number of intravitreal injections (Akula et al. 1994).

8.2.2.3 Treatment of Proliferative Vitreo Retinopathy

To reduce proliferative vitreo retinopathy (PVR) in rabbit eyes, PLA or PLGA microspheres loaded with 5-fluorouracil, adriamycin, or retinoic acid were used (Moritera et al. 1991, 1992; Giordano et al. 1993; Peyman et al. 1992). PLA microspheres containing 10 µg adriamycin significantly reduced the rate of PVR formation. PLGA microspheres releasing retinoic acid remained in the vitreous for 40 days and also reduced the extent of PVR in treated rabbit eyes. In these experiments, the microparticles mostly settled on the inferior quadrants of the retina and induced a localized multinuclear giant cell reaction. More recently, PEI/oligonucleotide complexes forming nanoparticles were injected in the rat vitreous to evaluate the possibility of inhibiting TGFb2 expression. Interestingly, these complexes localized in RMG cells (Gomes dos Santos, Bochot, Tsapsis et al. 2006).

8.2.2.4   Treatment of Intraocular Inflammation

As previously mentioned, particulate systems are ideal for delivering anti-inflammatory agents to resident macrophages or infiltrating cells because those cells naturally have a high phagocytic activity and accumulate the particles intracellularly at a high rate. Moreover, due to tissue barrier breakdown, particles penetrate quickly in the subretinal space and are undertaken by activated microglia. On the other hand, no local treatment can be expected due to ocular barrier breakdown and to the circulation of inflammatory cells containing particles in the cervical draining nodes and in the spleen.

De Kozak et al. have evaluated the efficacy of tamoxifen, a nonsteroidal estrogen-receptor modulator, in PEG-coated nanoparticles for the treatment of experimental autoimmune uveoretinitis. Intravitreal injection in a rat model performed 1–2 days before expected disease onset in controls significantly inhibited the disease owing to a shift in the immune response from a Th1- to a Th2-type response (de Kozak et al. 2004). The more classic immunosuppressive and hydrophobic drug, cyclosporine, was formulated in PLGA microspheres, 50 μm in diameter, for the treatment of uveitis (He et al. 2006) but the cyclosporine release monitored following intravitreal injections was performed in healthy rabbits with a maintenance of therapeutic concentrations for at least 65 days in the choroid–retina and iris–ciliary body.

Dexamethasone PLGA microspheres (20–53 μm in diameter) were used to reduce ocular inflammation in a rabbit model of uveitis elicited by intravitreal lipopolysaccharide injection. Both the short-term (15 days in length) and long-term study (33 days in length) demonstrated reduced inflammation by clinical evaluation, electroretinography, and histopathologic evaluation (Barcia et al. 2009).

8.3   FUTURE DIRECTIONS TOWARD CLINICAL APPLICATIONS

Many studies have been performed by different groups; they can hardly be compared because particle composition and physical properties and experimental methods vary. However, they have permitted us to break some old dogma and helped us to understand the dynamic aspect of transretinal migration processes. There is no doubt that nanoparticulate drug delivery systems, particularly functionalized ones, will be used to deliver drugs to specific retinal cells in the future. The larger-size polymeric particles could be used as injectable sustained drug release systems in the vitreous. But other long-lasting polymers should be used to design those particles. There is still a long way to go to gain more knowledge in this field in order to build guidelines for designing particulate systems adapted for specific durations of drug release, specific vitreous distributions and retinal cell targeting, depending on the particle size and composition, charge, structure, and surface modifications. More complex systems of release of nanoparticles could also be designed to combine sustained release in the vitreous and intraretinal cell targeting.

REFERENCES

Akula SK, Ma PE, Peyman GA, Rahimy MH, Hyslop NE Jr, Janney A, Ashton P. Treatment of cytomegalovirus retinitis with intravitreal injection of liposome encapsulated ganciclovir in a patient with AIDS. *Br J Ophthalmol.* 1994; 78: 677–680.
Andrieu-Soler C, Aubert-Pouëssel A, Doat M, Picaud S, Halhal M, Simonutti M, Venier-Julienne MC, Benoit JP, Behar-Cohen F. Intravitreous injection of PLGA microspheres

encapsulating GDNF promotes the survival of photoreceptors in the rd1/rd1 mouse. *Mol Vis*. 2005; 17 (11): 1002–1011.

Andrieu-Soler C, Bejjani RA, de Bizemont T, Normand N, BenEzra D, Behar-Cohen F. Ocular gene therapy: A review of nonviral strategies. *Mol Vis*. 2006; 30 (12): 1334–1347. Review.

Barcia E, Herrero-Vanrell R, Diez A, Alvarez-Santiago C, Lopez I, Calonge M. Downregulation of endotoxin-induced uveitis by intravitreal injection of polylactic-glycolic acid (PLGA) microspheres loaded with dexamethasone. *Exp Eye Res*. 2009; 89: 238–245.

Behar-Cohen FF, El Aouni A, Gautier S, David G, Davis J, Chapon P, Parel JM. Transscleral Coulomb-controlled iontophoresis of methylprednisolone into the rabbit eye: Influence of duration of treatment, current intensity and drug concentration on ocular tissue and fluid levels. *Exp Eye Res*. 2002; 74 (1): 51–59.

Bejjani RA, BenEzra D, Cohen H, Rieger J, Andrieu C, Jeaanny J-C, Golomb G, Behar-Cohen FF. Nanoparticles for gene delivery to retinal pigment epithelial cells. *Mol Vis*. 2005; 11: 124–132.

Bishop PN. Structural macromolecules and supramolecular organisation of the vitreous gel. *Prog Retin Eye Res*. 2000; 19 (3): 323–344.

Bishop PN. Vitreous as a substrate for vitreolysis. *Dev Ophthalmol*. 2009; 44: 7–19.

Bourges JL, Bloquel C, Thomas A, Froussart F, Bochot A, Azan F, Gurny R, BenEzra D, Behar-Cohen F. Intraocular implants for extended drug delivery: Therapeutic applications. *Adv Drug Deliv Rev*. 2006; 58 (11): 1182–1202.

Bourges JL, Gautier SE, Delie F, Bejjani RA, Jeanny JC, Gurny R, BenEzra D, Behar-Cohen FF. Ocular drug delivery targeting the retina and retinal pigment epithelium using polylactide nanoparticles. *Invest Ophthalmol Vis Sci*. 2003; 44: 3562–3569.

Cai X, Conley SM, Nash Z, Fliesler SJ, Cooper MJ, Naash MI. Gene delivery to mitotic and postmitotic photoreceptors via compacted DNA nanoparticles results in improved phenotype in a mouse model of retinitis pigmentosa. *FASEB J*. 2010 Apr; 24 (4): 1178–1191.

Camelo S, Lajavardi L, Bochot A, Goldenberg B, Naud MC, Brunel N, Lescure B et al. Protective effect of intravitreal injection of vasoactive intestinal peptide-loaded liposomes on experimental autoimmune uveoretinitis. *J Ocul Pharmacol Ther*. 2009 Feb; 25 (1): 9–21.

Camelo S, Lajavardi L, Bochot A, Goldenberg B, Naud MC, Fattal E, Behar-Cohen F, de Kozak Y. Ocular and systemic bio-distribution of rhodamine-conjugated liposomes loaded with VIP injected into the vitreous of Lewis rats. *Mol Vis*. 2007 Dec 7; 13: 2263–2274.

Camelo S, Lajavardi L, Bochot A, Goldenberg B, Naud MC, Fattal E, Behar-Cohen F, de Kozak Y. Drainage of fluorescent liposomes from the vitreous to cervical lymph nodes via conjunctival lymphatics. *Ophthalmic Res*. 2008; 40 (3–4): 145–50. Epub 2008 Apr 18.

Chen J, Patil S, Seal S, McGinnis JF. Rare earth nanoparticles prevent retinal degeneration induced by intracellular peroxides. *Nat Nanotechnol*. 2006; 1: 142–150.

Chen J, Patil S, Seal S, McGinnis JF. Nanoceria particles prevent ROI-induced blindness. *Adv Exp Med Biol*. 2008; 613: 53–55.

de Kozak Y, Andrieux K, Villarroya H, Klein C, Thillaye-Goldenberg B, Naud MC, Garcia E, Couvreur P. Intraocular injection of tamoxifen-loaded nanoparticles: A new treatment of experimental autoimmune uveoretinitis. *Eur J Immunol*. 2004; 34: 3702–3712.

Dib E, Maia M, Longo-Maugeri IM, Martins MC, Mussalem JS, Squaiella CC, Penha FM, Magalhães O Jr, Rodrigues EB, Farah ME. Subretinal bevacizumab detection after intravitreous injection in rabbits. *Invest Ophthalmol Vis Sci*. 2008 Mar; 49 (3): 1097–1100.

Ebrahim S, Peyman GA, Lee PJ. Applications of liposomes in ophthalmology. *Surv Ophthalmol*. 2005; 50: 167–182.

Edelhauser HF, Boatright JH, Nickerson JM. Drug delivery to posterior intraocular tissues: Third Annual ARVO/Pfizer Ophthalmics Research Institute Conference. *Invest Ophthalmol Vis Sci*. 2008; 49: 4712–4720.

Eter N. Molecular imaging in the eye. *Br J Ophthalmol.* 2010 Nov; 94 (11): 1420–1426.

Farjo R, Skaggs J, Quiambao AB, Cooper MJ, Naash MI. Efficient nonviral ocular gene transfer with compacted DNA nanoparticles. *PLoS ONE* 2006; 1 e38.

Giordano GG, Refojo MF, Arroyo MH. Sustained delivery of retinoic acid from microspheres of biodegradable polymer in PVR. *Invest Ophthalmol Vis Sci.* 1993; 34: 2743–2751.

Goldenberg DT, Giblin FJ, Cheng M, Chintala SK, Trese MT, Drenser KA, Ruby AJ. Posterior vitreous detachment with microplasmin alters the retinal penetration of intravitreal bevacizumab (Avastin) in rabbit eyes. *Retina.* 2011 Feb; 31 (2): 393–400.

Gomes dos Santos AL, Bochot A, Doyle A, Tsapis N, Siepmann J, Siepmann F, Schmaler J, Besnard M, Behar-Cohen F, Fattal E. Sustained release of nanosized complexes of polyethylenimine and anti-TGF-beta 2 oligonucleotide improves the outcome of glaucoma surgery. *J Control Release.* 2006 May 30; 112 (3): 369–381. Epub 2006 Mar 6.

Gomes dos Santos AL, Bochot A, Tsapis N, Artzner F, Bejjani RA, Thillaye-Goldenberg B, de Kozak Y, Fattal E, Behar-Cohen F. Oligonucleotide-polyethylenimine complexes targeting retinal cells: Structural analysis and application to anti-TGFbeta-2 therapy. *Pharm Res.* 2006 Apr; 23 (4): 770–781.

He Y, Liu Y, Liu Y, Wang J, Zhang X, Lu W, Ma Z, Zhu X, Zhang Q. Cyclosporine-loaded microspheres for treatment of uveitis: *In vitro* characterization and *in vivo* pharmacokinetic study. *Invest Ophthalmol Vis Sci.* 2006; 47: 3983–3988.

Hindson VJ, Gallagher JT, Halfter W, Bishop PN. Opticin binds to heparan and chondroitin sulfate proteoglycans. *Invest Ophthalmol Vis Sci.* 2005; 46 (12): 4417–4423.

Jager RD, Aiello LP, Patel SC, Cunningham ET Jr. Risks of intravitreous injection: A comprehensive review. *Retina.* 2004 Oct; 24 (5): 676–698.

Jiang C, Moore MJ, Zhang X, Klassen H, Langer R, Young M. Intravitreal injections of GDNF-loaded biodegradable microspheres are neuroprotective in a rat model of glaucoma. *Mol Vis.* 2007; 13: 1783–1792.

Kim H, Robinson SB, Csaky KG. Investigating the movement of intravitreal human serum albumin nanoparticles in the vitreous and retina. *Pharm Res.* 2009 Feb; 26 (2): 329–337.

Lajavardi L, Bochot A, Camelo S, Goldenberg B, Naud MC, Behar-Cohen F, Fattal E, de Kozak Y. Downregulation of endotoxin-induced uveitis by intravitreal injection of vasoactive intestinal peptide encapsulated in liposomes. *Invest Ophthalmol Vis Sci.* 2007 Jul; 48 (7): 3230–3238.

Lajavardi L, Camelo S, Agnely F, Luo W, Goldenberg B, Naud MC, Behar-Cohen F, de Kozak Y, Bochot A. New formulation of vasoactive intestinal peptide using liposomes in hyaluronic acid gel for uveitis. *J Control Release.* 2009 Oct 1; 139 (1): 22–30. Epub 2009 May 28.

Laude A, Tan LE, Wilson CG, Lascaratos G, Elashry M, Aslam T, Patton N, Dhillon B. Intravitreal therapy for neovascular age-related macular degeneration and inter-individual variations in vitreous pharmacokinetics. *Prog Retin Eye Res.* 2010 Nov; 29 (6): 466–475.

Li H, Tran VV, Hu Y, Mark Saltzman W, Barnstable CJ, Tombran-Tink J. A PEDF N-terminal peptide protects the retina from ischemic injury when delivered in PLGA nanospheres. *Exp Eye Res.* 2006; 83 (4): 824–833.

Liu X, Li SK, Jeong EK. Ocular pharmacokinetic study of a corticosteroid by 19F MR. *Exp Eye Res.* 2010 Sep; 91 (3): 347–352.

London NJ, Chiang A, Haller JA. The dexamethasone drug delivery system: Indications and evidence. *Adv Ther.* 2011 May; 28 (5): 351–366. Epub 2011 Mar 12.

Maurice DM. Injection of drugs into the vitreous body. In Leopold I, Burns R, eds. *Symposium on Ocular Therapy.* New York: John Wiley & Sons; 1976: 59–72

Maurice DM, Mishima S. Ocular pharmacokinetics. In Sears ML, ed. *Handbook of Experimental Pharmacology: Pharmacology of the Eye.* Berlin: Springer-Verlag; 1984: 19–116.

Merodio M, Irache JM, Valamanesh F, Mirshahi M. Ocular disposition and tolerance of ganciclovir-loaded albumin nanoparticles after intravitreal injection in rats. *Biomaterials*. 2002; 23: 1587–1594.

Moritera T, Ogura Y, Honda Y, Wada R, Hyon SH, Ikada Y. Microspheres of biodegradable polymers as a drug-delivery system in the vitreous. *Invest Ophthalmol Vis Sci*. 1991; 32: 1785–1790.

Moritera T, Ogura Y, Yoshimura N, Honda Y, Wada R, Hyon SH, Ikada Y. Biodegradable microspheres containing adriamycin in the treatment of proliferative vitreoretinopathy. *Invest Ophthalmol Vis Sci*. 1992; 33: 3125–3130.

Normand N, Valamanesh F, Savoldelli M, Mascarelli F, BenEzra D, Courtois Y, Behar-Cohen F. VP22 light controlled delivery of oligonucleotides to ocular cells *in vitro* and *in vivo*. *Mol Vis*. 2005; 11: 184–191.

Ogura Y, Kimura H. Biodegradable polymer microspheres for targeted drug delivery to the retinal pigment epithelium. *Surv Ophthalmol*. 1995; 39 (Suppl 1): S17–24.

Peyman GA, Conway M, Khoobehi B, Soike K. Clearance of microsphere-entrapped 5-fluorouracil and cytosine arabinoside from the vitreous of primates. *Int Ophthalmol*. 1992; 16: 109–113.

Read SP, Cashman SM, Kumar-Singh R. POD nanoparticles expressing GDNF provide structural and functional rescue of light-induced retinal degeneration in an adult mouse. *Mol Ther*. 2010 Nov; 18 (11): 1917–1926.

Sakai T, Kuno N, Takamatsu F, Kimura E, Kohno H, Okano K, Kitahara K. Prolonged protective effect of basic fibroblast growth factor-impregnated nanoparticles in royal college of surgeons rats. *Invest Ophthalmol Vis Sci*. 2007; 48 (7): 3381–3387.

Sakurai E, Ozeki H, Kunou N, Ogura Y. Effect of particle size of polymeric nanospheres on intravitreal kinetics. *Ophthalmic Res*. 2001; 33: 31–36.

Sebag J, Balazs EA. Morphology and ultrastructure of human vitreous fibers. *Invest Ophthalmol Vis Sci*. 1989; 30: 1867–1871.

Shahar J, Avery RL, Heilweil G, Barak A, Zemel E, Lewis GP, Johnson PT, Fisher SK, Perlman I, Loewenstein A. Electrophysiologic and retinal penetration studies following intravitreal injection of bevacizumab (Avastin). *Retina*. 2006 Mar; 26 (3): 262–269.

Tabata Y, Ikada Y. Macrophage phagocytosis of biodegradable microspheres composed of L-lactic acid/glycolic acid homo- and copolymers. *J Biomed Mater Res*. 1988; 22: 837–858.

Veloso AA Jr., Zhu Q, Herrero-Vanrell R, Refojo MF. Ganciclovir-loaded polymer microspheres in rabbit eyes inoculated with human cytomegalovirus. *Invest Ophthalmol Vis Sci*. 1997; 38: 665–675.

Ward MS, Khoobehi A, Lavik EB, Langer R, Young MJ. Neuroprotection of retinal ganglion cells in DBA/2J mice with GDNF-loaded biodegradable microspheres. *J Pharm Sci*. 2007; 96 (3): 558–568.

Zavaglia D, Normand N, Brewis N, O'Hare P, Favrot MC, Coll JL. VP22-mediated and light-activated delivery of an anti-c-raf1 antisense oligonucleotide improves its activity after intratumoral injection in nude mice. *Mol Ther*. 2003 Nov; 8 (5): 840–845.

Zhang L, Li Y, Zhang C, Wang Y, Song C. Pharmacokinetics and tolerance study of intravitreal injection of dexamethasone-loaded nanoparticles in rabbits. *Int J Nanomedicine*. 2009; 4: 175–183.

9 Transscleral and Suprachoroidal Drug Delivery

Damian E. Berezovsky and Henry F. Edelhauser

CONTENTS

9.1   INTRODUCTION

While pathologic conditions can affect all tissues of the eye, those affecting the posterior segment—especially the central retina—cause the greatest burden in quality of life as well as treatment costs. Conditions such as age-related macular degeneration (AMD), diabetic retinopathy, and glaucoma affect an estimated 7 million people in the United States alone and are expected to rise in prevalence in the coming years.[1] These diseases present an obvious target for novel drug development, and several highly effective agents have been developed over the past decade. Delivering these medications to their site of action in a safe, predictable, and effective manner, however, remains a challenge. Drug delivery to the anterior part of the eye (cornea, iris, and trabecular meshwork) can be achieved using topical eyedrops, and most current treatments for anterior segment disease use this delivery method. Some diseases that affect the posterior segment, such as glaucoma, have also been effectively treated through topical treatment. For other posterior segment diseases like AMD diabetic retinopathy, and some types of intraocular tumors, topical delivery does not result in therapeutic drug levels.

Drug delivery to the posterior ocular tissues has traditionally been approached in four ways, with varying success. The four proposed routes are topical, systemic, intraocular, and periocular (subconjunctival, sub-Tenon, or retrobulbar). While topical (eyedrop) delivery has been very successful in the treatment of anterior ocular conditions (dry eye, corneal infections, and glaucoma), early studies have determined that less than 5% of the applied drug penetrates the intraocular tissues; an even smaller proportion is expected to reach any specific target tissue within the posterior segment. The major factors against the penetration of drug into the posterior segment are the corneal epithelial tight junctions, tear drainage and turnover, and clearance by conjunctival blood and lymph vessels. Drugs injected into the systemic circulation may reach intraocular tissues, but large doses need to be administered to reach therapeutic levels within the eye. Exposure of multiple organ systems to large doses of medication may result in serious adverse effects, making this an undesirable approach for most ocular conditions, with the notable exception of intraocular tumors such as retinoblastoma or melanoma. Additionally, drugs administered systemically need to cross the blood–retinal barrier to reach the retina; this often requires extensive modification of existing drugs.[2]

Currently, the most commonly used approach to drug delivery to the posterior segment is the intravitreal injection. It has a key advantage over systemic and topical approaches in that it targets the posterior segment specifically, resulting in a large dose delivered in proximity to the retina. It is an office-based procedure generally well tolerated under topical anesthesia. Numerous studies have established the efficacy of intravitreal injections for the treatment of posterior segment diseases; many recent trials have centered around antiangiogenic therapy for AMD and diabetic retinopathy.[3,4] This technique does, however, have several drawbacks. The injection procedure has a small risk of retinal detachment, hemorrhage, or endophthalmitis. While these adverse events are rare, they often have devastating effects on patients' visual potential. Additionally, while the intravitreal half-life of most agents is longer than that of systemic injections, treatment regimens for chronic conditions often involve monthly injections. Intravitreal inserts can extend the half-life of intravitreal therapeutics, although they often require more extensive placement and removal procedures.

A transscleral approach to drug delivery provides an alternative that is less invasive and potentially better targeted to the tissues of the posterior segment. Transscleral drug delivery does not involve piercing the sclera; thus, there is a decreased risk of injection-related endophthalmitis. Because it does not rely on the vitreous for drug distribution, age-related vitreous liquefaction (syneresis) or previous vitrectomy are not likely to alter tissue drug levels.[5]

9.2 · TRANSSCLERAL DRUG DELIVERY

The human sclera is composed of multiple overlapping layers of collagen with fibrils of varying diameter, such that the resulting matrix does not allow light to pass through. The collagen matrix is supported by various proteoglycans within an aqueous environment. A dynamic system, the ability of molecules to traverse the sclera, is influenced by intraocular pressure and tissue hydration, as well as molecular characteristics. A large surface area (approximately 17 cm^2 in humans), over 15 times greater than that of the cornea, makes it an ideal medium for delivering molecules to large areas of the posterior segment. Mean scleral thickness ranges from 0.4 ± 0.2 mm near the equator to near 1 mm around the optic nerve.[6]

Although the sclera is largely avascular, blood vessels supporting the uvea and the retina pierce the sclera at various locations around the globe; there is evidence that outward fluid movement around these vessels plays a role in the regulation of intraocular pressure.[7] How this fluid movement may affect transscleral drug delivery, however, is not well known. In areas away from these vessels, the relatively low permeability of the sclera (which helps maintain normal intraocular pressure) sets up a hydrostatic gradient against the entry of material into the globe. Episcleral blood and lymph vessels crisscross the outer surface of the sclera; the highly vascular uvea, composed of the choroid and ciliary body, covers its inner surface. Although both of these vascular beds present barriers to drug penetration, experiments have shown that the episcleral vasculature plays a greater role in drug clearance during transscleral delivery.[8] Beyond the choroid, the retinal pigment epithelium (RPE) also represents a significant barrier to drug penetration; studies have shown the RPE barrier is especially significant when delivering lipophilic and large-molecular-weight products.

Despite the significant barriers to transscleral drug movement, numerous studies have successfully delivered various small- and large-molecular-weight compounds through human and animal sclera. Table 9.1, compiled from a number of *ex vivo* studies using isolated human sclera, shows the molecular weight and permeability constant (K_{trans}) of various agents.[9–15] Figure 9.1 shows a typical *in vitro* scleral diffusion chamber used for these experiments.

TABLE 9.1

Diffusion Coefficient of Various Molecules, as Established *In Vitro* Using Isolated Human Sclera

Drug or Molecule	Molecular Mass (Da)	K_{trans} (cm/s)	Reference
Polymyxin B	1800	3.90×10^{-7}	9
Doxil	580	4.74×10^{-7}	10
Vancomycin (BODIPY-tagged)	1723	6.66×10^{-7}	9
Single-stranded oligonucleotide (fluorescein-tagged)	7998	7.67×10^{-7}	11
Dexamethasone (fluorescein-tagged)	8414	1.64×10^{-6}	12
Rhodamine	479	1.86×10^{-6}	12
Penicillin G	661	1.89×10^{-6}	9
Methotrexate (fluorescein-tagged)	979	3.36×10^{-6}	13
Doxorubicin	580	3.50×10^{-6}	10
Doxorubicin (in nanoparticles)	580	4.97×10^{-6}	10
Fluorescein	332	5.21×10^{-6}	12
Cisplatin (in collagen matrix)	300	8.30×10^{-6}	14
Carboxyfluorescein	317	9.93×10^{-6}	13
Carboplatin (in fibrin sealant)	371	1.37×10^{-6}	15
Cisplatin	300	2.0×10^{-5}	13
Carboplatin	371	2.7×10^{-5}	15
Water	18	5.2×10^{-5}	14

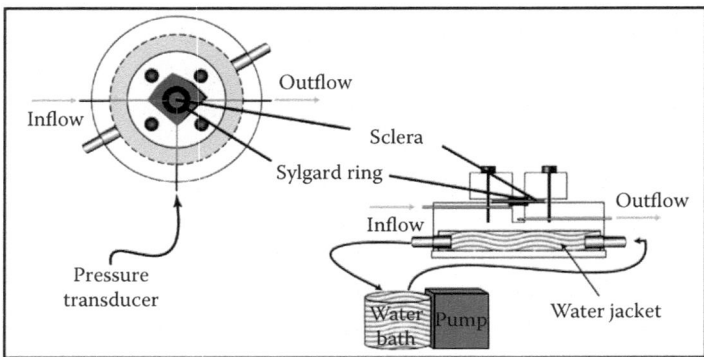

FIGURE 9.1 Typical setup for *in vitro* transscleral drug delivery experiments. Isolated sclera is sandwiched between Sylgard rings for a tight seal, then placed (choroidal side down) within a specially designed acrylic chamber. This results in a donor chamber above the sclera, where a drug or molecule of interest can be placed, and a recipient chamber, which can be sampled and measured for drug concentration. Regulation of recipient chamber inflow and outflow allows for the simulation of intraocular pressure as well as automated sampling. An underlying water jacket keeps the experimental apparatus close to physiologic temperature.

A number of early *in vitro* studies helped identify factors that affect the transscleral permeability of drugs and other agents. A general inverse relationship exists between a molecule's molecular weight and its K_{trans}, although detailed studies by Prausnitz et al. have established that molecular radius, rather than molecular weight, is best correlated to transscleral permeability.[16] Increasing drug lipophilicity has been shown to limit transscleral diffusion, an effect attributed to the choroid and Bruch's membrane rather than to the sclera itself.[17] Small (molecular weights less than 1000), hydrophilic molecules cross the sclera most readily; studies simulating elevated intraocular pressures have found that the sclera remains highly permeable to these agents at pressures up to 60 mm Hg.[16]

Drug molecules entering the sclera from the episcleral (outer) surface of the eye may move directly in the direction of the choroid, Bruch's membrane, and the retina, but may also move in a lateral direction or remain bound to components of the scleral matrix. An *in vitro* study by Jiang et al. using human tissue has shown that sulforhodamine can move laterally from a reservoir to a distance of 10 mm within 3 days.[18] In recent work from our laboratory, the episcleral surface of isolated human sclera was exposed to a prostaglandin analog for 15 minutes, and drug levels were measured from the opposite (inner) side of the sclera through 48 hours. Measurable drug levels were identified starting 15 minutes after exposure and remained well above detectable levels through 24 hours.[19] These results show that the sclera can absorb and retain drugs, releasing them over time and thereby acting as a basic form of sustained release. It is theorized that this phenomenon may contribute to the sustained (24 hours) drop in intraocular pressure, which follows a single dose of prostaglandin analogs in glaucoma patients.

It should also be noted that following the placement of drug on the episcleral surface of the eye, a variable period of time is needed before detectable levels of the drug reach the choroidal surface and inner ocular tissues. For example, carboxyfluo-rescein does not reach the opposite side of isolated human sclera until 20 minutes after exposure.[16] In an *in vivo* scenario, this lag time would allow episcleral blood flow to begin to clear the periocular drug depot—particularly if the drug was deliv-ered in aqueous solution. Because drug movement across the sclera is largely a pas-sive diffusive phenomenon, researchers have sought to maximize the amount of drug that can cross the sclera while minimizing the drug's clearance from the episcleral depot. It was hypothesized that a sustained-release device or a formulation placed on the episcleral surface could reach steady state with the underlying sclera, driving a maximum amount of drug deep into intraocular tissues.

Triamcinolone acetonide (TA), a synthetic steroid with a molecular mass of 434.5 Da, has been successfully delivered across isolated human sclera by Mora et al., with a scleral permeability coefficient of $1.47 \pm 0.17 \times 10^{-5}$ cm/s.[20] Subsequent *in vivo* experiments in rabbits have confirmed the presence of TA within the vitre-ous following a single sub-Tenon injection.[8,21] Robinson et al. selectively ablated the choroidal circulation at the injection location to further characterize the barriers to *in vivo* transscleral drug delivery. They found that the clearance of the drug from the sub-Tenon depot by episcleral blood plays a major role in limiting transscleral dif-fusion; in contrast, choroidal circulation was found to have a modest effect on drug entry into the vitreous.[8]

To maximize the contact time between the episcleral depot and the sclera, Cruysberg et al. delivered dexamethasone and methotrexate across isolated human sclera using a fibrin sealant and found that this approach delivered the drug in a slower, more sustained manner when compared to saline-based delivery.[22] Fibrin sealant solidifies at normal body temperature and can be overloaded with various drugs; these characteristics maximize the contact time between the episcleral depot and the underlying sclera, allowing for greater passive diffusion across the remain-ing tissues. The same principle was applied to carboplatin, a drug commonly used for the treatment of retinoblastoma, with similar results.[15] In an *in vivo* rabbit model, carboplatin-loaded fibrin was injected into the subconjunctival space, and intra-ocular drug levels were measured up to 2 weeks postinjection. Carboplatin levels remained detectable through 2 weeks after injection, suggesting that fibrin sealant is a favorable vehicle for sustained carboplatin delivery.

In an effort to evaluate the safety of carboplatin delivered through a subconjuncti-val fibrin sealant vehicle, Pardue et al. performed electroretinography (ERG) readings up to 3 weeks following such injections; they observed a transient suppression of the dark-adapted b-wave amplitude at 2 days, but no structural or functional changes were observed at 3 weeks postinjection.[23] Further studies were undertaken by Van Quill et al. in a mouse model of retinoblastoma, which showed intraocular tumor regression in 10 of 11 eyes injected with subconjunctival carboplatin (also in fibrin sealant).[24] No toxicity was observed on histological examination. Studies using topotecan-loaded fibrin in a similar mouse retinoblastoma model showed similar results.[25] Interestingly, the effects of topotecan were observed in the opposite eye as well, suggesting that the amount of drug absorbed into the systemic circulation reached therapeutic levels.

While small-molecule drugs such as carboplatin and topotecan have been shown to cross the sclera in *in vivo* animal models, larger molecules (such as proteins) may not be as easily delivered from a transscleral depot. Clinically important protein-based agents, such as bevacizumab (Avastin) and ranibizumab (Lucentis), are only available as intravitreal injections at this time. To assess whether these molecules could be delivered transsclerally, Kim et al. performed *in vitro* experiments using donor human scleral tissue, delivering Avastin (bevacizumab) and fluoresceinated rat immunoglobulin (Ig). Samples were taken from the recipient chamber every 2 hours, and the peak concentrations were found to be 1.3 µg/mL (for bevacizumab) and 1.47 µg/mL (for rat Ig) 10 hours after initial scleral exposure to the drugs.[26] This study shows that a protein such as bevacizumab can be delivered from a transscleral depot.

These initial *ex vivo* experiments prompted an interest in animal models of transscleral drug delivery; most *in vivo* animal studies utilized the rabbit eye and a variety of quantification techniques. In addition to the previously discussed studies, molecules successfully delivered across the sclera in live animal studies include fluorescein, fluoresceinated dextrans up to 70 kDa in molecular mass, and Ig. Ghate et al. compared several injection techniques (retrobulbar, sub-Tenon, and intravenous) using sodium fluorescein (484 Da) and fluorophotometry for quantification.[27] They observed detectable levels of fluorescein following all injections, with rapid clearance and a return to baseline fluorescence by 48 hours. Berezovsky et al. injected large-molecular-mass (40 and 70 kDa) fluoresceinated dextrans into live rabbits, followed by ocular fluorophotometry. Both of the large dextrans entered reached the posterior retina/choroid, reaching peak concentrations at 24 hours and a return to baseline fluorescence by 48 (40 kDa) and 72 (70 kDa) hours.[28] These results are in agreement with earlier experiments by Ambati et al., in which immunoglobulin G (IgG) was successfully delivered across the living rabbit sclera using an implantable osmotic pump.[29]

Based on findings from *in vitro* and *in vivo* transscleral delivery experiments, it has been postulated that a nonbiodegradable, unidirectional delivery device would provide optimal diffusion into the eye while avoiding rapid clearance by the episcleral circulation. Several such devices are currently at various stages of development; however, no transscleral implantable device has been approved for clinical use at this time.

9.3 SUPRACHOROIDAL DRUG DELIVERY

The suprachoroidal space (SCS) is, under physiologic conditions, a *potential* space located at the interface between sclera and choroid. This area of weak adhesions may become a true space under various conditions, both iatrogenic and naturally occurring. Well known to ophthalmic surgeons due to the risk of hemorrhage, the SCS has recently been proposed as a location for drug delivery in the treatment of posterior segment disease.

The SCS also referred to as the suprachoroid or suprachoroidea, describes the region where the innermost sclera meets the outer aspect of the choroid. It is made up of a loose collection of elastic fibers, arranged in layers such that they can slide past each other and provide channels along which fluid can move. It is fairly well defined at its outer face as it meets the lamina lutea (melanocytic layer covering the inner surface of the sclera); its inner face, its crisscrossing laminar fibers, blends into the body of the choroid. The SCS is held closed by a combination of intraocular pressure and weak adhesions between adjacent layers of fibers. In addition, the SCS becomes nonexistent at several locations around the globe, where the choroid firmly attaches to the sclera. These locations are the scleral spur (anteriorly), the optic disc (posteriorly), and the ampullae of the four vortex veins. Limited by these boundaries, the total surface area of the SCS has been estimated at 16 cm^2.[30] In normal individuals, the SCS is estimated to hold around 10 μL of aqueous fluid, which is believed to act as a lubricant for the choroid to glide over the inner face of the sclera.[31]

The SCS has a central role in the formation of a suprachoroidal hemorrhage, also called choroidal or ciliochoroidal detachment. The vast majority of suprachoroidal hemorrhages occur in the intraoperative or postoperative setting and are thought to form through the combination of hypotony (with subsequent suprachoroidal effusion) and rupture of a long ciliary artery.[32] In an effort to understand the pathophysiology of choroidal effusion and hemorrhage, a number of animal experiments have involved the injection of materials into the SCS. Observation of the effusion's natural history has revealed important information regarding the movement of fluids and molecules out of the SCS. Injected materials have included fibrinogen, autologous serum, Ringer's solution, and silicone oil. Small (0.25 mL) injections of human fibrinogen into normal rabbit eyes resulted in wide, flat choroidal detachments extending over most of the SCS. In animals that were sacrificed 21 days after injection, the suprachoroidal fibrin clot had disappeared, leaving no scar.[33] When a large (1 mL) amount of Ringer's solution or autologous serum was introduced into the SCS, it was easily visible as a four-lobed choroidal detachment immediately following injection, but became flat and ophthalmoscopically invisible within 2 days.[31]

There are several potential advantages to delivering drugs directly into the SCS. Because of its proximity to the choroid, Bruch's membrane, and RPE, drugs delivered into the SCS may reach these tissues at higher levels than drugs delivered outside the sclera or in the middle of the vitreous. Because suprachoroidal injections do not rely on (or disturb) the vitreous body, injections are less likely to cause significant disruptions in a patient's vision; additionally, changes in the state of the vitreous body (age-related changes or previous vitrectomy) are less likely to affect drugs delivered into the SCS. Given these potential advantages, several studies have specifically explored its potential in drug delivery. Fibrin glue and poly(ortho) ester were injected into living rabbit eyes as well as indocyanine green dye and TA in pigs.[34] The injection technique in all these studies involved a scleral incision and advancement of a cannula toward the posterior pole. Despite the invasive nature of this technique, however, no significant functional or structural damage was reported in animals that received a successful injection.[35–37]

Perhaps the greatest advantage of suprachoroidal injections, when compared to intravitreal or transscleral techniques, is the ability to deliver large amounts of drug to the choroid and Bruch's membrane. This is particularly relevant to the treatment of AMD, since the thickening of Bruch's membrane and subsequent drusen formation are believed to be some of the earliest events in the pathogenesis of the disease.[38]

Current therapies for neovascular AMD and diabetic retinopathy rely on antiangio-genic protein agents; delivering them directly adjacent to the choroid, as opposed to the central vitreous cavity, may greatly increase the efficacy of treatment. As new therapeutic agents are discovered or designed, the ability to deliver them directly to the choroid may be of critical importance.

Hou et al. injected fibrin glue directly into the SCS of live rabbits using a scleral incision and 26-gauge cannula; the rabbits underwent slit-lamp examina-tion, indirect ophthalmoscopy, fundus photography, fluorescein angiography, and electroretinography.[36] Through 90 days postinjection, the authors noted only minor hemorrhaging of the choroid 1 day after injection; all other parameters remained within normal limits, and no injection-related complications were noted. Upon his-tological examination, inflammation was observed within the area of fibrin glue injection; inflammatory cells disappeared within 7 days, and fibrosis was observed by 30 days. These findings suggest that cannula-based suprachoroidal delivery is well tolerated, with minor structural and no apparent functional changes. Olsen et al. compared the pharmacokinetic profiles of bevacizumab delivered through conventional intravitreal or cannula-based suprachoroidal injection in a live pig model.[39] While bevacizumab injected into the SCS showed greater direct dif-fusion into the RPE and photoreceptor outer segments, drug levels fell below detectable levels more quickly when compared to similar doses injected into the vitreous cavity.

Microneedle-based suprachoroidal delivery was conceived as a less-invasive alternative to cannula-based techniques, eliminating the need for scleral incision and passage of plastic tubing across choroidal tissue. Patel et al. experimented on *ex vivo* rabbit, pig, and human eyes to determine whether a single hollow microneedle could successfully deliver solutions and particle suspensions to the back of the eye (Figure 9.2).[40] Furthermore, the authors characterized several parameters, such as needle length, fluid pressure, and the size of the particles injected, which determine the success of suprachoroidal injections. Particles up to 1000 nm in diameter were successfully delivered using 800–1000-μm hollow microneedles. To characterize the surface area (as measured along the inner sclera) covered by a single supracho-roidal injection, donor human eyes were injected with 50, 100, or 150 μL of a latex solution; the eyes were dissected and the surface covered by latex was determined. It was found that 50, 100, and 150 μL injections covered an average of 17%, 39%, and 40% of the total sclera surface area.[41]

In live animals, microneedle-based suprachoroidal injections have been suc-cessfully used to deliver aqueous solutions of fluorescein and fluoresceinated dextrans, as well as aqueous suspensions of TA and polystyrene particles, into live rabbits.[42–44] *In vivo* fluorophotometry of these animals showed that aqueous solutions remained within the SCS for short periods (24 hours or less) before fall-ing below detectable levels. Polystyrene particles, however, were detected in the SCS of live rabbits more than 60 days after injection. Histological examination showed the particles in the SCS and also embedded in the choroidal tissue; no active inflammation or scar tissue was observed.[43]

Suprachoroidal drug delivery for posterior segment disease remains a novel approach, with clear advantages over the current intravitreal injection. These include

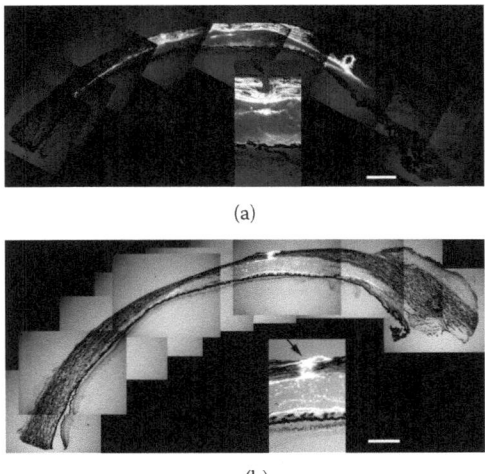

(a)

(b)

FIGURE 9.2 **(See color insert.)** Fluorescence imaging of porcine (a) and human (b) eyes following *ex vivo,* microneedle-based suprachoroidal injection of fluorescent nanoparticles. Particle sizes: 500 nm (a) and 1 μm (b). Fluorescence can be seen within the sclera near the site of injection (inset, black arrow) as well as within the choroid and SCS. The absence of fluorescence within the retina and vitreous shows that the microneedle did not penetrate Bruch's membrane or the retinal pigment epithelium. The white scale bar represents 500 μm. (Adapted from Patel SR et al., *Pharm Res*, 28, 166–176, 2011.)

the specific placement of drug adjacent to the site of disease, the lack of reliance on the vitreous for drug distribution, and the minimally invasive nature of the procedure (when using microneedles). As research continues in this area, particular attention should be directed at the development of sustained-release formulations for delivery to the SCS. The ability to deliver a therapeutic agent directly to the choroid provides a unique opportunity to halt disease progression at an early stage, and maintaining adequate drug levels over time will prove critical to the success of such a therapy.

ACKNOWLEDGMENTS

Supported in part by R24EY017045, T32EY7092, and Research to Prevent Blindness.

REFERENCES

1. Congdon N, O'Colmain B, Klaver CC et al. Causes and prevalence of visual impairment among adults in the United States. *Arch Ophthalmol.* Apr 2004;122(4):477–485.
2. Tachikawa M, Ganapathy V, Hosoya K-i. Systemic route for retinal drug delivery: Role of the blood-retinal barrier. In *Drug Product Development for the Back of the Eye*, Kompella UB, Edelhauser HF, eds. Vol 2, 85–109, Springer: New York, 2011.

3. Abouammoh M, Sharma S. Ranibizumab versus bevacizumab for the treatment of neovascular age-related macular degeneration. *Curr Opin Ophthalmol.* May 2011;22(3):152–158.

4. Ozturk BT, Kerimoglu H, Bozkurt B, Okudan S. Comparison of intravitreal bevacizumab and ranibizumab treatment for diabetic macular edema. *J Ocul Pharmacol Ther.* Aug 2011;27(4):373–377.

5. Lee SS, Ghosn C, Yu Z et al. Vitreous VEGF clearance is increased after vitrectomy. *Invest Ophthalmol Vis Sci.* Apr 2010;51(4):2135–2138.

6. Olsen TW, Aaberg SY, Geroski DH, Edelhauser HF. Human sclera: Thickness and surface area. *Am J Ophthalmol.* Feb 1998;125(2):237–241.

7. Alm A, Nilsson SF. Uveoscleral outflow—A review. *Exp Eye Res.* Apr 2009;88(4): 760–768.

8. Robinson MR, Lee SS, Kim H et al. A rabbit model for assessing the ocular barriers to the transscleral delivery of triamcinolone acetonide. *Exp Eye Res.* Mar 2006;82(3):479–487.

9. Kao JC, Geroski DH, Edelhauser HF. Transscleral permeability of fluorescent-labeled antibiotics. *J Ocul Pharmacol Ther.* Feb 2005;21(1):1–10.

10. Kim ES, Durairaj C, Kadam RS et al. Human scleral diffusion of anticancer drugs from solution and nanoparticle formulation. *Pharm Res.* May 2009;26(5):1155–1161.

11. Shuler RK, Jr., Dioguardi PK, Henjy C, Nickerson JM, Cruysberg LP, Edelhauser HF. Scleral permeability of a small, single-stranded oligonucleotide. *J Ocul Pharmacol Ther.* Apr 2004;20(2):159–168.

12. Cruysberg LP, Nuijts RM, Geroski DH, Koole LH, Hendrikse F, Edelhauser HF. *In vitro* human scleral permeability of fluorescein, dexamethasone-fluorescein, methotrexate-fluorescein and rhodamine 6G and the use of a coated coil as a new drug delivery system. *J Ocul Pharmacol Ther.* Dec 2002;18(6):559–569.

13. Gilbert JA, Simpson AE, Rudnick DE, Geroski DH, Aaberg TM, Jr., Edelhauser HF. Transscleral permeability and intraocular concentrations of cisplatin from a collagen matrix. *J Control Release.* May 20 2003;89(3):409–417.

14. Rudnick DE, Noonan JS, Geroski DH, Prausnitz MR, Edelhauser HF. The effect of intraocular pressure on human and rabbit scleral permeability. *Invest Ophthalmol Vis Sci.* Nov 1999;40(12):3054–3058.

15. Simpson AE, Gilbert JA, Rudnick DE, Geroski DH, Aaberg TM, Jr., Edelhauser HF. Transscleral diffusion of carboplatin: An *in vitro* and *in vivo* study. *Arch Ophthalmol.* Aug 2002;120(8):1069–1074.

16. Prausnitz MR, Noonan JS. Permeability of cornea, sclera, and conjunctiva: A literature analysis for drug delivery to the eye. *J Pharm Sci.* Dec 1998;87(12):1479–1488.

17. Cheruvu NP, Kompella UB. Bovine and porcine transscleral solute transport: Influence of lipophilicity and the Choroid-Bruch's layer. *Invest Ophthalmol Vis Sci.* Oct 2006;47(10): 4513–4522.

18. Jiang J, Geroski DH, Edelhauser HF, Prausnitz MR. Measurement and prediction of lateral diffusion within human sclera. *Invest Ophthalmol Vis Sci.* Jul 2006;47(7):3011–3016.

19. Jani AR, Kim ES, Berezovsky DE, Kadam R, Kompella UB, Edelhauser HF. Scleral uptake and extended release of prostaglandin analogs: Bimatoprost, latanoprost, and travoprost in an *in vitro* model. *Invest Ophthalmol Vis Sci.* Apr, 2009;50(5):5967.

20. Mora P, Eperon S, Felt-Baeyens O et al. Trans-scleral diffusion of triamcinolone acetonide. *Curr Eye Res.* May 2005;30(5):355–361.

21. Lee SJ, Kim ES, Geroski DH, McCarey BE, Edelhauser HF. Pharmacokinetics of intraocular drug delivery of Oregon green 488-labeled triamcinolone by subtenon injection using ocular fluorophotometry in rabbit eyes. *Invest Ophthalmol Vis Sci.* Oct 2008;49(10):4506–4514.

22. Cruysberg LP, Nuijts RM, Gilbert JA, Geroski DH, Hendrikse F, Edelhauser HF. *In vitro* sustained human transscleral drug delivery of fluorescein-labeled dexamethasone and methotrexate with fibrin sealant. *Curr Eye Res.* Aug 2005;30(8):653–660.

23. Pardue MT, Hejny C, Gilbert JA, Phillips MJ, Geroski DH, Edelhauser HF. Retinal function after subconjunctival injection of carboplatin in fibrin sealant. *Retina.* Oct 2004;24(5):776–782.

24. Van Quill KR, Dioguardi PK, Tong CT et al. Subconjunctival carboplatin in fibrin sealant in the treatment of transgenic murine retinoblastoma. *Ophthalmology.* Jun 2005;112(6):1151–1158.

25. Tsui JY, Dalgard C, Van Quill KR et al. Subconjunctival topotecan in fibrin sealant in the treatment of transgenic murine retinoblastoma. *Invest Ophthalmol Vis Sci.* Feb 2008;49(2):490–496.

26. Kim ES, Berglin L, Morohoshi K, Geroski DH, Edelhauser HF. *In vitro* diffusion of avastin (bevacizumab) across human sclera. *ARVO Meeting Abstracts.* Apr 11, 2009;50(5):5988.

27. Ghate D, Brooks W, McCarey BE, Edelhauser HF. Pharmacokinetics of intraocular drug delivery by periocular injections using ocular fluorophotometry. *Invest Ophthalmol Vis Sci.* May 2007;48(5):2230–2237.

28. Berezovsky DE, Patel SR, McCarey BE, Edelhauser HF. *In vivo* ocular fluorophotometry: Delivery of fluoresceinated dextrans via transscleral diffusion in rabbits. *Invest Ophthalmol Vis Sci.* 2011;52(10):7038–7045.

29. Ambati J, Gragoudas ES, Miller JW et al. Transscleral delivery of bioactive protein to the choroid and retina. *Invest Ophthalmol Vis Sci.* Apr 2000;41(5):1186–1191.

30. Brubaker RF, Pederson JE. Ciliochoroidal detachment. *Surv Ophthalmol.* Mar–Apr 1983;27(5):281–289.

31. Pederson JE, Gaasterland DE, MacLellan HM. Experimental ciliochoroidal detachment: Effect on intraocular pressure and aqueous humor flow. *Arch Ophthalmol.* Mar 1979;97(3):536–541.

32. Chu TG, Green RL. Suprachoroidal hemorrhage. *Surv Ophthalmol.* May–Jun 1999;43(6):471–486.

33. Edmund J, Gregersen E. Experimental detachment of the choroid: A new possibility of treatment in retinal detachment. *Acta Ophthalmol (Copenh).* 1964;42:269–276.

34. Alexandrakis G, Chaudhry NA, Liggett PE, Weitzman M. Spontaneous suprachoroidal hemorrhage in age-related macular degeneration presenting as angle-closure glaucoma. *Retina.* 1998;18(5):485–486.

35. Einmahl S, Savoldelli M, D'Hermies F, Tabatabay C, Gurny R, Behar-Cohen F. Evaluation of a novel biomaterial in the suprachoroidal space of the rabbit eye. *Invest Ophthalmol Vis Sci.* May 2002;43(5):1533–1539.

36. Hou J, Tao Y, Jiang YR, Wang K. *In vivo* and *in vitro* study of suprachoroidal fibrin glue. *Jpn J Ophthalmol.* Nov 2009;53(6):640–647.

37. Olsen TW, Feng X, Wabner K et al. Cannulation of the suprachoroidal space: A novel drug delivery methodology to the posterior segment. *Am J Ophthalmol.* Nov 2006;142(5):777–787.

38. Booij JC, Baas DC, Beisekeeva J, Gorgels TG, Bergen AA. The dynamic nature of Bruch's membrane. *Prog Retin Eye Res.* Jan 2010;29(1):1–18.

39. Olsen TW, Feng X, Wabner K, Csaky K, Pambuccian S, Cameron JD. Pharmacokinetics of pars plana intravitreal injections versus microcannula suprachoroidal injections of bevacizumab in a porcine model. *Invest Ophthalmol Vis Sci.* Jun 2011;52(7):4749–4756.

40. Patel SR, Lin AS, Edelhauser HF, Prausnitz MR. Suprachoroidal drug delivery to the back of the eye using hollow microneedles. *Pharm Res.* Jan 2011;28(1):166–176.

41. Patel SR, Bergman L, Berglin LC et al. Surface area coverage of suprachoroidal injections using a hollow microneedle in fresh human globes. *ARVO Meeting Abstracts.* Apr 22, 2011;52(6):3255.

42. Mansoor S, Patel SR, Tas C, Grossniklaus HE, Edelhauser HF, Prausnitz MR. Pharmacokinetics and biodistribution of triamcinolone acetonide following suprachoroidal injection into the rabbit eye *in vivo* using a microneedle. *ARVO Meeting Abstracts.* Apr 22, 2011;52(6):6585.

43. Berezovsky DE, Patel SR, McCarey BE, Grossniklaus HE, MR, Edelhauser HF. *In vivo* evaluation of fluorescent microparticles injected in the rabbit suprachoroidal space using hollow microneedles. *ARVO Meeting Abstracts.* Apr 22, 2011;52(6):2058.

44. Patel SR, Berezovsky D, McCarey BE, Nickerson JM, Edelhauser HF, Prausnitz MR. Intraocular pharmacokinetics of suprachoroidal drug delivery administered using hollow microneedles. *Invest Ophthalmol Vis Sci.* Apr, 2010;51(5):3796.

10 Protein Drug Delivery to Retina and Choroid

Hongwen M. Rivers and Patrick M. Hughes

CONTENTS

10.1 OVERVIEW

Protein and peptide drugs possess many advantages over their small-molecule counterparts: they have a relatively high specificity and low toxicity as well as unique mechanisms of action. The value of antibodies in ocular therapy is clearly evident in the clinical success of Lucentis for treating wet age-related macular degeneration (AMD). Antibodies, antibody fragments, and proteins are showing promise in clinical and preclinical stages of development for treating a wide variety of ophthalmic conditions (Tables 10.1 and 10.2). Obviously, proteins represent a valuable class of therapeutic agents for the eye. However, the clinical application of protein and peptide drug substances is limited due to poor bioavailability and disposition. Protein molecules are often hydrophilic large molecules, impermeable to most membranes, and have poor systemic absorption. Intravenous and subcutaneous delivery is the most common route of administration for protein drugs. Direct intravenous administration of protein therapeutics mitigates the issues of poor bioavailability; however, proteins often suffer from short plasma half-lives and poor penetration into the target tissues. This results in a requirement for frequent high-dose administrations often leading to poor compliance and potential toxic side effects.

Local sustained delivery of proteins and peptides could potentially alleviate many of these issues, circumventing barriers to productive absorption, protecting drugs from systemic degradation, reducing the total dose administered and subsequent

TABLE 10.1

Protein/Antibody Drug Candidates for Ocular Indications

	Patent Number	Publication Date	Claims	Patent Assignee
Recent patents	WO2009048537, WO2009048538, WO2009048539	4/16/2009	Anti-Aβ antibodies for retinal conditions	AC Immune SA/ Genentech
	WO2009040336	4/2/2009	Aβ-binding proteins for AMD and glaucoma	GlaxoSmithKline plc
	WO2008048675	4/24/2008	sCR1 for wet AMD	Celldex Therapeutics Inc
	WO2009061910	5/14/2009	Anti-factor B antibodies for CNV, AMD	Genentech Inc
	WO2009046405	6/10/2008	Anti-HtrA1 antibodies for AMD, DR, retinopathy of prematurity (ROP), or macular edema	University of Utah
	WO2006113311	10/26/2006	Galectin proteins for dry eye	Tufts University

	Compound	Type	Indications	Company/Institution
Preclinical	Nerve growth factor (NGF)[90,91]	Protein	Glaucoma	University of Rome
	Glial cell line-derived neutrophic factor (GDNF)[92,93]	Protein	Glaucoma	Harvard Medical School, Massachusetts Institute of Technology, University of California, Irvine, Tulane University, Otto-von-Guericke University
	Ciliary-derived neutrophic factor (CNTF)[94]	Protein	Glaucoma	Johns Hopkins University School of Medicine, Louisiana State University Health Sciences Center, University of Florida
	Neurotrophin-4/5 (NT-4/5)[95]	Protein	Glaucoma	University of Western Australia
	Neurotrophin-1 (NT-1)[96]	Protein	Glaucoma	University of Pennsylvania, Harvard Medical School, University Eye Hospital Tuebingen, Human Genome Sciences, Inc

TABLE 10.1   (*Continued*)
Protein/Antibody Drug Candidates for Ocular Indications

	Compound	Type	Indications	Company/Institution
	Brain-derived neurotrophic factor (BDNF)[97]	Protein	Glaucoma	University of Western Australia
	Erythropoietin (EPO)[97,98]	Protein	Glaucoma	Albert Einstein College of Medicine
	CCR3-neutralizing antibody[99]	Antibody	AMD	University of Kentuky, Nagoya City University, University of North Carolina Chapel Hill, University of Cincinnati, University of Utah School of Medicine, Veteran Affairs Salt lake City Healthcare Systems, Harvard Medical School, Oregon Health and Sciences University, University of Luebeck
	AVT-101 (anti-inflammatory protein)	Protein	AMD	Wellstat Ophthalmics
Ph. I/II planned	Anti-factor D	Antibody fragment	AMD	Genentech

Aβ, β-amyloid; AMD, age-related macular degeneration; CNV, choroidal neovascularization; DR, diabetic retinopathy; ROP, retinopathy of prematurity.

toxicities, and allowing for improved patient compliance. Moreover, the use of sustained, controlled, or targeted delivery systems can optimize the clinical performance of protein drugs. Local and sustained ocular drug delivery for anterior and posterior segment diseases is feasible and often preferred.[1,2]

The topical ocular route of administration is the preferred route of dosing for ocular delivery. It is noninvasive and has a relative ease of administration. Unfortunately, bioavailability is low, even to the anterior chamber with small molecules. Absorption from topical delivery can occur by the corneal and noncorneal (conjunctival/scleral) pathways. The barriers to ocular bioavailability from topical administration are well known and include precorneal drainage, lacrimation and tear dilution, tear turnover, conjunctival absorption, and low corneal epithelial permeability. Typically, less than 1–5% of the instilled dose reaches the aqueous humor, with further penetration into the posterior segment negligible. Noncorneal penetration into the anterior segment has been described. However, further penetration into the posterior segment with

TABLE 10.2
Protein/Antibody Drugs for Ocular Indications in Clinical Development or on the Market

Clinical	Compound	Type	Current Indications	Company/ Institution	Potential New Indications[108]
Ph. I	Murine nerve growth factor (mNGF)[100–102]	Protein	Neurotrophic keratitis	University of Rome	n/a
Ph. I	RN6G (Anti-amyloid β Ab)	Antibody	AMD	Pfizer	
Ph. I completed	iSONEP™ (Sphingomab™)	Antibody	AMD, glaucoma, diabetic retinopathy	Lpath	
Ph. I/IIa completed	MP0112 (VEGF-A antagonist)	Protein	AMD, DME	Molecular Partners	
Ph. II	iCo-008 (anti-eotaxin; Bertilimumab)[103]	Antibody	Allergic conjunctivitis	iCo Therapeutics	
Ph. II/III	VEGF Trap-Eye[104,105]	Protein	AMD, CRVO, DME	Regeneron/Bayer HealthCare	
On the market	Lucentis (anti-VEGF-A; ranibizumab)	Antibody fragment	AMD CNV, macular edema	Genentech/Novartis	Diabetic retinopathy
Investigated for ocular use	Avastin (anti-VEGF-A; bevacizumab)	Antibody	AMD CNV, macular edema	Genentech	Neovascular glaucoma
	Humira (anti-TNF-α; adalimumab)[106–108]	Antibody	Uveitis and inflammatory diseases	Abbott	AMD and macular edema
	Zenapax (anti-CD25; daclizumab)[106]	Antibody	Uveitis	Roche	AMD
	Enbrel (etanercept; TNF-α and β-inhibitor)[106,107,109]	Protein	Uveitis and inflammatory diseases	Amgen	AMD and macular edema
	Remicade (anti-TNF-α; infliximab)[106,107,109]	Antibody	Uveitis and inflammatory diseases	Centocor	AMD and macular edema
	Rituxan (anti-CD20; rituximab)[106,107,110]	Antibody	Lymphoma and inflammatory diseases	Genentech	AMD

AMD, age-related macular degeneration; CRVO, central retinal vein occlusion; DME, diabetic macular edema; TNF, tumor necrosis factor; VEGF, vascular endothelial growth factor.

diffusion to the macula is highly unlikely. Much of the literature describing topical delivery to the retina involves research on rodents or rabbits where vitreous volumes and diffusional path lengths are much smaller than humans.

Significant anatomic and physiologic barriers to posterior segment drug delivery exist and will be extensively reviewed in Chapters 2, 6, and 7 of this book. Briefly, the inner (endothelial cells of the retinal vessels) and outer (tight junctions of the retinal pigment epithelium [RPE]) blood–retinal barriers (BRB) greatly limit systemic bioavailability to the posterior segment. For a drug to cross the BRB, it should either exhibit optimum membrane partition characteristics or be a substrate for one of the membrane transporters present in the RPE or the endothelium of retinal blood vessels. The overwhelming majority of drugs are unable to achieve therapeutic concentrations in the posterior segment following systemic delivery. As a result of this, direct intraocular administration has become the preferred route of delivery for posterior segment diseases.

For highly soluble peptide and protein drugs, the high vitreal drug concentrations required for therapeutic effect[3] can only be achieved by local drug administration such as multiple intravitreal or periocular injections. Direct intravitreal injection is currently being used for the administration of drugs such as Macugen® (pegaptanib), Avastin® (bevacizumab injection), and Lucentis® (ranibizumab injection). One biopharmaceutical advantage of macromolecules relative to intravitreal administration is that their vitreal half-life is significantly greater than small molecules. Lucentis and Avastin have vitreal half-lives of 2.9 and 4.3 days, respectively, in the rabbit (MW 48 and 149 KDa).[4,5] The vitreal half-life of Lucentis in humans has been estimated to be approximately 9 days. However, even with these prolonged vitreal half-lives, these agents must be injected once every 4 weeks to maintain the effect. Multiple intraocular injections of this sort may lead to poor patient compliance and also increases the risk of intravitreal hemorrhages, retinal and vitreous detachment, cataract, and endophthalmitis.[3,6] Sustained and controlled delivery directly to the vitreous addresses this and also tempers the potential high peak drug levels achieved through frequent pulsed dosing. For this reason, there is a very significant unmet medical need for sustained intraocular macromolecule delivery systems. Subconjunctival/sub-Tenon's, episcleral, suprachoroidal, retrobulbar, and direct intravitreal implants have been explored for ocular drug delivery.[2] Limited success has been achieved with sustained delivery of some peptides. Table 10.3 lists the sustained release formulations for all ocular or all protein/peptide drugs on the market. However, protein delivery systems for ocular indications are still lacking.

Developing protein and peptide delivery systems is not a trivial task. These compounds often require intact quaternary structures for their biologic activity. Maintaining the compound's structural integrity within the formulation, during the manufacturing processes, and throughout the performance of the sustained delivery system is quite complex. Proteins can be denatured by heat, shear forces, pH extremes, organic solvents, hydrophobic interfaces, freezing, and drying.[7] Proteins or peptides can also be susceptible to damage from irradiation utilized in the terminal sterilization of the final drug product. Proteins or peptides may interact with many of the hydrophobic polymers used in the fabrication of sustained delivery systems,

TABLE 10.3

Sustained Release Formulations for All Ocular or All Protein/Peptide Drugs on the Market

	All Ocular Drugs (Small Molecule)	All Peptide/Protein Drugs
Microspheres	—	Lupron Depot® for prostate cancer
		Sandostatin LAR® for acromegaly
		Suprecur MP® for prostate cancer
		Decapeptyl® for prostate cancer
		Somatuline® LA for acromegaly
		Trelstar™ Depot for prostate cancer
		Bydureon for type II diabetes
Nondegradable implants	Vitrasert®[10,111] (approved in 1996)	Viadur® for prostate cancer[80,115]
	Retisert®[10,112] (approved in 2005)	
	Iluvien®[113] (phase III)	
Biodegradable implants	Ozurdex®[114] (approved in 2009)	Zoladex® for prostate and breast cancer[116,117]
In situ-forming implants	—	Eligard® for prostate cancer
Liposomes	Visudyne[17] (approved in 2000)	

becoming adsorbed, degraded, aggregated, or denatured. This may lead to loss of activity and immunogenicity.[7–9]

Some progress has been made in the delivery of peptide drugs. Short peptides are simple molecules relative to proteins and a few have been successfully developed as sustained-release drug products. Attempts at sustained protein delivery, on the other hand, have been met with much less success. In this chapter, we will cover the current successes of protein and peptide delivery, both preclinically and clinically. The overall focus will be on protein and peptide delivery to the eye with nanoparticulates. Because much of the nanoparticulate work is still in its nascent stage, microspheres will also be discussed, as the data may apply to particulates in general and lays the foundation for nanoparticulate protein delivery. The focus will include manufacturing processes, formulation excipients, stability issues, polymers used in the delivery systems, and drug release control. Liposomal delivery of proteins and peptides will also be reviewed. It should be emphasized, however, that what has worked for one combination of protein and delivery vehicle may not be expected to work for another combination.

10.2 MICROPARTICLES

Microparticles, microcapsules, and microspheres are small drug-loaded particles with diameters in the range from 1 to 250 μm. Microcapsules contain an inner drug core surrounded by a polymer shell, and microspheres contain a drug dissolved in or homogeneously dispersed throughout a polymer matrix. Microparticles can also have a drug adsorbed to their surface. Microparticles can be suspended in suitable buffers and easily administered by injection through narrow-gauge needles, making them suitable for ocular applications. Microparticles are often composed of biodegradable and biocompatible polymers such as poly(D,L-lactic-co-glycolide) (PLGA), PLGA-polyethylene glycol (PLGA-PEG), PLGA-poly(ε-caprolactone) (PLGA-PCL), and PEG-poly(butylene terephthalate) (PEG-PBT). Microparticles prepared with these and similar polymers can protect a protein therapeutic against enzymatic degradation and allow for sustained release of the drug.

Much of the microparticulate literature relating to proteins, peptides, and ocular drug delivery involve the use of microspheres, and consequently, this will be the focus, unless otherwise noted. Microspheres have been investigated for intraocular drug delivery to treat vitreoretinal diseases[6] such as uveitis[10] and proliferative vitreoretinopathy[11] as well as for transscleral delivery of ocular drugs to suppress choroidal neovascularization (CNV).[12,13] PLGA microspheres were shown to deliver compounds to the RPE cells through phagocytosis.[6,14] Poly(lactic acid) (PLA) microspheres were shown to remain in the vitreous for 1.5 months in normal rabbit eyes and 2 weeks in vitrectomized eyes after intravitreal injection.[15] Studies have shown that PLGA microspheres injected into the vitreous displayed no adverse effects on the ocular tissues.[6,15,16] Transparent and biocompatible viscous vehicles such as hyaluronic acid (HA) or hydroxypropyl methylcellulose (HPMC) can be used to improve the syringability of PLGA microspheres for intraocular delivery.[6] Numerous articles have been published on the use of microspheres for posterior segment delivery; however, to date, use of microspheres for ocular drug delivery is still limited to preclinical evaluations.[17,18]

10.2.1 PEPTIDE AND PROTEIN MICROSPHERE DELIVERY SYSTEMS

The success of microsphere protein or peptide delivery systems depends on several critical factors: the stability of the protein or peptide throughout the manufacturing processes and *in vivo* drug release, obtaining an appropriate drug release profile, and the tolerability of the microspheres. Ideally, the polymers used should be bioerodible to obviate the need for removal after drug release.

To date, there has been only one marketed protein microsphere drug, Nutropin Depot® (somatropin [rDNA origin] for injectable suspension). There are several peptide microsphere drugs approved by the FDA, most delivering luteinizing hormone-releasing hormone (LHRH) analogues or somatostatin analogues (octreotide acetate and lanreotide). Virtually all marketed peptide-loaded microspheres are composed of PLGA. Hydrophilic peptides or proteins exhibit a triphasic release from PLGA matrices. The first phase involves diffusion of the drug from the surface; in the second phase the drug is released through diffusion, with concomitant polymer hydrolysis and reduction of polymer molecular weight; and the tertiary phase represents drug release as a result of dissolution of low-molecular-weight polymer fragments and erosion of the bulk of the polymer.[19] Protein stability and drug release from microspheres can be affected by the manufacturing processes employed, formulation excipients, and the polymers used. All these must be tailored to the properties of the specific proteins being delivered.

10.2.2 Manufacturing Strategies

Considerable effort has been expended to stabilize protein or peptide drugs during the microsphere manufacturing process and to optimize drug release profiles. Drug-loaded microsphere manufacturing methods include emulsification, solvent evaporation, phase separation, and spray drying. To control the size range of microspheres, a precise microsieve™ (Nanomi Oldenzaal, the Netherlands.) can be used with the emulsification processes. Except for spray drying, microspheres are formed in the liquid phase. Double emulsification such as water/oil/water (W/O/W) is commonly used for loading hydrophilic drugs. However, this method could denature the protein due to the presence of a W/O interface. Disaccharides such as trehalose and mannitol or nonionic surfactant polymers such as PEG[20] and poloxamer 188[21] have been shown to help stabilize proteins in these systems.

The ProLease® delivery system from Alkermes is a nonaqueous encapsulation process in which proteins are stabilized with zinc, micronized, lyophilized, and loaded into PLGA microspheres by a spray-freeze-drying process.[22,23] This process eliminates the need for elevated temperatures, high concentrations of surfactants, or mixtures of organic and aqueous solvents, thus minimizing the chance of protein denaturation or degradation. A microsphere formulation of recombinant human growth hormone (hGH) (22 KDa), Nutropin Depot® (somatropin [rDNA origin] for injection suspension, Genentech/Alkermes), was commercialized using ProLease as a bimonthly injection for the treatment of hGH deficiency in children. This drug product was later discontinued due to the extensive resources required for manufacturing and commercialization.

Another alternative process designed to maintain the stability of protein drugs during microsphere manufacturing involves the use of a solid-in-oil-in-water emulsion process. This process involves lyophilizing the protein with stabilizing PEG and subsequently suspending the solid in organic solvents containing biodegradable polymers. Microspheres are then manufactured through a solvent evaporation process.[24] The method was successfully used to stabilize γ-chymotrypsin during encapsulation. The emulsifying agent polyvinyl alcohol (PVA) was replaced with PEG to further reduce protein instability and aggregation at the oil–water interface during encapsulation and protect the protein during release.[25]

10.2.3 Protein and Peptide Microsphere Formulation Approaches

In addition to the manufacturing processes and polymers used, formulation excipients have an effect on protein and peptide performance in these systems. An example of this is sucrose acetate isobutyrate (SAIB). It has been shown that SAIB embedded in PLGA microspheres not only increased protein loading efficiency by suppressing the leakage of proteins during the secondary emulsification but also prolonged the release of protein.[26] The authors further demonstrated that the addition of a nonionic surfactant, sorbitan monooleate 80, to the SAIB-incorporated PLGA microspheres helped to achieve a near-zero-order protein release over 60 days with a minimal initial burst.[26] In another study, urea was used to suppress protein aggregation and reduce its nonspecific adsorption to the matrix of degrading polymers.[27]

Composite systems have been designed to better preserve protein stability in microsphere formulations. A polymer-alloy method has been devised to create a reservoir type of microsphere in which hydrophilic proteins localize in a PLGA-rich core and are surrounded by a PLA-rich shell. Bovine superoxide dismutase was tested as a model protein for this method, and near-linear release over 28 days was achieved.[24]

In other composite systems, proteins are first formulated into particles or hydrogels with a hydrophilic matrix and then these formulations are incorporated into PLGA microspheres to achieve sustained release of protein. PLGA microspheres loaded with hydrophilic gelatin nanoparticles containing proteins have been made using a phase separation or solvent extraction method. Sustained release of protein was achieved with this method.[28] The gelatin-particle PLGA microspheres stabilized basic fibroblast growth factor (bFGF) over standard PLGA microspheres and were able to release the protein over a much longer duration.[29] Interferon-*f*-2a-loaded microspheres comprising calcium alginate cores surrounded by a PLA-PEG (PELA) coating showed a reduced initial burst (from 31% to 14%) when compared to conventional PLA or PLA-PEG microspheres.[30] PLGA-PVA composite microspheres were also shown to stabilize bovine serum albumin (BSA) during its release over more than 50 days.[31]

A recent example of a composite system is PLGA microspheres loaded with 1–4-µm glassy AqueSpheres®. AqueSpheres are protein-containing polysaccharide particles.[32] AqueSpheres are produced by a freezing-induced phase separation process during which the delicate structures as well as the biological activities of proteins can be preserved. Unlike microspheres made with conventional W/O/W methods, microspheres containing AqueSpheres are manufactured by a solid/oil/water (S/O/W) process and have shown significant improvement in the release of proteins. With this system, model proteins such as myoglobin and BSA can be released continuously for up to 50 days with very low initial burst.[32]

10.2.4 Polymers Used in Peptide and Protein Microspheres

Polymers utilized in protein and peptide drug delivery systems are listed in Table 10.4. PLGA polymers are generally biocompatible, biodegradable, proven through clinical experience and marketed drug products, and possess tunable physical and chemical properties to accommodate different requirements for specific drugs. However, PLA and PLGAs are hydrophobic polyesters and pose challenges for sustained release of hydrophilic proteins and peptides. In general, high drug loadings and longer durations of drug release become more difficult for hydrophilic drugs. Proteins can aggregate and also be reversibly or irreversibly adsorbed or bonded to the polymer matrix. Additionally, proteins or peptides may not remain physically or chemically intact when concentrated in a hydrated or acidified polymeric matrix.[8,33–35] Proteins may become denatured, aggregated, degraded, or adsorbed onto the polymer during fabrication or drug delivery, resulting in incomplete drug release.[7–9,33,36–38]

Many strategies have been pursued to improve the ability to formulate proteins and peptides in PLGA-based polymers. Hydrophilicity can be introduced into these polymers with PVA-based branched graft polyesters bearing PLGA side chains. The PLGA chain length and the MW of the PVA backbone can be varied to adjust glass transition

TABLE 10.4
Examples of Materials Explored for the Sustained
Release of Proteins

Microspheres	PLGA (the most common)
	Hydroxylated aliphatic polyester[118]
	PVA-PLGA[39–41], PELA[30,42]
	PEG and poloxamers (PROMAXX technology)
	PEG polybutylene terephthalate (PBT) (PolyActive)
	Cross-linked dextran (OctoDEX)
	PLA, PGA, PCL, and PEG (SynBiosys)
	Silk proteins[53–55]
	Polyanhydrides[119]
Nanoparticles	PLGA (the most common)
	PLGA-PEG[71,120]
	Polybutylcyanoacrylate (PBCA)[121]
	PLGA-amino cyclodextrins[45]
	Poly L-glutamate grafted with vitamin E (Medusa)[80]
	Cholesterol-bearing pullulan (CHP)[81]
	Chitosan[66,73,74]
	Solid lipid[75–79]
Nondegradable implants	Polyvinyl alcohol (PVA)[111–113]
	Ethylene vinyl acetate (EVA)[111,112]
Biodegradable implants	PLGA (the most common)
	Polyanhydrides[122]
	Poly(ortho ester) (POE)[48–50]
	PLA, PGA, PCL, and PEG (SynBiosys)
	Polysaccharides (the Eureka DUET family)
	Silk fibroin[51,52]
In situ-forming systems	PLGA
	Sucrose acetate isobutyrate (SAIB)[123]
In situ-forming hydrogels	PLGA-PEG-PLGA or PEG-PLGA-PEG (ReGel®)[124–126]
	Poloxamers[127,128]
	PEO-PLA,[129] PLA/PEO/PLA triblock[130]
	Modified Pluronic multiblock[131–133]
	MPEG-PCL diblock[134]
	Polyphosphazenes[135–137]
	PNIPAAm[56,138,139]
	Chitosan[59,60]
	Alginate[61]
	Elastin-like peptides[56,57]
	Poly(PEG/PPG/PHB) urethane copolymer[140]
Liposomes	Phospholipid[82,87,88]

temperatures (T_g), thermomechanical properties, as well as the degradation and erosion kinetics of the polymers. Microspheres consisting of these polymers showed linear release of hydrophilic proteins such as BSA with low initial burst.[39–41] Unfortunately, drug encapsulation efficiency decreased significantly when drug loading exceeded 5% w/w.[41] PEG has also been incorporated into the hydrophobic PLGA polyester backbone as a hydrophilic stabilizing agent to increase the affinity of the polymer matrix for proteins. PELA was used to encapsulate glucose oxidase into microspheres and was effective in reducing the specific activity loss of the protein.[42] The local low pH generated during the degradation of PLGA microspheres significantly stresses incorporated proteins and peptides. Proteins can aggregate at low pH on their own or through noncovalent or covalent interactions with polymers, undergo hydrolytic degradation at susceptible peptide bonds such as ASP-X, or experience chemical modifications such as deamination or acylation.[7,8,33,34,36,38] Basic salts such as $Mg(OH)_2$ have been used to neutralize acidic PLGA degradants.[36,43,44] In addition, PLGA has been conjugated to two amino-cyclodextrins, mono(6-(2-aminoethyl)amino-6-deoxy)-β-cyclodextrin and ethylene-diamino bridged bis (β-cyclodextrin), to create a less acidic and more hydrophilic microenvironment.[45] Both the entrapment efficiency and the structural stability of the released protein were improved using these polymers.

In addition to the PLGA polymers, other biodegradable synthetic polymers have been explored for protein-delivering microspheres. PolyActive® from OctoPlus is a series of poly(ether ester) multiblock copolymers consisting of hydrophilic PEG and hydrophobic PBT blocks. During the double emulsion manufacturing process of microspheres from PolyActive, the integrity of proteins can be protected through the presence of PEG in the polymer backbone and the surfactant properties of the polymers. The PEG content and chain length in PolyActive can be tailored to achieve release of protein drugs over weeks to months. Lysozyme was encapsulated in PolyActive microspheres as a model protein and shown to retain full activity when released over 40 days *in vitro* (www.octoplus.nl). Locteron®, a PolyActive microsphere formulation for recombinant interferon α2b (IFN-α2b), is in clinical development for hepatitis C.[46] Octoplus has also developed OctoDEX™ based on cross-linked dextran microspheres prepared without organic solvents. OctoDEX was developed for the sustained delivery of larger proteins (www.octoplus.nl). SynBiosys™ polymers developed by InnoCore are proprietary multiblock biodegradable polymers for delivering peptide and potentially protein drugs. In these poly(ether ester) copolymers, DL-lactide, glycolide, ε-caprolactone, and PEG are arranged into soft hydrophilic and rigid hydrophobic blocks.[47] By varying these two blocks, the hydrophilicity, swelling properties, and polymer degradation profile can be tuned to achieve sustained peptide drug release for up to 6 months (www.innocore.nl).

Poly(ortho esters) (POE) (Biochronomer Technology, AP Pharma) polymers have been tested as an alternative to PLGA for protein drug delivery in implants.[48–50] POE polymers are biocompatible and degrade into relatively harmless compounds through hydrolysis. Unlike PLGA, POE polymers undergo surface—rather than bulk—erosion and therefore have the potential of zero-order drug release. BSA, as a model protein, has been continuously and completely delivered over a 2-month period from POE implants.[48] The addition of PEG in the polymer structure further reduced the lag time in early release.

It is possible to incorporate proteins into carrier-free protein microspheres with water-soluble polymers. Utilizing the PROMAXX® technology (Baxter), microspheres of 1–2 μm diameter loaded with more than 90% of protein can be produced through a controlled phase separation process that involves cooling a highly concentrated protein solution together with aqueous polymers such as PEG and poloxamer. This technology has emerged as a promising platform for pulmonary delivery of proteins including insulin and hGH. The system is useful for immediate or controlled release and could potentially be applied to ocular drug delivery.

Natural degradable polysaccharides such as the Eureka™ DUET family were developed for protein delivery. Implants fabricated from these polymers are degraded *in vivo* by enzyme-mediated surface erosion, and the release of large proteins in the range from 30 to 150 KDa can be controlled. The polymer breakdown products are natural protein stabilizers and as such protein functionality can be maintained for up to 6 months *in vivo* (www.surmodics.com).

Silk fibroin can also be made into protein-embedded films for the sustained release of proteins.[51,52] The release of the protein, though, can be incomplete due to interactions between the protein and the silk fibroin as well as the molecular weight of the protein. The release of nerve growth factor (NGF) over 3 weeks was reported.[51] Microspheres of naturally derived silk proteins have been tested for protein drug delivery.[53–55] The degradation rate of silk fibroin can be adjusted from weeks to months *in vivo* by controlling the silk crystallinity. Proteins including horseradish peroxidase (HRP) and insulin-like growth factor-I (IGF-I) were encapsulated into silk fibroin microspheres, and their release was found to be affected by both silk–drug interactions and the molecular weights of the drugs. With theoretical drug loadings at 0.14% (w/w), microspheres prepared from aqueous solutions with laminar jet breakup technology were used to continuously release biologically active IGF-I over 7 weeks.[55] Entrapment efficiency was greater than 96%.

10.2.5 POLYMERS USED IN GELLING PROTEIN DELIVERY SYSTEMS

Another area of interest in protein delivery is the use of gels or thermogelling systems. The polymers utilized in these systems may have value in particulate delivery on their own or in composite systems. Elastin-like peptides (ELPs) are thermogelling biopolymers that consist of a repeating pentapeptide sequence derived from native elastin. ELPs are soluble in aqueous solutions and become insoluble and precipitate at body temperatue.[56,57] This forms a drug depot from which the polymer will resolubilize over time. ELPs are biocompatible and nonimmunogenic and could potentially be used in delivering protein or peptide drugs.

Natural polymers such as chitosan have been used for manufacturing protein delivery systems. Chitosan is a linear amino-polysaccharide mucoadhesive polymer. It is polycationic, biocompatible, and possesses antimicrobial and wound-healing properties.[58,59] Chitosan can degrade *in vivo* through chitosanase and lysozyme activity. Chitosan-based injectable thermosensitive hydrogels were formed when PEG was grafted into chitosan chains through covalent bonding and then cross-linked with genipin. These gels were shown to release BSA as a model protein continuously for over 40 days.[60]

Alginate is a linear copolysaccharide and represents another natural polymer that is biocompatible with low toxicity. Alginate is not degradable but can be dissolved into the surrounding media when divalent cations such as Ca^{++} are lost. Alginates were approved by the FDA as wound-dressing material.[61] Interestingly, sodium alginate can gel in the eye without external calcium ions and has been proposed as an in situ–forming ophthalmic drug delivery system.[61] Besides alginate, pectins can also gel in the presence of calcium ions. GelSite® polymers made with aloe vera pectin have been shown to form in situ gels and deliver functional bFGF.[62]

Polymers derivatized from dextran have been used to make vehicles for protein delivery. In one case, a dextran backbone is grafted with D- or L-oligo-lactate and gelation occurs in situ when the two isomers are mixed and physically cross-linked. This gel is biocompatible and biodegradable and can release a protein such as interleukin-2 (IL-2) for days or weeks.[63] In another case, oppositely charged hydroxyethyl methacrylate-derivatized dextran microspheres assemble into an injectable hydrogel through electrostatic attractions. Proteins of different sizes or isoelectric points can be loaded in and/or between the microspheres, maintain their active structures, and be continuously released by diffusion. A sustained drug release for up to 60 days has been demonstrated *in vitro* with model proteins.[64]

10.3 NANOPARTICLES

Nanoparticles are solid, colloidal matrices with diameters ranging from 10 to 1000 nm. Therapeutic molecules can be trapped or encapsulated into nanoparticles and released by diffusion, desorption, or as the polymeric matrix degrades. Nanoparticles can enhance cellular entry. Studies using negatively charged fluorescent polystyrene particles ranging in size from 20 nm to 2 μm showed that smaller particles were taken up by retinal pigment epithelial cells *in vitro* (ARPE-19 cells)[65] better than the larger particles due to an endocytosis. On the other hand, nanoparticles >200 nm and microspheres appear to be more valuable in extending retinal drug release.[65] Nanoparticles have been studied as topical suspensions or local injectable systems for conditions such as ocular inflammation, infection, and glaucoma.[17] For topical delivery, it was found that particles of ≤100 nm were able to penetrate across the corneal barrier[66] and that modification of nanoparticles with either PEG or chitosan improves mucoadhesion and drug permeation.[66] Intravitreal injection of PLA nanoparticles resulted in their enrichment in the RPE. The nanoparticles were detectable within the RPE for 4 months after a single injection, suggesting that sustained release of an ocular drug is possible.[67] Nanoparticles loaded with red nuclear fluorescent protein were also observed in RPE cells following intravitreal administration.[68] Finally, nanoparticles are being tested for periocular delivery such as subconjunctival administration.[3]

Nanoparticles can be made with synthetic polymers such as PLGA, PLA-PEG, and PCL[66,69] or natural polymers such as chitosan, gelatin,[66] and lipids. Proteins or peptides can be loaded into PLGA nanoparticles through a double emulsion evaporation method. However, due to their small sizes and short diffusional path lengths, nanoparticles tend to exhibit extremely low protein-loading efficiency, high initial burst, and relatively short release durations. Thus, in contrast to microspheres, less progress on the sustained delivery of proteins using nanoparticles has been reported in the literature. In one study, about 90% recombinant human granulocyte colony-stimulating factor (rhG-CSF) was encapsulated into PLGA nanoparticles with a single oil-in-water emulsion process and released over a 1-week period.[70] PEGylated PLGA nanoparticles (stealth nanoparticles) were tested for improving protein release *in vivo*.[71] To improve protein drug loading, PEG/PLA nanoparticles were made with a semicontinuous compressed CO_2 antisolvent precipitation method. In this system, insulin release was found to take place by a diffusion mechanism and was affected by PEG concentration and molecular weight.[72]

Chitosan is an example of a natural polymer for nanoparticle fabrication. Cationic chitosan can bind to negatively charged mucins in the tear film, creating a muco-adhesive delivery system, improving the precorneal residence time, and enhancing drug uptake by ocular epithelia.[66] Protein-loaded chitosan nanoparticles can be obtained by a simple ionotropic gelation process using polyanionic tripolyphosphate (TPP). However, the release of protein from this system could be diminished due to interactions between the chitosan and the protein molecule.[73,74]

Solid lipid nanoparticles have also been extensively investigated for their application in delivering peptides and proteins.[75] Lipid molecules can form solid matrices, and proteins can be incorporated into solid lipid particles by microemulsion, high-pressure homogenization, solvent emulsification-evaporation, and solvent emulsification-diffusion or supercritical fluid technologies. These particles have the advantage of delivering drugs across cell membranes or the blood-brain barrier.[76,77] Unfortunately, drug release durations are generally short, limited to hours or days.[75,78,79]

Nanoparticles made with semisynthetic materials have also been investigated for protein and peptide delivery. The Medusa® polymers (Flamel Technologies) consisting of a poly L-glutamate backbone grafted with hydrophobic α-tocopherol molecules (vitamin E) can be used to form stable nanoparticles (10–50 nm) through self-assembly in water.[80] Proteins or peptides bind to the hydrophobic nanodomains within the nanoparticles and are slowly released when being displaced by endogenous proteins present in the body fluid. Since both the capture and release of proteins are nondenaturing, the protein's integrity can be preserved. Formulations can be manufactured under mild conditions without the use of organic solvents, and the polymers are biocompatible and biodegradable. Proteins including human IFN-α2b were shown to be released for days with the technology, and almost all released protein remained active.[80] Proteins such as IFN-α2b and insulin delivered with the Medusa system are currently in various clinical studies. It remains to be seen how this system works in the eye.

Hydrophobic polysaccharides such as cholesterol-bearing pullulan (CHP) have been found to form relatively monodispersed and colloidally stable hydrogel nanoparticles in water upon self-aggregation. A variety of molecules including proteins and peptides were encapsulated and released from such nanogels. CHP nanogels are thought to have a "molecular chaperone" activity that can help protein refold into its active conformation. Drug-loaded nanogels are extremely simple to make, but may not be stable *in vivo* since serum proteins such as BSA can induce the rapid release of the drug from the complex.[81]

10.4 LIPOSOMES

Liposomes are vesicles composed of phospholipid bilayers surrounding an aqueous core and can be of various sizes and morphologies. Liposomes are a well-established drug delivery platform and have been successfully utilized to commercialize small molecules such as doxorubicin HCl to achieve localized and controlled drug delivery. Liposomes are well tolerated and have low toxicity. They can be taken up by phagocytic cells such as the RPE cells, thus enabling intracellular drug delivery.[18] Due to the presence of their aqueous core, liposomes can potentially deliver hydrophilic compounds including proteins or peptides. The compositions of liposomes can be optimized to release drugs such as IFN-γ over periods up to several days.[82,83]

To date, Visudyne® (verteporfin injection, Novartis Pharmaceuticals) is the only ocular liposomal drug approved for clinical use.[17] While Visudyne is approved for ocular indications, wet AMD (predominantly classic subfoveal CNV), and pathologic myopia or presumed ocular histoplasmosis, it is systemically administered. Visudyne is given by infusion followed by external activation with photodynamic therapy. Rostaporfin (Photrex®, Miravant Medical Technologies) is a similar liposomal photosensitizing agent that was under development for treating AMD.[17] Miravant filed an new drug application (NDA) for Photrex in 2004 for which they received an approvable letter from the FDA requesting a confirmatory clinical study. Enrollment in the confirmatory clinical study was suspended in February 2006 due to delays and poor enrollment.

Preclinical research has been conducted on direct intravitreal administration of liposomes. These studies indicated that intravitreal injection of drug-loaded liposomes did not cause toxic effects to the retina and further reduced the toxicity of the loaded drug.[84,85] Liposomes have also been investigated for topical delivery.[3] Topically applied positively charged liposomes were found to bind preferentially to the negatively charged corneal surface. Theoretically, this may allow for direct transfer of a drug from a liposome into the corneal epithelial cell membranes.[86] The main limitation was the instability of the liposomes, with lipids aggregating on the surface[66] mucin of the cornea.

DepoFoam™ (Pacira Pharmaceuticals) is a multivesicular liposome formulation from which drug release can be controlled by a temperature-sensitive diffusion through the phospholipid bilayers.[87] DepoFoam releases encapsulated drugs over 1–14 days and can be administered through 27–31-gauge needles. This technology has enabled the commercialization of drug products (DepoCyt® and DepoDur®) and has been shown to be feasible for the relatively short-term delivery of protein molecules including IFN-α and erythropoietin (EPO) (www.pacira.com).

Unfortunately, despite considerable effort in the past two decades, and in contrast to the achievements on delivering small molecules, liposomal protein formulations have yet to enter advanced clinical trials or be commercialized for ocular use. Issues with liposomal protein formulations that still need resolution include protein denaturation in the organic solvents and detergents used in the manufacturing processes, low encapsulation efficiencies,[88] and premature release of the drug *in vivo*.[75,89]

10.5 FUTURE OUTLOOK

Several issues surrounding the development of protein and peptide delivery systems to the eye remain, including maintaining protein stability throughout the manufacture of the system and drug release, achieving a suitable drug loading and release duration, and ensuring the safety and tolerability of the system. Innovation and necessity will continue to drive progress in protein and peptide drug delivery. Less-invasive administration, high ocular tolerability, local delivery to the vitreous and retina, and longer and more linear protein and peptide release will continue to be the goals for new nanoparticulate protein and peptide delivery systems. New polymers are being engineered to not only stabilize proteins and peptides and improve drug loading but also to achieve long-term drug release. The desirable polymers should be bioerodible, biocompatible, and possess suitable chemical, physical, and mechanical properties to allow for optimal protein delivery. With more antibody, protein, and peptide drugs entering the development pipeline, we anticipate significant progress in sustained and local ocular delivery systems for these biological therapeutics.

REFERENCES

1. Parkinson J. The promise of ocular implants. *Retinal Physician.* 2005; 2: 79–84.
2. Short BG. Safety evaluation of ocular drug delivery formulations: Techniques and practical considerations. *Toxicol Pathol.* 2008; 36: 49–62.
3. Gaudana R, Jwala J, Boddu SHS, Mitra AK. Recent perspectives in ocular drug delivery. *Pharm Res.* 2009; 26: 1197–1216.
4. Bakri SJ, Snyder MR, Reid JM, Pulido JS, Singh RJ. Pharmacokinetics of intravitreal bevacizumab (Avastin). *Ophthalmology.* 2007; 114: 855–859.
5. Bakri SJ, Snyder MR, Reid JM, Pulido JS, Ezzat MK, Singh RJ. Pharmacokinetics of intravitreal ranibizumab (Lucentis). *Ophthalmology.* 2007; 114: 2179–2182.
6. Herrero-Vanrell R, Refojo MF. Biodegradable microspheres for vitreoretinal drug delivery. *Adv Drug Deliv Rev.* 2001; 52: 5–16.
7. van de Weert M, Hennink WE, Jiskoot W. Protein instability in poly(lactic-co-glycolic acid) microparticles. *Pharm Res.* 2000; 17: 1159–1167.
8. Houchin ML, Topp EM. Chemical degradation of peptides and proteins in PLGA: A review of reactions and mechanisms. *J Pharm Sci.* 2008; 97: 2395–2404.
9. Crotts G, Park TG. Protein delivery from poly(lactic-co-glycolic acid) biodegradable microspheres: Release kinetics and stability issues. *J Microencapsul.* 1998; 15: 699–713.
10. Barcia E, Herrero-Vanrell R, Díez A, Alvarez-Santiago C. Downregulation of endotoxin-induced uveitis by intravitreal injection of polylactic-glycolic acid (PLGA) microspheres loaded with dexamethasone. *Exp Eye Res.* 2009; 89: 238–245.
11. Moritera T, Ogura Y, Yoshimura N, Honda Y, Wada R, Hyon SH, Ikadat Y. Biodegradable microspheres containing adriamycin in the treatment of proliferative vitreoretinopath. *Invest Ophthalmol Vis Sci.* 1992; 33: 3125–3130.
12. Saisbin Y, Silva RL, Saisbin Y, Callaban K, Schoch C, Ablbeim M, Lai H et al. Periocular injection of microspheres containing PKC412 inhibits choroidal neovascularization in a porcine model. *Invest Ophthalmol Vis Sci.* 2003; 44: 4989–4993.
13. Garrasquillo KG, Ricker JA, Rigas IK, Miller JW, Gragoudas ES, Adamis AP. Controlled delivery of the antiVEGF aptemer EYE001 with poly(lactic-co-glycolic) acid microspheres. *Invest Ophthalmol Vis Sci.* 2003; 44: 290–299.

14. Moritera T, Ogura Y, Yoshimura N, Kuriyama S, Honda Y, Tabata Y, Ikada Y. Feasibility of drug targeting to the retinal pigment epithelium with biodegradable microspheres. *Curr Eye Res.* 1994; 13: 171–176.

15. Moritera T, Ogura Y, Honda Y, Wada R, Hyon SH, Ikadat Y. Microspheres of biodegradable polymers as a drug-delivery system in the vitreous. *Invest Ophthalmol Vis Sci.* 1991; 32: 1785–1790.

16. Mordenti J, Thomsen K, Licko V, Berleau L, Kahn JW, Cuthbertson RA, Duenas ET et al. Intraocular pharmacokinetics and safety of a humanized monoclonal antibody in rabbits after intravitreal administration of a solution or a PLGA microsphere formulation. *Toxicol Sci.* 1999; 52: 101–106.

17. Eljarrat-Binstock E, Pe'er J, Domb AJ. New techniques for drug delivery to the posterior eye treatment. *Pharm Res.* 2010; 27: 530–543.

18. del Amo EM, Urtti A. Current and future ophthalmic drug delivery systems. A shift to the posterior segment. *Drug Discov Today.* 2008; 13: 135–143.

19. Sanders LM, Kent JS, McRae GI, Vickery BH, Tice TR, Lewis DH. Controlled release of a luteinizing hormone-releasing hormone analogue from poly(d,l-lactide-co-glycolide) microspheres. *J Pharm Sci.* 1984; 73: 1294–1297.

20. Kang F, Singh J. Effects of additives on the release of model protein from PLGA microspheres. *AAPS PharmSciTech.* 2001; 2:1–7.

21. Schwendeman SP, Tobio M, Joworowicz M, Alonso MJ, Langer R. New strategies for the microencapsulation of tetanus vaccine. *J Microencapsul.* 1998; 15: 299–318.

22. Bartus RT, Tracey MA, Emerich DE, Zale SE. Sustained delivery of proteins for novel therapeutic products. *Science.* 1998; 281: 1161–1162.

23. Herbert P, Murphy K, Johnson O, Dong N, Jaworowicz W, Tracey MA, Cleland H, Putney SD. A large-scale process to produce microencapsulated proteins. *Pharm Res.* 1998; 15: 357–361.

24. Morita T, Sakamura Y, Horikiri Y, Sukuki T, Yoshino H. Protein encapsulation into biodegradable microspheres by a novel S/O/W emulsion method using poly(ethylene glycol) as a protein micronization adjuvant. *J Control Release.* 2000; 69: 435–444.

25. Castellanos IJ, Crespo R, Griebenow K. Poly(ethylene glycol) as stabilizer and emulsifying agent: A novel stabilization approach preventing aggregation and inactivation of proteins upon encapsulation in bioerodible polyester microsphere. *J Control Release.* 2003; 88: 135–145.

26. Lee ES, Kwon MJ, Lee H, Na K, Kim JJ. *In vitro* study of lysozyme in poly(lactide-co-glycolide) microspheres with sucrose acetate isobutyrate. *Eur J Pharm Sci.* 2006; 29: 435–441.

27. Nam YS, Song SH, Choi JY, Park TG. Lysozyme microencapsulation within biodegradable PLGA microspheres: Urea effect on protein release and stability. *Biotechnol Bioeng.* 2000; 70: 270–277.

28. Li JK, Wang N, Wu XS. A novel biodegradable system based on gelatin nanoparticles and poly(lactic-co-glycolic acid) microspheres for protein and peptide drug delivery. *J Pharm Sci.* 1997; 86: 891–895.

29. Li, SH, Cai SX, Liu B, Ma KW, Wang ZP, Li XK. *In vitro* characteristics of poly (lactic-co-glycolic acid) microspheres incorporating gelatin particles loading basic fibroblast growth factor. *Acta Pharmacol Sin.* 2006; 27: 754–759.

30. Zhou S, Deng X, He S, Li X, Jia W, Wei D, Zhang Z, Ma J. Study on biodegradable microspheres containing recombinant interferon-f-2α. *J Pharm Pharmacol.* 2002; 54: 1287–1292.

31. Wang N, Wu XS, Li JK. A heterogeneously structured composite based on poly (lactic-co-glycolic acid) microspheres and poly (vinyl alcohol) hydrogel nanospheres for long-term protein drug delivery. *Pharm Res.* 1999; 16: 1430–1435.

32. Yuan W, Wu F, Jin T. Microencapsulation of protein-loaded polysaccharide particles within poly(D, L-lactic-co-glycolic acid) microspheres using S/O/W: Characterization and release studies. *Polym Adv Technol.* 2009; 20: 834–842.

33. Estey T, Kang J, Schwendeman SP, Carpenter JF. BSA degradation under acidic conditions: A model for protein instability during release from PLGA delivery systems. *J Pharm Sci.* 2006; 95: 1626–1639.

34. Lucke A, Göpferich A. Acylation of peptides by lactic acid solutions. *Eur J Pharm Biopharm.* 2003; 55: 27–33.

35. Mariette B, Coudane J, Vert M, Gautier J-C, Moneton P. Release of the GRF29NH2 analog of human GRF44NH2 from a PLA/GA matrix. *J Control Release.* 1993; 24: 237–246.

36. Giteau A, Venier-Julienne MC, Aubert-Pouëssel A, Benoit JP. How to achieve sustained and complete protein release from PLGA-based microparticles? *Int J Pharm.* 2008; 350: 14–26.

37. Kim HK, Park TG. Microencapsulation of human growth hormone within biodegradable polyester microspheres: Protein aggregation stability and incomplete release mechanism. *Biotechnol Bioeng.* 1999; 65: 659–667.

38. Houchin ML, Heppert K, Topp EM. Deamination, acylation and proteolysis of a model peptide in PLGA films. *J Control Release.* 2006; 112: 111–119.

39. Breitenbach A, Kissel T. Biodegradable comb polyesters: Part I. Synthesis, characterization and structural analysis of poly(lactide) and poly(lactide-co-glycolide) grafted onto water-soluble poly(vinyl alcohol) as backbone. *Polymer.* 1998; 39: 3261–3271.

40. Breitenbach A, Pistel KF, Kissel T. Biodegradable comb polyesters: Part II. Erosion and release properties of poly(vinyl alcohol)-g-poly(lactic-co-glycolic acid. *Polymer.* 2000; 41: 4781–4792.

41. Pistel KF, Breitenbach A, Regina Z-V, Kissel T. Brush-like branched biodegradable polyesters: Part III. Protein release from microspheres of poly (vinyl alcohol)-grafted-poly(D,L-lactic-co-glycolic acid). *J Control Release.* 2001; 73: 7–20.

42. Li X, Zhang Y, Yan R, Jia W, Yuan M, Deng X, Huang Z. Influence of process parameters on the protein stability encapsulated in poly-DL-lactide-poly(ethylene glycol) microspheres. *J Control Release.* 2000; 68: 41–52.

43. Zhu G, Schwendeman SP. Stabilization of proteins encapsulated in cylindrical poly(lactide-co-glycolide) implants: Mechanism of stabilization by basic additives. *Pharm Res.* 2000; 17: 351–357.

44. Zhu G, Mallery SR, Schwendeman SP. Stabilization of proteins encapsulated in injectable poly(lactide-co-glycolide). *Nat Biotechnol.* 2000; 18: 52–57.

45. Gao H, Wang YN, Fan YG, Ma JB. Conjugates of poly(DL-lactide-co-glycolide) on amino cyclodextrins and their nanoparticles as protein delivery system. *J Biomed Mater Res A.* 2006; 80: 111–122. doi: 10.1002.

46. De Leede LG, Humphries JE, Bechet AC, Van Hoogdalem EJ, Verrijk R, Spencer DG. Novel controlled-release Lemna-derived IFN-alpha2b (Locteron): Pharmacokinetics, pharmacodynamics, and tolerability in phase I clinical trial. *J Interferon Cytokine Res.* 2008; 28: 113–122.

47. Steendam R, van der Laan A, Hissink D. Bioresorbable drug-eluting stent coating formulations based on SynBioysys biodegradable multi-block copolymers. *J Control Release.* 2006; 116: e94–e95.

48. Rothen-Weinhold A, Schwach-Abdellaoui K, Barr J, Ng SY, Shen HR, Gurny R, Heller J. Release of BSA from poly(ortho ester) estruded thin strands. *J Control Release.* 2001; 71: 31–37.

49. van de Weert M, van Steenbergen MJ, Cleland JL, Heller J, Hennink WE, Crommelin DJA. Semisolid, self-catalyzed poly(ortho ester) as controlled-release systems: Protein release and protein stability issues. *J Pharm Sci.* 2002; 91: 1065–1074.

50. Heller J, Barr J. Poly(ortho esters)-from concept to reality. *Biomacromolecules*. 2004; 5: 1625–1632.

51. Uebersax L, Mattotti M, Papaloïzos M, Merkle HP, Gander B, Meinel L. Silk fibroin matrices for the controlled release of nerve growth factor (NGF). *Biomaterials*. 2007; 28: 4449–4460.

52. Hofmann S, Foo CT, Rossetti F, Textor M, Vunjak-Novakovic G, Kaplan DL, Merkle HP, Meinel L. Silk fibroin as an organic polymer for controlled drug delivery. *J Control Release*. 2006; 111: 219–227.

53. Wang X, Yucel T, Lu Q, Hu X, Kaplan DL. Silk nanospheres and microspheres from silk/pva blend films for drug delivery. *Biomaterials*. 2010; 31: 1025–1035.

54. Wang X, Wenk E, Matsumoto A, Meinel L, Li C, Kaplan DL. Silk microspheres for encapsulation and controlled release. *J Control Release*. 2007; 117: 360–370.

55. Wenk E, Wandrey AJ, Merkle HP, Meinel L. Silk fibroin spheres as a platform for controlled drug delivery. *J Control Release*. 2008; 132: 26–34.

56. Bikram M, West JL. Thermo-responsive systems for controlled drug delivery. *Expert Opin Drug Deliv*. 2008; 5: 1077–1091.

57. Betre H, Liu W, Zalutsky MR, Chilkoti A, Kraus VB, Setton LA. A thermally responsive biopolymer for intra-articular drug delivery. *J Control Release*. 2006; 115: 175–182.

58. Mundada AS, Avari, JG. In situ gelling polymers in ocular drug delivery systems: A review. *Crit Rev Ther Drug Carrier Syst*. 2008; 26: 85–118.

59. Wadhwa S, Paliwal R, Paliwal SR,Vyas SP. Chitosan and its role in ocular therapeutics. *Mini Rev Med Chem*. 2009; 9: 1639–1647.

60. Bhattarai N, Ramay HR, Gunn J, Matsen FA, Zhang M. PEG-grafted chitosan as an injectable thermosensitive hydrogel for sustained protein release. *J Control Release*. 2005; 103: 609–624.

61. Cohen S, Lobel E, Trevgoda A, Peled Y. A novel in situ-forming ophthalmic drug delivery system from alginate undergoing gelation in the eye. *J Control Release*. 1997; 44: 201–208.

62. Yawei N, Kenneth MY. In-situ gel formation of pectin. US patent no. 6,777,000 B2. 2004.

63. Bos GW, Jacobs JJL, Koten JW, Van Tomme S, Veldhuis T, van Nostrum CF, Otter WD, Hennink WE. In situ crosslinked biodegradable hydrogels loaded with IL-2 are effective tools for local IL-2 therapy. *Eur J Pharm Sci*. 2004; 21: 561–567.

64. Van Tomme SR, De Geest BG, Braeckmans K, De Smedt SC, Siepmann F, Siepmann J, van Nostrum CF, Hennink WE. Mobility of model proteins in hydrogels composed of oppositely charged dextran microspheres studied by protein release and fluorescence recovery after photobleaching. *J Control Release*. 2005; 110: 67–78.

65. Amrite AC, Kompella UB. Nanoparticles and microparticles: Particle engineering, cell uptake, *in vivo* disposition & efficacy. *Drug Deliv Technol*. 2007; 7: 52–56.

66. Nagarwal RC, Kant S, Singh PN, Maiti P, Pandit JK. Polymeric nanoparticulate system: A potential approach for ocular drug delivery. *J Control Release*. 2009; 136: 2–13.

67. Bourges JL, Gautier SE, Delie F, Bejjani RA, Jeanny JC, Gurny R, BenEzra D, Behar-Cohen FF. Ocular drug delivery targeting the retina and retinal pigment epithelium using polylactide nanoparticles. *Invest Ophthalmol Vis Sci*. 2003; 44: 3562–3569.

68. Bejjani RA, BenEzra D, Cohen H, Rieger J, Andrieu C, Jeanny JC, Gollomb G, Behar-Cohen FF. Nanoparticles for gene delivery to retinal pigment epithelial cells. *Mol. Vis.* 2005; 11: 124–132.

69. Balasubramanian V, Onaca O, Enea R, Hughes DW, Palivan CG. Protein delivery: From conventional drug delivery carriers to polymeric nanoreactors. *Expert Opin Drug Deliv*. 2010; 7: 63–78.

70. Choi SH, Park TG. G-CSF loaded biodegradable PLGA nanoparticles prepared by a single oil-in-water emulsion method. *Int J Pharm*. 2006; 311: 223–228.

71. Li YP, Pei YY, Zhang XY, Gu ZH, Zhou ZH, Yuan WF, Zhou JJ, Zhu JH, Gao XJ. PEGylated PLGA nanoparticles as protein carriers: Synthesis, preparation and biodistribution in rats. *J Control Release.* 2001; 71: 203–211.

72. Caliceti P, Salmaso S, Elvassore N, Bertucco A. Effective protein release from PEG/PLA nano-particles produced by compressed gas anti-solvent precipitation techniques. *J Control Release.* 2004; 94: 195–205.

73. Gan Q, Wang T. Chitosan nanoparticle as protein delivery carrier-Systemic examination of fabrication conditions for efficient loading and release. *Colloids Surf B Biointerfaces.* 2007; 59: 24–34.

74. Pan Y, Li Y, Zhao H, Zheng J, Xu H, Wei G, Hao J, Cui F. Bioadhesive polysaccharide in protein delivery system: Chitosan nanoparticles improve the intestinal absorption of insulin *in vivo*. *Int J Pharm.* 2002; 249: 139–147.

75. Almeida AJ, Souto E. Solid lipid nanoparticles as a drug delivery system for peptides and proteins. *Adv Drug Deliv Rev.* 2007; 59: 478–490.

76. Fundarò A, Cavalla R, Bargoni A, Vighetto D, Zara GP, Gasco MR. Non-stealth and stealth solid lipid nanoparticles (SLN) carrying doxorubicin: Pharmacokinetics and tissue distribution after i.v. administration to rats. *Pharm Res.* 2000; 42: 337–343.

77. Zara GP, Cavalli R, Bargoni A, Fundarò A, Vighetto D, Gasco MR. Intravenous administration to rabbits of non-stealth and stealth doxorubicin-loaded solid lipid nanoparticles at increasing concentrations of stealth agent: Pharmacokinetics and distribution of doxorubicin in brain and other tissues. *J Drug Target.* 2007; 10: 327–335.

78. Reithmeier H, Herrmann J, Göpferich A. Lipid microparticles as a parental controlled release device for peptides. *J Control Release.* 2001; 73: 339–350.

79. Reithmeier H, Herrmann J, Göpferich A. Development and characterization of lipid microparticles as a drug carrier for somatostatin. *Int J Pharm.* 2001; 218: 133–143.

80. Chan YP, Meyrueix R, Kravtzoff R, Nicolas F, Lundstrom K. Review on Medusa®: A polymer-based sustained release technology for protein and peptide drugs. *Expert Opin Drug Deliv.* 2007; 4: 441–451.

81. Shimizu T, Kishida T, Hasegawa U, Ueda Y, Imanishi J, Yamagishi H, Akiyoshi K, Otsuji E, Mazda O. Nanogel DDS enables sustained release of IL-12 for tumor immunotherapy. *Biochem Biophys Res Commun.* 2008; 367: 330–335.

82. Van Slooten ML, Boerman O, RomØren K, Kedar E, Crommelin DJ, Storm G. Liposomes as sustained release system for human interferon-gamma: Biopharmaceutical aspects. *Biochim Biophys Acta.* 2001; 1530: 134–145.

83. Allen TM, Mehra T, Hansen C, Chin YC. Stealth liposomes: An improved sustained release system for 1-beta-D-arabinofuranosylcytosine. *Cancer Res.* 1992; 52: 2431–2439.

84. Wiechens B, Neumann D, Grammer JB, Pleyer U, Hedderich J, Duncker GIW. Retinal toxicity of liposome-incorporated and free ofloxacin after intravitreal injection in rabbit eyes. *Int Ophthalmol.* 1999; 22: 133–143.

85. Zhang R, He R, Qian J, Guo J, Xue K, Yuan Y. Treatment of experimental autoimmune uveoretinitis with intravitreal injection of Tacrolimus (FK506) encapsulated in liposomes. *Invest Ophthalmol Vis Sci.* 2010; 51: 3575–3582.

86. Sahoo SK, Dilnawaz F, Krishnakumar S. Nanotechnology in ocular drug delivery. *Drug Discov Today.* 2008; 13: 144–151.

87. Mantripragada S. A lipid based Depot (DepoFoam® technology) for sustained release drug delivery. *Prog Lipid Res.* 2002; 41: 392–406.

88. Yatuv R, Robinson M, Dayan I, Baru M. Enhancement of the efficacy of therapeutic proteins by formulation with PEGylated liposomes: A case of FVIII, FVIIa and G-CSF. *Expert Opin Drug Deliv.* 2010; 7: 187–201.

89. Lee KY, Yuk SH. Polymeric protein delivery systems. *Prog Polym Sci.* 2007; 32: 669–697.

90. Baltmr A, Duggan J, Nizari S, Salt TE, Cordeiro MF. Neuroprotection in glaucoma—Is there a future role? *Exp Eye Res.* 2010; 91: 554–566.

91. Lambiase A, Aloe L, Centofanti M, Parisi V, Mantelli F, Colafrancesco V, Manni GL, Bucci MG, Bonini S, Lei-Montalcini R. Experimental and clinical evidence of neuroprotection by nerve growth factor eye drops: Implications for glaucoma. *Proc Natl Acad Sci USA.* 2009; 106: 13469–13474.

92. Jiang C, Moore MJ, Zhang X, Klassen H, Langer R, Young M. Intravitreal injections of GDNF-loaded biodegradable microspheres are neuroprotective in a rat model of glaucoma. *Mol Vis.* 2007; 13: 1783–1792.

93. Naskar R, Vorwerk C, Dreyer EB. Concurrent downregulation of a glutamate transporter and receptor in glaucoma. *Invest Ophthalmol Vis Sci.* 2000; 41: 1940–1944.

94. Pease ME, Zack DJ, Berlinicke C, Bloom K, Cone F, Wang Y, Klein EL, Hauswirth WW, Quigley HA. Effect of CNTF on retinal ganglion cell survival in experimental glaucoma. *Invest Ophthalmol Vis Sci.* 2009; 50: 2194–2200.

95. Cui Q, Harvey AR. At least two mechanisms are involved in the death of retinal ganglion cells following target ablation in neonatal rats. *J Neurosci.* 1995; 15: 8143–8155.

96. Schuettauf F, Zurakowski D, Quinto K, Varde MA, Besch D, Laties A, Anderson R, Wen R. Neuroprotective effects of cardiotrophin-like cytokine on retinal ganglion cells. *Graefes Arch Clin Exp Ophthalmol.* 2005; 243: 1036–1042.

97. Digicaylioglu M, Lipton SA. Erythropoietin-mediated neuroprotection involves crosstalk between Jak2 and NF-κB signaling cascade. *Nature.* 2001; 412: 641–647.

98. Junk AK, Mammis A, Savitz SI, Singh M, Roth S, Malhotra S, Rosenbaum PS, Cerami A, Brines M, Rosenbaum DM. Erythropoietin administration protects retinal neurons from acute ischemia-reperfusion injury. *Proc Natl Acad Sci USA.* 2002; 99: 10659–10664.

99. Takeda A, Baffi JZ, Kleinman ME, Cho WG, Nozaki M, Yamada K, Kaneko H et al. CCR3 is a target for age-related macular degeneration diagnosis and therapy. *Nature.* 2009; 460: 225–231.

100. Lambiase A, Mantelli F, Bonini S. Nerve growth factor eye drops to treat glaucoma. *Drug News Perspect.* 2010; 23: 361–367.

101. Bonini S, Lambiase A, Rama P, Caprioglio G, Aloe L. Topical treatment with nerve growth factor for neurotrophic keratitis. *Ophthalmology.* 2000; 107: 1347–1351.

102. Lambiase A, Rama P, Bonini S, Caprioglio G, Aloe L. Topical treatment with nerve growth factor for corneal neurotrophic ulcers. *N Engl J Med.* 1998; 338: 1174–1180.

103. Origlieri C, Bielory L. Emerging drugs for conjunctivitis. *Expert Opin Emerg Drugs.* 2009; 14: 523–536.

104. Ciulla TA, Rosenfeld PJ. Antivascular endothelial growth factor therapy for neovascular age-related macular degeneration. *Curr Opin Ophthalmol.* 2009; 20: 158–165.

105. Chappelow AV, Kaiser PK. Neovascular age-related macular degeneration: Potential therapies. *Drugs.* 2008; 68: 1029–1036.

106. Rodrigues EB, Farah ME, Maia M, Penha FM, Regatieri C, Melo GB, Pinheiro MM, Zanetti CR. Therapeutic monoclonal antibodies in ophthalmology. *Prog Retin Eye Res.* 2009; 28: 117–144.

107. Lim L, Suhler EB, Smith JR. Biologic therapies for inflammatory eye disease. *Clin Experiment Ophthalmol.* 2006; 34: 365–374.

108. Penha FM, Rodrigues EB, Maia M, Furlani BA, Regatieri C, Melo GB, Magalhães O Jr., Manzano R, Farah ME. Retinal and ocular toxicity in ocular application of drugs and chemicals—Part II: Retinal toxicity of current and new drugs. *Opthalmic Res.* 2010; 44: 205–224.

109. Saurenmann RK, Levin AV, Rose JB, Parker S, Rabinovitch T, Tyrrell PN, Feldman BM, Laxer RM, Schneider R, Silverman ED. Tumour necrosis factor alpha inhibitors in the treatment of childhood uveitis. *Rheumatology (Oxford).* 2006; 45: 982–989.

110. Shome D, Esmaeli B. Targeted monoclonal antibody therapy and radioimmunotherapy for lymphoproliferative disorders of the ocular adnexa. *Curr Opin Ophthalmol.* 2008; 19: 414–421.

111. Bourges JL, Bloquel C, Thomas A, Froussart F, Bochot A, Azan F, Gurny R, BenEzra D, Behar-Cohen F. Intraocular implants foe extended drug delivery: Therapeutic applications. *Adv Drug Deliv Rev.* 2006; 58: 1182–1202.

112. Jaffe GJ, Martin D, Callanan D, Pearson PA, Levy B, Comstock T. Fluocinolone acetonide implant (Retisert) for noninfectious posterior uveitis: Thirty-four-week results of a multicenter randomized clinical study. *Ophthalmology.* 2006; 113: 1020–1027.

113. Kane FE, Burdan J, Cutino A, Green KE. Iluvien: A new sustained delivery technology for posterior eye disease. *Expert Opin Drug Deliv.* 2008; 5: 1039–1046.

114. Kuppermann BD, Blumenkranz MS, Haller JA, Williams GA, Weinberg DV, Chou C, Whitcup SM. Randomized controlled study of an intravitreous dexamethasone drug delivery system in patients with persistent macular edema. *Arch Ophthalmol.* 2007; 125: 309–317.

115. Wright JC, Tao Leonard S, Stevenson CL, Beck JC, Chen G, Jao RM, Johnson PA, Leonard J, Skowronski RJ. An *in vivo/in vitro* comparison with a leuprolide osmotic implant for the treatment of prostate cancer. *J Control Release.* 2001; 75: 1–10.

116. Hutchinson FG, Furr BJA. Biodegradable polymers for the sustained release of peptides. *Biochem Soc Trans.* 1985; 13: 502–523.

117. Furr BJA, Hutchinson FG. A biodegradable delivery system for peptides: Preclinical experience with the gonadotrophin-releasing hormone agonist. *J Control Release* 1992; 21: 117–128.

118. Ghassemi AH, Steenbergen MJV, Talsma H, van Nostrum CF, Jiskoot W, Crommelin DJA, Hennink WE. Preparation and characterization of protein loaded microspheres based on a hydroxylated aliphatic polyester, poly(lactic-co-hydroxymethyl glycolic acid). *J Control Release.* 2009; 138: 57–63.

119. Sun L, Zhou S, Wang W, Su Q, Li X, Weng J. Preparation and characterization of protein-loaded polyanhydrides microspheres. *J Mater Sci Mater Med.* 2009; 20: 2035–2042.

120. Dziubla TD, Shuvaev VV, Hong NK, Hawkins BJ, Madesh M, Takano H, Simone E et al. Endothelial targeting of semi-permeable polymer nanocarriers for enzyme therapies. *Biomaterials.* 2008; 29: 215–227.

121. Hasadsri L, Kreuter J, Hattori H, Iwasaki T, George JM. Functional protein delivery into neurons using polymeric nanoparticles. *J Biol Chem.* 2009; 284: 6972–6981.

122. Leong KW, D'Amore PD, Marletta M, Langer R. Bioerodible polyanhydrides as drug-carrier matrices. II. Biocompatibility and chemical reactivity. *J Biomed Mater Res.* 1986; 20: 51–64.

123. Rathbone MJ. *Modified-Release Drug Delivery Technology*, 2nd ed., vol. 2, *Drugs and the Pharmaceutical Sciences.* New York: Informa Healthcare, 2008, 155.

124. Zenter GM, Rathi R, Shih C, McRea JC, Seo MH, Oh H, Rhee BG et al. Biodegradable block copolymers for delivery of proteins and water-insoluble drugs. *J Control Release.* 2001; 72: 203–215.

125. Qiao M, Chen D, Hao T, Zhao X, Hu H, Ma X. Injectable thermosensitive PLGA-PEG-PLGA triblock copolymers-based hydrogels as carriers for interleukin-2. *Pharmazie.* 2008; 63: 27–30.

126. Kim, YJ, Choi S, Koh JJ, Lee M, Ko, KS, Kim SW. Controlled release of insulin from injectable biodegradable triblock copolymer. *Pharm Res.* 2001; 18: 548–550.

127. Zentner GM. Biodegradable, thermally reversible gels for drug delivery. In *The 9th International Symposium on Recent Advances in Drug Delivery Systems*, Salt Lake City, UT, 22/22/99-2/25/99.

128. Al-Tahami K, Singh J. Smart polymer based delivery systems for peptides and proteins. *Recent Pat Drug Deliv Formul.* 2007; 1: 65–71.

129. Singh S, Webster DC, Singh J. Thermosensitive polymers: Synthesis, characterization, and delivery of proteins. *Int J Pharm.* 2007; 341: 68–77.
130. Molina I, Li SM, Martinez MB, Vert M. Protein release from physically cross-linked hydrogels of the PLA/PEO/PLA triblock copolymer-type. *Biomaterials.* 2001; 22: 363–369.
131. Wang B, Zhu W, Zhang Y, Yang ZG, Ding JD. Synthesis of a chemically cross-linked thermo-sensitive hydrogel film and in situ encapsulation of model protein drugs. *React Funct Polym.* 2006; 66: 509–518.
132. Park SY, Chung HJ, Lee Y, Park TG. Injectable and sustained delivery of human growth hormone using chemically modified Pluronic copolymer hydrogels. *Biotechnol J.* 2008; 3: 1–7.
133. Chung HJ, Lee Y, Park TG. Thermo-sensitive and biodegradable hydrogels based on stereocomplexed pluronic multi-block copolymers for controlled protein delivery. *J Control Release.* 2008; 127: 22–30.
134. Hyun H, Kim YH, Song IB, Lee JW, Kim MS, Khang G, Park K, Lee HB. *In vitro* and *in vivo* release of albumin using a biodegradable MPEG-PCL diblock copolymer as an in situ gel-forming carrier. *Biomacromolecules.* 2007; 8: 1093–1100.
135. Nguyen MK, Lee DS. Injectable biodegradable hydrogels. *Macromol Biosci.* 2010; 10: 563–579.
136. Seong JY, Jun YJ, Kim BM, Park YM, Sohn YS. Synthesis and characterization of biocompatible poly(organophosphazenes) aiming for local delivery of protein drugs. *Int J Pharm.* 2006; 314: 90–96.
137. Kang GD, Song SC. Effect of chitosan on the release of protein from thermosensitive poly(organophosphazene) hydrogels. *Int J Pharm.* 2008; 349: 188–195.
138. Kang Derwent JJ, Mieler WF. Thermosensitive hydrogels as a new ocular drug delivery platform to the posterior segment of the eye. *Tans Am Ophthalmol Soc.* 2008; 106: 206–213.
139. Cao Y, Zhang C, Shen W, Cheng Z, Yu L, Ping Q. Poly(N-isopropylacrylamide)-chitosan as thermosensitive in situ gel-forming system for ocular drug delivery. *J Control Release.* 2007; 120: 186–194.
140. Loh XJ, Goh SH, Li J. Hydrolytic degradation and protein release studies of thermogelling polyurethane copolymers consisting of poly[(*R*)-3-hydroxybutyrate], poly(ethylene glycol), and poly(propylene glycol). *Biomaterials.* 2007; 28: 4113–4123.

11 Transscleral Drug Delivery to the Posterior Segment of the Eye

Bernard F. Godley, Cheryl L. Rowe-Rendleman,
Ed Kraft, and Gabriella Kulp

CONTENTS

11.1 INTRODUCTION

11.1.1 BASIC NEEDS OF OCULAR DRUG DELIVERY

Our current understanding of ocular anatomy and physiology has impeded the development of a truly effective noninvasive method to deliver drugs to the posterior segment. The drug delivery platforms such as drops, inserts, and injections that are currently used in the patient care setting are each effective and safe for a limited number of drug entities; however, a gold standard method that is well tolerated, generalizable to a large range of molecule species, and has a capacity for sustained and targeted delivery has not yet been achieved.

11.1.2 Continued Need for Improvement

The use of topical drops for anterior segment diseases is generally regarded as safe, but the effective delivery of drugs with drops is diminished by the need for solubility, penetrance, and increased contact time with the ocular surface. For the past 30 years, the aim of these methods has been to accomplish the safe and effective delivery of drugs to the posterior segment without the clinical burden of injections or surgical implantations. In 1980, David Maurice stated that "an understanding of the movement from an eye drop to its target can be based only on a consideration of tissue barriers and depots and of the fluid drainage systems that stand between them" (Maurice 1980). Citing a two-phase model to describe the passive permeation of drugs through corneal absorption, Maurice's work set the stage for the exploration of methods that would result in facilitated delivery of topical drugs to the posterior segment. In the past 3 years, at least four topical eye drop formulations have been evaluated in early phase 1 and 2 clinical trials for the treatment of posterior segment diseases (Campochiaro et al. 2010; GlaxoSmithKline 2010; Palanki et al. 2008; Sternberg et al. 2010). However, the degree of corneal penetration of most topical drugs intended for the treatment of posterior segment diseases has been disappointing. Consequently, the use of eye drops to treat vitreoretinal disease is extremely limited because the static and dynamic barriers of the eye are not easily breached with current therapy.

For many years, researchers have attempted to develop ocular drugs based on the premise that productive absorption from the ocular surface to the posterior chamber can occur by extra-corneal permeation. Definitive work by Ahmed et al. (1987) and others (Doane et al. 1978; Edelhauser and Maren 1988) showed that after topical administration of some drugs, high concentrations could be found in the tissues of the iris and ciliary body, presumably as a result of scleral absorption. Using labeled inulin and timolol, Ahmed and Patton (1985) demonstrated that a noncorneal route might contribute significantly to the overall penetration of topically applied drugs in rabbits. In an experiment in which access to the cornea was physically blocked by a glass cylinder that was glued to the surface of the globe, the degree of absorption of a labeled drug by the sclera and conjunctiva was equivalent whether or not corneal access was permitted. It was determined that for some drugs, scleral and conjunctival routes of biodistribution were important. Accumulating data from several animal models allowed Mizuno et al. (2009) to conclude that there are three possible local penetration routes from the ocular surface to the posterior segment (Figure 11.1). These include (1) periocular and transposterior scleral route (conjunctival cul-de-sac → periocular Tenon tissue → posterior sclera → posterior choroid, and then retina); (2) transvitreal route (cornea → anterior chamber → vitreous → posterior retina); and (3) uveal route (cornea → anterior chamber → anterior choroid → posterior choroid → retina).

Accumulating evidence from Edelhauser and Maren (1988) and others (Ambati, Canakis et al. 2000; Amaral et al. 2005) suggests that there is less resistance in the sclera to the penetration of drugs than in corneal tissue. With its large surface area and high degree of hydration, the sclera is permeable to diverse molecules of different

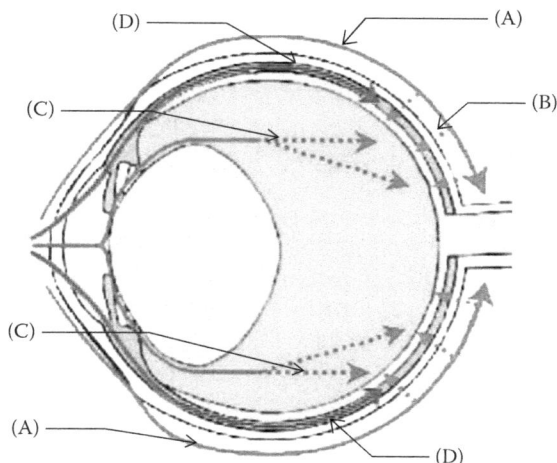

FIGURE 11.1 Three possible local penetration routes of topically instilled drugs are shown: periocular and transposterior scleral route (A and B), transvitreal route (C), and uveal route (D). (Adapted from Mizuno et al., Investigative Ophthalmology and Visual Science 2009;50.)

weights and sizes. By analyzing rabbit sclera *in vitro* in a diffusion chamber, Ambati, Canakis et al. (2000) demonstrated that molecules as large as 150 kDa could permeate the tissue. As expected, the rate of passive diffusion through the sclera declined exponentially with increasing molecular weight and molecular radius. A secondary analysis of the data revealed that molecular radius was a better predictor of permeability in the sclera than molecular weight.

Testing in animal models indicated that the sclera could be used to deliver bioactive molecules to specific targets in the eye. When fluorescein isothiocyanate-tagged immunoglobulin (FITC-IgG, $MW_r \sim 150$ kDa) was delivered by osmotic pump to the orbital surface of a rabbit sclera, a spatial gradient of fluorescence was observed in the sclera, retina, and choroid. Remarkably, no fluorescence was detected in the plasma or the ipsilateral eye. In the hemisphere of the choroid that was proximal to the pump, the concentration of IgG delivered by the osmotic pump reached a plateau of 6% (Ambati, Gragoudas et al. 2000). The deposition of IgG in the tissue of the posterior segment suggests that passive delivery through the scleral route can be a productive means of delivering large molecules. It is commonly thought and was demonstrated in recent work by Jiang et al. (2009) that after forming a drug depot within the scleral matrix, large molecular entities diffuse to neighboring tissues such as the choroid, retina, or ciliary body.

11.2 METHODS OF APPROACHING SCLERAL DELIVERY

Even molecules that passively penetrate the sclera may have little opportunity to reach targets in the posterior segment because of diffusion away from the scleral surface, drainage into the lacrimal gland, and absorption into the systemic circulation. This problem is not adequately remedied by most scleral delivery methods. Devices and formulations that control the direction of drug flow after topical application to the sclera are needed.

11.2.1 JUXTA-SCLERAL DELIVERY

The juxta-scleral injection of anecortave acetate for macular degeneration was a technique that placed the drug on the surface of the sclera over the macula. Unlike injections, juxta-scleral administration is not associated with a risk of endophthalmitis, retinal detachment, or increase in intraocular pressure because the globe is not penetrated during administration. A special curved cannula was designed to follow the curvature of the globe and allow the healthcare provider to place the drug underneath the conjunctiva and Tenons's capsule on bare sclera in the back of the eye (Augustin 2006). Although preclinical studies in animals indicated that this approach was capable of achieving therapeutic levels of anecortave acetate in the retina and choroid for several months (Clark 2007; Dahlin and Rahimy 2007; Jockovich et al. 2007), the results of human trials were disappointing, with no clear strong antiangiogenic effect demonstrated in humans with neovascular age-related macular degeneration (Csaky 2009). However, it was not possible to determine whether the lack of effectiveness seen in these trials was due to the drug or the mode of application (Csaky 2009).

11.2.2 EPISCLERAL DELIVERY

One approach to improving the efficiency of injections into the periocular space is permitting a drug to flow in only one direction after application to the sclera. As a result of this concept, investigators designed techniques to evaluate unidirectional transscleral drug delivery. When Kato et al. (2004) sutured an episcleral implant into place that was loaded with a lipophilic fluorophore, 6-carboxyfluorescein diacetate (6-CFDA), as a drug marker, the diffusion into the sclera was limited to the area around the implantation site. Likewise, Pontes de Carvalho et al. (2006) demonstrated that a similar device called an exoplant delivered significantly higher levels of sodium fluorescein ($MW_r = 376$ Da) to the retina and posterior vitreous at all time points as compared to periocular injection of fluorescein. Upon analysis of the pattern of distribution, peak fluorescence was measured in the retina and posterior vitreous and spread to the anterior chamber over time. Although these methods support a valuable proof of concept, commercial execution of the exoplant is not straightforward.

11.2.3 INTRASCLERAL DELIVERY

Although somewhat more invasive than episcleral delivery, intrascleral drug delivery involves penetration of the scleral tissue with a microneedle. Our group (CR-R) described a microsurgical approach to pierce the episclera with a microneedle and deposit steroids directly in the sclera for diffusion to the choroid, retinal pigment epithelium (RPE), and retina (Rowe-Rendleman et al. 2002). When perfusion pressure was maintained by a micropump, adenoviral, oligonucleotide, and microparticle entities could be introduced into the sclera and subsequently diffused throughout the scleral layers and the suprachoroidal space (Shuler et al. 2004). Prolonged diffusion was aided by the development of a depot. Diffusion out of the eye was prevented but not eliminated by hydrostatic pressure.

A study by Jiang et al. (2009) demonstrated that hollow microneedles that are allowed to penetrate the sclera can infuse solutions containing soluble molecules, nanoparticles, and microparticles in a minimally invasive manner. Microneedles penetrate the sclera to a depth of a few hundredths of a micrometer and do not penetrate the ocular tunic. The distribution and clearance of molecules after intrascleral delivery is not well understood, but choroidal circulation and clearance from the suprachoroidal space is a dynamic factor in the distribution of a drug following intrascleral delivery. Currently, all methods to create intrascleral injections with microneedles are experimental (Jiang et al. 2009).

11.2.4 Transscleral Delivery

11.2.4.1 Iontophoresis

Iontophoresis was initially developed to induce the permeation of chemical entities across the epidermis and into the deep layers of the skin. The methods were adapted for use in the eye by applying an electrical current to the ocular surface. In brief, a donor electrode containing the drug to be delivered is placed over the cornea. To complete the circuit, another electrode is placed at a distal site on the body. According to the work summarized by Hao et al. (2009), the mechanism of iontophoretic transport of molecules across the ocular surface is believed to involve three properties: (1) direct interaction of the electric field with the charge of the ionic permeant drug, (2) convective solvent flow that affects the transport of both neutral and ionic compounds, and (3) electric field–induced pore formation in the membrane. Hughes and Maurice (1984) described that the main factors influencing transscleral iontophoretic drug delivery were current density, duration of treatment, drug concentration, pH, and the permeability capacity of the tissue for the drug molecules.

As demonstrated by several toxicity and safety studies, iontophoresis is not without risk of complications. In a study examining the iontophoretic transfer of aspirin across the sclera, Voight et al. (2002) explained that transscleral iontophoresis may lead to retinal and choroidal cell damage, resulting in thinning and disorganization of retinal layers. When tissues are damaged by heat, the impedance changes with time, resulting in variable electrical fields (volts per square centimeter), which affects the iontophoretic drug transfer characteristics. This can occur at the epithelial surface (e.g., conjunctiva) when high current densities are applied. To avoid these problems, modern ocular iontophoresis employs a constant electrical field across the conjunctival epithelium barrier. The device uses negative pressure to secure the electrodes to the tissue. Under these conditions, the threshold for avoiding ocular toxicity due to transscleral iontophoresis has been determined to be a current density of 500 mA cm^{-2} for 5-minute duration (Voight et al. 2002).

In early ocular ionotophoresis studies, Francine Behar Cohen et al. (1997) demonstrated that the transport of dexamethasone across the cornea and sclera was just as effective as systemic administration for the treatment of experimentally induced uveitis. In recent clinical studies sponsored by EyeGate Pharmaceuticals, transscleral delivery of corticosteroids in patients suffering from anterior segment diseases such as graft rejection, uveitis, and chronic dry eye has been found to be safe (Patane et al. 2010). Although iontophoresis seems effective for the delivery of drugs to the front of the eye, there are little data to support its use for delivery of drugs to the posterior segment.

11.2.4.2 Photokinetic Drug Delivery

Recently, we presented evidence that supports the theory that energy from light can induce the translocation of drugs across the sclera (Godley et al. 2010). Parameters described by Kausar et al. (2009) demonstrated that upon photostimulation at a specific wavelength, a molecule of azobenzene could be caused to convert rapidly between *cis* and *trans* forms, an effect that can be used to produce work. In the Kausar experiments, it was demonstrated that coherent light from an argon laser at $\lambda = 488$ nm could induce the gross organization of azobenzene molecules that were entrapped in a soft liquid crystalline matrix. The stimulation of azobenzene in the matrix caused the molecules to rapidly change conformation, which subsequently induced the gross movement of microscopic glass rods that were sprinkled over the surface of the film. In the experiment, the glass rods moved unidirectionally at a speed of 173 μm min^{-1} away from the source of light. The evidence suggests that upon absorption of a photon with a particular wavelength, excited molecules are promoted to a higher unstable energy level that can be exploited to do work. Left to itself, the higher energy state will decay back to the ground state. Dissipation of the excitation energy can occur by one of three mechanisms: vibration, rotation, and/or translocation (Figure 11.2).

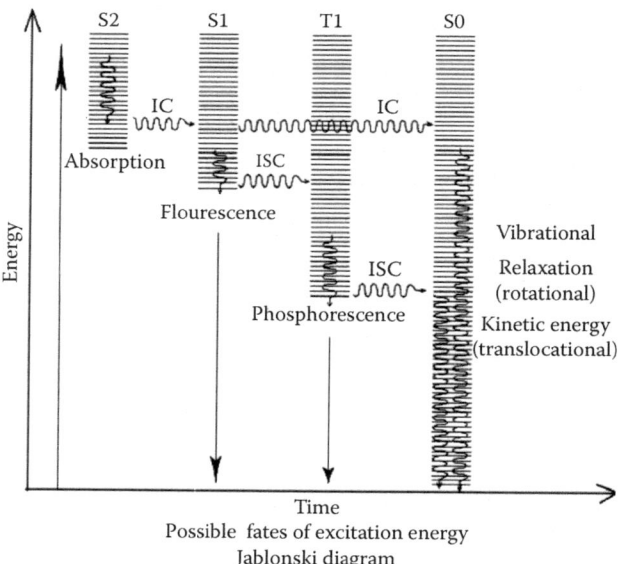

Possible fates of excitation energy
Jablonski diagram

FIGURE 11.2 The illumination of biologically active molecules with certain wavelengths of light results in excitation of the molecule. Depending on the excited state achieved, the pathway to molecular relaxation can involve fluorescence; phosphorescence (emitting decay); or vibration, rotation, and translocation (nonemitting decay). Jablonski diagram. (Adapted from Olympus microscopy. Accessed at http://www.olympusmicro.com/primer/java/jablonski/jabintro/.)

In related work on molecules with rotaxane substructures, Sebastian (2004) proposed that if a molecule is placed in the presence of light of a certain wavelength, it is possible to generate a net unidirectional motion (rotational or translational). In our laboratory, photokinetic drug delivery was performed with different biologically active molecules that did not have rotaxane substructures. These were applied directly to split-thickness skin *in vitro*. Kraft and coworkers observed that energy imparted by a specific wavelength of pulsed light could induce the permeation of molecules into the tissue and away from the light. It was theorized that incoherent (nonlaser) light could be used to excite different types of molecules in such a way that upon relaxation there was a significant increase in skin permeation and deposition in the dermal epithelium (Kraft and Kulp 2009). Further studies demonstrated that the light would not denature the drug molecule or harm the skin itself (Koutrouvelis et al. 2008). To apply the biophysical observations made by Kraft and others to a biological system such as the eye, we focused on certain key parameters, including (1) delivery of light energy without causing damage to the eye, (2) generalization of the concept to all drug molecules, and (3) development of a device to deliver the light energy to the eye.

11.3 DELIVERY OF LIGHT WITHOUT CAUSING DAMAGE TO THE EYE

The use of bright light therapy in or around the eye has been a subject of research because the visible blue light wavelengths can cause retinal damage (Lack and Wright 2007). Our group (B.G.) has previously described that exposure of human primary retinal pigmented epithelial cell cultures to visible light in the blue-green range (390–550 nm) for up to 6 hours at 2.8 mW cm^{-2} induces cellular damage primarily through reactive oxygen species (ROS) (Godley et al. 2005). In that work, it was concluded that ROS such as the hydroxyl radical, superoxide anion, and singlet oxygen are produced when visible light excites putative cellular photosensitizers (e.g., retinoids, melanin, and lipofuscin). Unlike cells in a culture dish, the eye possesses a number of survival factors and antioxidative mechanisms that defend against phototoxicity (LaVail et al. 1998). The data presented by LaVail and coworkers clearly demonstrate that retinal degeneration can be induced by extended exposure to ambient light in genetically susceptible strains of rodents. The toxicity seen in these animals is thought to result from transpupillary exposure. Certain risk factors such as the use of medications that sensitize the eye to damage by visible light pose important considerations for patients.

In contrast, phototoxicity caused by exposure of the sclera to a discrete wavelength of light in the visible range and at low power has not been described. Based on the evidence from animal studies, scleral cross-linking with riboflavin and blue light ($\lambda = 465$ nm) has a stiffening effect on the sclera but leaves no tissue damage as revealed by histology (Iseli et al. 2008). Previous work performed by Kulp et al. (2010) on skin demonstrated that the application of a pulsed narrow wavelength of incoherent light from a light-emitting diode (LED) enhances the permeation of insulin eightfold but does not damage the epidermis. Accordingly, we have developed a transscleral method of photokinetic drug delivery by focusing light from an LED onto a solution containing a responsive drug in contact with the sclera.

11.4 GENERALIZATION OF THE PHOTOKINETIC CONCEPT TO WIDE CLASSES OF MOLECULES

The range of experimental candidates for scleral drug delivery includes prodrugs, nanoparticles, antiangiogenic macromolecules, and gene therapy. Translational research is needed to bridge the gap between experimental investigation and therapeutic drug candidates. In an overview of the photoelectric effect of light, Siu et al. (2008) stated that when a photon strikes an atom, a quantum of energy, directly proportional to the frequency and inversely proportional to the wavelength of the light, is absorbed. Whereas the energy can be dissipated as reemission of light, it can also be converted into vibrational or kinetic energy.

The repertoire of drugs that can be delivered with photokinetic therapy is broad and includes organic compounds with chromophores of discernible activity in the visible and nonvisible ranges of light. Ongoing tests reveal that the methods can be applied to a variety of small molecules, proteins, and peptides. Photokinetic drug delivery uses an incoherent light source with output in the visible range that can be pulsed at a specific frequency measured in cycles per second (cps) to induce the molecular flexion of molecules. Flexion is translated into motion, which supports the translocation of molecules across a permeable surface.

In vitro studies indicate that a range of molecules can be induced to permeate the sclera extending in MW_r from 450 to 1600 KDa (Figure 11.3), which is a broader range than that described by Ambati. Despite this increased capability relative to

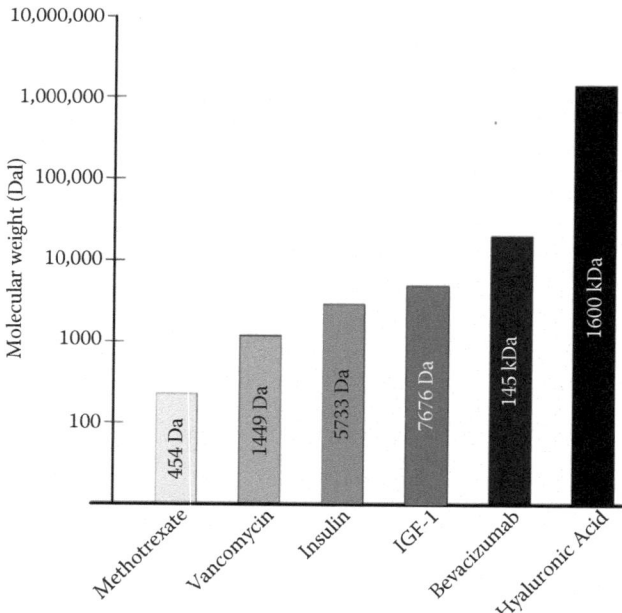

FIGURE 11.3 A wide variety of drugs, ranging from $MW_r = 454$ Da (methotrexate) to 1600 kDa (hyaluronic acid), were evaluated *in vitro*. We were not able to reach a molecular weight limit above which photokinetic-induced permeation does not work.

diffusion, it is also important to recognize that some molecular entities are not good candidates for photokinetic delivery, including certain fluorescent molecules, because the absorbed photonic energy is directed to fluorescence rather than kinetic (translocational) energy (Figure 11.2).

Our data indicated that three variables are needed to describe the permeation of molecules: wavelength, pulse rate, and time. An example of this is demonstrated by the graph in Figure 11.4 representing the permeation of EMLA (AstraZeneca, Wilmington, DE), a eutectic mixture of lidocaine and prilocaine developed to anesthetize intact skin. Using split-thickness skin in a traditional Franz cell, which was adapted to receive LEDs on top of the donor compartment, tests were performed to determine the optimum wavelength of light for lidocaine- and prilocaine-enhanced photokinetic delivery over 24 hours. In this study, LEDs were used with a peak emission at $\lambda = 390, 405, 436,$ and 450 nm (with 50% radiance output of $1.3–2.5$ mW). The LEDs were driven by a square wave pulse generator providing electric energy set at 24 cps with a 50% duty cycle. As shown in Figure 11.4, three of the four wavelengths tested induced increased permeation of prilocaine and the two highest wavelengths induced significantly ($p < .001$) increased permeation of lidocaine (Koutrouvelis et al. 2008). The amount of drug that crossed the skin was dependent on the wavelength of light and the cps. To more fully demonstrate how this property can be applied to

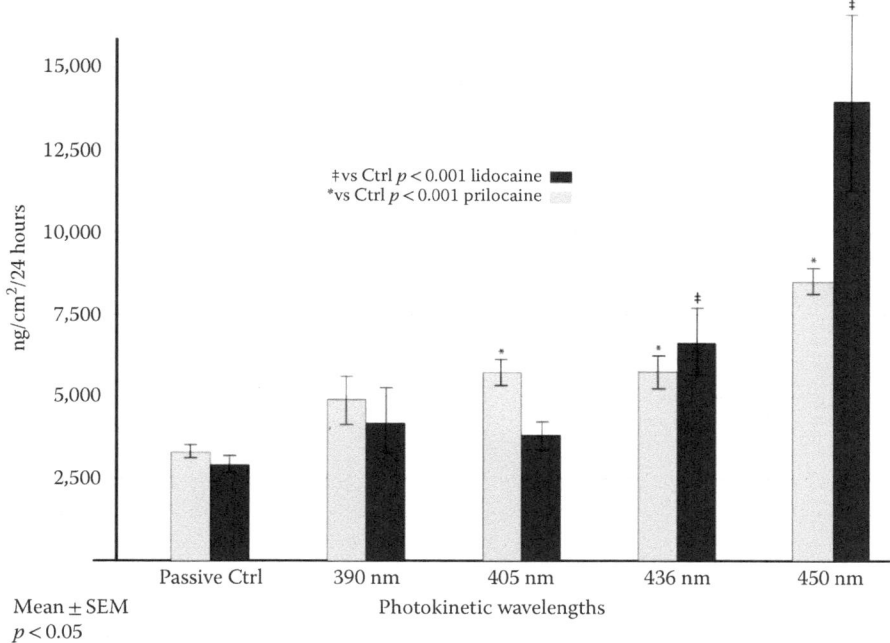

FIGURE 11.4 Light induces the permeation of anesthetics in skin. A commercial mixture of 2.5% lidocaine and 2.5% prilocaine hydrochloride was applied to split-thickness skin from cadaver donors (thickness 0.6 mm) in a Franz chamber. The amount of drug permeating the dermal tissue is dependent on the wavelength and pulse rate of light. The graph shows permeation profiles of lidocaine and prilocaine under different photokinetic conditions.

molecules on the surface of the sclera, we describe our *in vitro* results with three model candidates: methotrexate, vancomycin, and insulin-like growth factor-1 (IGF-1).

11.4.1 METHOTREXATE

Studies were conducted to examine the flux of methotrexate (MTX, $MW_r = 454$ Da) across sheep sclera *in vitro*. In clinical care, high doses of intraocular methotrexate are preferred to treat intraocular lymphoma because concentrations of the drug sufficient to treat persistent cancer cells in the vitreous humor are not obtained with systemic therapy alone (Batchelor et al. 2003). In addition to intraocular malignancies, the immunosuppressive nature of methotrexate is useful for the treatment of various ocular inflammatory conditions such as uveitis. So far, intraocular injections are the primary route to administer methotrexate to the eye. A noninvasive system that delivers the drug may be more desirable to some patients and physicians than injections.

A traditional Franz cell apparatus (PermeGear, Inc, Bethlehem, PA) was modified to allow placement of the LEDs into the donor cell (Figure 11.5). Ovine eyes

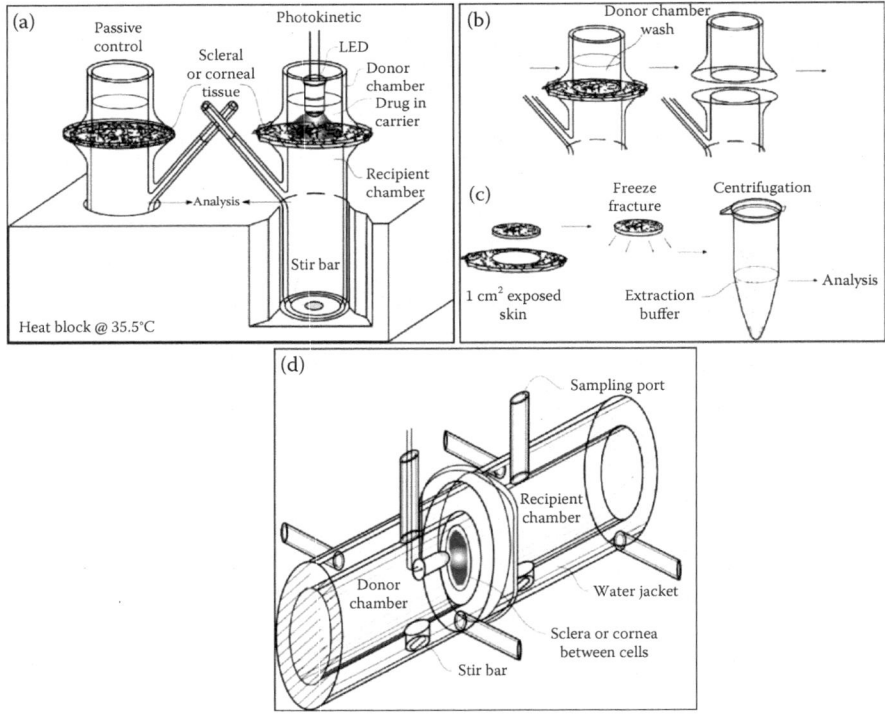

FIGURE 11.5 (a) Experiments were conducted in Franz cells held at a constant temperature of 35.5°C. (b) Skin or sclera was exposed to a solution of the drug in the donor chamber. Permeation was allowed to proceed in the presence or absence of pulsed incoherent light. (c) After exposure, the area of tissue under the donor solution was washed with recipient buffer, frozen in liquid nitrogen, pulverized in a mechanical cryogenic pulverizer, and extracted in a suitable buffer for each drug and submitted for analysis (d) Expanded view of modified Franz cell.

were procured through a tissue-sharing program from other investigators conducting animal studies that did not affect the eye. Eyes were enucleated within minutes of euthanasia, placed immediately in RPMI 1640 medium with antibiotics and antimycotics and refrigerated until use. Peri-equatorial circumferential sections of the eye were dissected. The sclera was removed from the sections and sealed between the donor and recipient chambers of the Franz cell apparatus. Scleral tissue was used within 36 hours of enucleation.

The area of sclera that was exposed to the drug was 1 cm^2. The Franz cells were placed in an aluminum block and maintained at a constant temperature of 35.5°C. The recipient fluids were constantly mixed with a magnetic stirrer. Methotrexate was formulated in high-performance liquid chromatography (HPLC)-grade water or a carrier containing preservatives and 0.2–1.0% hyaluronic acid ($MW_r = 1600$ kDa) as a thickening agent. The formulation was introduced to the donor chamber. Either HPLC-grade water or buffer was used in the recipient chamber for ease of analysis. Samples from the recipient fluid were taken through the side port at various time points. Control cells were set up the same way with ambient room light and appropriate experimental temperatures. They were not exposed to pulsed incoherent light.

In the Franz donor chamber, 2.5 mg mL^{-1} methotrexate was illuminated with pulsed light at $\lambda = 350 \pm 5$ and 370 ± 5 nm. Permeation of sheep sclera was allowed to proceed for 1 hour. Nonilluminated controls were monitored for passive diffusion for 1 hour. LEDs with a radiant output up to 1.2 mW were pulsed at 24 cps. Methotrexate was analyzed by HPLC on an XTerra RP-C18, 150 × 40-mm, 3-μm column. The drug was eluted with a gradient mobile phase from 25% to 30% methanol in HPLC-grade water with 0.1% trifluoroacetic acid (TFA) at a flow rate of 0.7 mL min^{-1}. The drug was detected at 260 nm. Of the two wavelengths tested, the peak permeation was achieved at $\lambda = 370$ nm and 24 cps (Figure 11.6). Active permeation was higher than passive after 15, 30, or 60 minutes of exposure (paired t-test, $p = .003$, $p = .007$, and $p = .006$ respectively).

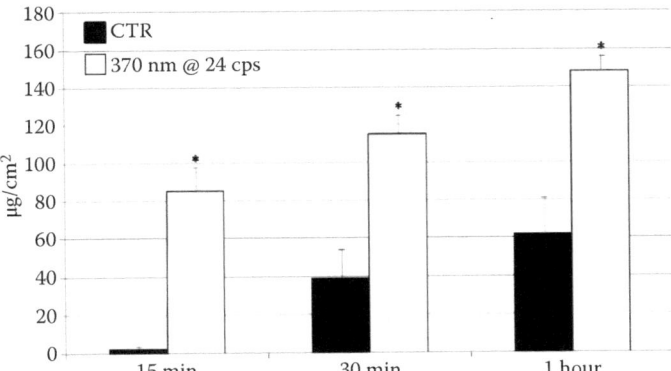

FIGURE 11.6 Methotrexate was added to the donor chamber of a Franz cell. LEDs with an output of 370 nm were pulsed at 24 cps. Illuminated and nonilluminated controls (CTR) were monitored for 15 minutes, 30 minutes, or 1 hour. Although there was measurable flux of methotrexate under control conditions, light evoked a more robust and time-dependent response. Analysis by paired t = test, significance indicated by * ; 15 min $p = .03$, 30min $p = .07$, 1 hour $p = .006$.

The rate of permeation of the two concentrations of methotrexate was compared. Under passive diffusion conditions, the amount of methotrexate that permeated the sclera was proportional to the concentration of the drug in the donor chamber. Compared to controls, light increased the permeation of 1% methotrexate by 2.5-fold, as indicated by calculation of the area under the curve (AUC). The AUC_{0-3hr} of the 1% and 2% control solutions were respectively 687.5 and 1593.1 hr^{-1}, which reflects a proportional 2.3-fold increase in the extent of passive diffusion. The AUC_{0-3hr} of the 1% and 2% photokinetic-treated solutions were respectively 1750 hr^{-1} and 2593.75 hr^{-1}. Light-induced permeation of the methotrexate solutions did not increase proportionately with the concentration of the drug in the donor chamber, suggesting that concentration gradient is not the main driving factor of photokinetic permeation (compare Figure 11.7a and b).

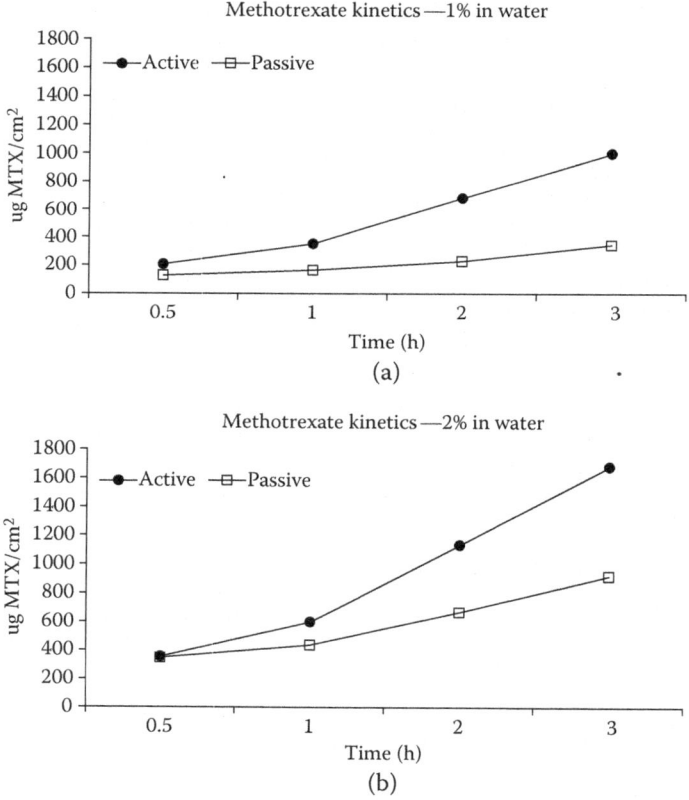

FIGURE 11.7   Kinetic measurements for methotrexate permeation were tested at two concentrations. Solutions of methotrexate were prepared at 1% and 2% in water. In both cases, the AUC for active permeation was higher than that for passive diffusion. In the passive diffusion curves, the concentration gradient is the main driving factor for permeation. (a) The AUC of the 2% solution is 2× greater than the 1% solution. (b) In contrast, with photokinetic treatment, the AUC of the 2% solution is only 1.6× greater than the 1% solution. The data suggest that concentration is not the main driving force in photokinetic drug delivery.

11.4.2 VANCOMYCIN

The recommended treatment of endophthalmitis includes intravitreal injections of broad spectrum antibiotics. Vancomycin ($MW_r = 1449$ Da) is the drug of choice for Gram-positive bacteria such as coagulase-negative staphylococci, *Staphylococcus aureus*, and streptococci (Lifshitz et al. 2000). The commonly recommended vancomycin dose of 1 mg is extremely high because it results in vitreous concentrations that are many times higher than recommended to treat severe extraocular infections such as endocarditis and meningitis (Gan et al. 2001). Since the antimicrobial effect of vancomycin depends on the length of time that the drug exceeds the minimum inhibitory concentration for the susceptible bacteria, efficacious therapy of the antibiotic depends on frequent intraocular dosing.

A 1% solution of vancomycin was introduced to the donor chamber of a Franz cell. The permeation of sheep sclera was allowed to proceed for 24 hours with photokinetic treatment at $\lambda = 405$ or 450 nm, with pulses at 24 or 100 cps. Vancomycin was analyzed, on a Nucleosil C18, 250×46-mm, 5-μm column. The drug was eluted with a mobile phase consisting of 85% acetonitrile (ACN) in HPLC-grade water with 0.1% TFA at a flow rate of 1 mL min^{-1}. Vancomycin was detected at 230 nm.

At both of the wavelengths tested, photokinetic treatment permitted more permeation of vancomycin through sheep sclera than passive controls (Figure 11.8). At 405 nm, the slower pulse at 24 cps was more effective at promoting permeation than the faster pulse at 100 cps (control vs. treated, paired t-test, $p = .043$). Under these light conditions, the permeation of vancomycin was twice as high as that seen in controls. This result indicates that the system is sensitive to the light wavelength as well as the pulse rate of the light. At the longer wavelength, the effect of pulses at 24 and 100 cps was equivalent.

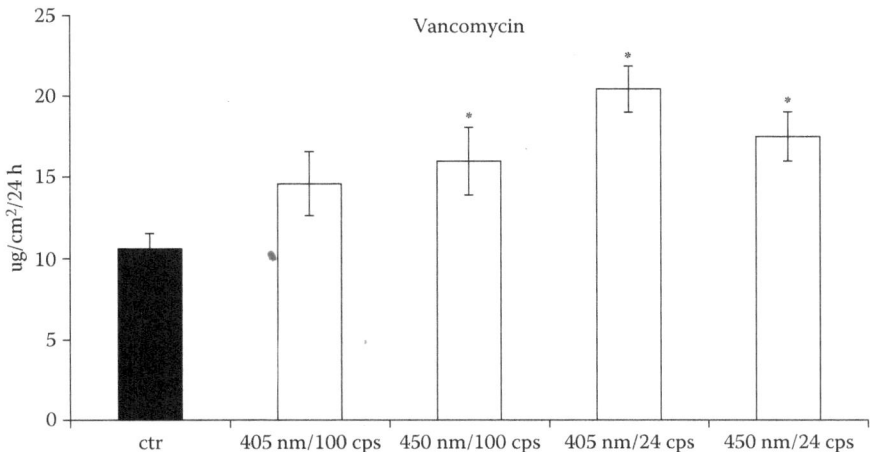

FIGURE 11.8 Permeation of vancomycin through sclera was tested at 405 and 450 nm and 24 and 100 cps. The amounts recovered from recipient chamber were compared to passive controls (by Student's t-test). Active permeation was significantly higher than passive controls at three of the conditions tested. Analysis by paired t = test, significance indicated by * ; control vs. 450 nm at 100 and 24 cps, $p = .015$; control vs 405 nm at 24 cps, $p = .001$. CTR = nonilluminated control.

The extent of permeation was significantly higher than controls (paired t-test, control vs. 450 nm at 100 cps, $p = .015$; control vs. 450 nm at 24 cps, $p = .001$; control vs. 450 nm at 24 cps, $p = .015$).

11.4.3 IGF-1

A third proof of concept for photokinetic drug delivery was demonstrated in ovine sclera with the delivery of IGF-1 ($MW_r = 7676$ Da), an angiogenic growth factor that is thought to play a critical role in diabetic retinopathy (Haurigot et al. 2009). In the normal retina, mRNA for IGF-I and its receptor, IGF-I receptor, is expressed throughout the neuroretinal layers, the RPE, and in some choriocapillary and retinal capillary endothelial cells. In transgenic mice, increased levels of IGF-1 lead to a diabetic-like phenotype in the eye (Ruberte et al. 2004). In humans with diabetic retinopathy, high levels of IGF-1 in the vitreous are significantly associated with the degree of neovascularization (Grant et al. 1986). Several mimetics that target the IGF-1 receptor have been developed for clinical trials in cancer (Gualberto et al. 2009). The targeting of IGF-1 or the IGF-1 receptor may be useful in the development of antiangiogenic therapy for ocular vasculopathies. Photokinetic drug delivery may be useful for administering these drugs to retinal targets.

Studies were completed using two LED sources that emitted light at $\lambda = 405$ or 450 nm. The lights were pulsed at either 24 or 100 cps. A solution of 1 mg mL^{-1} IGF-1 was added to the donor chamber of the Franz cell. After 24 hours, the sheep sclera and recipient compartments were analyzed using a commercially available ELISA kit (Diagnostic System Laboratories, Webster, TX). The tests were performed according to the manufacturer's instructions. Briefly, samples were incubated in 96-well plates coated with specific primary antibody. Positive reactions were developed with secondary horseradish peroxidase (HRP)-conjugated antibodies and visualized with a tetramethylbenzidine substrate. The plates were read on a microplate reader at 450 nm.

Unlike methotrexate and vancomycin, all samples that had been treated with light had levels of permeation that were at least 100-fold greater than passive permeation controls (Figure 11.9). The results of this experiment suggest that photokinetic treatment can facilitate the permeation of peptide molecules that would not ordinarily diffuse through the sclera.

11.5 DEVELOPMENT OF A DEVICE TO DELIVER LIGHT ENERGY

A device to deliver photokinetic therapy to the eye requires two elements; a narrow bandwidth light source with low radiance output that is capable of receiving an oscillating signal and responding with pulsing light. In our development of such a device, we employed an adjustable pulse rate square wave signal generator and LEDs ranging in output from 0.04 to 20 mW. In our preliminary experiments, we learned that the wavelength and not the intensity of light was the primary determinant for

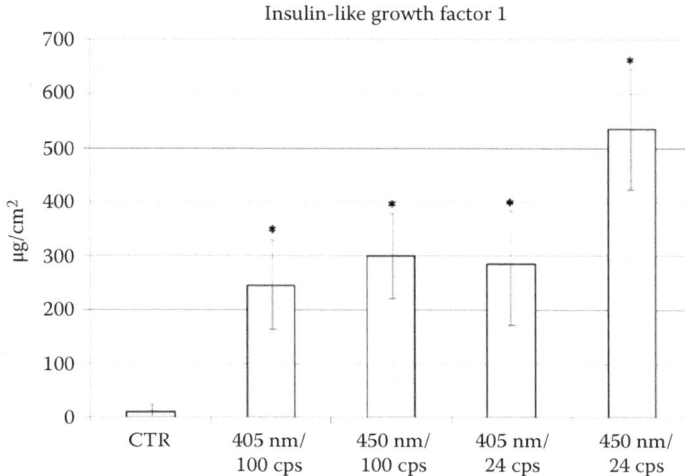

FIGURE 11.9 'Insulin-like growth factor-1 (IGF-1) was used as a model for a large peptide. All photokinetic treatment conditions tested revealed significantly higher amounts permeated versus passive controls (by one-way ANOVA vs. control with post hoc correction for unequal variance *$p < .05$). CTR = nonilluminated control.

the photokinetic effect. When one, two, or three LEDs (0.04, 0.08, or 0.12 mW cm^{-2}) were placed in the same donor chamber and permeation of insulin was allowed to proceed over 6 hours, there was no statistical difference between the active groups in the amount permeated ($p = .962$); however, all the amounts were significantly higher than passive controls, $p < .05$ (one-way ANOVA with post hoc correction vs. control, Dunn's method) (Sokal and Rohlf 1995).

Photokinetic drug delivery differs from two other therapeutic approaches in which light is also used as a triggering agent. As summarized by Alvarez-Lorenzo et al. (2009), photodynamic therapy as currently used in cancer and ophthalmology is intended to cause cell death and involves a photosensitizer, light, and oxygen in the tissue. Laser light stimulates the creation of singlet oxygen that reacts with nearby biomolecules, subsequently provoking destructive reactions. As used in the retinal disease setting, a photosensitive drug such as verteporfin is injected into the bloodstream, where it proceeds to diseased vasculature in the back of the eye. Subsequently, laser light aimed at the specific sites in the macula activates the photosensitive molecules, which destroy the leaky blood vessels by localized thrombosis and occlusion in patients with macular degeneration. Unlike photokinetic drug delivery, the interaction with light in photodynamic therapy causes the destruction of the blood vessels and does not provide the mode of transport for molecules.

The methods also differ from emerging technology that uses light to promote the fusion of drug-transporting liposomes with cells or tissues. Liposomes consist of concentric layers of phospholipids and/or other amphiphilic molecules. In this case, light-controlled disruption of the liposomes and drug delivery is permitted by incorporating a chromophore into the bilayer structure. With this technology, UV/visible or near-infrared light can cause reversible destabilization of the micelle and passive delivery of the incorporated drug to a cell or tissue. In contrast to photokinetic drug delivery, liposomal delivery relies on passive delivery of drugs to the target tissue.

Photokinetic drug delivery uses cyclic illumination of a drug molecule at a selected wavelength and pulse rate that causes gross unidirectional movement and migration of molecules across the sclera. The photokinetic force is different from diffusion, and drugs that are not permeable in the sclera under passive delivery conditions can cross with light stimulation. In all the experiments, the drive current for the LEDs was attenuated to a level below the manufacturer's specified rating. This was specifically designed to eliminate any secondary and unnecessary heat production from the LEDs. Also, the wavelengths and intensities of the light used specifically avoid the initiation of a photochemical reaction or damage from ionizing irradiation. Thus, there is improbable risk of photochemical and thermal damage to the surrounding tissues.

While light and pulse rate are two essential elements of photokinetic drug delivery, they are by no means sufficient for creating a biologically compatible device that could eventually be used in the clinic setting. Lessons from iontophoretic drug delivery have taught us that thorough study of the tissue must be conducted in living animal models (Figure 11.10). Specifically, neither the device nor the light that is delivered should evoke irreversible changes in the hydration level, mechanical organization, or cellular integrity of the sclera and underlying retinal tissues during treatment.

Since the placement of a foreign object on the ocular surface usually provokes an irritative response, the device must be engineered from materials that are compatible with the ocular surface, which in turn should prevent ocular irritation. This can be done by making sure that the interface between the device and the patient is well tolerated. In our initial studies, we encapsulated an LED in a medical-grade silicone

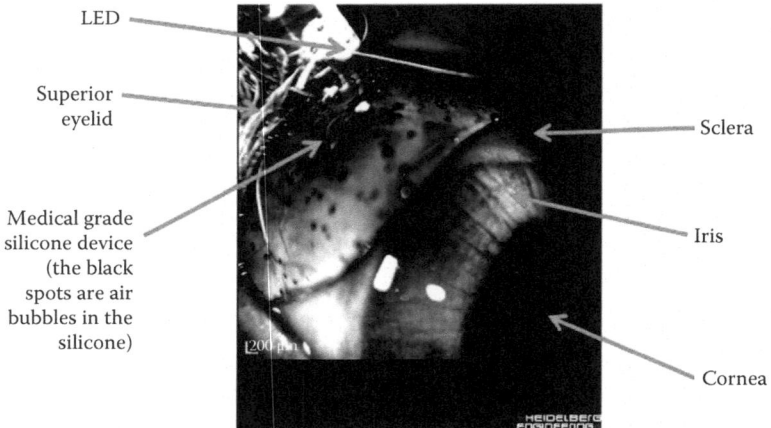

FIGURE 11.10   A beta-prototype device to deliver photokinetic therapy to the eye.

shell that conformed to the geometry of the rabbit eye, 10 mm × 6 mm in size, with a concave area to hold 150 μL of donor formulation (Figure 11.10). Unlike the exoplant devices, suturing of the device to the sclera is not necessary. Ocular irritation and tolerability studies are in progress.

11.6 CONCLUSION

Previous *in vitro* experiments have demonstrated that the human sclera is permeable to molecules as large as 70 kDa (Cruysberg et al. 2005), and Ambati, Canakis et al. (2000) showed that a 150-kDa IgG molecule can diffuse across rabbit tissue. In our studies, we showed that selected wavelengths of pulsed incoherent light have a unique action on molecules that would not cross the sclera under passive conditions. In this chapter, we reviewed studies that demonstrated that small molecules ranging MW_r from 400 to 7600 Da in hyaluronic acid (1600 kDa) cross sheep sclera following photic stimulation. Experience allows us to generalize across molecular classes that a particular wavelength and pulse may trigger the permeation of many different types of molecules that are related by similarities in structure. It may be possible to design drug candidates such that they are more likely to absorb light energy and release it as motion.

A number of new approaches to deliver drugs to the posterior segment are under consideration. Many of these have data that support their ocular safety, pharmacokinetics with ideal compounds, and sustained or controlled release characteristics. One limitation of this approach is that *in vitro* studies were performed with ovine sclera. Comparisons and interpolations from sheep to an *in vivo* rabbit model are not straightforward since there are few published studies describing the characteristics (e.g., thickness, cellularity, and water content) of the sheep sclera. The feasibility and ocular safety of photokinetic drug delivery is under investigation in rabbit models. In the ophthalmic setting, it is presumed that molecules will be translocated by photokinetic light as they come in contact with the sclera. Although we dissolved our test molecules in simple solutions of water and hyaluronic acid, more work will be needed to devise formulations that are compatible with the ocular surface and pulsed light.

At the molecular level, we have explained that pulsed incoherent light may induce reversible conformational changes within a molecule by raising the energy level to an excited state followed by relaxation and movement. Functional studies with insulin peptides indicate that the intrinsic activity of these biomolecules is not changed by photic stimulation. Bioassay methods will be needed to demonstrate that repeated activation and relaxation does not interfere with the bioactivity of selected molecules.

A need exists for a safe and efficient drug delivery method that eliminates side effects and damage to the barrier function or appearance of the patient's eye caused by drug administration. It is anticipated that for some molecules, photokinetic drug delivery will replace the invasive delivery of these drugs with needles. Methods are in early development to evaluate the potential of photokinetic drug delivery to treat posterior segment diseases noninvasively.

REFERENCES

Ahmed I., Gokhale R., Shah M. et al. 1987. Physicochemical determinants of drug diffusion across the conjunctiva, sclera and cornea. *J Pharm Sci.* 76:583–586.

Ahmed I., Patton T. 1985. Importance of the noncorneal absorption route in topical ophthalmic drug delivery. *Invest Ophthalmol Vis Sci.* 26:584–587.

Alvarez-Lorenzo., Bromberg L., Concheiro A. 2009. Light-sensitive intelligent drug delivery systems. *Photochem and Photobiol.* 85:848–860.

Amaral J., Fariss R., Campos M. et al. 2005. Trans-scleral-RPE permeability of PEDF and ovalbumin proteins: Implications for subconjunctival protein delivery. *Invest Ophthalmol Vis Sci.* 46:4385–4392.

Ambati J., Canakis C., Miller J. et al. 2000. Diffusion of high molecular weight compounds through sclera. *Invest Ophthalmol Vis Sci.* 41:1181–1185.

Ambati J., Gragoudas E., Miller J. et al. 2000. Transscleral delivery of bioactive protein to the choroid and retina. *Invest Ophthalmol Vis Sci.* 41:1186–1191.

Augustin A. 2006. Anecortave acetate in the treatment if age-related macular degeneration. *Clin Interv Aging.* 1:237–246.

Batchelor T., Kolak G., Ciordia R. et al. 2003. High-dose methotrexate for intraocular lymphoma. *Clin Cancer Res.* 9:711–715.

Behar-Cohen F., Parel J., Pouliquen Y. et al. 1997. Iontophoresis of dexamethasone in the treatment of endotoxin-induced-uveitis in rats. *Exp Eye Res.* 65:533–545.

Campochiaro P., Shah S., Hafiz G. et al. 2010. Topical mecamylamine for diabetic macular edema. *Am J Ophthalmol.* 149:839–851.

Clark A. 2007. Preclinical efficacy of anecortave acetate. *Surv Ophthalmol.* 52 (suppl. 1): S41–S48.

Cruysberg L., Nuijts R., Geroski D. et al. 2005. The influence of intraocular pressure on the transscleral diffusion of high-molecular-weight compounds. *Invest Ophthalmol Vis Sci.* 46:3790–3794.

Csaky K. 2009. Hurdles and prospects for episcleral drug delivery. *Ret. Phys* (July), http://www.retinalphysician.com/article.aspx?article=103201, accessed December 2, 2010.

Dahlin D., Rahimy M. 2007. Pharmacokinetics and metabolism of anecortave acetate in animals and humans. *Surv Ophthalmol.* 52 (suppl. 1):S49–S61.

Doane M., Jensen A., Dohlman G. 1978. Penetration routes of topically applied eye medications. *Am J Ophthalmol.* 85:383–386.

Edelhauser H., Maren T. 1988. Permeability of human cornea and sclera to sulfonamide carbonic anhydrase inhibitors. *Arch Ophthalmol.* 106:1110–1113.

Gan I., van Dissek J., Beekhuis W. et al. 2001. Intravitreal vancomycin and gentamicin concentrations in patients with postoperative endopthalmitis. *Br J Ophthalmol.* 85:1289–1293.

GlaxoSmithKline. 2010. Dose ranging study of pazopanib to treat neovascular age related macular degeneration. Bethesda, MD: US National Library of Medicine. 2000–2011. http://www.clinicaltrials.gov/ct2/show/NCT01134055, accessed December 2, 2010.

Godley B., Kulp G., Kraft E. et al. 2010. Photokinetic trans-scleral drug delivery—A novel platform for posterior segment drug delivery. *Invest Ophthalmol Vis Sci.* 51:E-abstract 5319.

Godley B., Shamsi F., Liang F. et al. 2005. Blue light induces mitochondrial DNA damage and free radical production in epithelial cells. *J Biol Chem.* 280:21061–21066.

Grant M., Russell B., Fitzgerald C. et al. 1986. Insulin-like growth factors in vitreous. Studies in control and diabetic subjects with neovascularization. *Diabetes.* 35:416–442.

Gualberto A., Pollak M. 2009. Clinical development of inhibitors of the insulin-like growth factor receptor in oncology. *Curr Drug Targets.* 10:923–936.

Hao J., Li S., Liu C. et al. 2009. Electrically assisted delivery of macromolecules into the corneal epithelium. *Exp Eye Res.* 89:934–941.

Haurigot V., Villacampa P., Ribera A. et al. 2009. Increased intraocular insulin-like growth factor-I triggers blood-retinal barrier breakdown. *J Biol Chem.* 284:22961–22969.

Hughes L., Maurice D. 1984. A fresh look at iontophoresis. *Arch Ophthalmol.* 102:1825–1829.

Iseli H., Spoerl E., Wiedemann P. et al. 2008. Efficacy and safety of blue-light scleral cross-linking. *J Refract Surg.* 24:S752–S755.

Jiang J., Moore J., Edelhauser H. et al. 2009. Intrascleral drug delivery to the eye using hollow microneedles. *Pharm Res.* 26:395–403.

Jockovich M., Murray T., Clifford P. et al. 2007. Posterior juxtascleral injection of anecortave acetate: Magnetic resonance and echographic imaging and localization in rabbit eyes. *Retina.* 27:247–252.

Kato A., Kimura H., Okabe K. et al. 2004. Feasibility of drug delivery to the posterior pole of the rabbit eye with an episcleral implant. *Invest Ophthalmol Vis Sci.* 45:238–244.

Kausar A., Nagano H., Ogata T. et al. 2009. Photocontrolled translational motion of a microscale solid object on azobenzene-dopped liquid crystalline films. *Angew Chem Int Ed Engl.* 121:2178–2181.

Koutrouvelis A., Kulp G., Kraft E. et al. 2008. Photokinetic transdermal delivery of a lidocaine-prilocaine eutectic mixture. In *Proceedings of the Controlled Release Society 35th Annual Symposium Abstract,* Publisher Controlled Release Society.

Kraft E., Kulp G. 2009. Photokinetic delivery of biologically active substances using pulsed incoherent light. U.S. Patent 20090156463, published June 18.

Kulp G., Urban R., Kraft E. et al. 2010. Photokinetic transdermal drug delivery (PYDD)—A novel platform technology for insulin delivery. In *Proceedings of the Diabetes Technology Society Meeting,* November 11–13, 2010, Bethesda, MD.

Lack L. C., Wright H. R. 2007. Clinical management of delayed sleep phase disorder. *Behav Sleep Med.* 1:57–76.

LaVail M., Yasumura D., Matthes M. et al. 1998. Protection of mouse photoreceptors by survival factors in retinal degenerations. *Invest Ophthalmol Vis Sci.* 39:592–602.

Lifshitz T., Lapid-Gortzak R., Finkelman Y. et al. 2000. Vancomycin and ceftazidime incompatibility upon intravitreal injection. *Br J Ophthalmol.* 84:117–121.

Maurice D. 1980. Structures and fluids involved in the penetration of topically applied drugs. *Int Ophthalmol Clin.* 20:7–20.

Mizuno K., Koide T., Shimada S. et al. 2009. Route of penetration of topically instilled nipradilol into the ipsilateral posterior retina. *Invest Ophthalmol Vis Sci.* 50:2839–2847.

Palanki M., Akiyama H., Campochiaro P. et al. 2008. Development of prodrug 4-chloro-3-(5-methyl-3-{[4-(2-pyrrolidin-1-ylethoxy) phenyl]amino}-1,2,4-benzotriazin-7-yl)phenyl benzoate (TG100801): A topically administered therapeutic candidate in clinical trials for the treatment of age-related macular degeneration. *J Med Chem.* 51:1546–1559.

Patane M., Cohen A., Sugarman J. et al. 2010. Randomized, double masked study of four iontophoresis dose levels of EGP-437 in non-infectious anterior segment uveitis patients. *Invest Ophthalmol Vis Sci.* 51:E-abstract 5263.

Pontes de Carvalho R., Krausse M., Murphree A. et al. 2006. Delivery from episcleral exoplants. *Invest Ophthalmol Vis Sci.* 47:4532–4539.

Rowe-Rendleman C., Higa A., Iwai S. et al. 2002. Prophylactic intra-scleral injection of steroid compounds in rabbit model of retinal neovascularization. *Invest Ophthalmol Vis Sci.* 43:E-Abstract 3872.

Ruberte J., Ayuso E., Navarro M. et al. 2004. Increased ocular levels of IGF-1 in transgenic mice leads to diabetes-like eye disease. *J Clin Invest.* 113:1149–1157.

Sebastian K. 2004. Theoretical design for a light-driven molecular motor based on rotaxanes. *Curr Sci.* 87:232–236.

Shuler R. Jr., Dioguardi P., Henjy C. et al. 2004. Scleral permeability of a small, single-stranded oligonucleotide. *J Ocul Pharmacol Ther.* 20:159–168.

Siu T., Morley J., Coroneo M. 2008. Toxicology of the retina: Advances in understanding the defense mechanisms and pathogenesis of drug- and light-induced retinopathy. *Clin Experiment Ophthalmol.* 36:176–185.

Sokal R., Rohlf F. 1995. *Biometry: The principles and practice of statistics in biological research*, 3rd ed. New York: W.H. Freeman and Co.

Sternberg P., Rosenfeld P., Slakter J. et al. 2010. A topical OT-551 for treating geographic atrophy: Phase II results. *Invest Ophthalmol Vis Sci.* 51:E-abstract 6416

Voight M., Kralinger M., Kieselbach G. et al. 2002. Ocular aspirin distribution: A comparison of intravenous, topical, and coulomb-controlled iontophoresis administration. *Invest Ophthalmol Vis Sci.* 43:3299–3306.

12 Drug Delivery to the Suprachoroidal Space

Sung Won Cho and Timothy W. Olsen

CONTENTS

12.1 INTRODUCTION

Age-related macular degeneration (AMD), retinal vascular disease, and diabetic macular edema (DME) represent leading causes of vision loss from diseases that affect the posterior segment of the eye in developed countries (Augustin and Offermann 2006; Campbell et al. 2010; DRCR Network et al. 2009, 2010; Ip et al. 2009; Nomoto et al. 2009; Ramchandran et al. 2008; Soheilian et al. 2009; Stahl et al. 2007). Other posterior segment disorders, such as uveitis, retinitis pigmentosa (RP), and optic nerve disease, also represent potentially treatable forms of blindness that involve disease processes that affect the posterior segment tissues of the eye, including the retina, choroid, and optic nerve (Cai et al. 2010; Giansanti et al. 2008; Gilger et al. 2010; Jacobson and Cideciyan 2010; Janoria et al. 2010; Kanamori et al. 2010). Administration or delivery of a specific drug to posterior segment tissues has gained significant attention in research and represents an important area of investigation. Our goal is to develop effective, targeted, and local delivery for the treatment of posterior segment diseases that affect vision. Recently, pharmaceutical innovation and development has produced new drugs for the treatment of the posterior segment, and novel drug delivery technologies are under investigation. Delivering drugs to the suprachoroidal space, a potential space between the choroid and the sclera, will be discussed and represents a new space that is gaining attention for novel drug delivery technology.

Traditionally, the most common and simplest form of drug delivery to the eye is to use either topical or systemic delivery. The advantages of topical drug delivery are most relevant to the treatment of anterior segment ocular disease and tissues such as the tear film (dry eye therapy), conjunctiva, cornea, iris, and ciliary body. Eye drops are convenient, usually painless, and this route avoids any form of "first-pass effect" through hepatic metabolism (Davies 2000). However, simple diffusional pharmacokinetics and aqueous outflow limit the ability of topical agents to reach posterior segment tissues with sufficient concentration to exert a therapeutic effect. In addition, there are several posterior segment diffusional barriers, such as the lens and the internal limiting membrane, that limit drug movement toward the macula and optic nerve. Other limitations include the following:

- As a drop is instilled into the eye, 50–80% of the medication is lost through the natural tear drainage routes.
- The remaining drug is diluted by the tear film and tear production (the turnover rate of tears is 16% volume per minute) even though many eye drops contain materials such as polymers and cellulose derivatives to optimize retention.
- There are corneal and conjunctival epithelial barriers, including the epithelium, that limit the diffusion of hydrophilic drugs.
- Drugs that penetrates the conjunctival epithelium may be absorbed into the systemic circulation through conjunctival blood vessels, thus removing them from a therapeutic effect.

- The corneal stroma is thick and is more permeable to hydrophilic drugs and less permeable to lipophilic moieties. Therefore, the cornea has biphasic resistance to both hydrophilic and lipophilic drugs.
- As the drug diffuses through the cornea and enters into the anterior chamber, it is diluted yet again with aqueous humor (1.5–4.0 µL/min with complete turnover of the aqueous every 100 minutes) (Brubaker 1982).

Consequently, the intraocular bioavailability of drug administration by the topical route to the vitreous is less than 5% of the given dose (Gaudana et al. 2010).

On the other hand, the surface area of the conjunctival epithelium is 17-fold larger than the cornea (Watsky et al. 1988) and is more permeable to drugs than corneal epithelium. Plus, the scleral stroma is hydrophilic and drugs are capable of penetrating the scleral tissues (based largely on the molecular size) (Olsen et al. 1995). These features suggest that a highly potent drug could still be effective using a topical approach.

Systemic administration, using either the oral or intravenous route, represents another means to deliver drugs to the posterior segment. With systemic administration, drugs arrive at the local ocular tissues through the bloodstream. The choroid consists of numerous blood vessels lined by endothelial cells that facilitate the exchange of nutrients, drugs, and waste products readily, as this vasculature is not part of the blood–retinal barrier (BRB). Systemic drug delivery could directly treat a pathologic condition of the choroid. However, the neurosensory retina is more difficult to treat due to the BRB. The outer BRB occurs at the zonula occludens at the retinal pigment epithelial (RPE) cell layer. The inner BRB resides at the tight junctions of the retinal capillaries. Normal capillary endothelial cells throughout the body are fenestrated, while the retinal capillary endothelium does not have these fenestrations. Multidrug efflux pumps, such as P-glycoproteins (P-gps), are present on the retinal vascular endothelium, while the RPE also contains these proteins plus multidrug resistance (MDR) proteins. These proteins resist the passage of substances from either the retinal vasculature or the choroidal side into the neurosensory retina, and further characteristics add to the barrier properties of the inner and the outer BRB (Duvvuri et al. 2003). Because of these blood–ocular barriers, only an estimated 1–2% of plasma drug concentration is achieved in the vitreous humor. To maintain a therapeutic effective concentration in the vitreous and the retina, frequent administration and high systemic levels of a drug are required. The variable absorption of a drug by other tissues may also lead to systemic side effects.

Periocular injection represents another commonly employed method of local drug delivery to deliver medication to the posterior segment tissues. These routes include subconjunctival, peribulbar, retrobulbar, and posterior sub-Tenon injections using either sharp needles or blunt cannulae. Periocular injection bypasses some of the surface barriers, as compared with topical delivery, and offers a reasonable and relatively safe route for drug delivery (Del Amo and Urtti 2008). Drugs delivered using this method reach the posterior segment of the eye by simple diffusion, generally through a transscleral route (Ghate and Edelhauser 2006). Nevertheless, three types of barriers, static, dynamic, and metabolic, still hinder the drug transfer to the retina (Kim, Lutz et al. 2007). Especially, drug clearance by the conjunctival lymphatics and blood flow in the episcleral veins, which is one of the dynamic barriers, plays a significant role in pharmacokinetics (Balachandran and Barocas 2008). Moreover, the transscleral route may be limited by suboptimal drug diffusion kinetics, especially for macromolecules or less-hydrophilic compounds (Olsen et al. 2006). For example, the intravitreal bioavailability of prednisolone administered periocularly has been reported to be in the range of 0.01–0.1% (Lee and Robinson 2004).

Since 2006, intravitreal injections have become an increasingly common route for pharmaceutical treatment of posterior disorders, especially exudative AMD. Intravitreal injections originated over 30 years ago for the treatment of acute intraocular diseases, such as the use of antibiotics for acute endophthalmitis. Intravitreal drugs may diffuse directly to many posterior segment tissues, including the macula, and minimize many systemic side effects. Because drugs injected intravitreally display first-order kinetics and show rapid clearance, the intravitreal route of delivery may be preferred for drugs with a higher molecular weight and a longer half-life (Janoria et al. 2007) and may require repeated administration, making this route less acceptable to patients. Local complications of the injection procedure include some pain and discomfort; transient elevations in intraocular pressure (IOP); cataract, subconjunctival, and occasionally vitreous hemorrhage; retinal detachment; and infection (Del Amo and Urtti 2008).

Finally, the suprachoroidal space is a potential space, limited anteriorly in the region of the scleral spur and posteriorly by the transscleral connections of the short posterior ciliary vessels to the choroid. There are well-established focal, equatorial connections at the vortex ampullae where venous blood exits the globe (Olsen et al. 2006) (Figure 12.1). The suprachoroidal space is considered a potential space to implant electrical retinal stimulators because mechanically induced retinal injury from direct contact with the electrodes and the heat generated by a device may

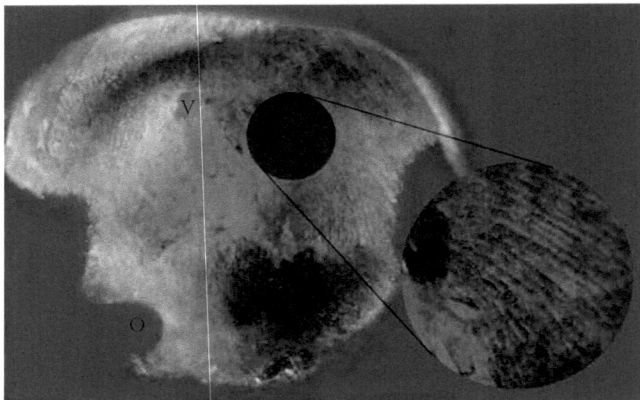

FIGURE 12.1 **(See color insert.)** Cast of the suprachoroidal space. Note the defects in the cast from the optic nerve (O) and vortex ampullae (V). The high-power view (circle) shows the fingerprint of large choroidal vessels.

damage the neurosensory retina. In addition, the suprachoroidal space may also be used for implantation of sustained-release drug delivery systems (Hou et al. 2009). During the past decade, the frequency of intravitreal injection has increased significantly (Campbell et al. 2010). Naturally, the incidence of intraocular side effects would also increase (Nelson et al. 2003; Ozkiriş and Erkiliç 2005; Parke 2003; Sutter and Gillies 2003). Drug delivery to the suprachoroidal space may be a reasonable alternative that may avoid some of the intraocular complications, avoid the visual axis with the more opaque medications, bypass the conjunctival clearance mechanism of periocular injection, optimize pharmacokinetics to macular tissue, and more directly target posterior segment tissues, specifically the macula, choroid, RPE, and optic nerve.

12.2 HISTORY/BACKGROUND

In 1997, Krohn et al. studied the suprachoroidal spaces in pig and human eyes using the corrosion cast method. The authors aimed to find the uveoscleral drainage route of the aqueous humor (Krohn and Bertelsen 1997a,b). These authors characterized the suprachoroidal space using Indian ink–stained gelatin and a light microscope (Krohn and Bertelsen 1998).

In 2002, Einmahl et al. investigated the feasibility and tolerance of suprachoroidal injections in the rabbit model. They injected a poly(ortho ester) using a solid metal, olive-tipped cannulae inserted into the suprachoroidal space. The authors demonstrated that the material remained in the suprachoroidal space for 3 weeks. However, there were RPE irregularities, presumably from the cannulae, associated with the injections (Einmahl et al. 2002).

In 2006, Gilger et al. investigated the effects of deep scleral lamellar administration of a cyclosporine A (CsA)-releasing device for treatment of equine recurrent uveitis (ERU). The device was a biodegradable reservoir impregnated with CsA. It was inserted under a partial a partial-thickness lamellar scleral flap, and the implant was shown to be highly effective in achieving therapeutic ocular drug concentrations and controlling the equine uveitis. The authors suggested the possibility that the deep scleral lamellar deposit may optimize the suprachoroidal space and allow more direct diffusion of the drug to bathe the external choroid instead of relying on slower full-thickness transscleral drug diffusion (Gilger et al. 2006).

In 2006, we described the use of a flexible, fiber-optic microcannula with the ability to directly access the suprachoroidal space in a safe, repeatable, and reliable manner. We demonstrated the methodology in the pig model. The microcannula was originally designed to access Schlemm's canal for circumferential viscodilation during canaloplasty surgery. In our studies, we delivered the drug directly into the suprachoroidal space, which was performed in 94 porcine eyes, and demonstrated sustained local delivery and excellent pharmacokinetics for triamcinolone that maintained local tissue levels for at least 120 days postinjection (Olsen et al. 2006).

In 2007, Kim et al. investigated the suprachoroidal space with gadolinium-diethylenetriaminopentaacetic acid (Gd-DTPA) infused in the intrascleral space using dynamic contrast-enhanced magnetic resonance imaging (MRI). An intrascleral injection using a catheter placed anteriorly was successful in transporting Gd-DTPA to the posterior segment from an anterior injection site. They also demonstrated that supra-choroidal space is an expandable conduit for drug transport to the posterior segment. However, the clearance rate of the Gd-DTPA followed first-order kinetics (Kim, Galbán et al. 2007). In 2009, Hou et al. tested the safety of suprachoroidal fibrin glue. They showed that suprachoroidal fibrin glue induces a localized inflammatory reaction in the early stage that turned into fibrotic tissue (Hou et al. 2009). Finally, in 2010, supracho-roidal drug delivery using a hollow microneedle was reported (Patel et al. 2010).

12.3 TECHNOLOGIES

12.3.1 Deep Scleral Dissection

The conventional method to approach the suprachoroidal space is through a deep scleral dissection for placing an implant device (Figure 12.2). Using the methodol-ogy documented for horses, after sterile ocular preparation, a conjunctival incision was made followed by a partial-thickness lamellar dissection, creating a scleral flap (90–95% depth centrally). The implant was placed under the flap, and the flap was closed with a suture. In these studies, few abnormalities were observed on ophthal-mic examination of these horses for up to 12 weeks after surgery and included mild conjunctival hyperemia over the surgical site for a week. No histopathologic abnor-malities were noted on the eyes of horses 12 weeks after implantation with the CsA device. However, the implant was encapsulated in a thin layer of fibrous tissue that may have prevented implant migration. Signs of retinal or uveal toxicity were not observed (Gilger et al. 2006).

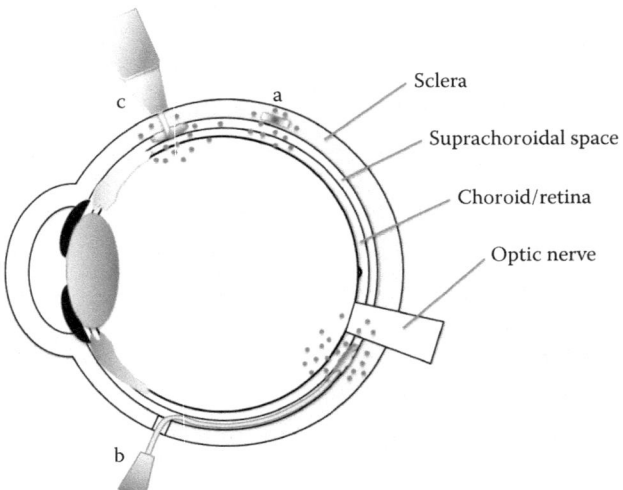

FIGURE 12.2 Schematic representation of various techniques of suprachoroidal drug delivery. (a) Deep lamellar scleral implant is placed under the partial-thickness scleral flap. (b) After full-thickness scleral incision, a solid cannula or catheter is inserted. The catheter allows easy access to the subperifoveal and peripapillary region. (c) A hollow microneedle, injected through the conjunctiva, is a minimally invasive method to access the suprachoroidal space.

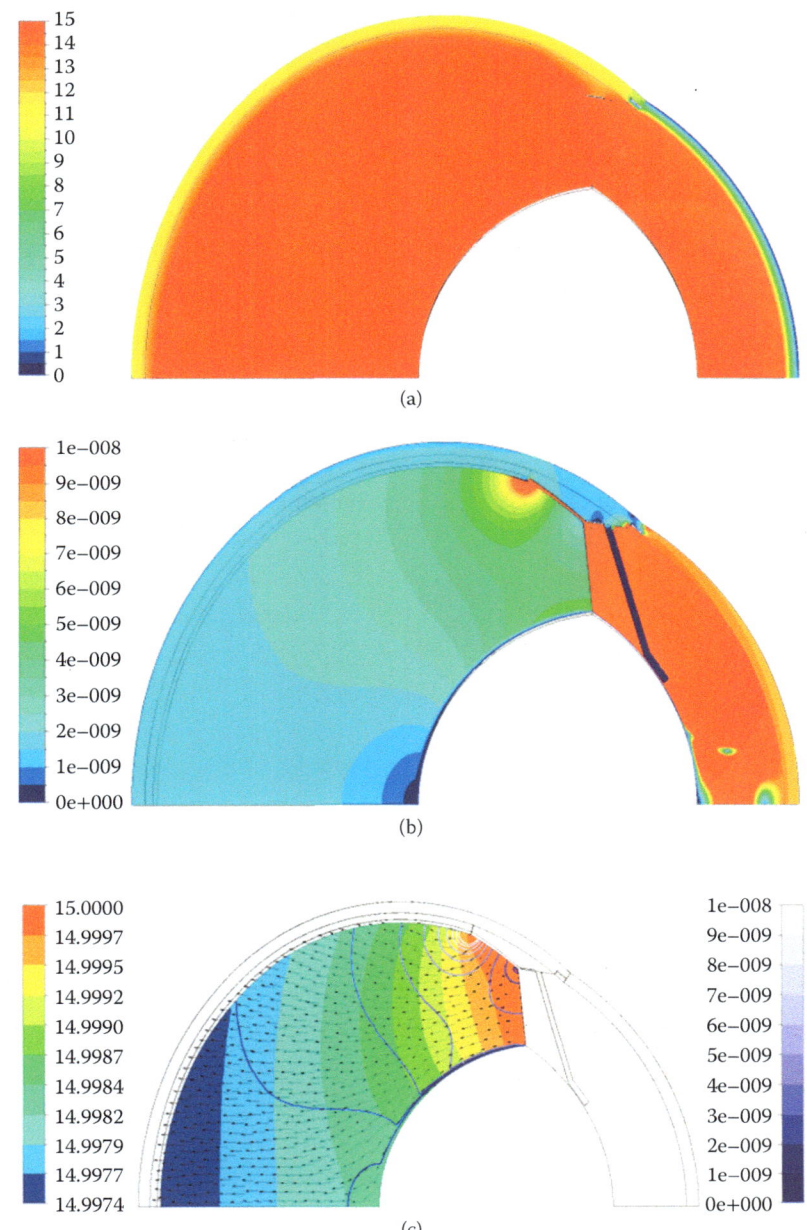

FIGURE 4.9

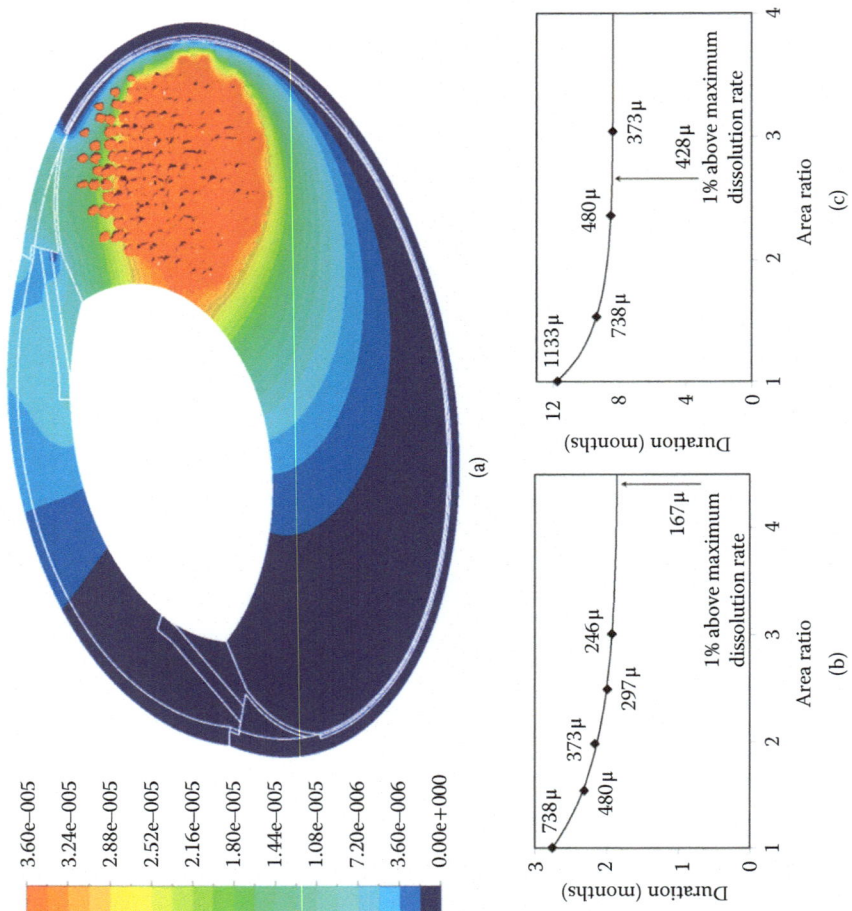

FIGURE 4.11

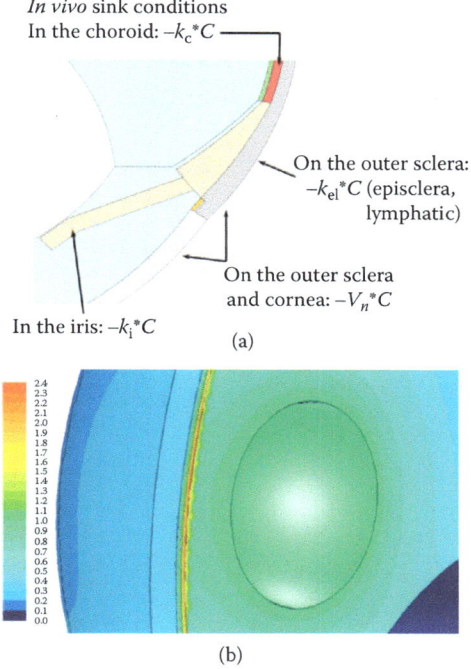

In vivo sink conditions
In the choroid: $-k_c{}^*C$

On the outer sclera:
$-k_{el}{}^*C$ (episclera,
lymphatic)

On the outer sclera
and cornea: $-V_n{}^*C$

In the iris: $-k_i{}^*C$

(a)

(b)

FIGURE 4.14

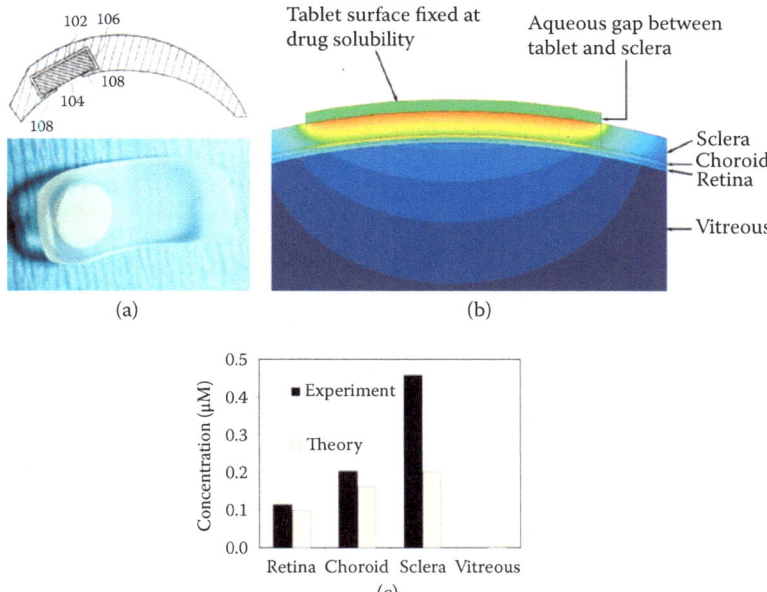

102 106

108

104

108

Tablet surface fixed at
drug solubility

Aqueous gap between
tablet and sclera

Sclera
Choroid
Retina

Vitreous

(a)   (b)

Concentration (μM)

■ Experiment

Theory

Retina Choroid Sclera Vitreous

(c)

FIGURE 4.15

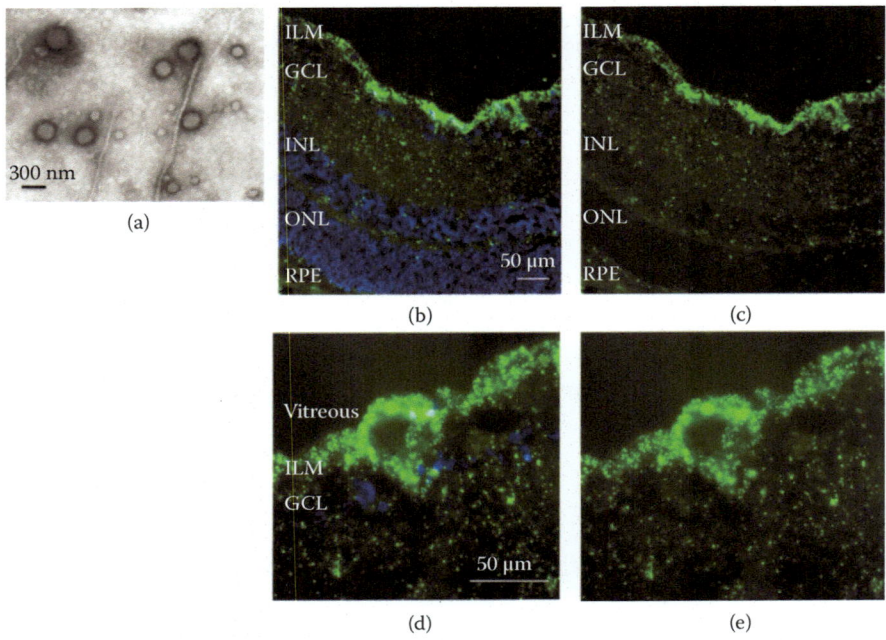

FIGURE 8.2

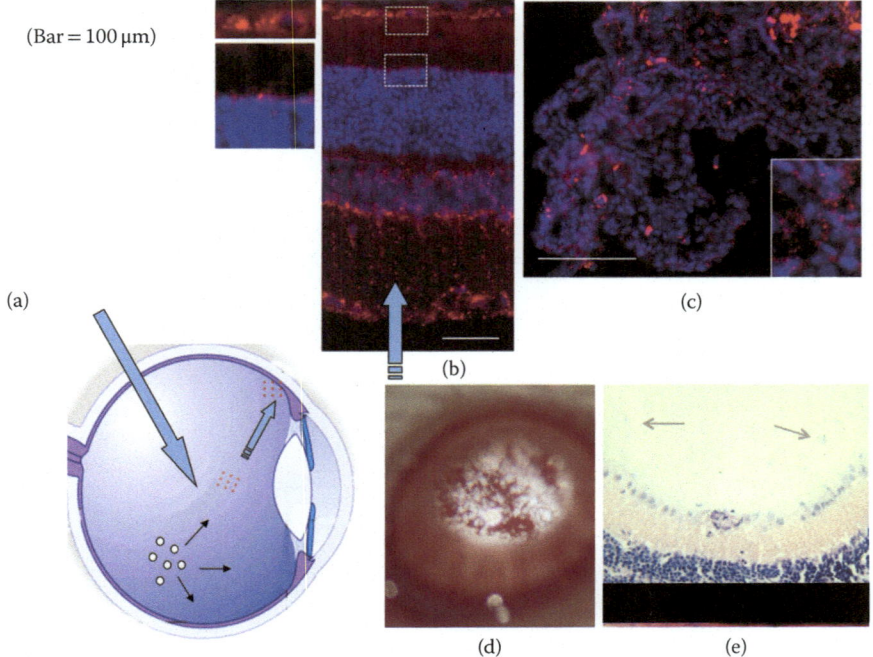

FIGURE 8.4

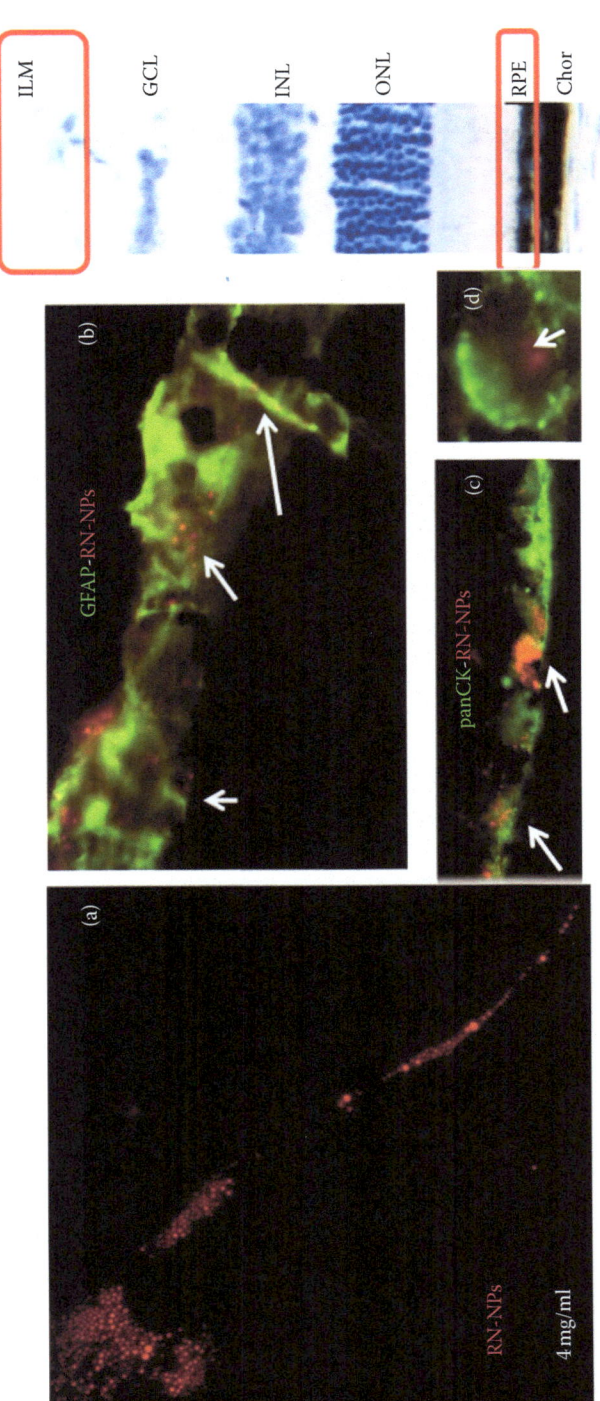

FIGURE 8.5

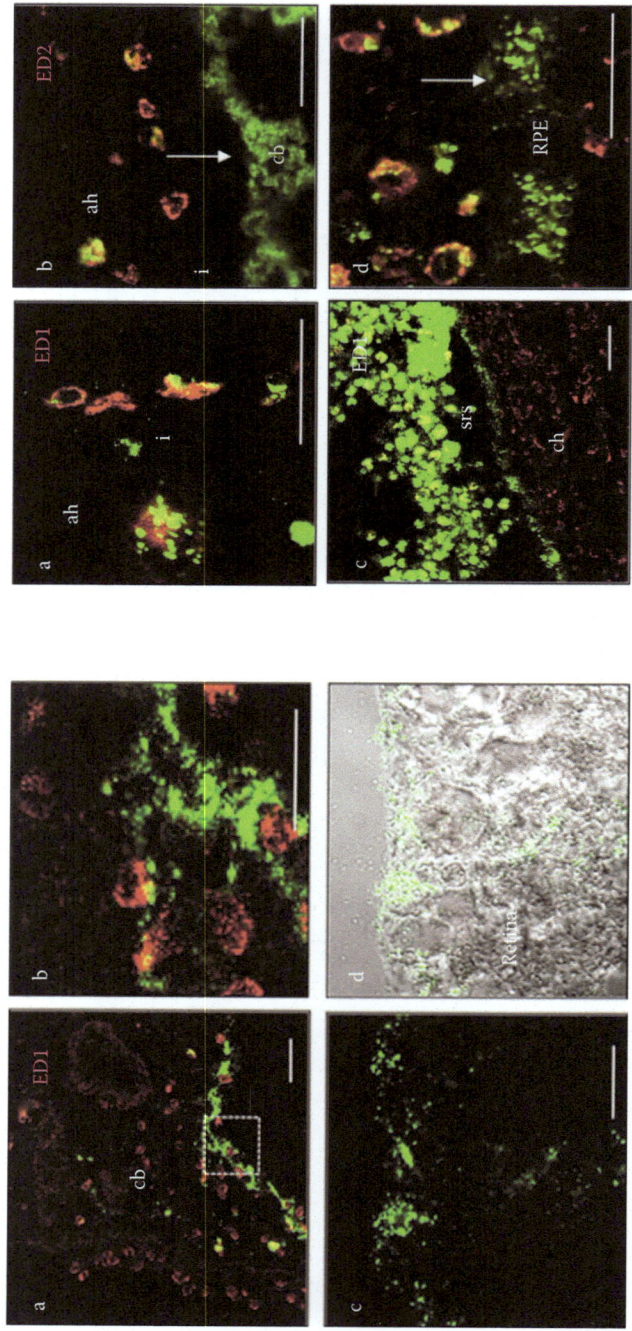

(left)

(right)

FIGURE 8.6

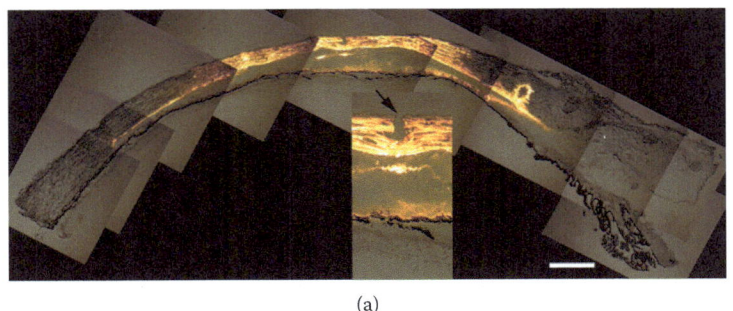

(a)

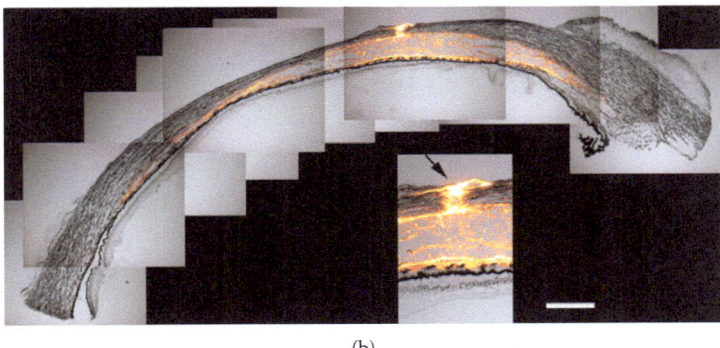

(b)

FIGURE 9.2

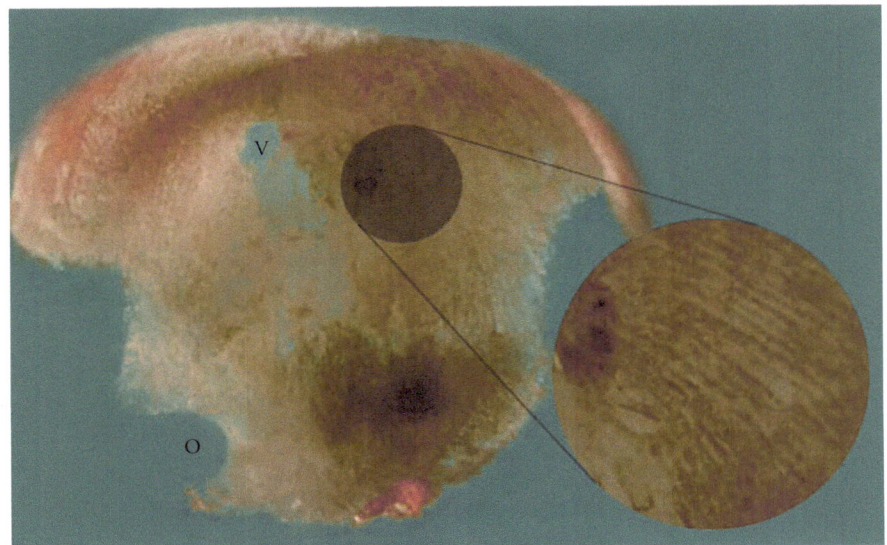

FIGURE 12.1

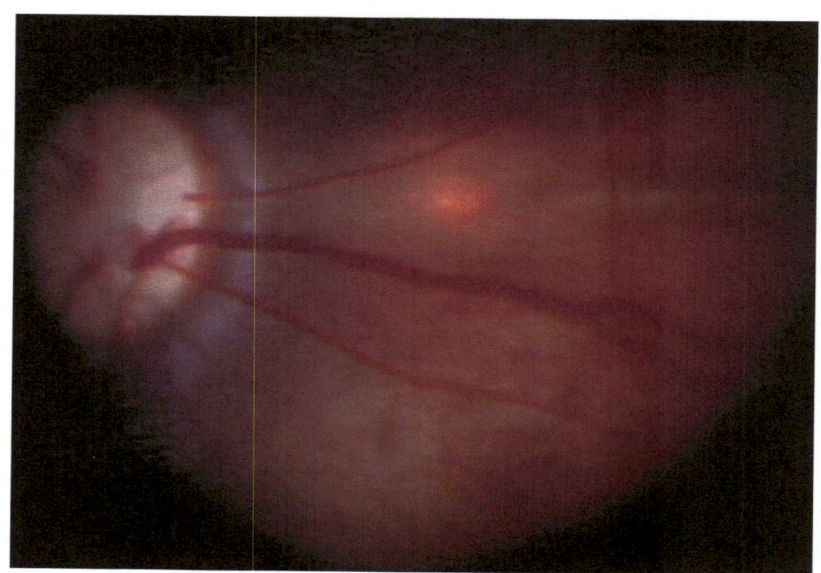

FIGURE 12.4

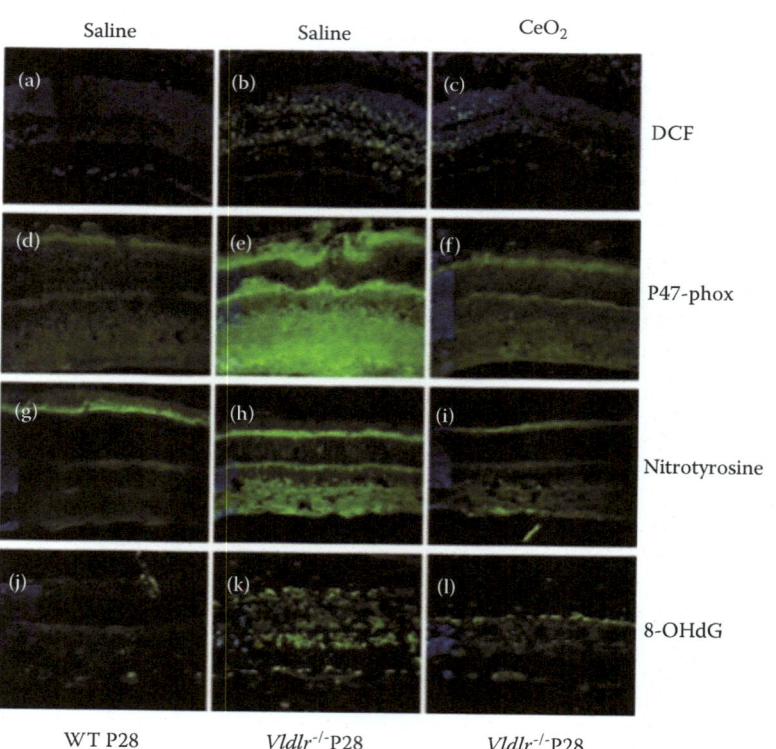

FIGURE 13.4

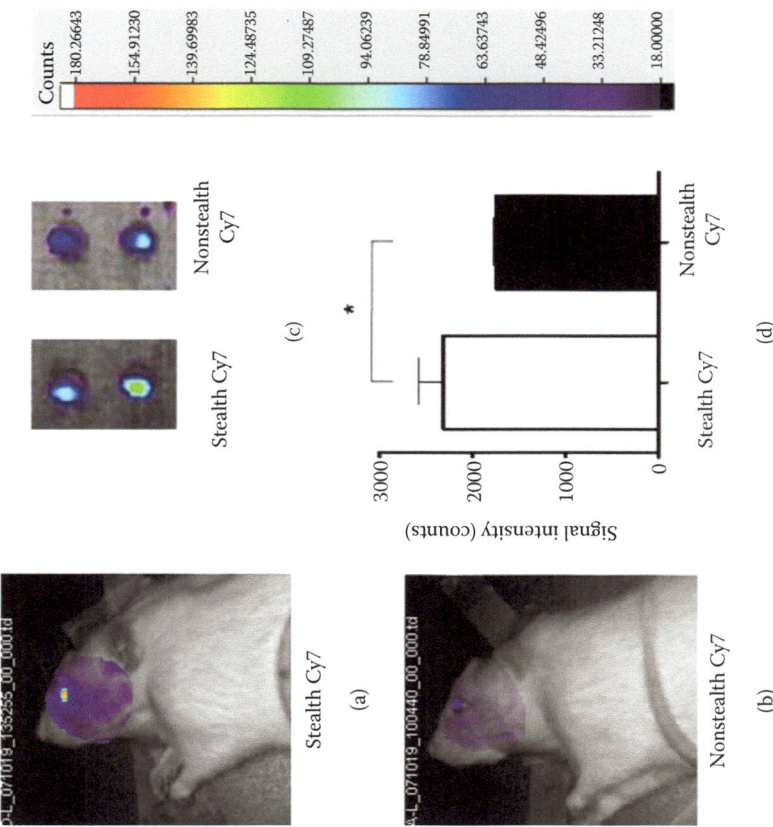

FIGURE 18.1

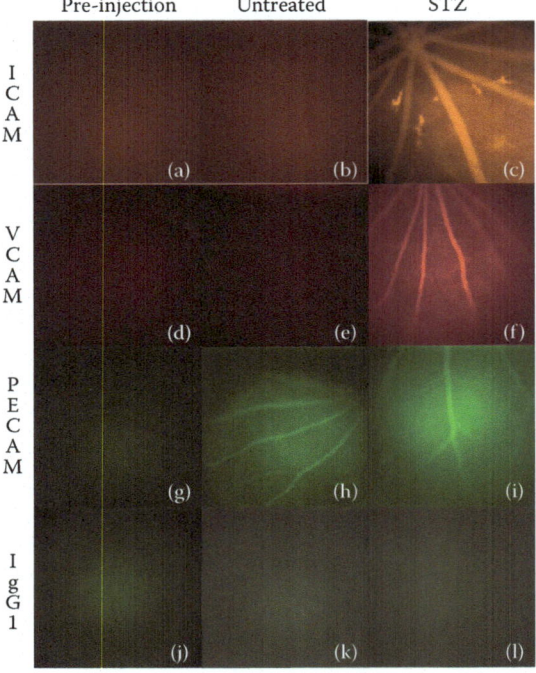

FIGURE 19.2

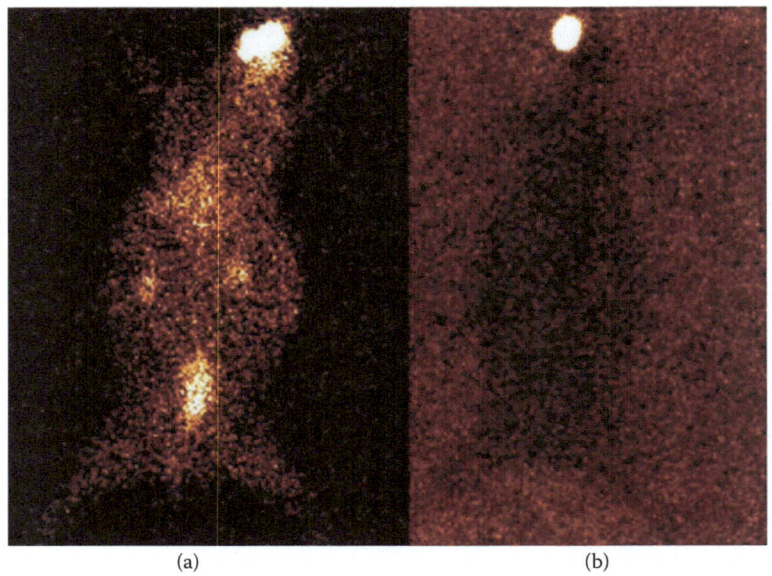

FIGURE 21.12

12.3.2 Solid Cannulae

Cannulae have been reported to serve as a viable conduit to the suprachoroidal space. In a rabbit study, the conjunctiva and Tenon capsule were opened, a small (2-mm) full-thickness scleral incision was created parallel to the limbus, and a scleral tunnelization was made, which gained access to the suprachoroidal space. A 100-μL injection was performed through a curved cannula with an olivary tip (0.6 mm of internal diameter). No reflux of material was observed through the sclerotomy and the incision was sutured (Einmahl et al. 2002).

In another recent study for rabbits, the authors reported the use of a "soft pipe." The sclera was exposed and a 2-mm scleral tunnel incision was made with a blade. Approximately 0.1–0.2 mL of aqueous humor was drained from the anterior chamber before the drug injection. The 26-gauge soft pipe was gently inserted through the tunnel into the posterior suprachoroidal space, advanced toward the posterior pole, and 100 μL of drug was injected. They reported that the suprachoroidal hemorrhage at postoperative 1 day in 37.5% of eyes cleared by 3 days postinjection (Hou et al. 2009).

12.3.3 Catheter

In 2006, we investigated the suprachoroidal drug delivery route using a microcannulation system (Figure 12.3; posterior delivery system [PDS], iScience Surgical, Menlo Park, CA). The advantageous characteristics of the cannula include a flexible, small-diameter cannula (325-μm outer diameter and 175-μm inner diameter cannula channel), a stainless steel wire shaft for optimizing the transition properties (flexibility with enough stiffness to be pushable through the tissues), a fiber-optic light source with a flashing tip or "beacon tip" (Figure 12.4), and a contoured tip design (360 μm maximum outer diameter) for maneuvering safely in the suprachoroidal space that does not tear or injure the overlying RPE and allows easy accessibility back to the submacular and peripapillary region. A brief technique description is as follows. An anterior conjunctival peritomy provided access to pre-equatorial sclera, where a radial incision was made to expose bare choroid. Next, viscoelastic was injected with the cannula to hydrodissect the entry site choroid from the sclera. The PDS was inserted using two non-toothed forceps and introduced parallel to the plane of the suprachoroidal space in a hand-over-hand manner, gently guiding the tip posteriorly. Care was taken to avoid snagging the choroid. When resistance was encountered, the cannula was minimally retracted. The wide-angle viewing system was used to visualize the PDS tip (red flashing light). Direct visualization was possible to guide the probe tip into the desired position, usually superior to the fovea. The drug was slowly injected using a mechanical viscous fluid injection system while the cannula tip remained in position for 1 minute to minimize reflux. The scleral and conjunctival incisions were closed with suture (Olsen et al. 2006).

12.3.4 Microneedle

The microneedle is a device originally developed for drug and vaccine delivery to the skin. Microneedles are solid or hollow needles that penetrate into tissue for targeted drug delivery that may occur in a minimally invasive manner. Hollow microneedles

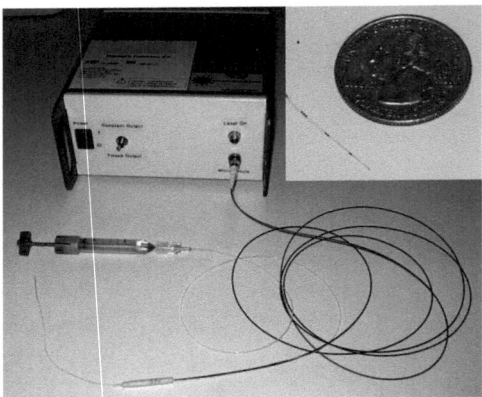

FIGURE 12.3   The catheter system for cannulation to the suprachoroidal space. It is composed of a flexible, small-diameter cannula and a fiber-optic light source. Compare the size (360 μm in diameter) with a quarter coin (upper right box).

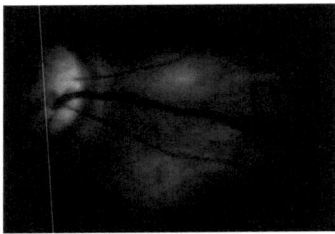

FIGURE 12.4   (See color insert.) Wide-angle surgical view of the posterior pole demonstrating the shaft of the cannula. The shaft of the posterior delivery system and red probe tip are visible in the suprachoroidal space. (Reprinted from *American Journal of Ophthalmology*, 142, Olsen, T.W. et al., Cannulation of the suprachoroidal space: A novel drug delivery methodology to the posterior segment, 777–787, © [2006], with permission from Elsevier.)

were proposed to deliver drugs to the suprachoroidal space. Glass microneedles were fabricated with borosilicate micropipette tubes (outer diameter of 1.5 mm, inner diameter of 0.86 mm, B150-86-15; Sutter Instrument, Novato, CA) that were fire-polished. These blunt-tip microneedles were beveled to a 20° tip angle using a glass grinder (BV-10, Sutter Instrument) and might be cleaned for reuse. A pen-like device was constructed using a threaded cap, fabricated to position the microneedle, and allowed precise length adjustment. This handheld device was attached to a micro-pipette holder (MMP-KIT, World Precision Instruments, Sarasota, FL) with tubing connected to a carbon dioxide gas cylinder for direct infusion pressure application. In that *ex vivo* study, the authors demonstrated that microneedles could deliver drugs into the suprachoroidal space of rabbit, pig, and human eyes. Optimization of the delivery device parameters showed that microneedle length, pressure, and particle size played an important role in determining successful delivery of the drug formulation into the suprachoroidal space (Jiang et al. 2009; Patel et al. 2010).

12.4 PHARMACOKINETICS

To understand the pharmacokinetics in suprachoroidal drug delivery, several factors should be considered. There are relatively few reports to date that carefully investigate the pharmacokinetics of suprachoroidal drug delivery. However, one could extrapolate from the pharmacokinetics of transscleral drug delivery, a route that has been more intensively investigated (Amrite et al. 2008; Balachandran and Barocas 2008; Barar et al. 2008; Kim et al. 2009; Nicoli et al. 2009; Nomoto et al. 2009; Robinson et al. 2006;). The following represents key factors that determine accurate pharmacokinetics.

12.4.1 DIFFUSION

The simple diffusion of a locally administered drug depot, injected directly into the target tissues, represents a primary means of drug distribution in the relevant tissues. A drug depot results in diffusional spread from the administration site to the surrounding tissues, down the concentration gradient, specifically from higher to lower concentration. The diffusional distribution of a drug into the retina is hindered by several barriers, classified as static, dynamic, or metabolic barriers (Kim, Lutz et al. 2007). In the transscleral system, the sclera acts as an important static barrier to the diffusion of drug molecules into the retina (Ambati, Canakis et al. 2000; Ambati, Gragoudas et al. 2000; Cruysberg et al. 2005; Jiang et al. 2006; Kim et al. 2009; Kim, Lutz et al. 2007; Olsen et al. 1995; Shuler et al. 2004). However, the scleral barrier of drug delivery to the choroid, RPE, and retina is bypassed by using suprachoroidal delivery. In fact, the scleral barrier may help minimize diffusion away from the target tissue or even act as a depot for the drug. Diffusion across the sclera may occur through perivascular spaces (i.e., the vortex ampullae), through the aqueous media of the gel-like mucopolysaccharides, and across the scleral matrix itself. Regarding scleral diffusion, low-molecular-weight, hydrophilic and negatively charged molecules will be cleared more rapidly than larger, high-molecular-weight, lipophilic, or positively charged drugs.

Next, the diffusion of the drug depot in the suprachoroidal space toward the choroid, RPE, and neurosensory retina should be considered. The choroid itself is very permeable and drugs easily diffuse within the choroidal stroma, as there are few structural barriers. Once the drug reaches the Bruch's membrane complex and the RPE layer, a significant barrier to diffusion occurs. Any drug residing within the choroidal stroma can then either diffuse through the Bruch's membrane or may also be swept away by the rapid flow of the choroidal vasculature, a vasculature that is devoid of tight junctions, especially at the choriocapillaris. Fewer studies have carefully examined this layer and the pharmacokinetics are not as well studied. The *in vitro* study (no blood flow) using bovine and porcine choroid-Bruch's layer suggested that this layer was a more significant barrier to drug transport than was the sclera. Specifically, this layer hindered the transport of lipophilic solutes, especially cationic solutes, more than hydrophilic solutes (Cheruvu and Kompella 2006). In a bovine RPE-choroid preparation, RPE was a major barrier and may be the rate-limiting barrier in the diffusion of hydrophilic drugs and macromolecules. For lipophilic molecules, the RPE-choroid layer demonstrated an equal barrier effect, as did the sclera. Interestingly, lipophilic drugs had an asymmetric permeability from inward to outward diffusion, suggesting a possible active transport mechanism (Pitkänen et al. 2005). In the aging eye, Bruch's membrane thickness increases by 135%, from 2 μm in the 1st decade to 4.7 μm in the 10th decade (Ramrattan et al. 1994). Using human tissues, the hydraulic conductivity of the choroid-Bruch's membrane complex also demonstrates an exponential decrease with increasing age (Moore et al. 1995) with the overall permeability of choroid-Bruch's membrane to serum proteins decreasing ten fold from the first to the ninth decade of life (Moore and Clover 2001).

12.4.2 SCLERAL DEPOT IN IONTOPHORESIS

In a study that investigated human scleral diffusion of vinblastine and doxorubicin, human sclera acted as a relative depot for free doxorubicin (Kim et al. 2009). Separately, methotrexate delivery using hydrogel iontophoresis showed that the drug accumulated in the sclera that in turn served as a concentrated "depot" (Eljarrat-Binstock et al. 2007). Iontophoresis seems to cause accumulation of the drug within the contact tissue. As stated, the primary factor that drives the drug into the intraocular tissues is its diffusional properties (Anderson et al. 2003). A study that investigated the pharmacokinetics of bevacizumab in rabbits suggested an extended half-life of the drug within the iris/ciliary body and retina/choroid after subconjunctival injection as compared to intravitreal injection (Nomoto et al. 2009). However, these studies should be interpreted with caution because the anatomy of the rabbit eye and the scleral thickness and size parameters are quite different from humans. In fact, the pig is probably the best mode to study pharmacokinetics due to the similarities in scleral thickness (Olsen et al. 1998), choroidal blood flow, RPE structure, and retinal vasculature as compared to humans (Olsen et al. 2002).

12.4.3 UVEOSCLERAL OUTFLOW

In 1965, Anders Bill was studying uveoscleral outflow and reported that albumin injected into suprachoroidal space disappeared by bulk flow and could penetrate the sclera both through the perivascular spaces and also directly through the scleral substance (Bill 1965).

Uveoscleral outflow is a major exit route for the aqueous humor and is believed to traverse the ciliary body face, and drains directly into the suprachoroidal space. The contribution of the uveoscleral pathway to the total aqueous outflow has been reported to be in the range of 40–50% in nonhuman primates. In the human eye, most data has been calculated indirectly; however, the results are similar (Alm and Nilsson 2009). As the eye ages, drainage of the aqueous humor through the uveoscleral outflow pathway decreases (Toris et al. 1999).

The uveoscleral outflow pathway is a key dynamic factor that influences the pharmacokinetics of drugs administered in the suprachoroidal space.

12.4.4 CHOROIDAL BLOOD FLOW

Drug clearance into the high-flow choroidal vasculature is considered another major dynamic factor that affects suprachoroidal pharmacokinetics. Choroidal blood flow is among the highest per unit volume of any tissue in the body, and the choriocapillaris endothelial barrier has large fenestrations (Bill et al. 1980). These features suggest that the choroidal flow acts as a sink for drug diffusion, down the the always-decreasing gradient from the local tissue to the extremely low systemic serum levels. However, there is very little data on the effect of choroidal clearance on transscleral delivery (Balachandran and Barocas 2008).

A rabbit study that used episcleral implants and dynamic three-dimensional MRI showed that *in vivo*, episcleral implant did not deliver a significant amount of Gd-DTPA into the vitreous. A 30-fold increase in vitreous Gd-DTPA concentration occurred in the enucleated eye, suggesting dynamic barriers to the movement of drugs from the episcleral space into the vitreous found only in the *in vivo* studies are very revealing. While the choroidal blood flow of the rabbit eye is different than the human eye, the sink effect of the choroid is clearly demonstrated (Kim et al. 2004). Similarly, Amrite et al. investigated the effect of circulation on the ocular tissue distribution of nanoparticles after periocular injection in a rat model. They noted that the tissue levels of particles in the sclera and choroid most adjacent to the site of administration were 19-fold higher in the dead animals than in the live animals. They speculated that the higher observed levels could be due to the absence of episcleral and/or choroid circulatory systems in addition to the absence of other periocular clearance mechanisms (Amrite et al. 2008). In our opinion, the mouse or rat model is a poor model for studying the pharmacokinetics of a drug that relates to the human condition.

Robinson et al. applied cryotherapy to treat the choroid of a rabbit model and assess for differential diffusion of the ocular barriers. Four weeks after cryotherapy, a mature chorioretinal scar developed with theoretic obliteration of the choroid and retinal vasculature. They compared vitreous drug levels following a sub-Tenon's injection of triamcinolone acetonide in treated versus control eyes. Interestingly, the injury to the choroid did not have detectable differences in the vitreous concentrations (Robinson et al. 2006). The results of a computer model of pharmacokinetics were in agreement with the rabbit study (Balachandran and Barocas 2008).

Our current studies have shown more rapid diffusion of drugs away from this space when they are formulated into a large biologic agent that is suspended in Healon (hyaluronic acid; Abbott Medical Optics, Abbott Park, IL). Sustained-release small molecules may be ideal candidates for suprachoroidal delivery, while large biologic agents have a more rapid clearance from the suprachoroidal space. The high-flow blood channels of the choroid and choriocapillaris are likely responsible for systemic clearance, but we cannot rule out alternate pathways such as the uveoscleral pathways (Olsen et al. 2011).

12.4.5 MELANIN

Ocular melanin is found in the uveal tract and the RPE. Various drugs bind to melanin and may be retained with these cells for extended periods of time. The significance of such binding could be that the melanin protects the pigmented cells and adjacent tissues by adsorbing potentially harmful substances (Larsson 1993). More basic pH and lipophilic drugs seem to have a higher affinity to melanin (Leblanc et al. 1998).

As an example, one drop of 0.25% timolol was applied to one eye of 10 subjects with brown iridies. No IOP reduction was obtained in either the treated or control eyes, presumably because the drug was bound to the iris pigment (Salminen et al. 1985). The drug permeability of the PE-choroid layer may have a greater lag time for the lipophilic beta-blockers than the hydrophilic beta-blockers, also perhaps due to the drug binding to melanin in the RPE and choroidal melanocytes (Pitkänen et al. 2005). Similarly, the transscleral drug delivery of lipophilic celecoxib was significantly lower in pigmented rats than in albino rats with the same rationale (Cheruvu et al. 2008). Consequently, the pigmentation of ocular tissues can act as an important barrier and may affect the delivery of drugs to the retina and vitreous, especially for lipophilic drugs (Amrite et al. 2010).

12.4.6 RPE BARRIER

The RPE serves as the outer BRB through the tight junction between neighboring RPE that acts as a static barrier, hindering diffusion. As mentioned, melanin pigment in the RPE acts as reservoir or barrier for lipophilic drugs. Finally, the RPE has a dynamic barrier effect through active cellular transport.

Active transport of fluorescein from the vitreous by the retinal vessels and retina was first observed by Cunha-Vaz and Maurice (1967). Since then, many investigators have reported active transport in the RPE-choroid layer *in vitro* using an Ussing chamber (Kimura et al. 1996; Koyano et al. 1993; Tsuboi 1987; Tsuboi et al. 1984).

Since P-gp was first identified in multidrug-resistant tumor cells in 1992, a number of membrane transporters have been discovered in various ocular tissues, such as the corneal epithelium, retinal capillary endothelial cells, RPE, ciliary nonpigmented epithelium, and iris and ciliary endothelial cells (Aukunuru et al. 2001; Gandhi et al. 2004; Kennedy and Mangini 2002; Ling 1992; Saha et al. 1998; Yang et al. 2000). Initially, the transporters were believed to be responsible for translocation of essential nutrients across the cell membrane, but later studies have shown a more active role in drug transport (Janoria et al. 2007). Cell membrane transporters in the retina are primarily located within the RPE cells as well as in the retinal vascular endothelium. A variety of transport proteins are expressed in the RPE cell surface, including drug efflux pumps, amino acid transporters, organic ion transporters, and various kinds of transporter proteins (Mannermaa et al. 2006, 2009; Senthilkumari et al. 2009).

12.5 CLINICAL APPLICATIONS

12.5.1 AMD

AMD is the leading cause of visual loss in people older than 50 years in developed countries, with a prevalence that continues to rise. The stages of AMD are best characterized in the Age-Related Eye Disease Study (AREDS Group 2001).

The Eye Diseases Prevalence Research Group estimates that in the year 2000, AMD affected more than 1.75 million individuals in the United States and is expected to increase to nearly 3 million by 2020 (Friedman et al. 2004).

The exact mechanism involved in the development of choroidal neovascularization is poorly understood. However, vascular endothelial growth factor (VEGF) clearly represents an important stimulus for neovascularization. VEGF is a potent endothelial cell-selective growth factor that in turn promotes degradation of the local extracellular matrix and facilitates endothelial cell migration. VEGF also mediates vascular hyperpermeability resulting in leakage of plasma proteins that serve as a substrate for endothelial cell growth with secondary retinal edema, bleeding, and eventual tissue destruction.

Currently, exudative AMD is treated primarily with intravitreal injections of anti-VEGF. The most successful results of an anti-VEGF drug to date are the intravitreal delivery of ranibizumab (Lucentis; Genentech, Inc, South San Francisco, CA). Ranibizumab has been shown to increase the chance of improving visual acuity by 15 letters (approximately 3 lines) in 25–40% of treated cases as compared with untreated controls (Brown et al. 2006; Rosenfeld et al. 2006). Bevacizumab is also used in an off-label manner with a similar anti-VEGF mechanism. Pegaptanib, an RNA aptamer, was the first anti-VEGF agent released with FDA approval for the treatment of exudative AMD. Currently, several other anti-VEGF agents are in various phases of clinical trials (Kuno and Fujii 2010).

EphB4/EphrinB2 is erythropoietin-producing hepatocellular receptor, and its ligands have been demonstrated to posses antiangiogenic effects on CNV (choroidal neovascularization) (Brar et al. 2010). Pazopanib is a small-molecule kinase inhibitor that blocks VEGFR1, 2, and 3 and also has substantial activity against PDGFR-alpha (platelet-derived growth factor receptor), PDGFR-beta, FGFR1 (fibroblast growth factor receptor), and FGFR3. It has shown strong antitumor and antiangiogenic activities in mice (Takahashi et al. 2009). The agent a36 is a rationally designed, 13-amino acid cyclic peptide that competitively inhibits urokinase plasminogen activator urokinase-type plasminogen activator (uPA) receptors and is expected to inhibit the migration of human endothelial cells (Falkenstein et al. 2010). EMD478761 is an integrin antagonist that significantly suppressed laser-induced choroidal neovascularization in an experimental rat model (Fu et al. 2007).

Rationally, since the primary pathologic origin of neovascular AMD is at the level of the choriocapillaris under the macula, suprachoroidal drug delivery may offer the most direct route to administer drugs into this layer. Therefore, perhaps, exudative AMD could be an excellent candidate for suprachoroidal drug delivery. Further study, especially around drug formulation, is necessary.

12.5.2 DME

According to the data from the 2007 National Diabetes Fact Sheet, diabetes affects approximately 23.5 million or 10.7% of U.S. adults. Diabetes is the leading cause of new cases of blindness among adults aged 20–74 years, and diabetic retinopathy causes 12,000–24,000 new cases of blindness each year in the United States. A major complication of diabetic retinopathy is DME. In a long-term U.S. study, the incidence of DME over a 10-year period was 20.1% for "younger-onset" patients (those diagnosed before 30 years of age), 25.4% for "older-onset" patients using insulin, and 13.9% for "older-onset" patients not using insulin (Klein et al. 1995).

DME is defined by accumulation of fluid and hard exudates that results from leakage of the retinal capillaries in the macular area. Tractional forces of the posterior hyaloid on the retinal surface along with an increase in vascular permeability due to numerous cytokines are known to contribute the development of DME. DME is commonly treated with focal laser photocoagulation. Diffuse DME results from a generalized breakdown of the inner BRB and responded less well to laser photocoagulation. VEGF is known to be a critical factor contributing to the formation of DME. Depot injections of corticosteroids and intravitreal injections of an anti-VEGF agent are increasingly used in the management of DME, usually in conjunction with laser photocoagulation. Diabetic Retinopathy Clinical Research (DRCR) Network reported 3-year follow-up of focal/grid laser versus intravitreal triamcinolone injection for the treatment of DME and showed that the long-term effect of laser photocoagulation was superior to intravitreal triamcinolone alone (DRCR Network et al. 2009). Another report showed that intravitreal ranibizumab with prompt or deferred laser was more effective, through at least 1 year, than laser alone for the treatment of DME involving the central macula (DRCR Network et al. 2010).

12.5.3 RETINAL VEIN OCCLUSION

Retinal vein occlusion (RVO) is the second most common retinal vascular disease, next to diabetic retinopathy. The prevalence of RVO is reported as 5 per 1000 (Rogers et al. 2010). The two types of RVO are central retinal vein occlusion (CRVO) and branch retinal vein occlusion (BRVO). CRVO is subdivided again into nonischemic and ischemic types. The main cause of vision loss from RVO is from either ischemic damage to the macula or due to the development of macular edema.

The Branch Vein Occlusion Study reported that grid laser photocoagulation reduced macular edema and improved visual acuity in BRVO patients (BVOS 1986). Intravitreal steroids or anti-VEGF is another method for the treatment of macular edema. The SCORE study (Standard Care vs. Corticosteroid for Retinal Vein Occlusion) reported no difference in visual acuity by 12 months for the standard care group compared with the triamcinolone groups; however, the rates of adverse events, particularly elevated IOP and cataract, were highest in the 4-mg group (Scott et al. 2009). Since it understood that VEGF is an important factor in the pathophysiology of RVO (Adamis and Shima 2005), numerous clinical studies have used intravitreal anti-VEGF agents. Both ranibizumab and bevacizumab had long-term benefits to patients with macular edema from RVO (Campochiaro et al. 2010; Kriechbaum et al. 2009).

12.5.4 UVEITIS

Uveitis is a general term for inflammation involving uveal tissues. Uveal tissues include the iris, ciliary body, and choroid. Common types of posterior uveitis include toxoplasmosis, pars planitis, CMV (cytomegalovirus) retinitis, and histoplasmosis. The treatment of posterior uveitis is challenging to most ophthalmologists because drug delivery to the retina and choroid is difficult, the course of disease is chronic and recurrent, and the etiology can be difficult to identify. Topical drug administration for posterior uveitis is limited by diffusion of drug into the vitreous and is usually insufficient to treat posterior uveitis. Chronic systemic administration of corticosteroids is associated with many systemic side effects, including hypertension; hyperglycemia; and increased susceptibility to infection, peptic ulcers, and psychosis. Besides corticosteroids, various immunosuppressive drugs, such as cyclophosphamide, chlorambucil, methotrexate, azathioprine, cyclosporin A, bromocriptine, dapsone, and colchicine, form the mainstays in the treatment of uveitis. Viral uveitis is treated with antiviral agents such as acyclovir, ganciclovir, and famciclovir.

12.5.5 Glaucoma

Glaucoma is a progressive optic neuropathy characterized by gradual degeneration of neuronal tissue due to retinal ganglion cell (RGC) loss and an accompanying visual field loss (Gupta et al. 2006). RGCs gather together at the optic disc to form the optic nerve. In Western developed countries, glaucoma causes 18% of total blindness and represents the second most frequent cause of visual loss, following AMD (Resnikoff et al. 2004). Primary glaucoma is classified as open angle or closed angle. Open angle glaucoma (OAG) again can be subdivided as primary open angle glaucoma (POAG), accompanied by increased IOP, and normal tension glaucoma (NTG), characterized by a normal IOP.

The mainstay of treatment for OAG is either medical or surgical lowering of the IOP. More recently, neuroprotection has become a focus as a possible modality to reduce vision loss and minimize degeneration of RGC. An extensive review of a potential neuroprotective agent and its mechanism is nicely reported by Baltmr et al. (2010).

12.6 SUBSTANCES

12.6.1 Chemical Compounds

The most traditionally targeted drugs for suprachoroidal delivery include antibiotics, anti-inflammatory agents, immunosuppressive agents, and antimetabolites. These pharmaceutical compounds usually have a low molecular weight as compared to "biologics" that usually are larger peptides or proteins. So far, there have been relatively few investigations of biologic drugs in the suprachoroidal space. Our previous study showed triamcinolone acetonide administered in pig's suprachoroidal space (SCS) was maintained until at least 120 days (Olsen et al. 2006). To improve the bioavailability of "chemical drugs," optimization of the formulation is key. Optimization of lipophilicity, reduction of molecular weight, modification of molecular charge, and enhancement of aqueous solubility are some of the strategies for improvement in drug bioavailability (Shirasaki 2008).

12.6.2 Biomolecules

Biomolecules or biologics include any organic molecules, large polymers, proteins, polysaccharides, or nucleic acids. The molecular weights range from 1000 to 5000 kDa. Clinically useful agents in this category include monoclonal antibodies such as ranibizumab, bevacizumab, and infliximab. The exact pharmacokinetics of these large molecules in the SCS has not been studied. In an upcoming report from our lab, we noted a very rapid clearance time (<7 days) when bevacizumab was injected directly into the SCS of pig eye (Olsen et al. 2011). We believe that high-molecular-weight biological drugs tend to clear rapidly from the SCS, and so drug manufacturers should consider slower-release formulations of drugs.

Biomolecular drugs are currently being investigated for the treatment of RP. Various neurotrophic and growth factors have demonstrated a protective effect on photoreceptor and RPE loss in several animal models of degeneration (Sakai et al. 2007; Sieving et al. 2006). Ciliary neurotrophic factors (CNTF) are currently in two-phase—two to three clinical trials for the treatment of visual loss associated with RP—one consisting of patients with earlier-stage disease (60 patients) and the second consisting of patients with later-stage disease (60 patients).

12.6.3 GENE-BASED DRUGS

Gene-based therapies are also currently in human clinical trials. The suprachoroidal delivery route offers a unique avenue for accessing the photoreceptors more directly than from intravitreal injections. There are many potential vectors to transport gene-based drugs to target cells, including viral-mediated and nonvirally mediated vectors. Recently, the chemical method has been described by using nanoparticle technology (de la Fuente et al. 2008; Ding et al. 2009; Singh et al. 2009).

12.6.4 TRANSPORTER-TARGETED PRODRUGS

Prodrugs are structurally modified drugs with specific substrates that enhance the solubility, stability, and permeability of the parent compound. Recently, membrane transporters have been identified that may help facilitate prodrugs, and prodrugs targeting tissues through specific cellular transporters are under investigation (Cheng et al. 2004; Chong et al. 2009; Janoria et al. 2010; Kansara et al. 2007).

12.7 NOVEL DRUG DELIVERY SYSTEMS

12.7.1 POLYMERS

Polymers may be useful to optimize pharmacokinetics and bioavailability using suprachoroidal drug delivery. A polymer is generally thought of as a large molecule composed of multiple repeating structural subunits or monomers. Such monomers are typically connected by covalent chemical bonds. Recently, various polymers have been under investigation and serve as a reservoir or matrix for controlled-release drug delivery systems. The polymers are either biodegradable or nonbiodegradable, with the latter being used primarily for sustained-release implants such as the Vitrasert®, Retisert®, Iluvien™, and a metal device, the helical-shaped I-vation™ (Barnett 2009; Jaffe et al. 2005; Kane et al. 2008; Musch et al. 1997). Although nonbiodegradable systems have the advantage of sustained delivery with more stable drug tissue concentrations, the devices may need to be retrieved or exchanged.

Biodegradable polymers includes poly(ortho ester), polylactide, poly(lactic-co-glycolic acid) (PLGA), polyvinyl alcohol, PGLC (glycolide-co-lactide-co-caprolactone copolymer), and so on. They can be made into a more solid pellet (implantable), a semisolid gel (injectable), or a colloidal state (also injectable). Solid state or semisolid gel carriers degrade over time and do not need to be retrieved. However, the drug release kinetics are more variable, with an initial burst, slow release, and then a second burst occurring during the final dissolution phase. A semisolid gel-like Poly(ortho esters) (POE) was injected into SCS and has been reported to have detectable tissue levels for at least 6 months postinjection (Heller 2005; Einmahl et al. 2003).

12.7.2 COLLOIDAL SYSTEMS

A variety of submicron-sized colloidal carriers are being developed, including nanoparticulate or microparticulate systems that provide sustained and controlled release of the drug at the targeted site with reduced frequency of administration and the capability to penetrate blood–ocular barriers (Gaudana et al. 2010). Such a formulation may be ideal for suprachoroidal drug delivery.

12.7.3 NANOPARTICLES OR MICROPARTICLES

Nanoparticles or microparticles, classified by their size, are polymeric colloidal particles where the drug molecule is encapsulated within the polymeric matrix or simply adsorbed or conjugated onto the particle surface. Nanoparticles have diameters that range from <1 to 1000 nm. They can be further classified into matrix-type nanospheres or reservoir-type nanocapsules filled with either oil or water (Rabinovich-Guilatt et al. 2004). In the matrix system, the drug and nanospheres are combined with the drug being released through diffusion from the polymer matrix with simultaneous polymer degradation. The reservoir-type nanocapsule system has encapsulated drugs within a polymeric shell (Shah et al. 2010). An *ex vivo* study using a hollow microneedle revealed that the microneedle could deliver nanoparticle and microparticle suspensions into the suprachoroidal space of rabbit, pig, and human eyes (Patel et al. 2010).

12.7.4 VESICULAR SYSTEMS

Liposomes are a vesicular system and are composed of multiple phospholipid *bilayers* surrounding a water compartment. Liposomes may allow for the delivery of a wide variety of drug molecules such as protein, nucleotides, and even plasmids. Using liposomes, hydrophilic, lipophilic, and amphiphilic drugs can be encapsulated and delivered directly through cell membranes.

Micelles are also a vesicular system, with a smaller diameter in the range of 10–100 nm. They are composed of lipid *monolayer* that encapsulates a lipophilic drug within its core.

Niosomes are *bilayered* vesicles made up of nonionic surfactants that are biodegradable and relatively nontoxic. They are capable of encapsulating both lipophilic and hydrophilic compounds and are a relatively stable and inexpensive alternative to liposomes that release the drug independent of pH and potential for enhanced ocular bioavailability (Azeem et al. 2009).

12.8 SUMMARY

Drug delivery to the suprachoroidal space may be a reasonable alternative that avoids some intraocular complications in intravitreal injection; bypasses the conjunctival clearance of periocular injection; and focuses drug delivery to the choroid, optic nerve, and retina. Such methodology may be used to treat various posterior segment diseases such as AMD, DME, uveitis, RVO, glaucoma, and optic nerve disease and may deliver various forms of drugs such as small molecules, larger biologics, various prodrugs, or even gene-based vectors. Further investigation will improve our ability to deliver drugs in a minimally invasive manner and optimize safety. Novel drug formulations could potentially increase the sustained effect of locally delivered drugs in the suprachoroidal space.

ACKNOWLEDGMENT

Financial support by an unrestricted departmental grant from Research to Prevent Blindness.

REFERENCES

Adamis, A.P., and D.T. Shima. 2005. The role of vascular endothelial growth factor in ocular health and disease. *Retina* 25 (2):111–8.

Alm, A., and S.F.E. Nilsson. 2009. Uveoscleral outflow—A review. *Exp Eye Res* 88 (4):760–8.

Ambati, J., C.S. Canakis, J.W. Miller, E.S. Gragoudas, A. Edwards, D.J. Weissgold, I. Kim, F.C. Delori, and A.P. Adamis. 2000. Diffusion of high molecular weight compounds through sclera. *Invest Ophthalmol Vis Sci* 41 (5):1181–5.

Ambati, J., E.S. Gragoudas, J.W. Miller, T.T. You, K. Miyamoto, F.C. Delori, and A.P. Adamis. 2000. Transscleral delivery of bioactive protein to the choroid and retina. *Invest Ophthalmol Vis Sci* 41 (5):1186–91.

Amrite, A.C., H.F. Edelhauser, S.R. Singh, and U.B. Kompella. 2008. Effect of circulation on the disposition and ocular tissue distribution of 20 nm nanoparticles after periocular administration. *Mol Vis* 14:150–60.

Amrite, A., V. Pugazhenthi, N. Cheruvu, and U. Kompella. 2010. Delivery of celecoxib for treating diseases of the eye: Influence of pigment and diabetes. *Expert Opin Drug Deliv* 7 (5):631–45.

Anderson, C.R., R.L. Morris, S.D. Boeh, P.C. Panus, and W.L. Sembrowich. 2003. Effects of iontophoresis current magnitude and duration on dexamethasone deposition and localized drug retention. *Phys Ther* 83 (2):161–70.

AREDS Group, Age-Related Eye Disease Study Research. 2001. The Age-Related Eye Disease Study system for classifying age-related macular degeneration from stereoscopic color fundus photographs: The Age-Related Eye Disease Study Report Number 6. *Am J Ophthalmol* 132 (5):668–81.

Augustin, A.J., and I. Offermann. 2006. Emerging drugs for age-related macular degeneration. *Expert Opin Emerg Drugs* 11 (4):725–40.

Aukunuru, J.V., G. Sunkara, N. Bandi, W.B. Thoreson, and U.B. Kompella. 2001. Expression of multidrug resistance-associated protein (MRP) in human retinal pigment epithelial cells and its interaction with BAPSG, a novel aldose reductase inhibitor. *Pharm Res* 18 (5):565–72.

Azeem, A., M.K. Anwer, and S. Talegaonkar. 2009. Niosomes in sustained and targeted drug delivery: Some recent advances. *J Drug Target* 17 (9):671–89.

Balachandran, R.K., and V.H. Barocas. 2008. Computer modeling of drug delivery to the posterior eye: Effect of active transport and loss to choroidal blood flow. *Pharm Res* 25 (11):2685–96.

Baltmr, A., J. Duggan, S. Nizari, T.E. Salt, and M.F. Cordeiro. 2010. Neuroprotection in glaucoma—Is there a future role? *Exp Eye Res* 91 (5):554–66.

Barar, J., A.R. Javadzadeh, and Y. Omidi. 2008. Ocular novel drug delivery: Impacts of membranes and barriers. *Expert Opin Drug Deliv* 5 (5):567–81.

Barnett, P.J. 2009. Mathematical modeling of triamcinolone acetonide drug release from the I-vation intravitreal implant (a controlled release platform). *Conf Proc IEEE Eng Med Biol Soc* 2009:3087–90.

Bill, A. 1965. Movement of albumin and dextran through the sclera. *Arch Ophthalmol* 74:248–52.

Bill, A., P. Törnquist, and A. Alm. 1980. Permeability of the intraocular blood vessels. *Trans Ophthalmol Soc U K* 100 (3):332–6.

Brar, M., L. Cheng, R. Yuson, F. Mojana, W.R. Freeman, and P.S. Gill. 2010. Ocular safety profile and intraocular pharmacokinetics of an antagonist of EphB4/EphrinB2 signalling. *Br J Ophthalmol* 94 (12):1668–73.

Brown, D.M., P.K. Kaiser, M. Michels, G. Soubrane, J.S. Heier, R.Y. Kim, J.P. Sy, S. Schneider, and ANCHOR Study Group. 2006. Ranibizumab versus verteporfin for neovascular age-related macular degeneration. *N Engl J Med* 355 (14):1432–44.

Brubaker, R.F. 1982. The flow of aqueous humor in the human eye. *Trans Am Ophthalmol Soc* 80:391–474.

BVOS. 1986. Argon laser scatter photocoagulation for prevention of neovascularization and vitreous hemorrhage in branch vein occlusion. A randomized clinical trial. Branch Vein Occlusion Study Group. *Arch Ophthalmol* 104 (1):34–41.

Cai, X., S.M. Conley, and M.I. Naash. 2010. Gene therapy in the retinal degeneration slow model of retinitis pigmentosa. *Adv Exp Med Biol* 664:611–9.

Campbell, R.J., S.E. Bronskill, C.M. Bell, J.M. Paterson, M. Whitehead, and S.S. Gill. 2010. Rapid expansion of intravitreal drug injection procedures, 2000 to 2008: A population-based analysis. *Arch Ophthalmol* 128 (3):359–62.

Campochiaro, P.A., G. Hafiz, R. Channa, S.M. Shah, Q.D. Nguyen, H. Ying, D.V. Do et al. 2010. Antagonism of vascular endothelial growth factor for macular edema caused by retinal vein occlusions: Two-year outcomes. *Ophthalmology* 117 (12):2387–94.e5.

Cheng, L., K.Y. Hostetler, J. Lee, H.J. Koh, J.R. Beadle, K. Bessho, M. Toyoguchi, K. Aldern, J. Bovet, and W.R. Freeman. 2004. Characterization of a novel intraocular drug-delivery system using crystalline lipid antiviral prodrugs of ganciclovir and cyclic cidofovir. *Invest Ophthalmol Vis Sci* 45 (11):4138–44.

Cheruvu, N.P.S., A.C. Amrite, and U.B. Kompella. 2008. Effect of eye pigmentation on transscleral drug delivery. *Invest Ophthalmol Vis Sci* 49 (1):333–41.

Cheruvu, N.P.S., and U.B. Kompella. 2006. Bovine and porcine transscleral solute transport: Influence of lipophilicity and the choroid-Bruch's layer. *Invest Ophthalmol Vis Sci* 47 (10):4513–22.

Chong, D.Y., M.W. Johnson, T.H. Huynh, E.F. Hall, G.M. Comer, and D.N. Fish. 2009. Vitreous penetration of orally administered famciclovir. *Am J Ophthalmol* 148 (1):38–42.e1.

Cruysberg, L.P.J., R.M.M.A. Nuijts, D.H. Geroski, J.A. Gilbert, F. Hendrikse, and H.F. Edelhauser. 2005. The influence of intraocular pressure on the transscleral diffusion of high-molecular-weight compounds. *Invest Ophthalmol Vis Sci* 46 (10):3790–4.

Cunha-Vaz, J.G., and D.M. Maurice. 1967. The active transport of fluorescein by the retinal vessels and the retina. *J Physiol* 191 (3):467–86.

Davies, N.M. 2000. Biopharmaceutical considerations in topical ocular drug delivery. *Clin Exp Pharmacol Physiol* 27 (7):558–62.

de la Fuente, M., B. Seijo, and M.J. Alonso. 2008. Novel hyaluronic acid-chitosan nanoparticles for ocular gene therapy. *Invest Ophthalmol Vis Sci* 49 (5):2016–24.

Del Amo, E.M., and A. Urtti. 2008. Current and future ophthalmic drug delivery systems. A shift to the posterior segment. *Drug Discov Today* 13 (3–4):135–43.

Ding, X.-Q., A.B. Quiambao, J.B. Fitzgerald, M.J. Cooper, S.M. Conley, and M.I. Naash. 2009. Ocular delivery of compacted DNA-nanoparticles does not elicit toxicity in the mouse retina. *PLoS One* 4 (10):e7410.

DRCR Network, Diabetic Retinopathy Clinical Research Network, R.W. Beck, A.R. Edwards, L.P. Aiello, N.M. Bressler, F. Ferris, A.R. Glassman et al. 2009. Three-year follow-up of a randomized trial comparing focal/grid photocoagulation and intravitreal triamcinolone for diabetic macular edema. *Arch Ophthalmol* 127 (3):245–51.

DRCR Network, Diabetic Retinopathy Clinical Research Network, M.J. Elman, L.P. Aiello, R.W. Beck, N.M. Bressler, S.B. Bressler, A.R. Edwards et al. 2010. Randomized trial evaluating ranibizumab plus prompt or deferred laser or triamcinolone plus prompt laser for diabetic macular edema. *Ophthalmology* 117 (6):1064–77.e35.

Duvvuri, S., S. Majumdar, and A.K. Mitra. 2003. Drug delivery to the retina: Challenges and opportunities. *Expert Opin Biol Ther* 3 (1):45–56.

Einmahl, S., S. Ponsart, R.A. Bejjani, F. D'Hermies, M. Savoldelli, J. Heller, C. Tabatabay, R. Gurny, and F. Behar-Cohen. 2003. Ocular biocompatibility of a poly(ortho ester) characterized by autocatalyzed degradation. *J Biomed Mater Res A* 67 (1):44–53.

Einmahl, S., M. Savoldelli, F. D'Hermies, C. Tabatabay, R. Gurny, and F. Behar-Cohen. 2002. Evaluation of a novel biomaterial in the suprachoroidal space of the rabbit eye. *Invest Ophthalmol Vis Sci* 43 (5):1533–9.

Eljarrat-Binstock, E., A.J. Domb, F. Orucov, J. Frucht-Pery, and J. Pe'er. 2007. Methotrexate delivery to the eye using transscleral hydrogel iontophoresis. *Curr Eye Res* 32 (7–8):639–46.

Falkenstein, I.A., L. Cheng, T.R. Jones, W.R. Freeman, B. Babson, I. Kozak, A.M. Tammewar, and E.C. Barron. 2010. Intraocular properties of a repository urokinase receptor antagonist a36 Peptide in rabbits. *Curr Eye Res* 35 (8):742–50.

Friedman, D.S., B.J. O'Colmain, B. Muñoz, S.C. Tomany, C. McCarty, P.T.V.M. de Jong, B. Nemesure, P. Mitchell, J. Kempen, and Eye Diseases Prevalence Research Group. 2004. Prevalence of age-related macular degeneration in the United States. *Arch Ophthalmol* 122 (4):564–72.

Fu, Y., M.L. Ponce, M. Thill, P. Yuan, N.S. Wang, and K.G. Csaky. 2007. Angiogenesis inhibition and choroidal neovascularization suppression by sustained delivery of an integrin antagonist, EMD478761. *Invest Ophthalmol Vis Sci* 48 (11):5184–90.

Gandhi, M.D., D. Pal, and A.K. Mitra. 2004. Identification and functional characterization of a Na(+)-independent large neutral amino acid transporter (LAT2) on ARPE-19 cells. *Int J Pharm* 275 (1–2):189–200.

Gaudana, R., H.K. Ananthula, A. Parenky, and A.K. Mitra. 2010. Ocular drug delivery. *AAPS J* 12 (3):348–60.

Ghate, D., and H.F. Edelhauser. 2006. Ocular drug delivery. *Expert Opin Drug Deliv* 3 (2):275–87.

Giansanti, F., M. Ramazzotti, L. Vannozzi, E. Rapizzi, T. Fiore, B. Iaccheri, D.D. Innocenti, D. Moncini, and U. Menchini. 2008. A pilot study on ocular safety of intravitreal infliximab in a rabbit model. *Invest Ophthalmol Vis Sci* 49 (3):1151–6.

Gilger, B.C., J.H. Salmon, D.A. Wilkie, L.P.J. Cruysberg, J. Kim, M. Hayat, H. Kim et al. 2006. A novel bioerodible deep scleral lamellar cyclosporine implant for uveitis. *Invest Ophthalmol Vis Sci* 47 (6):2596–605.

Gilger, B.C., D.A. Wilkie, A.B. Clode, R.J. McMullen, M.E. Utter, A.M. Komaromy, D.E. Brooks, and J.H. Salmon. 2010. Long-term outcome after implantation of a suprachoroidal cyclosporine drug delivery device in horses with recurrent uveitis. *Vet Ophthalmol* 13 (5):294–300.

Gupta, N., L.-C. Ang, L. Noël de Tilly, L. Bidaisee, and Y.H. Yücel. 2006. Human glaucoma and neural degeneration in intracranial optic nerve, lateral geniculate nucleus, and visual cortex. *Br J Ophthalmol* 90 (6):674–8.

Heller, J. 2005. Ocular delivery using poly(ortho esters). *Adv Drug Deliv Rev* 57 (14):2053–62.

Hou, J., Y. Tao, Y.-R. Jiang, and K. Wang. 2009. *In vivo* and *in vitro* study of suprachoroidal fibrin glue. *Jpn J Ophthalmol* 53 (6):640–7.

Ip, M.S., N.L. Oden, I.U. Scott, P.C. VanVeldhuisen, B.A. Blodi, M. Figueroa, A. Antoszyk, M. Elman, and SCORE Study Investigator Group. 2009. SCORE Study report 3: Study design and baseline characteristics. *Ophthalmology* 116 (9):1770–7.e1.

Jacobson, S.G., and A.V. Cideciyan. 2010. Treatment possibilities for retinitis pigmentosa. *N Engl J Med* 363 (17):1669–71.

Jaffe, G.J., R.M. McCallum, B. Branchaud, C. Skalak, Z. Butuner, and P. Ashton. 2005. Long-term follow-up results of a pilot trial of a fluocinolone acetonide implant to treat posterior uveitis. *Ophthalmology* 112 (7):1192–8.

Janoria, K.G., S.H.S. Boddu, S. Natesan, and A.K. Mitra. 2010. Vitreal pharmacokinetics of peptide-transporter-targeted prodrugs of ganciclovir in conscious animals. *J Ocul Pharmacol Ther* 26 (3):265–71.

Janoria, K.G., S. Gunda, S.H.S. Boddu, and A.K. Mitra. 2007. Novel approaches to retinal drug delivery. *Expert Opin Drug Deliv* 4 (4):371–88.

Jiang, J., D.H. Geroski, H.F. Edelhauser, and M.R. Prausnitz. 2006. Measurement and prediction of lateral diffusion within human sclera. *Invest Ophthalmol Vis Sci* 47 (7):3011–6.

Jiang, J., J.S. Moore, H.F. Edelhauser, and M.R. Prausnitz. 2009. Intrascleral drug delivery to the eye using hollow microneedles. *Pharm Res* 26 (2):395–403.

Kanamori, A., M.-M. Catrinescu, A. Mahammed, Z. Gross, and L.A. Levin. 2010. Neuroprotection against superoxide anion radical by metallocorroles in cellular and murine models of optic neuropathy. *J Neurochem* 114 (2):488–98.

Kane, F.E., J. Burdan, A. Cutino, and K.E. Green. 2008. Iluvien: A new sustained delivery technology for posterior eye disease. *Expert Opin Drug Deliv* 5 (9):1039–46.

Kansara, V., Y. Hao, and A.K. Mitra. 2007. Dipeptide monoester ganciclovir prodrugs for transscleral drug delivery: Targeting the oligopeptide transporter on rabbit retina. *J Ocul Pharmacol Ther* 23 (4):321–34.

Kennedy, B.G., and N.J. Mangini. 2002. P-glycoprotein expression in human retinal pigment epithelium. *Mol Vis* 8:422–30.

Kim, E.S., C. Durairaj, R.S. Kadam, S.J. Lee, Y. Mo, D.H. Geroski, U.B. Kompella, and H.F. Edelhauser. 2009. Human scleral diffusion of anticancer drugs from solution and nanoparticle formulation. *Pharm Res* 26 (5):1155–61.

Kim, H., M.R. Robinson, M.J. Lizak, G. Tansey, R.J. Lutz, P. Yuan, N.S. Wang, and K.G. Csaky. 2004. Controlled drug release from an ocular implant: An evaluation using dynamic three-dimensional magnetic resonance imaging. *Invest Ophthalmol Vis Sci* 45 (8):2722–31.

Kim, S.H., C.J. Galbán, R.J. Lutz, R.L. Dedrick, K.G. Csaky, M.J. Lizak, N.S. Wang, G. Tansey, and M.R. Robinson. 2007. Assessment of subconjunctival and intrascleral drug delivery to the posterior segment using dynamic contrast-enhanced magnetic resonance imaging. *Invest Ophthalmol Vis Sci* 48 (2):808–14.

Kim, S.H., R.J. Lutz, N.S. Wang, and M.R. Robinson. 2007. Transport barriers in transscleral drug delivery for retinal diseases. *Ophthalmic Res* 39 (5):244–54.

Kimura, M., M. Araie, and S. Koyano. 1996. Movement of carboxyfluorescein across retinal pigment epithelium-choroid. *Exp Eye Res* 63 (1):51–6.

Klein, R., B.E. Klein, S.E. Moss, and K.J. Cruickshanks. 1995. The Wisconsin Epidemiologic Study of Diabetic Retinopathy. XV. The long-term incidence of macular edema. *Ophthalmology* 102 (1):7–16.

Koyano, S., M. Araie, and S. Eguchi. 1993. Movement of fluorescein and its glucuronide across retinal pigment epithelium-choroid. *Invest Ophthalmol Vis Sci* 34 (3):531–8.

Kriechbaum, K., F. Prager, W. Geitzenauer, T. Benesch, C. Schütze, C. Simader, and U. Schmidt-Erfurth. 2009. Association of retinal sensitivity and morphology during antiangiogenic treatment of retinal vein occlusion over one year. *Ophthalmology* 116 (12):2415–21.

Krohn, J., and T. Bertelsen. 1997a. Corrosion casts of the suprachoroidal space and uveoscleral drainage routes in the human eye. *Acta Ophthalmol Scand* 75 (1):32–5.

Krohn, J., and T. Bertelsen. 1997b. Corrosion casts of the suprachoroidal space and uveoscleral drainage routes in the pig eye. *Acta Ophthalmol Scand* 75 (1):28–31.

Krohn, J., and T. Bertelsen. 1998. Light microscopy of uveoscleral drainage routes after gelatine injections into the suprachoroidal space. *Acta Ophthalmol Scand* 76 (5):521–7.

Kuno, N., and S. Fujii. 2010. Biodegradable intraocular therapies for retinal disorders: Progress to date. *Drugs Aging* 27 (2):117–34.

Larsson, B.S. 1993. Interaction between chemicals and melanin. *Pigment Cell Res* 6 (3):127–33.

Leblanc, B., S. Jezequel, T. Davies, G. Hanton, and C. Taradach. 1998. Binding of drugs to eye melanin is not predictive of ocular toxicity. *Regul Toxicol Pharmacol* 28 (2):124–32.

Lee, T.W.-Y., and J.R. Robinson. 2004. Drug delivery to the posterior segment of the eye III: The effect of parallel elimination pathway on the vitreous drug level after subconjunctival injection. *J Ocul Pharmacol Ther* 20 (1):55–64.

Ling, V. 1992. Charles F. Kettering Prize. P-glycoprotein and resistance to anticancer drugs. *Cancer* 69 (10):2603–9.

Mannermaa, E., K.-S. Vellonen, T. Ryhänen, K. Kokkonen, V.-P. Ranta, K. Kaarniranta, and A. Urtti. 2009. Efflux protein expression in human retinal pigment epithelium cell lines. *Pharm Res* 26 (7):1785–91.

Mannermaa, E., K.-S. Vellonen, and A. Urtti. 2006. Drug transport in corneal epithelium and blood-retina barrier: Emerging role of transporters in ocular pharmacokinetics. *Adv Drug Deliv Rev* 58 (11):1136–63.

Moore, D.J., and G.M. Clover. 2001. The effect of age on the macromolecular permeability of human Bruch's membrane. *Invest Ophthalmol Vis Sci* 42 (12):2970–5.

Moore, D.J., A.A. Hussain, and J. Marshall. 1995. Age-related variation in the hydraulic conductivity of Bruch's membrane. *Invest Ophthalmol Vis Sci* 36 (7):1290–7.

Musch, D.C., D.F. Martin, J.F. Gordon, M.D. Davis, and B.D. Kuppermann. 1997. Treatment of cytomegalovirus retinitis with a sustained-release ganciclovir implant. The Ganciclovir Implant Study Group. *N Engl J Med* 337 (2):83–90.

Nelson, M.L., M.T.S. Tennant, A. Sivalingam, C.D. Regillo, J.B. Belmont, and A. Martidis. 2003. Infectious and presumed noninfectious endophthalmitis after intravitreal triamcinolone acetonide injection. *Retina* 23 (5):686–91.

Nicoli, S., G. Ferrari, M. Quarta, C. Macaluso, and P. Santi. 2009. *In vitro* transscleral iontophoresis of high molecular weight neutral compounds. *Eur J Pharm Sci* 36 (4–5):486–92.

Nomoto, H., F. Shiraga, N. Kuno, E. Kimura, S. Fujii, K. Shinomiya, A.K. Nugent, K. Hirooka, and T. Baba. 2009. Pharmacokinetics of bevacizumab after topical, subconjunctival, and intravitreal administration in rabbits. *Invest Ophthalmol Vis Sci* 50 (10):4807–13.

Olsen, T.W., S.Y. Aaberg, D.H. Geroski, and H.F. Edelhauser. 1998. Human sclera: Thickness and surface area. *Am J Ophthalmol* 125 (2):237–41.

Olsen, T.W., H.F. Edelhauser, J.I. Lim, and D.H. Geroski. 1995. Human scleral permeability. Effects of age, cryotherapy, transscleral diode laser, and surgical thinning. *Invest Ophthalmol Vis Sci* 36 (9):1893–903.

Olsen, T.W., X. Feng, K. Wabner, S.R. Conston, D.H. Sierra, D.V. Folden, M.E. Smith, and J.D. Cameron. 2006. Cannulation of the suprachoroidal space: A novel drug delivery methodology to the posterior segment. *Am J Ophthalmol* 142 (5):777–87.

Olsen, T.W., X. Feng, K. Wabner, K. Csaky, S. Pambuccian, and J.D. Cameron. 2011. Pharmacokinetics of pars plana intravitreal injections versus microcannula suprachoroidal injections of bevacizumab in a porcine model. *Invest Ophthalmol Vis Sci* 52 (7): 4749–56.

Olsen, T.W., S. Sanderson, X. Feng, and W.C. Hubbard. 2002. Porcine sclera: Thickness and surface area. *Invest Ophthalmol Vis Sci* 43 (8):2529–32.

Ozkiriş, A., and K. Erkiliç. 2005. Complications of intravitreal injection of triamcinolone acetonide. *Can J Ophthalmol* 40 (1):63–8.

Parke, D.W. 2003. Intravitreal triamcinolone and endophthalmitis. *Am J Ophthalmol* 136 (5):918–9.

Patel, S.R., A.S.P. Lin, H.F. Edelhauser, and M.R. Prausnitz. 2010. Suprachoroidal drug delivery to the back of the eye using hollow microneedles. *Pharm Res* 28 (1): 166–76.

Pitkänen, L., V.-P. Ranta, H. Moilanen, and A. Urtti. 2005. Permeability of retinal pigment epithelium: Effects of permeant molecular weight and lipophilicity. *Invest Ophthalmol Vis Sci* 46 (2):641–6.

Rabinovich-Guilatt, L., P. Couvreur, G. Lambert, and C. Dubernet. 2004. Cationic vectors in ocular drug delivery. *J Drug Target* 12 (9–10):623–33.

Ramchandran, R.S., S. Fekrat, S.S. Stinnett, and G.J. Jaffe. 2008. Fluocinolone acetonide sustained drug delivery device for chronic central retinal vein occlusion: 12-month results. *Am J Ophthalmol* 146 (2):285–91.

Ramrattan, R.S., T.L. van der Schaft, C.M. Mooy, W.C. de Bruijn, P.G. Mulder, and P.T. de Jong. 1994. Morphometric analysis of Bruch's membrane, the choriocapillaris, and the choroid in aging. *Invest Ophthalmol Vis Sci* 35 (6):2857–64.

Resnikoff, S., D. Pascolini, D. Etya'ale, I. Kocur, R. Pararajasegaram, G.P. Pokharel, and S.P. Mariotti. 2004. Global data on visual impairment in the year 2002. *Bull World Health Organ* 82 (11):844–51.

Robinson, M.R., S.S. Lee, H. Kim, S. Kim, R.J. Lutz, C. Galban, P.M. Bungay et al. 2006. A rabbit model for assessing the ocular barriers to the transscleral delivery of triamcinolone acetonide. *Exp Eye Res* 82 (3):479–87.

Rogers, S., R.L. McIntosh, N. Cheung, L. Lim, J.J. Wang, P. Mitchell, J.W. Kowalski, H. Nguyen, T.Y. Wong, and International Eye Disease Consortium. 2010. The prevalence of retinal vein occlusion: Pooled data from population studies from the United States, Europe, Asia, and Australia. *Ophthalmology* 117 (2):313–9.e1.

Rosenfeld, P.J., D.M. Brown, J.S. Heier, D.S. Boyer, P.K. Kaiser, C.Y. Chung, R.Y. Kim, and MARINA Study Group. 2006. Ranibizumab for neovascular age-related macular degeneration. *N Engl J Med* 355 (14):1419–31.

Saha, P., J.J. Yang, and V.H. Lee. 1998. Existence of a p-glycoprotein drug efflux pump in cultured rabbit conjunctival epithelial cells. *Invest Ophthalmol Vis Sci* 39 (7):1221–6.

Sakai, T., N. Kuno, F. Takamatsu, E. Kimura, H. Kohno, K. Okano, and K. Kitahara. 2007. Prolonged protective effect of basic fibroblast growth factor-impregnated nanoparticles in royal college of surgeons rats. *Invest Ophthalmol Vis Sci* 48 (7):3381–7.

Salminen, L., G. Imre, and R. Huupponen. 1985. The effect of ocular pigmentation on intraocular pressure response to timolol. *Acta Ophthalmol Suppl* 173:15–8.

Scott, I.U., M.S. Ip, P.C. VanVeldhuisen, N.L. Oden, B.A. Blodi, M. Fisher, C.K. Chan et al. 2009. A randomized trial comparing the efficacy and safety of intravitreal triamcinolone with standard care to treat vision loss associated with macular edema secondary to branch retinal vein occlusion: The Standard Care vs Corticosteroid for Retinal Vein Occlusion (SCORE) study report 6. *Arch Ophthalmol* 127 (9):1115–28.

Senthilkumari, S., T. Velpandian, N.R. Biswas, A. Bhatnagar, G. Mittal, and S. Ghose. 2009. Evidencing the modulation of P-glycoprotein at blood-ocular barriers using gamma scintigraphy. *Curr Eye Res* 34 (1):73–7.

Shah, S.S., L.V. Denham, J.R. Elison, P.S. Bhattacharjee, C. Clement, T. Huq, and J.M. Hill. 2010. Drug delivery to the posterior segment of the eye for pharmacologic therapy. *Expert Rev Ophthalmol* 5 (1):75–93.

Shirasaki, Y. 2008. Molecular design for enhancement of ocular penetration. *J Pharm Sci* 97 (7):2462–96.

Shuler, R.K., P.K. Dioguardi, C. Henjy, J.M. Nickerson, L.P.J. Cruysberg, and H.F. Edelhauser. 2004. Scleral permeability of a small, single-stranded oligonucleotide. *J Ocul Pharmacol Ther* 20 (2):159–68.

Sieving, P.A., R.C. Caruso, W. Tao, H.R. Coleman, D.J.S. Thompson, K.R. Fullmer, and R.A. Bush. 2006. Ciliary neurotrophic factor (CNTF) for human retinal degeneration: Phase I trial of CNTF delivered by encapsulated cell intraocular implants. *Proc Natl Acad Sci USA* 103 (10):3896–901.

Singh, S.R., H.E. Grossniklaus, S.J. Kang, H.F. Edelhauser, B.K. Ambati, and U.B. Kompella. 2009. Intravenous transferrin, RGD peptide and dual-targeted nanoparticles enhance anti-VEGF intraceptor gene delivery to laser-induced CNV. *Gene Ther* 16 (5):645–59.

Soheilian, M., A. Ramezani, A. Obudi, B. Bijanzadeh, M. Salehipour, M. Yaseri, H. Ahmadieh et al. 2009. Randomized trial of intravitreal bevacizumab alone or combined with triamcinolone versus macular photocoagulation in diabetic macular edema. *Ophthalmology* 116 (6):1142–50.

Stahl, A., H. Agostini, L.L. Hansen, and N. Feltgen. 2007. Bevacizumab in retinal vein occlusion-results of a prospective case series. *Graefes Arch Clin Exp Ophthalmol* 245 (10):1429–36.

Sutter, F.K.P., and M.C. Gillies. 2003. Pseudo-endophthalmitis after intravitreal injection of triamcinolone. *Br J Ophthalmol* 87 (8):972–4.

Takahashi, K., Y. Saishin, Y. Saishin, A.G. King, R. Levin, and P.A. Campochiaro. 2009. Suppression and regression of choroidal neovascularization by the multitargeted kinase inhibitor pazopanib. *Arch Ophthalmol* 127 (4):494–9.

Toris, C.B., M.E. Yablonski, Y.L. Wang, and C.B. Camras. 1999. Aqueous humor dynamics in the aging human eye. *Am J Ophthalmol* 127 (4):407–12.

Tsuboi, S. 1987. Measurement of the volume flow and hydraulic conductivity across the isolated dog retinal pigment epithelium. *Invest Ophthalmol Vis Sci* 28 (11):1776–82.

Tsuboi, S., T. Fujimoto, Y. Uchihori, K. Emi, S. Iizuka, K. Kishida, and R. Manabe. 1984. Measurement of retinal permeability to sodium fluorescein *in vitro*. *Invest Ophthalmol Vis Sci* 25 (10):1146–50.

Watsky, M.A., M.M. Jablonski, and H.F. Edelhauser. 1988. Comparison of conjunctival and corneal surface areas in rabbit and human. *Curr Eye Res* 7 (5):483–6.

Yang, J.J., K.J. Kim, and V.H. Lee. 2000. Role of P-glycoprotein in restricting propranolol transport in cultured rabbit conjunctival epithelial cell layers. *Pharm Res* 17 (5):533–8.

13 Use of Nanoparticles in the Treatment of Age-Related Macular Degeneration, Glaucoma, and Other Degenerative Retinal Diseases

*Marco A. Zarbin, Carlo Montemagno,
James F. Leary, and Robert Ritch*

CONTENTS

13.1 INTRODUCTION

Nanotechnology provides an important new set of tools for the diagnosis and treatment of ocular diseases, a concept that has been reviewed previously.[1–4] Nanoparticles include colloidal carrier systems that can improve the efficacy of drug delivery by overcoming diffusion barriers, permitting reduced dosing

259

(through more efficient tissue targeting) and sustained delivery. Viruses also can be considered a type of nanoparticle. Nanoparticle biodistribution is affected by particle size, shape, and surface properties.[5] Particle size, for example, influences whether the particle is internalized through phagocytosis, caveolin-mediated endocytosis, or clathrin-mediated endocytosis, which in turn results in exposure of the nanoparticle to different intracellular environments.[6–8] Trans-activating transcriptional activator peptide favors macropinocytosis. Compacted polylysine DNA nanoparticles are taken into cells and transported directly to the nucleus by the cell surface receptor nucleolin.[9]

Because some of these features can be manipulated with nanoengineering, nanoparticles offer attractive opportunities to improve drug, growth factor, and virus-mediated treatment of chronic conditions such as glaucoma,[10] uveitis,[11] or retinal edema (due to venous occlusion or choroidal neovascularization) as well as treatment of intraocular tumors and other conditions associated with cell proliferation such as capsular fibrosis after cataract surgery, ocular neovascularization, and proliferative vitreoretinopathy. In this chapter, we consider various types of nanoparticles and their applications to the treatment of glaucoma (Table 13.1), age-related macular degeneration (AMD), and retinal degenerative diseases such as retinitis pigmentosa (RP) (Table 13.2).

13.2   NANOVESICLES

13.2.1   Liposomes, Niosomes, and Discomes

A liposome is a lipid bilayer that can be multilamellar or unilamellar. Liposomes can carry hydrophobic or hydrophilic cargo. Niosomes are nonionic surfactant-based liposomes that are formed primarily by incorporation of cholesterol as an excipient. Niosomes may have better chemical stability and lower production costs compared to liposomes. Discomes are nonionic surfactant-based discoidal vesicles.[48] For cell targeting, nanovesicles can be coated with ligands that direct them to specific cell surface receptors as well as with polymers that prolong their half-life in the circulatory system. Opsonization by immunoglobulin and/or complement proteins can lead to recognition of the nanoparticle as foreign and induce a hypersensitivity reaction.[49,50] Polyethylene glycol (PEG) can be conjugated with different molecules to enhance solubility and stability in plasma and to reduce immunogenicity.[5] Coating nanoparticles, including liposomes, with albumin and/or PEG can create a hydrophilic surface that temporarily resists protein adsorption, thus imparting longer bioavailability to the particle.[5,51,52]

13.3   NANOPARTICLES

Nanoparticles may be more stable in storage media than vesicular systems. Their pharmacokinetics is modulated by the carrier material as well as particle size and charge. Nanoparticles can be used to deliver growth and neurotrophic factors to cells,[53] and polymeric nanoparticles can be used to chelate metals.[54]

TABLE 13.1

Nanotechnology for Glaucoma Therapy

Application	Device	Agent	In Vitro Effect	In Vivo Effect
Anti-Glaucoma Drug Delivery				
	Liposome	Acetazolamide[12]	Sustained release	Decreased IOP (rabbits)
		Acetazolamide[13]	Sustained release	Decreased IOP (rabbits)
	Niosome	Acetazolamide[14,15]	Sustained release	Decreased IOP (rabbits and swine)
		Timolol[16]	Sustained release	Higher aqueous humor concentration (rabbits)
	Discome	Timolol[17]	Sustained release	Sustained release
	Nanovesicle	Brimonidine[18,19]	Sustained release	Decreased IOP (rabbits)
	PLGA-PLA microspheres	Timolol[20]	Sustained release	
	$(Ca)_3(PO_4)_2$ nanoparticle	Methazolamide[21]	Sustained release	Decreased IOP (rabbits)
	Solid lipid nanoparticle	Methazolamide[22]	Sustained release	Decreased IOP (rabbits)
	Chitosan nanoparticle	Timolol + dorzolamide[23] (HA-modified)	Sustained release	Decreased IOP (rabbits)
	Poly(amidoamine) dendrimer	Pilocarpine[24]	Sustained release	Prolonged miosis (rabbits)
Inhibit Apoptosis				
	PLGA nanospheres and microspheres	CNTF[25]	Sustained release	
	Viral transfection	Ad-CNTF[26] (ON transfection)		Sustained CNTF production and increased RGC survival (rats)
		AAV-CNTF[27] (laser-induced glaucoma)		Increased CNTF production and RGC survival (rats)
	HSPs induction	EMZF-SP nanoparticles[28]		

(Continued)

TABLE 13.1 (Continued)
Nanotechnology for Glaucoma Therapy

Application	Device	Agent	In Vitro Effect	In Vivo Effect
Inhibit Scarring				
	PLA Disc	5-FU[29]	Sustained release	Decreased IOP, prolonged bleb persistence (rabbits)
	PLA nanoparticles	MMC[30]	Sustained release	
	CS-*g*-(PEI-*b*-mPEG) nanoparticles	IκB kinase beta targeting siRNA[31]	Decreased fibroblast activation	Prolonged bleb persistence (monkey)
	Viral transfection	Adp21(WAF-1/Cip-1)[32]	Inhibits fibroblast growth	Prolonged bleb persistence (rabbits)
		Adp27(KIP1)[33]		Reduced fibroblast proliferation, prolonged bleb persistence (rabbits)
Inhibit Steroid-Induced IOP Increase				
	Viral transfection	AdhGRE.MMP1[34]	Steroid-induced increase in MMP1 expression and collagen-1 degradation by human TM cells	

IOP, intraocular pressure; PLGA-PLA, poly(lactic-co-glycolic acid)/poly(lactic acid); HA, hyaluronic acid; Ad, adenovirus; RGC, retinal ganglion cell; ON, optic nerve; AAV, adeno-associated virus; CNTF, ciliary neurotrophic factor; BDNF, brain-derived neurotrophic factor; HSPs, heat shock proteins; EMZF-SP, engineered superparamagnetic manganese-zinc-iron oxide; 5-FU, 5-fluorouracil; MMC, mitomycin C; MMP, matrix metalloproteinase; TM, trabecular meshwork.

TABLE 13.2
Nanotechnology for Retinal Degenerative Disease

Application	Device	Agent	In Vitro Effect	In Vivo Effect
Growth Factor Delivery/Inhibition				
	Nanoparticle	Gelatin-bFGF[35] Dendrimer-anti-VEGF oligonucleotide[36]		Increased PR survival in RCS rats Inhibited laser-induced CNV in rats
Antioxidant Delivery				
	Nanoparticle	Nanoceria[37,38]		Increased PR survival in light-damaged rats and inhibited NV in *Vldlr* knockout mice
Gene Delivery				
	Polyplex	PEG-DNA-rds[39,40]		Rescued PRs in *rds* knockout mice
	Virus	rAAV-ChR2[41-43]		Restored light sensitivity in *rd1* mice and RCS rats
		rAAV-HaloR[44]		Restored light sensitivity in *rd1* mice
		rAAV (tyrosine mutated capsid)-ChR2[45]		Induced light sensitivity in ON bipolar cells in multiple mouse models of RP
		rAAV-ChR2-ankyrin, rAAV-HaloR-PSD-95[46]	Created excitatory center, antagonistic surround in rabbit RGCs	
		rAAV-HaloR (human rhodopsin, human red opsin, and mouse cone arrestin promoters)		Restored light sensitivity in cones from *rd1* and *Cnga3,Rho* double-knockout mice
		rAAV-LiGluR[47]		Restored light sensitivity and visually guided behavior in *rd1* mice

bFGF, basic fibroblast growth factor; PR, photoreceptor; RCS, Royal College of Surgeons; CNV, choroidal new vessels; NV, neovascularization; PEG, polyethylene glycol; *Rds*, retinal degeneration slow (peripherin) gene; rAAV, recombinant adeno-associated virus; ChR2, channelrhodopsin-2; HaloR, halorhodopsin; RP, retinitis pigmentosa; RGCs, retinal ganglion cells; LiGluR, light-activated ionotropic glutamate receptor; VEGF, vascular endothelial growth factor.

13.3.1 CHITOSAN

Chitosan is a polycationic polymer containing glucosamine. It is mucoadhesive, bio-degradable, and biocompatible. Chitosan can interact with the polyanionic mucosal surface through ionic and hydrogen bonding, which enhances drug residence time.[55] Hyaluronic acid is a polysaccharide whose mucoadhesion is mediated through CD44 receptor binding.[56] Combining chitosan nanoparticles with hyaluronic acid improves their mucoadhesion, which increases drug delivery time and corneal penetration after application to the ocular surface. Wadhwa et al.,[23] for example, used hyaluronic acid–modified chitosan nanoparticles loaded with timolol maleate and dorzolamide hydro-chloride to lower intraocular pressure (IOP) in rabbits (Figures 13.1 through 13.3, Table 13.1). Compared to plain drug solution, the nanoparticle formulation induced greater IOP lowering.

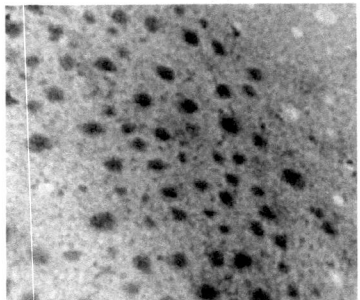

FIGURE 13.1 Transmission electron microscope image of hyaluronic acid–modified chi-tosan nanoparticles loaded with dorzolamide hydrochloride and timolol maleate. The image demonstrates that the particles are almost spherical. (Reproduced from Wadhwa, S. et al., *J Drug Target*, 18, 292–302, 2010. With permission.)

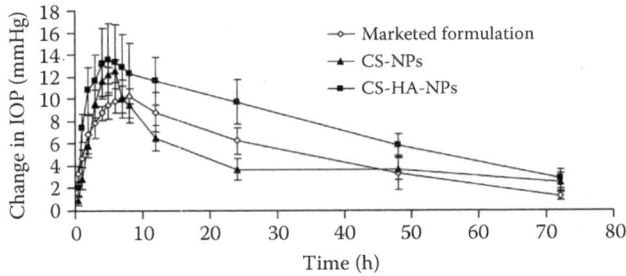

FIGURE 13.2 *In vivo* pharmacodynamic effect on intraocular pressure (IOP) in treated rabbit eyes using different preparations of dorzolamide and timolol. Results are expressed as mean ± standard deviation (*n* = 6). CS-HA-NPs produced prolonged IOP lowering compared to the marketed formulations. CS-NPs, chitosan nanoparticles; CS-HA-NPs, hyaluronic acid–modified chitosan nanoparticles. (From Wadhwa, S. et al., *J Drug Target*, 18, 292–302, 2010. With permission.)

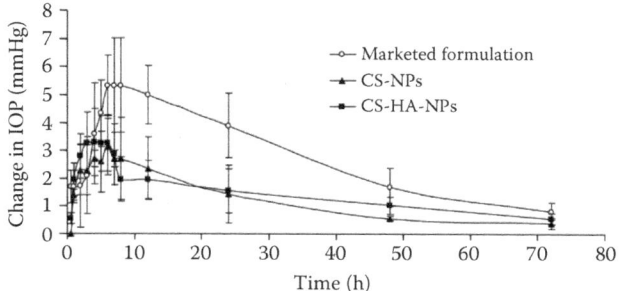

FIGURE 13.3 *In vivo* pharmacodynamic effect on intraocular pressure (IOP) in untreated fellow eyes of rabbits shown in Figure 13.2 using different preparations of dorzolamide and timolol. Results are expressed as mean ± standard deviation ($n = 6$). Marketed drug formulations induced more prominent IOP lowering in control eyes than CS-NPs, indicating greater systemic drug absorption with the marketed formulation. CS-NPs, chitosan nanoparticles; CS-HA-NPs, hyaluronic acid–modified chitosan nanoparticles. (From Wadhwa, S. et al., *J Drug Target*, 18, 292–302, 2010. With permission.)

13.3.2 COLLAGEN

Intravitreal nanoparticle-mediated basic fibroblast growth factor (bFGF) delivery provides sustained retinal rescue in Royal College of Surgeons (RCS) rats.[35] RCS rats have a mutation in a tyrosine kinase (Mertk) that prevents proper outer segment phagocytosis by retinal pigment epithelial (RPE) cells, which results in progressive rod and cone photoreceptor degeneration.[57] Some forms of RP in humans arise from this same mutation.[58–60] Sakai et al.[35] prepared bFGF nanoparticles using acidic gelatin isolated from bovine bone collagen by an alkaline process and human recombinant bFGF. Gelatin nanoparticles were cross-linked through a dehydrothermal process and ultraviolet irradiation of preprepared non-cross-linked gelatin particles. The nanoparticle diameter, assessed using dynamic light scattering, was ~585 nm. The bFGF was incorporated into the gelatin nanoparticles by dropping 5 mg/mL bFGF solution (20 μL) onto 2 mg freeze-dried gelatin nanoparticles.

13.3.3 POLY(LACTIC-CO-GLYCOLIC ACID)

Intravitreal glial cell line–derived neurotrophic factor (GDNF)-loaded biodegradable poly(lactic-co-glycolic acid) (PLGA) microspheres provided sustained ganglion cell protection in a rodent model of glaucoma.[61] Microspheres (average 8 μm diameter) containing GDNF were fabricated using a modification of a spontaneous emulsion technique.[62] In view of the fact that adeno-associated virus (AAV)-mediated GDNF secretion from glia delays retinal degeneration in a rat model of RP,[63] it seems likely that nanoparticle-mediated GDNF delivery could be applied to treating RP-like diseases also.

13.3.4 NANOCERIA

A fundamental property of nanomaterials is that their surface-area-to-volume ratio is relatively high. Alteration in the oxidation state of cerium oxide (CeO_2) nanoparticles ("nanoceria") creates defects in their lattice structure through loss of oxygen or its electrons. As their size decreases, nanoceria (3–5 nm diameter) exhibit more oxygen vacancies in their crystal structure.[64,65] Vacancy-engineered nanoceria may function as highly effective antioxidants. Diverse retinal diseases including AMD, RP, diabetic retinopathy, and retinopathy of prematurity are characterized, in part, by the presence of oxidative damage.[66–71]

Chen et al.[37] posited that engineered nanoceria can scavenge reactive oxygen intermediates and demonstrated that intravitreal injection of nanoceria prevents light-induced photoreceptor damage in rodents, even if injected after the initiation of light damage. Vacancy-engineered nanoceria also inhibit the development of and promote regression of pathological retinal neovascularization in the *Vldlr* knockout mouse, which carries a loss-of-function mutation in the *very low density lipoprotein receptor* gene and whose phenotype resembles a clinical entity known as retinal angiomatous proliferation (Figures 13.4 and 13.5).[38,72] This regression occurs even if intravitreal nanoceria treatment is administered after the mutant retinal phenotypes are established (Figure 13.6). A single injection has a prolonged effect (weeks) since nanoceria are both a catalytic and a regenerative antioxidant. Nanoceria inhibit development of increased vascular endothelial growth factor (VEGF) levels in this model,[38] which may mean this nanoparticle will be effective in treating macular edema in diabetic eyes and choroidal neovascularization-induced retinal edema in AMD eyes.[73–75]

13.3.5 DENDRIMERS

Dendrimers are synthesized, highly branched polymers that have precisely controllable nanoscale scaffolding and nanocontainer properties.[76] These globular macromolecules can deliver therapeutic agents that are incorporated into the scaffold core or are attached to surface terminal groups. The diameter of a poly(amidoamine) dendrimer ranges from 1.5 to 14.5 nm.[77] As generation (G) number increases, the number of active terminal groups doubles. G3 dendrimers, for example, contain 32 terminal groups, and G4 dendrimers contain 64 terminal groups. In poly(amidoamine) dendrimers, full generations (e.g., G3) have terminal amine or hydroxyl groups, while half-generation dendrimers (e.g., G3.5) have carboxylic acid terminal groups. Because dendrimers contain surface functional groups and void spaces within and between their branches, they can serve as delivery vehicles for therapeutic modalities such as carboplatin.[78]

Dendrimers have been explored as vehicles for controlled drug delivery, including cancer therapy, pilocarpine, gatifloxacin, and verteporfin-photodynamic therapy, and for VEGF inhibition.[78,79,36,24,80] Marano et al.[36] used a lipophilic amino acid dendrimer to deliver an anti-VEGF oligonucleotide into rat eyes with laser-induced choroidal new vessels. The dendrimer–oligonucleotide conjugate inhibited choroidal new vessel development by up to 95% for 4–6 months. The dendrimer–oligonucleotide conjugate was well tolerated *in vivo* with excellent biodistribution and no observable increase in inflammation-associated antigens. Ideta et al.[80] used dendrimer porphyrin encapsulated by a polymeric micelle to treat laser-induced choroidal new vessels in rodents and found significant enhancement of photodynamic therapy efficacy.

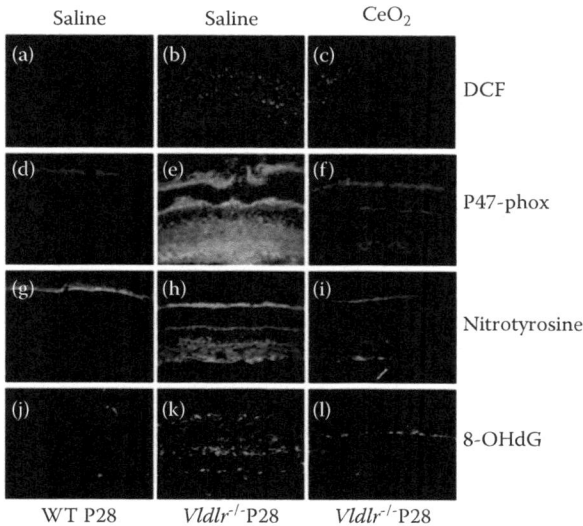

Saline   Saline   CeO$_2$

DCF

P47-phox

Nitrotyrosine

8-OHdG

WT P28   *Vldlr*$^{-/-}$P28   *Vldlr*$^{-/-}$P28

FIGURE 13.4 (**See color insert.**) Nanoceria reduce oxidative stress in the *Vldlr*$^{-/-}$ retina. Retinal sections from saline-injected wild-type (WT) mice (a, d, g, j); saline-injected *Vldlr*$^{-/-}$ mice (b, e, h, k); and CeO$_2$-injected (c, f, i, l) *Vldlr*$^{-/-}$ mice are shown as imaged by confocal microscopy. The 2′,7′-dicholoro-dihydro-fluorescein-diacetate (DCF) assay (a–c) visualizes reactive oxygen species (ROS) as punctuate fluorescence and demonstrates a very low level of ROS in the normal (a), a considerable amount in the *Vldlr*$^{-/-}$ (b), and a greatly reduced amount in the retina of the *Vldlr*$^{-/-}$ mice injected with CeO$_2$ (c). Similar results were obtained with the other three assays. NADPH-oxidase (P47-phox; d–f), a major producer of ROS, was very high in the *Vldlr*$^{-/-}$ retina and almost reduced to control levels in the CeO$_2$-injected mice. Nitrotyrosine (g–i), a reflection of oxidative activity due to increases in nitric oxide concentration, was highest in the *Vldlr*$^{-/-}$ retina and significantly reduced in the nanoceria-injected mice. ROS-mediated damage to DNA was indicated by the labeling of the retina with an antibody against a DNA adduct, 8-hydroxy-29-deoxyguanosine (8-OHdG; j–l), which showed little labeling in the control, significant labeling in the saline-injected *Vldlr*$^{-/-}$ retina, and a greatly reduced amount in the nanoceria-treated retina. DAPI (4′,6-diamidino-2-phenylindole) was used to visualize the nuclei. (From Zhou, X. et al., *PLoS One*, 6, e16733, 2011; Zarbin M.A. et al., *Wiley Interdiscip Rev Nanomed Nanobiotechnol* 4, 113–37, 2012. With permission.)

Shaunak et al.[81] used anionic, polyamidoamine, generation 3.5 dendrimers to make novel water-soluble conjugates of D(+)-glucosamine and D(+)-glucosamine 6-sulfate with immunomodulatory and antiangiogenic properties, respectively. Dendrimer glucosamine inhibited Toll-like receptor 4–mediated lipopolysaccharide-induced synthesis of pro-inflammatory chemokines (i.e., macrophage inflammatory protein [MIP]-1α, MIP-1β, and interleukin [IL]-8) and pro-inflammatory cytokines (i.e., tumor necrosis factor-α, IL-1β, and IL-6) primarily from immature human monocyte-derived dendritic cells and monocyte-derived macrophages, but allowed upregulation of the costimulatory molecules CD25, CD80, CD83, and CD86. Dendrimer glucosamine 6-sulfate blocked FGF-2-mediated human umbilical vein endothelial cell proliferation (but not VEGF-mediated proliferation) and neoangiogenesis in human Matrigel and placental angiogenesis assays. When dendrimer

Saline   Saline   CeO$_2$

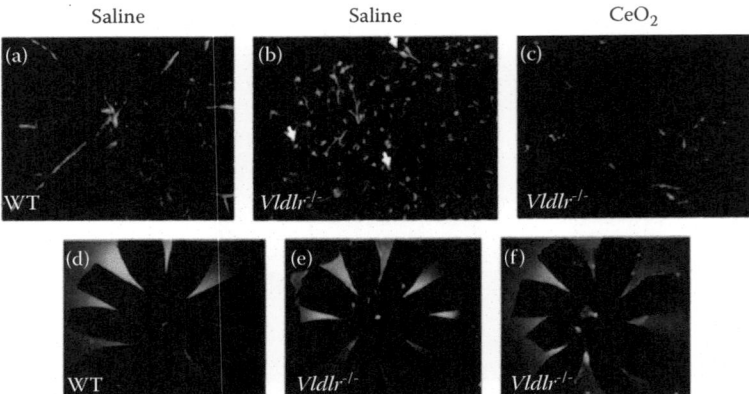

FIGURE 13.5 Nanoceria inhibit the development of pathologic intraretinal and subretinal vascular lesions in the *Vldlr$^{-/-}$* retina. Photomicrographs of whole-mount retinas (a–c) and eyecups (retinal pigment epithelium, choroid, and sclera) (d–f) from P28 animals are shown. All retinal blood vessels were labeled by the vascular filling assay. Wild-type (WT) retinas (a) showed the normal weblike retinal vasculature, whereas those from the *Vldlr$^{-/-}$* mice (b) showed numerous intraretinal vascular lesions or "blebs" (IRN blebs). See white arrows for example. A single injection of nanoceria at P7 inhibited (c) the appearance of these lesions. Eyecups from WT mice (d) showed no subretinal neovascular (SRN) "tufts" but those from *Vldlr$^{-/-}$* mice (e) had many bright SRN tufts. A single injection of nanoceria on P7 inhibited the appearance of these SRN tufts (f). (From Zhou, X. et al., *PLoS One*, 6, e16733, 2011; Zarbin M.A. et al., *Wiley Interdiscip Rev Nanomed Nanobiotechnol* 4, 113–37, 2012. With permission.)

glucosamine and dendrimer glucosamine 6-sulfate were used together in a clinically relevant rabbit model of scar tissue formation after glaucoma filtration surgery, they increased the long-term success of the surgery from 30% to 80% ($P = 0.029$).[82,83] A clinical trial exploring this technology, however, was not successful (Robert Ritch, MD and Peng Khaw, MD, personal communication).

13.3.6 CS-*G*-(PEI-*B*-ᴍPEG)

Ye et al.[31] used cationic nano-copolymer CS-*g*-(PEI-*b*-mPEG)-mediated IκB kinase beta (IKBKB) targeting small interfering RNA (siRNA) to modulate wound healing in a monkey model of glaucoma filtration surgery. IKBKB phosphorylates IκB and activates the inflammatory cascade mediated by NF-κB (nuclear factor **kappa**-light-chain-enhancer of activated **B** cells). These investigators posited that blocking NF-κB signaling would modulate wound healing. The siRNA targeting IKKβ (IKKβ-siRNA) was designed and delivered into human Tenon's fibroblasts using a ternary cationic copolymer called CS-*g*-(PEI-*b*-mPEG) as the vehicle. *In vitro* studies carried out by this group demonstrated that expression of IKKβ was downregulated, and the activation of NF-κB in the fibroblasts could be inhibited with this approach.[84] Blockade of the NF-κB pathway suppressed Tenon fibroblast proliferation. Although bleb survival was prolonged compared to control, mitomycin C treatment resulted in better bleb survival and lower IOP.

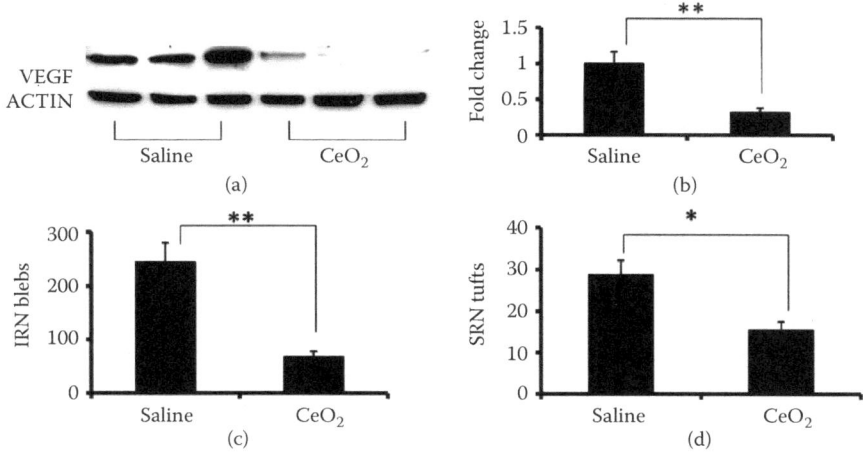

FIGURE 13.6 Retinal vascular lesions in *Vldlr*−/− retinas require continual production of excess reactive oxygen species (ROS). *Vldlr*−/− mice were injected at P28 with saline or nanoceria and killed 1 week later on P35. Analysis of vascular endothelial growth factor (VEGF) levels by Western blots (a) showed a fourfold reduction (b) within 1 week of nanoceria injection. The numbers of IRN blebs (c) and subretinal neovascular (SRN) tufts (d) were also dramatically reduced. *p = .05; **p = .01. (From Zhou, X. et al., *PLoS One*, 6, e16733, 2011; Zarbin M.A. et al., *Wiley Interdiscip Rev Nanomed Nanobiotechnol* 4, 113–37, 2012. With permission.)

13.4 NONVIRUS-MEDIATED GENE DELIVERY

This approach has been reviewed in detail elsewhere[1,3,4] as well as in Chapter 17. Viral vectors deliver genes efficiently but can be associated with risks such as immunogenicity and insertional mutagenesis. Nonviral vectors (e.g., polymers and lipids) and other methods (e.g., electroporation and nucleofection) have high gene-carrying capacity, low risk of immunogenicity, relatively low cost, and greater ease of production.[85] Nanoparticles can deliver genes efficiently to stem cells[86] and have been explored as a means for gene delivery in the diagnosis and treatment of ocular disease.[87–90]

Electrostatic interaction of cationic polymers with negatively charged DNA/RNA molecules results in condensation of the material into nanoparticles, protection of the genes from enzymes, and mediation of cellular entry.[91,92] Complexes of cationic polymers and plasmid DNA, termed polyplexes, can have transfection efficiency comparable to adenoviral vectors.[93] Nanometer-size polyplexes have large vector capacity, are stable in nuclease-rich environments, and have relatively high transfectivity for both dividing and nondividing cells.[90,93] Nanoparticles compacted with a lysine 30-mer linked to 10-kDa PEG-containing CMV-CFTR cDNA (cytomegalovirus-cystic fibrosis transmembrane conductance regulator complimentary DNA) were used successfully in a phase I/II clinical trial for treatment of cystic fibrosis.[94] Some particles, however, have low transfection efficiency, and the duration of gene expression can be short. The toxicity of polyplexes and nanoparticles is a reflection of their chemistry.[91]

Compacted DNA nanoparticles can be targeted to different tissues in the eye by varying the injection site (e.g., intravitreal injection can target the cornea, trabecular meshwork, lens, and inner retina; subretinal injection can target the outer retina and RPE).[90] Nanoparticle size and charge influence migration through the vitreous cavity.[95] Additional specificity in the locus of gene expression can be achieved by choosing promoters that are cell-specific. The rhodopsin promoter, for example, drives expression in rod photoreceptors, and the human red opsin promoter drives expression in cone photoreceptors.[39,96,97] Interphotoreceptor retinoid-binding protein drives expression in both rods and cones.[98] The vitelliform macular dystrophy promoter drives expression in RPE cells.[99] Farjo et al.[90] demonstrated that after subretinal injection of compacted lysine 30-mer DNA nanoparticles, gene expression extends through the retina and is not confined to the site of the injection.

Cai et al.[39,40] used DNA nanoparticles consisting of single molecules of DNA compacted with 10-kDa PEG-substituted lysine 30-mer peptides containing the wild-type (WT) retinal degeneration slow (*Rds*) gene, peripherin/rds, to induce cone photoreceptor rescue in an animal model (*rds*[+/-]) of human RP. These compacted plasmid DNA nanoparticles are small (8–20 nm) and have a large carrying capacity (at least up to 20 kb).[90,40] PLGA nanoparticles can deliver genes to RPE cells *in vitro* and *in vivo* with reasonable efficiency and safety.[100] PLGA DNA nanoparticles can be associated with long-term gene expression, perhaps due to sustained cytosolic plasmid release.[101] PLGA DNA nanoparticles tend to be larger than polylysine DNA nanoparticles,[102,103] which may affect the cellular uptake mechanism and delivery to the nucleus. PLGA DNA nanoparticles might be used to deliver therapeutic genes for conditions associated with RPE gene mutations, for example, Best disease[104] and a form of Leber congenital amaurosis.[105–107]

There are some concerns involving nonviral gene delivery. Although the immune response to polylysine-based nanoparticles seems to be less than that for capsid proteins, the efficiency of gene transfer is not as high since most are degraded in the endosomal complexes.[108] Thus, one may have to use large numbers of nanoparticles, which might generate an immune response nonetheless. Also, the immune response to both nanoparticles and viruses varies from one species to another, and the apparent low immunogenicity observed in murine models of RP may not be observed in human patients.[108]

13.5   VIRUS-MEDIATED GENE DELIVERY

Due to their relatively low immunogenicity, ability to target many nondividing cells, and capacity for sustained efficient therapeutic gene expression after a single treatment,[109] recombinant adeno-associated virus (rAAV) vectors have been used to treat eye diseases in preclinical models, including in preclinical models of human retinal disease[100,111] and also have been used to treat humans with Leber congenital amaurosis.[105,106,112] Nonetheless, important areas in which to improve gene therapy remain, including (1) vector uptake, transport, and uncoating; (2) vector genome persistence; (3) sustained transcriptional expression; (4) the host immune response; and (5) insertional mutagenesis and cancer.[108,113,114] We have reviewed elsewhere[4] the nanoengineering of the viral capsid and transgene that has been undertaken to address these issues. Modifications of the virus to improve clinical effectiveness provide a good example of some of the nanoengineering strategies that have been employed in this area. AAVs are small (4.7 kb carrying capacity), nonpathogenic, single-stranded DNA parvoviruses that can transduce dividing and nondividing cells.[115] The genes encoding replication and capsid proteins from the WT AAV genome are replaced by a promoter-therapeutic transgene cassette flanked by the normal AAV inverted terminal repeats needed for packaging and replication in rAAVs. The capsid is critical for extracellular events related to the recognition of specific receptors, which influences cell tropism, as well as intracellular processes involving AAV trafficking and uncoating, which influences transduction kinetics and efficiency of transgene expression.[109,116] Two nanoengineering techniques have been applied to improve vector cellular tropism, transduction efficiency, and immunogenicity: directed evolution and site-directed mutagenesis.

Directed evolution of AAV capsids has generated vectors that are highly resistant to neutralizing antibodies.[117,118] With one mutagenesis and three selection steps, Maheshri et al.[118] generated mutant capsids with a threefold improved neutralizing antibody titer (vs. WT capsid) and a ~7.5% infectivity at serum levels that completely neutralized WT infectivity.[118] Directed evolution has been used to generate AAV variants that transduce Muller cells after intravitreal injection,[119,120] which may provide a means to deliver growth factors to photoreceptors and RPE cells. These growth factors retard the progression of retinal degeneration in preclinical models of RP[121,122] and possibly in human patients also.[123]

Site-directed mutagenesis technology is being applied to the treatment of degenerative retinal disease in preclinical models. Vectors containing point mutations in surface-exposed capsid tyrosine residues in AAV serotypes 2, 8, and 9 display strong and widespread transgene expression in retinal cells after intravitreal or subretinal delivery.[124] Petrs-Silva et al.[124] demonstrated that tyrosine-to-phenylalanine capsid scAAV2 mutants showed much greater transduction efficiency of the entire retina after intravitreal injection compared to scAAV with WT capsids. Mutants of scAAV2, scAAV8, and scAAV9 also enhanced transduction of retinal ganglion cells compared to WT AAV2. (Previously only AAV2 could transduce retinal ganglion cells.) Intravitreal delivery may offer an important clinical advantage over subretinal delivery. Subretinal virus delivery, which has been used in clinical studies,[105–107] requires pars plana vitrectomy surgery in the operating room and has a higher likelihood of complications (e.g., retinal tear) than intravitreal delivery, which can be done in an office setting under topical anesthesia. However, Li et al.[125] have shown that a humoral immune response against AAV2 capsid proteins occurs after intravitreal but not after subretinal vector delivery. (Subretinal injection of one of the mutant scAAVs also transduced Muller cells.) This finding may reflect the relative immune privilege of the subretinal space.[126] These studies demonstrate two strategies for reducing the immune response to viral vectors through site-directed mutagenesis: increasing transduction efficiency (which permits lower doses of vector) and creation of multiple effective serotypes, which can be used sequentially for subsequent therapy.

13.5.1 Optogenetics

Optogenetics involves the use of viral particles to induce light sensitivity in cells or parts of cells that normally are not light sensitive. This concept has been explored in detail elsewhere.[2–4,127] Although rewiring of inner retinal circuits and inner retinal neuronal degeneration occur in association with photoreceptor degeneration in RP,[128,129] one can create visually useful percepts by stimulating retinal ganglion cells electrically.[130–133] Use of light-sensitive ion channels, rather than electrodes, to stimulate retinal ganglion cells provides an alternative approach to retinal cell stimulation.[134–137] Induced light sensitivity has the potential for noninvasive neuronal stimulation with high spatial resolution.

Channelopsin-2 is a light-gated, blue light-sensitive ion channel that is derived from green algae. Reversible photoisomerization of its attached chromophore, all-*trans* retinaldehyde, induces a conformational change in channelopsin-2 that alters its permeability to monovalent and divalent cations.[138] Channelrhodopsin-2 (ChR2) is the complex of channelopsin-2 and all-*trans* retinal. One can use an AAV delivery system (AAV serotype-2) in *rd1* mice, which have a null mutation in a cyclic GMP (cyclic guanosine monophosphate) phosphodiesterase (PDE6b), and in RCS rats to induce ChR2 expression in inner retinal neurons (primarily ON and OFF retinal ganglion cells). After ChR2 transfection, these neurons respond to light with membrane depolarization.[41,42,39,140,141] ChR2 can also restore the ability of the animals to encode light signals in the retina and transmit them to the visual cortex. *Rd1* mice and RCS rats each harbor mutations that cause RP in humans.

Inner retinal neuron transfection with halorhodopsin (HaloR), a yellow light-activated chloride ion pump from halobacteria, converts them into OFF cells.[44] Zhang et al.[44] showed that HaloR was ~20-fold less sensitive to light than ChR2. If cells express both HaloR and ChR2, then they can produce ON, OFF, and ON-OFF responses, depending on the illumination wavelength.[44] In these preclinical models, the kinetics of ChR2- and HaloR-mediated light responses is compatible with the retina's temporal information-processing requirements. ChR2 and HaloR both exhibit low light sensitivity, however, with threshold activation light intensities ~5–6 log units higher than those of cones.[41,44] Also, the light intensity operating range of microbial rhodopsins is 2–3 log units, whereas the normal retinal dynamic range is 10 log units. Recombinant AAV vector packaged in a tyrosine-mutated capsid has been used to achieve stable and specific ChR2 in ON bipolar cells,[142] and in these experiments, light levels that elicited visually guided behaviors were within the physiological range of cone photoreceptors. Signal convergence from bipolar cells onto retinal ganglion cells may mean that targeting ChR2 to rod bipolar cells will enhance light sensitivity and spatial resolution, but the alterations in synaptic circuitry that accompany photoreceptor degeneration might compromise this approach.[128,129,143–145]

Greenberg et al.[46] reconstructed an excitatory center and an antagonistic surround by targeting humanized ChR2 to the somata and enhanced HaloR to the dendrites of retinal ganglion cells. This approach, in contrast to insertion of optical neuromodulators, retains crucial information processing (edge detection) while being independent of the state of inner retinal circuit remodeling during degeneration. Since ankyrins couple sodium channels to the spectrin–actin network, fusion of the humanized ChR2 to ankyrin$_G$ polypeptide localized this opsin to the soma and proximal dendrites. Fusion of enhanced HaloR to PSD-95 (postsynaptic density protein-95) protein targeted this opsin to retinal ganglion cell dendrites. Depending on which opsin is fused to ankyrin$_G$ and which to PDS-95, both ON- and OFF-center ganglion cells could be created. Because of the nonphysiological center-surround dimensions created by this transfection strategy, Greenberg et al.[46] preprocessed the visual image with Gaussian blurring, such that when convolved with the dimensions of the soma and dendrites, the Gaussians approximated the relative dimensions of the ganglion cells' center and the surround receptive fields. Thus, extraction of edge information was obtained artificially. At this time, it seems that ChR2/HaloR-based retinal ganglion cell prosthetics will require image preprocessing to perform light amplification, dynamic range compression, and local gain control operations.[46]

In typical RP, cone degeneration occurs after rod photoreceptor degeneration.[129] Cone cell bodies and inner segments remain for a time after the outer segments have been lost. Using AAV transfection, Busskamp et al.[146] restored light sensitivity in mouse models of RP (i.e., the *rd1* mouse, which models fast forms of retinal degeneration, and *Cnga3*$^{-/-}$; *Rho*$^{-/-}$ double-knockout, which models a slow form of retinal degeneration) by enhancing HaloR expression in light-insensitive cones. Targeted expression of enhanced HaloR in photoreceptors was achieved using human rhodopsin, human red opsin, and mouse cone arrestin promoters. The resensitized cones activated all retinal cone pathways, drove directional selectivity, activated cortical circuits (in *rd1* mice), and mediated visually guided behaviors. Light-stimulated HaloR-transfected photoreceptors seemed to convey information through bipolar cells to retinal ganglion cells, including both ON and OFF pathways. These effects were obtained even at times when only ~25% of cone cell bodies remained and despite the synaptic reorganization of the inner retina that occurs with RP.

One might consider the use of ChR2 and HaloR as molecules to re-engineer cells and their behavior. One also can use bionanotechnology to re-engineer proteins first and to re-engineer cell behavior subsequently. For example, synthesis of light-sensitive ion channels has been achieved by coupling naturally occurring ion channels with molecules (e.g., azobenzene) whose photoisomerization results, ultimately, in reversible activation of the ion channel.[134,136,138,147] In the case of azobenzene, one end of the molecule is covalently tethered to the ion channel and to the other end is attached an "active moiety," for example, an agonist, antagonist, or pore-blocking agent. Light absorption by azobenzene creates a conformational change in the molecule that alters the relationship of its active moiety to the ion channel. In the case of azobenzene, the thermally relaxed *trans* isomer is more extended (~0.7 nm longer) than the higher-energy *cis* isomer. The active moiety can interact with the ion channel in only one of the isomeric states. For example, if light induces the *cis* conformation and if a pore-blocking agent is brought into position in the *cis* state, then light exposure will block ion movement into the transfected cell. Acrylamide azobenzene quaternary ammonium is a variant of the maleimide-azobenzene-quaternary ammonium molecule that permits affinity labeling of endogenous potassium channels without the need for receptor mutagenesis or genetic manipulation of the target cells (e.g., ganglion cells).[148] A genetically and chemically engineered light-gated mammalian ion channel, the light-activated glutamate receptor (LiGluR), has been expressed selectively in the retinal ganglion cells of the *rd1* mouse.[149] In these mice, the LiGluR restores light sensitivity to the retinal ganglion cells, reinstates light responsiveness to the primary visual cortex, and restores both the pupillary reflex and a natural light-avoidance behavior.

13.6 CONCLUSION

Nanoparticles can be designed to facilitate treatment of retinal and optic nerve diseases by improving the action of existing therapeutic modalities as well as by enabling the development of novel treatments. Nanoparticle chemistry can be modified, for example, to increase drug delivery to target tissue, increase drug half-life in target tissue, improve tissue and cell tropism, and even confer novel physiological functions to cells. In view of the fact that some of the major causes of blindness in the industrialized world are chronic diseases of the retina and optic nerve (e.g., AMD, diabetic retinopathy, and glaucoma), these features of nanoparticles are potentially quite important and portend a central role for this technology in the twenty-first century ophthalmic pharmacopeia.

REFERENCES

1. Zarbin MA, Montemagno C, Leary JF, Ritch R. Nanomedicine in ophthalmology: The new frontier. *Am J Ophthalmol* 2010;150:144–162 e2.
2. Zarbin MA, Montemagno C, Leary JF, Ritch R. Nanotechnology in ophthalmology. *Can J Ophthalmol* 2010;45:457–76.
3. Zarbin M, Montemagno C, Leary J, Ritch R. Artificial vision. *Panminerva Med* 2011;53:167–77.
4. Zarbin MA, Montemagno C, Leary JF, Ritch R. Regenerative nanomedicine and the treatment of degenerative retinal diseases. *Wiley Interdiscip Rev Nanomed Nanobiotechnol* 2012;4:113–37.
5. Petros RA, DeSimone JM. Strategies in the design of nanoparticles for therapeutic applications. *Nat Rev Drug Discov* 2010;9:615–27.
6. Rejman J, Oberle V, Zuhorn IS, Hoekstra D. Size-dependent internalization of particles via the pathways of clathrin- and caveolae-mediated endocytosis. *Biochem J* 2004;377:159–69.
7. Bareford LM, Swaan PW. Endocytic mechanisms for targeted drug delivery. *Adv Drug Deliv Rev* 2007;59:748–58.
8. Torchilin VP. Cell penetrating peptide-modified pharmaceutical nanocarriers for intracellular drug and gene delivery. *Biopolymers* 2008;90:604–10.
9. Chen X, Kube DM, Cooper MJ, Davis PB. Cell surface nucleolin serves as receptor for DNA nanoparticles composed of pegylated polylysine and DNA. *Mol Ther* 2008;16:333–42.
10. Chu TC, He Q, Potter DE. Biodegradable calcium phosphate nanoparticles as a new vehicle for delivery of a potential ocular hypotensive agent. *J Ocul Pharmacol Ther* 2002;18:507–14.
11. Kassem MA, Abdel Rahman AA, Ghorab MM, Ahmed MB, Khalil RM. Nanosuspension as an ophthalmic delivery system for certain glucocorticoid drugs. *Int J Pharm* 2007;340:126–33.
12. El-Gazayerly ON, Hikal, AH. Preparation and evaluation of acetazolamide liposomes as an ocular delivery system. *Int J Pharm* 1997;158:121–7.
13. Hathout RM, Mansour S, Mortada ND, Guinedi AS. Liposomes as an ocular delivery system for acetazolamide: *In vitro* and *in vivo* studies. *AAPS PharmSciTech* 2007;8:1.
14. Aggarwal D, Garg A, Kaur IP. Development of a topical niosomal preparation of acetazolamide: Preparation and evaluation. *J Pharm Pharmacol* 2004;56:1509–17.
15. Guinedi AS, Mortada ND, Mansour S, Hathout RM. Preparation and evaluation of reverse-phase evaporation and multilamellar niosomes as ophthalmic carriers of acetazolamide. *Int J Pharm* 2005;306:71–82.

16. Kaur IP, Aggarwal D, Singh H, Kakkar S. Improved ocular absorption kinetics of timolol maleate loaded into a bioadhesive niosomal delivery system. *Graefes Arch Clin Exp Ophthalmol* 2010;248:1467–72.

17. Vyas SP, Mysore N, Jaitely V, Venkatesan N. Discoidal niosome based controlled ocular delivery of timolol maleate. *Pharmazie* 1998;53:466–9.

18. Prabhu P, Nitish KR, Koland M et al. Preparation and evaluation of nano-vesicles of brimonidine tartrate as an ocular drug delivery system. *J Young Pharm* 2010;2:356–61.

19. Maiti S, Paul S, Mondol R, Ray S, Sa B. Nanovesicular formulation of brimonidine tartrate for the management of glaucoma: *In vitro* and *in vivo* evaluation. *AAPS PharmSciTech* 2011;12:755–63.

20. Bertram JP, Saluja SS, McKain J, Lavik EB. Sustained delivery of timolol maleate from poly(lactic-co-glycolic acid)/poly(lactic acid) microspheres for over 3 months. *J Microencapsul* 2009;26:18–26.

21. Chen R, Qian Y, Li R et al. Methazolamide calcium phosphate nanoparticles in an ocular delivery system. *Yakugaku Zasshi* 2010;130:419–24.

22. Li R, Jiang S, Liu D et al. A potential new therapeutic system for glaucoma: Solid lipid nanoparticles containing methazolamide. *J Microencapsul* 2011;28:134–41.

23. Wadhwa S, Paliwal R, Paliwal SR, Vyas SP. Hyaluronic acid modified chitosan nanoparticles for effective management of glaucoma: Development, characterization, and evaluation. *J Drug Target* 2010;18:292–302.

24. Vandamme TF, Brobeck L. Poly(amidoamine) dendrimers as ophthalmic vehicles for ocular delivery of pilocarpine nitrate and tropicamide. *J Control Release* 2005;102:23–38.

25. Nkansah MK, Tzeng SY, Holdt AM, Lavik EB. Poly(lactic-co-glycolic acid) nanospheres and microspheres for short- and long-term delivery of bioactive ciliary neurotrophic factor. *Biotechnol Bioeng* 2008;100:1010–9.

26. Weise J, Isenmann S, Klocker N et al. Adenovirus-mediated expression of ciliary neurotrophic factor (CNTF) rescues axotomized rat retinal ganglion cells but does not support axonal regeneration *in vivo*. *Neurobiol Dis* 2000;7:212–23.

27. Pease ME, Zack DJ, Berlinicke C et al. Effect of CNTF on retinal ganglion cell survival in experimental glaucoma. *Invest Ophthalmol Vis Sci* 2009;50:2194–200.

28. Jeun M, Jeoung JW, Moon S et al. Engineered superparamagnetic Mn0.5Zn0.5Fe2O4 nanoparticles as a heat shock protein induction agent for ocular neuroprotection in glaucoma. *Biomaterials* 2011;32:387–94.

29. Cui LJ, Sun NX, Li XH, Huang J, Yang JG. Subconjunctival sustained release 5-fluorouracil for glaucoma filtration surgery. *Acta Pharmacol Sin* 2008;29:1021–8.

30. Hou Z, Wei H, Wang Q et al. New method to prepare mitomycin C loaded PLA-nanoparticles with high drug entrapment efficiency. *Nanoscale Res Lett* 2009;4:732–7.

31. Ye H, Qian Y, Lin M et al. Cationic nano-copolymers mediated IKKbeta targeting siRNA to modulate wound healing in a monkey model of glaucoma filtration surgery. *Mol Vis* 2010;16:2502–10.

32. Perkins TW, Faha B, Ni M et al. Adenovirus-mediated gene therapy using human p21WAF-1/Cip-1 to prevent wound healing in a rabbit model of glaucoma filtration surgery. *Arch Ophthalmol* 2002;120:941–9.

33. Yang JG, Sun NX, Cui LJ, Wang XH, Feng ZH. Adenovirus-mediated delivery of p27(KIP1) to prevent wound healing after experimental glaucoma filtration surgery. *Acta Pharmacol Sin* 2009;30:413–23.

34. Spiga MG, Borras T. Development of a gene therapy virus with a glucocorticoid-inducible MMP1 for the treatment of steroid glaucoma. *Invest Ophthalmol Vis Sci* 2010;51:3029–41.

35. Sakai T, Kuno N, Takamatsu F et al. Prolonged protective effect of basic fibroblast growth factor-impregnated nanoparticles in Royal College of Surgeons rats. *Invest Ophthalmol Vis Sci* 2007;48:3381–7.

36. Marano RJ, Toth I, Wimmer N, Brankov M, Rakoczy PE. Dendrimer delivery of an anti-VEGF oligonucleotide into the eye: A long-term study into inhibition of laser-induced CNV, distribution, uptake and toxicity. *Gene Ther* 2005;12:1544–50.
37. Chen J, Patil S, Seal S, McGinnis JF. Rare earth nanoparticles prevent retinal degeneration induced by intracellular peroxides. *Nat Nanotechnol* 2006;1:142–50.
38. Zhou X, Wong LL, Karakoti AS, Seal S, McGinnis JF. Nanoceria inhibit the development and promote the regression of pathologic retinal neovascularization in the vldlr knockout mouse. *PLoS One* 2011;6:e16733.
39. Cai X, Conley SM, Nash Z, Fliesler SJ, Cooper MJ, Naash MI. Gene delivery to mitotic and postmitotic photoreceptors via compacted DNA nanoparticles results in improved phenotype in a mouse model of retinitis pigmentosa. *FASEB J* 2010;24:1178–91.
40. Cai X, Nash Z, Conley SM, Fliesler SJ, Cooper MJ, Naash MI. A partial structural and functional rescue of a retinitis pigmentosa model with compacted DNA nanoparticles. *PLoS One* 2009;4:e5290.
41. Bi A, Cui J, Ma YP et al. Ectopic expression of a microbial-type rhodopsin restores visual responses in mice with photoreceptor degeneration. *Neuron* 2006;50:23–33.
42. Lagali PS, Balya D, Awatramani GB et al. Light-activated channels targeted to ON bipolar cells restore visual function in retinal degeneration. *Nat Neurosci* 2008;11:667–75.
43. Tomita H, Sugano E, Isago H et al. Channelrhodopsin-2 gene transduced into retinal ganglion cells restores functional vision in genetically blind rats. *Exp Eye Res* 2010;90:429–36.
44. Zhang Y, Ivanova E, Bi A, Pan ZH. Ectopic expression of multiple microbial rhodopsins restores ON and OFF light responses in retinas with photoreceptor degeneration. *J Neurosci* 2009;29:9186–96.
45. Doroudchi MM, Greenberg KP, Liu J et al. Virally delivered channelrhodopsin-2 safely and effectively restores visual function in multiple mouse models of blindness. *Mol Ther* 2011;19:1220–9.
46. Greenberg KP, Pham A, Werblin FS. Differential targeting of optical neuromodulators to ganglion cell soma and dendrites allows dynamic control of center-surround antagonism. *Neuron* 2011;69:713–20.
47. Caporale N, Kolstad KD, Lee T et al. LiGluR restores visual responses in rodent models of inherited blindness. *Mol Ther* 2011;19:1212–9.
48. Uchegbu IF, Bouwstra JA, Florence AT. Large disk-shaped structures (discomes) in nonionic surfactant vesicle to micelle transitions. *J Phys Chem* 1992;96:10548–53.
49. Moghimi SM, Hamad I, Andresen TL, Jorgensen K, Szebeni J. Methylation of the phosphate oxygen moiety of phospholipid-methoxy(polyethylene glycol) conjugate prevents PEGylated liposome-mediated complement activation and anaphylatoxin production. *FASEB J* 2006;20:2591–3.
50. Hamad I, Christy Hunter A, Rutt KJ, Liu Z, Dai H, Moein Moghimi S. Complement activation by PEGylated single-walled carbon nanotubes is independent of C1q and alternative pathway turnover. *Mol Immunol* 2008;45:3797–803.
51. Yokoe J, Sakuragi S, Yamamoto K et al. Albumin-conjugated PEG liposome enhances tumor distribution of liposomal doxorubicin in rats. *Int J Pharm* 2008;353:28–34.
52. Furumoto K, Yokoe J, Ogawara K et al. Effect of coupling of albumin onto surface of PEG liposome on its *in vivo* disposition. *Int J Pharm* 2007;329:110–6.
53. Ferreira L, Park H, Choe H, Kohane D, Langer R. Human embryoid bodies containing nano- and micro-particulate delivery vehicles. *Adv Mat* 2008;20:2285–91.
54. Liu G, Men P, Harris PL, Rolston RK, Perry G, Smith MA. Nanoparticle iron chelators: A new therapeutic approach in Alzheimer disease and other neurologic disorders associated with trace metal imbalance. *Neurosci Lett* 2006;406:189–93.
55. De Campos AM, Sanchez A, Alonso MJ. Chitosan nanoparticles: A new vehicle for the improvement of the delivery of drugs to the ocular surface. Application to cyclosporin A. *Int J Pharm* 2001;224:159–68.

56. de la Fuente M, Seijo B, Alonso MJ. Bioadhesive hyaluronan-chitosan nanoparticles can transport genes across the ocular mucosa and transfect ocular tissue. *Gene Ther* 2008;15:668–76.

57. Vollrath D, Feng W, Duncan JL et al. Correction of the retinal dystrophy phenotype of the RCS rat by viral gene transfer of Mertk. *Proc Natl Acad Sci USA* 2001;98:12584–9.

58. Gal A, Li Y, Thompson DA et al. Mutations in MERTK, the human orthologue of the RCS rat retinal dystrophy gene, cause retinitis pigmentosa. *Nat Genet* 2000;26:270–1.

59. Charbel Issa P, Bolz HJ, Ebermann I, Domeier E, Holz FG, Scholl HP. Characterisation of severe rod-cone dystrophy in a consanguineous family with a splice site mutation in the MERTK gene. *Br J Ophthalmol* 2009;93:920–5.

60. Mackay DS, Henderson RH, Sergouniotis PI et al. Novel mutations in MERTK associated with childhood onset rod-cone dystrophy. *Mol Vis* 2010;16:369–77.

61. Jiang C, Moore MJ, Zhang X, Klassen H, Langer R, Young M. Intravitreal injections of GDNF-loaded biodegradable microspheres are neuroprotective in a rat model of glaucoma. *Mol Vis* 2007;13:1783–92.

62. Fu K, Harrell R, Zinski K et al. A potential approach for decreasing the burst effect of protein from PLGA microspheres. *J Pharm Sci* 2003;92:1582–91.

63. Dalkara D, Kolstad KD, Guerin KI et al. AAV mediated GDNF secretion from retinal glia slows down retinal degeneration in a rat model of retinitis pigmentosa. *Mol Ther* 2011;19:1602–8.

64. Deshpande S, Patil S, Kuchibhatla SV, Seal S. Size dependency variation in lattice parameter and valency states in nanocrystalline cerium oxide. *Appl Phys Lett* 2005;87:133113.

65. Tsunekawa S, Sahara R, Kawazoe Y, Ishikawa K. Lattice relaxation of monosize CeO2-x nanocrystalline particles. *Appl Surf Sci* 1999;152:53–6.

66. Brownlee M. A radical explanation for glucose-induced beta cell dysfunction. *J Clin Invest* 2003;112:1788–90.

67. Yorek MA. The role of oxidative stress in diabetic vascular and neural disease. *Free Radic Res* 2003;37:471–80.

68. Dugan LL, Lovett EG, Quick KL, Lotharius J, Lin TT, O'Malley KL. Fullerene-based antioxidants and neurodegenerative disorders. *Parkinsonism Relat Disord* 2001;7:243–6.

69. Shen JK, Dong A, Hackett SF, Bell WR, Green WR, Campochiaro PA. Oxidative damage in age-related macular degeneration. *Histol Histopathol* 2007;22:1301–8.

70. Komeima K, Rogers BS, Campochiaro PA. Antioxidants slow photoreceptor cell death in mouse models of retinitis pigmentosa. *J Cell Physiol* 2007;213:809–15.

71. Papp A, Nemeth I, Karg E, Papp E. Glutathione status in retinopathy of prematurity. *Free Radic Biol Med* 1999;27:738–43.

72. Truong SN, Alam S, Zawadzki RJ et al. High resolution fourier-domain optical coherence tomography of retinal angiomatous proliferation. *Retina* 2007;27:915–25.

73. Elman MJ, Aiello LP, Beck RW et al. Randomized trial evaluating ranibizumab plus prompt or deferred laser or triamcinolone plus prompt laser for diabetic macular edema. *Ophthalmology* 2010;117:1064–77 e35.

74. Rosenfeld PJ, Brown DM, Heier JS et al. Ranibizumab for neovascular age-related macular degeneration. *N Engl J Med* 2006;355:1419–31.

75. Brown DM, Kaiser PK, Michels M et al. Ranibizumab versus verteporfin for neovascular age-related macular degeneration. *N Engl J Med* 2006;355:1432–44.

76. Tomalia DA, Reyna LA, Svenson S. Dendrimers as multi-purpose nanodevices for oncology drug delivery and diagnostic imaging. *Biochem Soc Trans* 2007;35:61–7.

77. Hahn U, Gorka M, Vogtle F et al. Light-harvesting dendrimers: Efficient intra- and intermolecular energy-transfer processes in a species containing 65 chromophoric groups of four different types. *Angew Chem Int Ed Engl* 2002;41:3595–8, 3514.

78. Kang SJ, Durairaj C, Kompella UB, O'Brien JM, Grossniklaus HE. Subconjunctival nanoparticle carboplatin in the treatment of murine retinoblastoma. *Arch Ophthalmol* 2009;127:1043–7.
79. Durairaj C, Kadam RS, Chandler JW, Hutcherson SL, Kompella U. Nanosized dendritic polyguanidilyated translocators for enhanced solubility, permeability, and delivery of gatifloxacin. *Invest Ophthalmol Vis Sci* 2010;51:5804–16.
80. Ideta R, Tasaka F, Jang WD et al. Nanotechnology-based photodynamic therapy for neovascular disease using a supramolecular nanocarrier loaded with a dendritic photosensitizer. *Nano Lett* 2005;5:2426–31.
81. Shaunak S, Thomas S, Gianasi E et al. Polyvalent dendrimer glucosamine conjugates prevent scar tissue formation. *Nat Biotechnol* 2004;22:977–84.
82. Mead AL, Wong TT, Cordeiro MF, Anderson IK, Khaw PT. Evaluation of anti-TGF-beta2 antibody as a new postoperative anti-scarring agent in glaucoma surgery. *Invest Ophthalmol Vis Sci* 2003;44:3394–401.
83. Siriwardena D, Khaw PT, King AJ et al. Human antitransforming growth factor beta(2) monoclonal antibody—A new modulator of wound healing in trabeculectomy: a randomized placebo controlled clinical study. *Ophthalmology* 2002;109:427–31.
84. Duan Y, Guan X, Ge J et al. Cationic nano-copolymers mediated IKKbeta targeting siRNA inhibit the proliferation of human Tenon's capsule fibroblasts *in vitro*. *Mol Vis* 2008;14:2616–28.
85. Glover DJ, Lipps HJ, Jans DA. Towards safe, non-viral therapeutic gene expression in humans. *Nat Rev Genet* 2005;6:299–310.
86. Kutsuzawa K, Chowdhury EH, Nagaoka M, Maruyama K, Akiyama Y, Akaike T. Surface functionalization of inorganic nano-crystals with fibronectin and E-cadherin chimera synergistically accelerates trans-gene delivery into embryonic stem cells. *Biochem Biophys Res Commun* 2006;350:514–20.
87. Mo Y, Barnett ME, Takemoto D, Davidson H, Kompella UB. Human serum albumin nanoparticles for efficient delivery of Cu, Zn superoxide dismutase gene. *Mol Vis* 2007;13:746–57.
88. Prow T, Grebe R, Merges C et al. Nanoparticle tethered antioxidant response element as a biosensor for oxygen induced toxicity in retinal endothelial cells. *Mol Vis* 2006;12:616–25.
89. Cai X, Conley S, Naash M. Nanoparticle applications in ocular gene therapy. *Vision Res* 2008;48:319–24.
90. Farjo R, Skaggs J, Quiambao AB, Cooper MJ, Naash MI. Efficient non-viral ocular gene transfer with compacted DNA nanoparticles. *PLoS One* 2006;1:e38.
91. Ferreira L, Karp JM, Nobre L, Langer R. New opportunities: The use of nanotechnologies to manipulate and track stem cells. *Cell Stem Cell* 2008;3:136–46.
92. Pack DW, Hoffman AS, Pun S, Stayton PS. Design and development of polymers for gene delivery. *Nat Rev Drug Discov* 2005;4:581–93.
93. Incani V, Tunis E, Clements BA et al. Palmitic acid substitution on cationic polymers for effective delivery of plasmid DNA to bone marrow stromal cells. *J Biomed Mater Res* A 2007;81:493–504.
94. Konstan MW, Davis PB, Wagener JS et al. Compacted DNA nanoparticles administered to the nasal mucosa of cystic fibrosis subjects are safe and demonstrate partial to complete cystic fibrosis transmembrane regulator reconstitution. *Hum Gene Ther* 2004;15:1255–69.
95. Pitkanen L, Ruponen M, Nieminen J, Urtti A. Vitreous is a barrier in nonviral gene transfer by cationic lipids and polymers. *Pharm Res* 2003;20:576–83.
96. Flannery JG, Zolotukhin S, Vaquero MI, LaVail MM, Muzyczka N, Hauswirth WW. Efficient photoreceptor-targeted gene expression *in vivo* by recombinant adeno-associated virus. *Proc Natl Acad Sci USA* 1997;94:6916–21.

97. Li Q, Timmers AM, Guy J, Pang J, Hauswirth WW. Cone-specific expression using a human red opsin promoter in recombinant AAV. *Vision Res* 2008;48:332–8.

98. Porrello K, Bhat SP, Bok D. Detection of interphotoreceptor retinoid binding protein (IRBP) mRNA in human and cone-dominant squirrel retinas by in situ hybridization. *J Histochem Cytochem* 1991;39:171–6.

99. Esumi N, Oshima Y, Li Y, Campochiaro PA, Zack DJ. Analysis of the VMD2 promoter and implication of E-box binding factors in its regulation. *J Biol Chem* 2004;279:19064–73.

100. Bejjani RA, BenEzra D, Cohen H et al. Nanoparticles for gene delivery to retinal pigment epithelial cells. *Mol Vis* 2005;11:124–32.

101. Conley SM, Cai X, Naash MI. Nonviral ocular gene therapy: Assessment and future directions. *Curr Opin Mol Ther* 2008;10:456–63.

102. Conley SM, Naash MI. Nanoparticles for retinal gene therapy. *Prog Retin Eye Res* 2010;29:376–97.

103. Ravi Kumar MN, Bakowsky U, Lehr CM. Preparation and characterization of cationic PLGA nanospheres as DNA carriers. *Biomaterials* 2004;25:1771–7.

104. Marmorstein AD, Marmorstein LY, Rayborn M, Wang X, Hollyfield JG, Petrukhin K. Bestrophin, the product of the Best vitelliform macular dystrophy gene (VMD2), localizes to the basolateral plasma membrane of the retinal pigment epithelium. *Proc Natl Acad Sci USA* 2000;97:12758–63.

105. Maguire AM, Simonelli F, Pierce EA et al. Safety and efficacy of gene transfer for Leber's congenital amaurosis. *N Engl J Med* 2008;358:2240–8.

106. Bainbridge JW, Smith AJ, Barker SS et al. Effect of gene therapy on visual function in Leber's congenital amaurosis. *N Engl J Med* 2008;358:2231–9.

107. Cideciyan AV, Hauswirth WW, Aleman TS et al. Human RPE65 gene therapy for Leber congenital amaurosis: Persistence of early visual improvements and safety at 1 year. *Hum Gene Ther* 2009;20:999–1004.

108. Kay MA. State-of-the-art gene-based therapies: The road ahead. *Nat Rev Genet* 2011;12:316–28.

109 Surace EM, Auricchio A. Versatility of AAV vectors for retinal gene transfer. *Vision Res* 2008;48:353–9.

110. Acland GM, Aguirre GD, Bennett J et al. Long-term restoration of rod and cone vision by single dose rAAV-mediated gene transfer to the retina in a canine model of childhood blindness. *Mol Ther* 2005;12:1072–82.

111. Min SH, Molday LL, Seeliger MW et al. Prolonged recovery of retinal structure/function after gene therapy in an Rs1h-deficient mouse model of x-linked juvenile retinoschisis. *Mol Ther* 2005;12:644–51.

112. Hauswirth WW, Aleman TS, Kaushal S et al. Treatment of leber congenital amaurosis due to RPE65 mutations by ocular subretinal injection of adeno-associated virus gene vector: Short-term results of a phase I trial. *Hum Gene Ther* 2008;19:979–90.

113. Donsante A, Miller DG, Li Y et al. AAV vector integration sites in mouse hepatocellular carcinoma. *Science* 2007;317:477.

114. Hacein-Bey-Abina S, Von Kalle C, Schmidt M et al. LMO2-associated clonal T cell proliferation in two patients after gene therapy for SCID-X1. *Science* 2003;302:415–9.

115. Goncalves MA. Adeno-associated virus: From defective virus to effective vector. *Virol J* 2005;2:43.

116. Yang GS, Schmidt M, Yan Z et al. Virus-mediated transduction of murine retina with adeno-associated virus: Effects of viral capsid and genome size. *J Virol* 2002;76:7651–60.

117. Kwon I, Schaffer DV. Designer gene delivery vectors: Molecular engineering and evolution of adeno-associated viral vectors for enhanced gene transfer. *Pharm Res* 2008;25:489–99.

118. Maheshri N, Koerber JT, Kaspar BK, Schaffer DV. Directed evolution of adeno-associated virus yields enhanced gene delivery vectors. *Nat Biotechnol* 2006;24:198–204.

119. Koerber JT, Klimczak R, Jang JH, Dalkara D, Flannery JG, Schaffer DV. Molecular evolution of adeno-associated virus for enhanced glial gene delivery. *Mol Ther* 2009;17:2088–95.

120. Klimczak RR, Koerber JT, Dalkara D, Flannery JG, Schaffer DV. A novel adeno-associated viral variant for efficient and selective intravitreal transduction of rat Muller cells. *PLoS One* 2009;4:e7467.

121. LaVail MM, Yasumura D, Matthes MT et al. Protection of mouse photoreceptors by survival factors in retinal degenerations. *Invest Ophthalmol Vis Sci* 1998;39:592–602.

122. Chaum E. Retinal neuroprotection by growth factors: A mechanistic perspective. *J Cell Biochem* 2003;88:57–75.

123. Sieving PA, Caruso RC, Tao W et al. Ciliary neurotrophic factor (CNTF) for human retinal degeneration: Phase I trial of CNTF delivered by encapsulated cell intraocular implants. *Proc Natl Acad Sci USA* 2006;103:3896–901.

124. Petrs-Silva H, Dinculescu A, Li Q et al. High-efficiency transduction of the mouse retina by tyrosine-mutant AAV serotype vectors. *Mol Ther* 2009;17:463–71.

125. Li Q, Miller R, Han PY et al. Intraocular route of AAV2 vector administration defines humoral immune response and therapeutic potential. *Mol Vis* 2008;14:1760–9.

126. Streilein JW. Limitations in the study of immune privilege in the subretinal space of the rodent. *Invest Ophthalmol Vis Sci* 1999;40:3069.

127. Busskamp V, Picaud S, Sahel JA, Roska B. Optogenetic therapy for retinitis pigmentosa. *Gene Ther* 2012;19:169–75.

128. Jones BW, Watt CB, Frederick JM et al. Retinal remodeling triggered by photoreceptor degenerations. *J Comp Neurol* 2003;464:1–16.

129. Milam AH, Li ZY, Fariss RN. Histopathology of the human retina in retinitis pigmentosa. *Prog Retin Eye Res* 1998;17:175–205.

130. Lakhanpal RR, Yanai D, Weiland JD et al. Advances in the development of visual prostheses. *Curr Opin Ophthalmol* 2003;14:122–7.

131. Chen SJ, Mahadevappa M, Roizenblatt R, Weiland J, Humayun M. Neural responses elicited by electrical stimulation of the retina. *Trans Am Ophthalmol Soc* 2006;104:252–9.

132. Zrenner E. Will retinal implants restore vision? *Science* 2002;295:1022–5.

133. Weiland JD, Liu W, Humayun MS. Retinal prosthesis. *Annu Rev Biomed Eng* 2005;7:361–401.

134. Banghart M, Borges K, Isacoff E, Trauner D, Kramer RH. Light-activated ion channels for remote control of neuronal firing. *Nat Neurosci* 2004;7:1381–6.

135. Choi SY, Sheng Z, Kramer RH. Imaging light-modulated release of synaptic vesicles in the intact retina: Retinal physiology at the dawn of the post-electrode era. *Vision Res* 2005;45:3487–95.

136. Szobota S, Gorostiza P, Del Bene F et al. Remote control of neuronal activity with a light-gated glutamate receptor. *Neuron* 2007;54:535–45.

137. Volgraf M, Gorostiza P, Numano R, Kramer RH, Isacoff EY, Trauner D. Allosteric control of an ionotropic glutamate receptor with an optical switch. *Nat Chem Biol* 2006;2:47–52.

138. Nagel G, Szellas T, Huhn W et al. Channelrhodopsin-2, a directly light-gated cation-selective membrane channel. *Proc Natl Acad Sci USA* 2003;100:13940–5.

139. Tomita H, Sugano E, Yawo H et al. Restoration of visual response in aged dystrophic RCS rats using AAV-mediated channelopsin-2 gene transfer. *Invest Ophthalmol Vis Sci* 2007;48:3821–6.

140. Bowes C, Li T, Danciger M, Baxter LC, Applebury ML, Farber DB. Retinal degeneration in the rd mouse is caused by a defect in the beta subunit of rod cGMP-phosphodiesterase. *Nature* 1990;347:677–80.

141. D'Cruz PM, Yasumura D, Weir J et al. Mutation of the receptor tyrosine kinase gene Mertk in the retinal dystrophic RCS rat. *Hum Mol Genet* 2000;9:645–51.
142. Doroudchi MM, Greenberg KP, Liu J et al. Virally delivered channelrhodopsin-2 safely and effectively restores visual function in multiple mouse models of blindness. *Mol Ther* 2011;19:1220–9.
143. Marc RE, Jones BW, Watt CB, Vazquez-Chona F, Vaughan DK, Organisciak DT. Extreme retinal remodeling triggered by light damage: Implications for age related macular degeneration. *Mol Vis* 2008;14:782–806.
144. Strettoi E, Pignatelli V, Rossi C, Porciatti V, Falsini B. Remodeling of second-order neurons in the retina of rd/rd mutant mice. *Vision Res* 2003;43:867–77.
145. Gargini C, Terzibasi E, Mazzoni F, Strettoi E. Retinal organization in the retinal degeneration 10 (rd10) mutant mouse: A morphological and ERG study. *J Comp Neurol* 2007;500:222–38.
146. Busskamp V, Duebel J, Balya D et al. Genetic reactivation of cone photoreceptors restores visual responses in retinitis pigmentosa. *Science* 2010;329:413–7.
147. Gorostiza P, Isacoff EY. Nanoengineering ion channels for optical control. *Physiology (Bethesda)* 2008;23:238–47.
148. Fortin DL, Banghart MR, Dunn TW et al. Photochemical control of endogenous ion channels and cellular excitability. *Nat Methods* 2008;5:331–8.
149. Caporale N, Kolstad KD, Lee TD et al. LiGluR restores visual responses in rodent models of inherited blindness. *Mol Ther* 2011;19:1212–9.

Section IV

Drug Delivery Systems

14 Drug Molecule Characteristics and Their Impact on Anterior versus Posterior Segment Drug Delivery Strategies

Kay D. Rittenhouse

CONTENTS

14.1 INTRODUCTION

Now more than ever, focus has been intensified on understanding ocular physiology and factors important in the delivery of therapeutics to the eye. Every year, the number of conferences, papers, dedicated journals, book chapters, and textbooks that focus on drug delivery to the eye continues to increase. Using a Google search engine and the key words "ophthalmic drug delivery" or "ocular drug delivery" constrained by the year 2000 and beyond, the following statistics were returned: conferences, 1,700+ hits; book chapters or journals, 8,250 hits; papers, 14,800 hits; web sites, 71 hits. Statistics such as these reinforce the premise of the importance and high degree of interest for novel and improved drug delivery systems as an ever more important component of effective pharmacotherapy.

The rationale for the high demand for developing novel options for delivery of drugs to the eye is based on the need to progress from drug delivery concerns discovered in earlier research on topically administered drugs. Work by a number of researchers has identified external adnexa and intraocular regions that may impact drug disposition and are thus important for optimizing delivery of topically administered drugs (Wang et al. 1991; Lee et al. 1993; Morrison et al. 1996). The recent entry of locally administered pharmaceutical treatments targeting the posterior segment with vitreous placement, such as Vitrasert® (ganciclovir non-bioerodible intravitreal device, approved in 1996), Retisert® (fluocinolone acetonide non-bioerodible intravitreal device, approved in 2005), Macugen® (pegaptanib sodium, a pegylated aptamer administered intravitreally [IVT], approved in 2004 for wet age related macular edema [AMD]), Lucentis® (vascular endothelial growth factor-A [VEGF-A] Fab administered IVT, approved in 2005 for wet AMD, and in 2011 for macular edema, retinal vein occlusion [RVO], etc.), and Ozurdex® (a dexamethasone bioerodible IVT implant approved in 2009 for macular edema following branch retinal vein occlusion [BRVO] or central retinal vein occlusion [CRVO]), as well as off-label use of glucocorticoids, like IVT-administered triamcinolone acetonide, has renewed interest in pursuing sustained ocular delivery. The value of such an approach includes a reduction in invasive injection procedures and avoidance of intraocular infection and trauma to the eye as well as decreased treatment burden on treating physicians and patients by reducing the patient clinic visit frequency per year.

14.2   SOPHISTICATION OF DRUG MODALITIES

Other drivers for the explosion in the interest and pursuit of effective ocular drug delivery modalities are the innovative classes of molecules that have been identified and developed in recent years. Biologicals such as Mab's, Fab's, fusion proteins, peptides, aptamers, small interfering RNA (siRNA), locked nucleic acids (LNAs), ribozymes, DNA, and other macromolecules have changed the landscape regarding specific challenges to be surmounted in drug delivery to ocular disease targets. Before the arrival of approved IVT administered drugs for wet AMD, apart from anti-inflammatory or anti-infective small-molecule drugs to treat uveitis or endophthalmitis, drug delivery problem solving was largely focused on the rule of five issues (MW <500 Da, partition coefficient, polar surface area, aqueous solubility, charge, etc.) (Ghose et al. 1999; Lipinski et al. 2001) relevant for small molecules, or other physicochemical properties that offer challenges in oral administration, which may also be important for drug access to ocular surface regions such as the cornea or the conjunctiva following topical administration or to the blood–retinal barrier (BRB) or blood–aqueous barrier (BAB) through the systemic route. Moreover, advances in pharmacokinetic/pharmacodynamic (PK/PD) analysis, with greater understanding of both anatomic and physiological barriers encountered following ocular drug delivery (Rittenhouse et al. 1999; Robinson et al. 2007), have elucidated the importance of developing better drug delivery platforms that will provide sufficient target concentrations to relevant ocular regions and decrease invasive dosing procedures.

14.3 DRUG CHARACTERISTICS VERSUS ANATOMIC AND PHYSIOLOGICAL FACTORS IN OCULAR DRUG DELIVERY

14.3.1 ANTERIOR SEGMENT DRUG DELIVERY

Historically, ophthalmology topical therapies involved delivery of drops of solutions or suspensions to the corneal or conjunctival surfaces of the eye, and thus, a large body of work was developed to understand the intraocular disposition of drugs with a focus on anterior segment pharmacokinetics (Hussain et al. 1980; Maurice and Mishma 1984; Chiang and Shoenwald 1986; Lee 1993; Lee et al. 1993; Schoenberg 1993; Vouri et al. 1993; Ohtori et al. 1998; Rittenhouse et al. 1998; Rittenhouse et al. 1999). Species such as rabbits were used to model ocular disposition of agents that are topically delivered and whose pharmacologic action involved modulation of aqueous humor formation and/or turnover, exit pathways through the trabecular meshwork (Schmitt et al. 1981; Sato et al. 1996). Because rabbits do not have the uveoscleral route for aqueous humor exit, drugs affecting this pathway must be examined in alternative species such as cats or nonhuman primates (Toris 2008). An important factor that was characterized by such studies of drug delivery to the anterior segment includes the impact of hydrophilicity versus lipophilicity on ocular penetration of corneal versus conjunctival tissues of the ocular surface. Lee et al. (1993) demonstrated that a series of beta-blockers with known lipophilicity have log P-dependent differences in ocular bioavailability and systemic bioavailability following topical drug administration. Timolol is a classic example of a water-soluble beta-blocker with a known systemic impact and has a log P value in the middle of the range. Low- or high-log P compounds were less likely to have increased systemic exposure of drugs from a similar structural series.

A second aspect of topical drug delivery focused on how ocular surface barriers contribute to apparent absorption and elimination kinetics of drugs based on charge, the molecular weight of the molecule, and physiological factors such as vascular or aqueous perfusion rates and their proximity to the barriers. The recent ARVO-sponsored Summer Eye Research Conference of 2009 (ARVO 2009 SERC) focused on retinal drug delivery, and a considerable number of presenters addressed issues regarding ocular barriers and their impact on retinal drug disposition (Edelhauser et al. 2010). Lessons obtained from the study of ocular surface barrier function have been the bridge to advancing our understanding of ocular barriers that impact back-of-the-eye targets as well.

Limitations imposed by animal species on animal to human predictions remain a significant conundrum. Rabbit eyes are still considered a robust model for evaluating ocular drug disposition, especially for anterior segment pharmacokinetics (PK), largely due to their similar size to humans and accessibility (Edman 1993). However, there are ocular characteristics in rabbit eyes that are dissimilar to humans and contribute to artifacts that confound translation to human ocular PK. Such characteristics include differences in blink rates; tear turnover rates; corneal thickness;

corneal to conjunctival surface area differences (Edman 1993); and species-selective differences in responses to insult, inflammation, or infection (Rittenhouse et al. 1998, 1999). Alternative species such as dogs, rodents, pigs, and nonhuman primates have been used to evaluate ocular PK, and each one has specific anatomic or physiologic properties that must be taken into account to manage animal to human predictions.

Analogous to the importance of animal species selection for confidence in human translation for topical drug delivery to the eye is the recognition of the impact the pigment has on intraocular drug disposition. Historically, New Zealand White rabbits, an albino species, were routinely used for ocular PK studies, due to access and ease of visualizing the anterior segment or other associated regions. Studies by Larsson (1993) and others (Lee 1993; Acheampong et al. 1995; Rittenhouse et al. 1998) have shown that pigment binding of lipophilic compounds can greatly alter ocular disposition of drugs, prolonging the apparent elimination t ½ many folds, reducing C_{max}, and in some cases may impact on the tissue-specific concentrations of a drug retained in various pigmented tissues within the eye such as the iris-ciliary body and the choroid (Dayhaw-Barker 2002). Moreover, ocular pigments are subcategorized by type of melanin (i.e., eumelanin vs. pheomelanin) with associated beneficial or deleterious effects on ocular drug disposition (Dayhaw-Barker 2002). Anesthesia may introduce artifacts in ocular PK estimates. Rittenhouse et al. (1999) found that, following topical administration of a drug to rabbits, aqueous humor peak concentrations of the drug were higher in anesthetized animals than conscious animals. These results are consistent with a time dependency for peak tissue concentrations post-euthanasia, due to cessation of blood flow and thus the removal of perfusion barriers for greater tissue diffusion of drugs. Similar results were described by Robinson et al. when following a sub-Tenon's injection of triamcinolone to rabbits and selective blocking of relevant ocular vascular beds. At a 10-mg dose, no triamcinolone was detected in the vitreous in live animals; detectable levels were observed at 20 mg. Following euthanasia, and in contrast to the live animal results, detectable vitreous levels of triamcinolone were observed (Robinson et al. 2007) at 10 mg. These results are indicative of an ocular vascular perfusion barrier in animals that is eliminated after euthanasia. Factors such as these must be taken into account when comparing and evaluating animal ocular drug disposition data from the literature.

Ocular enzymatic activity and transport systems have been shown to alter the drug disposition of topically administered drugs. No more than ~2% of topically administered aqueous-based drug products will be bioavailable to the eye (Chastain 2003). Thus, the impact of ocular barriers, including enzymatic degradation and solute transport systems, is appreciable, further reducing intraocular access using the topical route. Researchers conducting studies using in vivo and in vitro models have identified enzymatic mediators and transport systems impactful on drug delivery to the anterior segment. Esterase activity was identified in cornea and conjunctiva and was employed for prodrug strategies to circumvent challenges in delivering drug candidates to the anterior segment (Bodor and Buchwald 2005). Drugs such as the isopropyl ester prodrug latanoprost, an F2alpha prostanoid (Xalatan®), and loteprednol etabonate (Lotemax® and Alrex®), a so-called soft steroid, are examples that have exploited esterase enzymatic degradation of molecules to reduce adverse effects of drugs or to increase drug penetration into the eye through specific ocular tissue barriers such as the cornea (Bodor and Buchwald 2005). Works by Hosoya et al. (2005) and Dey et al. (2003)

have extensively examined important transport systems resident in conjunctival and corneal tissues in the eye, such as ion-coupled solute transport systems that transport amino acids (large neutral amino acid transporter 1 [LAT1] and LAT2), glucose (glucose transporter 1 [GLUT1]), nucleosides, dipeptides (peptide transporter 1 [PEPT1]/ PEPT2), and multidrug resistance–associated transporter systems such as MRP and P-glycoprotein (PGP), contributing to the advance in the current understanding of specific mechanisms involved in modulating drug delivery to selected ocular targets.

14.3.2 Posterior Segment Drug Delivery

Development of drug therapies for retinopathies has provided the impetus for studies to understand anatomic, physiologic, and drug characteristics that a effect drug delivery to the back of the eye. Gerald Chader, one of the conference presenters at the ARVO 2009 SERC, summarized the core issues that need to be addressed for successful programs in posterior segment pharmaceutical intervention (Edelhauser et al. 2010): (1) development of effective products, (2) identification and implementation of the best drug delivery approach, (3) appropriate animal model selection for testing drug efficacy and safety, (4) identification of a patient sample (or population) with well-developed clinical study design and incorporation of relevant end points, and (5) commercial considerations—market partners and finance. Under this broader context, the specific aspects relevant for this successful strategy in the development of drugs for treating back of the eye diseases are discussed in the following sections.

14.3.2.1 Posterior Segment Physiological Considerations for Drug Delivery

The vitreous humor, a gel-like media composed of water (~99%), chondroitin sulfates, glycosaminoglycans, hyaluronic acid, and various electrolytes, is the matrix into which small molecular entities (SMEs) or large molecular entities (LMEs) are delivered through intravitreous injection for access to the retina (Duvvuri et al. 2005; Le Goff and Bishop 2008). Unlike the aqueous humor, the aqueous media is formed by the nonpigmented epithelial cells of the ciliary body, and it is not replenished (Bishop 2000). Simplistically modeled in the past as a virtual "unstirred" compartment (Figure 14.1), recent studies and sophisticated modeling (Stay et al. 2003; Wilson et al. 2011) have elucidated how the molecular structure of the vitreous undergoes age-related changes such as liquefaction, which significantly impacts intraocular drug disposition. Moreover, structural entities within the vitreous, such as type IX collagen molecules that comprise fibril-like entities and specific anatomic regions of vitreous–retinal adhesion, introduce further complexity to the accurate prediction of ocular pharmacokinetics of drugs, depending on the drug's physicochemical and structural characteristics (Le Goff and Bishop 2008).

14.3.2.2 Systemic versus Ocular PK

Anatomic factors influence intraocular exposure to drugs administered systemically. These include typical factors that modulate small molecule absorption and clearance: e.g., charge, molecular weight, or size; whether the molecule is a typical substrate for active transport or efflux systems such as PGP, MRP, or ion channels; and the metabolism pathway is primary or secondary, or whether the degree of lipophilicity

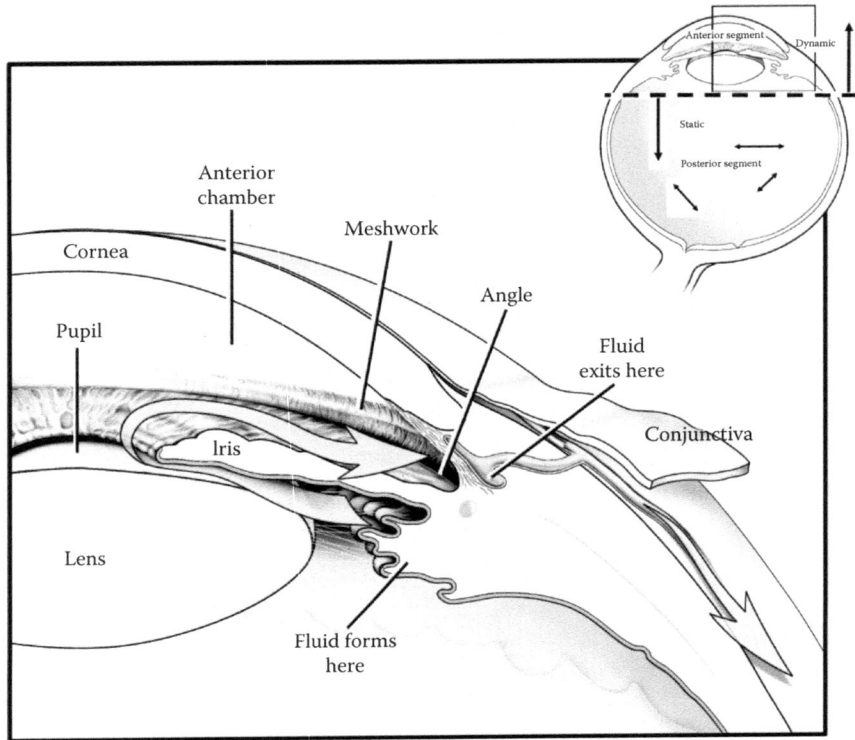

FIGURE 14.1 Diagram of the eye depicting the static posterior segment environment versus the dynamic anterior segment environment (upper right corner). Due to regeneration of the aqueous humor at the ciliary body and efflux from the anterior chamber through the trabecular meshwork or uveoscleral route, the aqueous flow influences the disposition of drugs through volume dilution, aqueous turnover, and active and passive transport systems. Although the vitreous is not regenerated once formed, the so-called static environment contains more complex structures than in the aqueous, such as glycosaminoglycans or collagen that may provide diffusional barriers or distribution channels within the vitreous. Illustrative adaptation. (From the National Eye Institute, National Institutes of Health Ref#: NEA11. With permission.)

directly affects ocular drug disposition. Drugs accessing ocular target tissues in the front versus back of the eye encounter a number of challenges for achieving pharmacologically relevant concentrations that depend on the route of local administration to the eye. Anatomic barriers such as BRB and BAB versus permissive barriers of vascular tissues such as fenestrated choriocapillaris (Edelhauser et al. 2010), are important determinants of tissue access for molecules.

14.3.2.3 Transport across Cell Membranes

Active transport involves drug transport against a concentration (nonelectrolyte) or electrochemical (electrolyte) gradient; it requires energy (ATP) and in some cases may involve symport-mediated (unidirectional) or antiport-mediated (opposite directional) travel into or out of the cell. Facilitated diffusion has a weak association of the carrier protein with small molecules, and generally, the direction for transport of the solute is based on the electrolyte gradient. In some cases, transport may go against a concentration gradient. Diffusion of solute is nonenergy dependent, going with the concentration (nonelectrolytes) or electrochemical gradients. Small molecule transport across cell membranes is governed by diffusion (Fick's law), active and facilitated transport, energy-dependent antiport and symport, or channels/gated transport across membranes. In the anterior segment, the ciliary body, a virtual syncytium, with the apical faces of nonpigmented and pigmented epithelial bilayers toward each other like mirror images, contain multiple transport systems. Representative transport systems are depicted in Figure 14.2, with heavier representation of transporters on the basolateral face of the bilayers. Both SMEs and LMEs can be endocytosed. Particle size dictates the class of endocytic pathway exploited. Phagocytosis can allow for ingestion of large particles, senescent cells, debris >250 nm. Typical cell populations that phagocytize cells are macrophages and neutrophils. Pinocytosis or

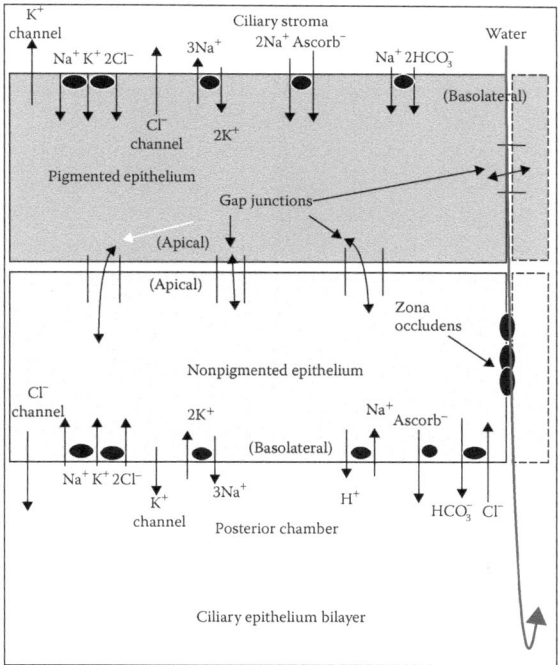

FIGURE 14.2 The ciliary epithelium bilayer is a unique anatomic tissue arrangement that has been described as a syncytium (McLaughlin et al. 1998). The unusual arrangement is such that the apical domains of pigmented versus nonpigmented epithelia face each other and contain gap junctions allowing communication between the layers. The orientation of and enrichment in the number of the transporters and channels localized on the basolateral face of the bilayer provide the electrolytic and concentration gradient that directs the flux of fluid from the ciliary stroma toward the posterior chamber. The arrow at the right shows the directionality of fluid flow down these gradients in conjunction with the hydrostatic pressure gradient and paracellular flow.

fluid ingestion can engulf <150 nm; macrophages can engulf up to 25% of their cell volume per hour (Helenius et al. 1983). Fibroblasts engulf at a lower rate. Receptor-mediated endocytic routes can involve clathrin-coated or caveolin-coated pits containing specific receptors recognizing as many as 1000 different substrates (Bretscher 1984).

A number of factors should be considered for optimizing the influx and egress of xenobiotics, including molecular size, transport systems, and route. For systemically administered drugs, blood vessel diameter limitations contribute to the influx and egress of xenobiotics. In the vascular endothelium, molecules with small pore sizes of 45 Å or less in diameter may involve paracellular routes of delivery. If molecular weight only was important, oligonucleotides could be transported through the paracellular route. Because of the contribution of the hydrodynamic radii of globular proteins, which are charged, such molecules are larger than uncharged polymers of similar mass. Thus, a dramatic decline in vascular permeability occurs with molecular size approaching ~5 nm or 50 Å (Juliano et al. 2009).

An important factor influencing the transit of macromolecules in blood or matrices of selected fluid-filled compartments or organs is the size of macromolecules. How molecular size impacts the transit of drugs in blood is a function of molecular or hydrodynamic radius (Armstrong et al. 2004), the conformability of molecule, vascular resistance/compromise of the capillary, the "stickiness" of the molecule in transit, and the rate of blood or fluid flow in transit (e.g., diapedesis, platelet aggregation, or enhancement of vascular permeability). Macromolecular clearance pathways or barriers to the entry of SMEs and LMEs may incorporate the body's processes of recognition such as opsonization (opsonins—immunoglobulins, complement components, and serum proteins), agglomeration, engulfment by specialized scavenger cells—for example, activated macrophages and immunity pathways—humoral and cellular, and specialized target organs with enhanced fenestration of vascular capillaries (100–200 nm diameter): liver, kidney, and spleen. The reticuloendothelial system (liver: Kupffer cells; kidney: proximal tubule cells; and spleen: macrophages) provides a rapid clearance mechanism for substrates >5 nm in diameter (Scherphof 1991). Figure 14.3 depicts blood component size parameters that elucidate specific physiological barriers to consider for optimizing drug transport for LMEs.

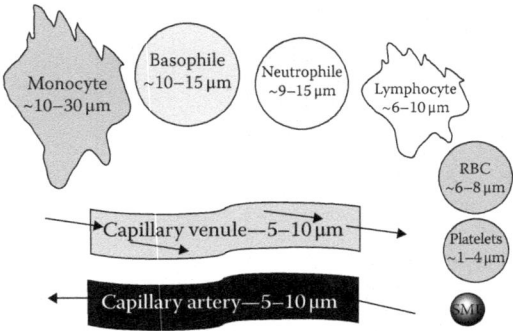

FIGURE 14.3 Size dimensions of various blood components and vessels.

14.4 RNA INTERFERENCE AND GENE DELIVERY CHALLENGES

Factors for increasing transfection efficiency involve specific optimization of endocytosis (Dalby et al. 2004). Relevant aspects of focus include cellular uptake, vesicular trafficking, and endosomal release. Each may control the downstream response of RNAi molecules such as oligonucleotides, for example, based on what is rate limiting in the process (Nguyen et al. 2008). Lack of lysosomal escape effectively can result in lysosomal degradation of the oligonucleotide if the molecule cannot escape endosomal compartments. RNAi onset and duration of action is a function of cell type, mitotic rate, and turnover (apoptosis) (Bartlet and Davis 2006; White et al. 2009). Ocular-specific cell types of relevance in the eye include neuronal cells such as those resident in the retina (ganglion cells, bipolars, amacrines, horizontal cells, and photoreceptors) that are terminally differentiated and not renewed, stromal cells such as the corneal endothelium, mucosal cells such as those of the conjunctiva and goblet cells, and the endothelial cells of the retinal and choroidal capillaries.

Typical onset of activity rates for various SMEs versus macromolecules are described in several publications (Jenne et al. 1998; O'Neill et al. 2009). Small molecules such as glucocorticoids and mechanisms of nongenomic versus genomic pathways may exhibit effects from within minutes (nongenomic) to days (genomic) (Croxtall et al. 2002; Revollo and Cidlowski 2009). Specific receptor-mediated endocytosis (RME) pathways have been examined, and the time frame for passage of substrate through the pathway can occur within ~20 minutes with exit from endosomes into the cytoplasm within 2 hours (Jenne 1998; O'Neill et al. 2009). Lipophilic SMEs' diffusion is rapid. Monoclonal antibody antagonists may reduce target circulating ligand or downregulate target receptors within a week time frame (Shah and Del Priore 2009). Shah and Del Priore conducted a post hoc analysis of wet AMD patients using sequential macular volume and central foveal point thickness (CFPT) measurements using optical coherence tomography (OCT) to estimate onset and duration of Lucentis, a Fab, versus Avastin, a monoclonal antibody, both against VEGF-A. Macular volume was reduced in these subjects following a single intravitreal injection of Lucentis as early as 2 weeks post-dosing, the earliest observation time point, confirming relatively rapid onset of action.

siRNA silencing of target genes has been documented to silence genes within hours to days (Rittenhouse et al. 2010). siRNA is an RNA duplex ranging from 15 to 23 nucleotides in length (Whitehead et al. 2009; Rittenhouse et al. 2010). Entry into cells by siRNA is theorized to occur through RME (Dalby et al. 2004). Because in general siRNAs are too large (~13 kDa or larger) and negatively charged for facile cell entry, transfection is facilitated by usage of positively charged lipids (Whitehead et al. 2009) or physical delivery techniques such as electroporation (SABiosciences Corporation et al. 2009), although some target tissues such as brain, lung, and eye have shown messenger RNA (mRNA) modulation following "naked" siRNA administration. In the cytosol of the cell, an enzymatic protein called Dicer cuts the double-stranded RNA (dsRNA) or siRNA from ~30 in length to 25 nt or less in length. A helicase called Ago2 located in the intracellular machinery called the RNA-induced silencing complex (RISC) unwinds the RNA duplex such that the antisense strand, which is thermodynamically more stable than the passenger or sense strand, can then access RISC. Upon association with the antisense strand, RISC becomes activated and results in downstream silencing of translation of the target protein (Figure 14.4).

Localization within the cell for the RISC complex is still under study. Some have proposed that RISC resides in the so-called P-bodies or "processing center" vacuoles within the cell, and others have suggested that the passenger or sense strand may not be destroyed and thus may also access the RISC machinery (Sen and Blau 2006). The antisense strand, after cell entry, self-associates to its complementary strand *in vivo* and thus also is processed through RISC (Sen and Blau 2006), which may provide rationale for the perceived less-efficient activity of antisense compared to siRNA (Elbashir et al. 2001).

Aspects of the RNAi molecule or approach that influence the onset and persistency of activity include stability of the substrate against enzymatic degradation, the transfection efficiency, the copy number available to silence, facility of endosomal/lysosomal escape, the relationship between the onset of gene silencing versus desired downstream phenotypic responses, and the apparent hysteresis—delay in phenotypic

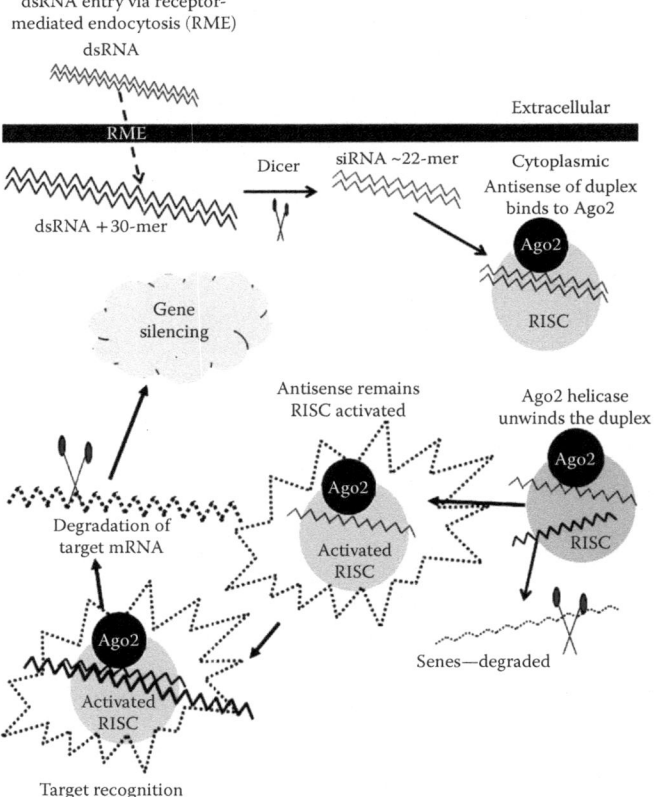

FIGURE 14.4 Diagram of the RNAi mechanisms used in the RISC, RNA-induced silencing complex.

outcomes. Bartlet and Davis (2006), in their studies on siRNA's kinetics, confirmed with bioluminescent imaging (firefly luciferase protein turnover), used to estimate transfection efficiency in rapidly versus nondividing cells, that in rapidly dividing cells, siRNA silencing of luciferase resulted in luciferase protein levels returning to pretreatment values in <1 week, but in nondividing fibroblasts, >3 weeks passed before return to basal levels of protein was observed. In their *in vivo* studies, gene knockdown in subcutaneous tumors of A/J mice lasted ~10 days; gene knockdown in nondividing hepatocytes (BALB/c mice) took up to 3–4 weeks. Based on their work, it appears that siRNA half-life does not dictate the duration of gene silencing; although chemical modification (2′F ribose) increased systemic enzymatic stability, this did not alter/increase the duration of gene silencing. Dilution from cell division dictates the persistence of silencing of target proteins. Thus, neuronal cells such as in the retina, which do not divide, should in theory continue to silence genes through siRNA mechanisms as long as RISC remains activated and accessible in these cells (Bartlet and Davis 2006).

One then must speculate as to why there continue to be challenges developing therapies using siRNA for wet AMD. One aspect of the debate regarding siRNA mechanisms is whether the so-called naked siRNAs can indeed enter cells without significant modifications such as use of transfecting agents, positively charged lipophilic media, PEGylation, or the introduction of electroporation or iontophoresis (Matsuda and Cepko 2004; SABiosciences Corporation et al. 2009). Preclinical studies have shown knockdown of relevant target genes in the anterior segment (Hou et al. 2009) and retina using RNAi (Chen et al. 2009; Hirano et al. 2010; Rittenhouse et al. 2010, 2011), supporting the position that the eye and its immune privilege characteristics make it a receptive target organ for siRNA and gene delivery. The specificity of gene silencing activity using siRNA has been challenged by recent papers (Jackson and Linsey 2010), where it appears that the efficacy of the phenotype may occur independent of a true siRNA mechanism, due to nonspecific toll-like receptor 3 (TLR3) activation (Kleinman et al. 2008). However, others have shown through robust studies that incorporated nonselective siRNA negative controls that siRNA selectively knocks down the gene of interest and results in efficacy in murine and nonhuman primates (Shen et al. 2006; Chen et al. 2009).

Gene delivery involves the safe and effective delivery of a specific, efficacious, and safe genomic package to the desired population of cells of sufficient quantity and potency to elicit robust expression, coupled with an efficient and sufficiently persistent level and duration of expression. During the course of investigating the mechanistic importance of target genes for various ocular diseases, relevant genes have been selected to employ gene delivery approaches for improving the prognosis of various diseases of the eye. Gene delivery is currently being explored in ophthalmology by many players for (1) replacing gene mutations (Bainbridge et al. 2006), (2) introducing genes that produce proteins that protect neurons (Van Adel et al. 2003), and (3) converting the function of specific retinal cell populations to perform the function of another group that is dysfunctional (Tomita et al. 2009). Important components for a successful gene therapeutic approach include (1) selection of the relevant gene target, (2) selection of an appropriate a gene delivery vector, (3) a gene delivery platform that will access the relevant ocular target tissue, (4) stability of gene expression (i.e., does not mutate), (5) sufficiency of the magnitude of gene expression, (6) persistency of gene expression, (7) translation of preclinical models to predict successful clinical application, and lastly and importantly, (8) the mitigation strategy for "self" versus "foreign" immunological responses to gene delivery for effects on efficacy and safety.

Viral and nonviral gene transfer vectors have been evaluated in preclinical and clinical studies of ophthalmic disease. Viral-derived vectors include adenoviral, adeno-associated viral (AAV), retroviral, lentiviral (a diploid subclass of retrovirus that is enveloped and surrounded by a capsid shell), alphaviral, herpes viral, and baculoviral vectors (Walther and Stein 2000; Van Adel et al. 2003). Although viral vectors are generally more efficient and can deliver greater DNA payloads than other delivery systems, the heightened concern for long-term safety and the problems encountered due to immunogenicity have not been overcome. Tropism, the specific feature of the capsid coat causing it to track to specific cell types of a target organ, has been comprehensively evaluated by Choi et al. (2005) and reviewed by Van Adel et al. (2003). For the AAV vectors, there are more than eight different AAV serotypes, with specialized tracking to target cells that have been enlisted for robust and persistent gene expression and profiles of differing kinetics of expression.

Nonviral vectors such as plasmids, cationic lipids, or polymers that complex with DNA, similar to those exploited for siRNA delivery enhancement, and that condense their size for access to the nucleus (<25-nm pore), like nanoparticles, or physical delivery modalities such as electroporation or pressure (Van Adel et al. 2003) have also been explored. Generally, nonviral vector gene transfer constructs are considered safer but much less efficient (Lee et al. 2009).

14.5 SUMMARY

Characteristics of SMEs versus LMEs provide significantly affects on the pharmacology of onset/duration by different mechanisms. Anatomic and physiological barriers against SME versus LME have differing affects on intraocular disposition. Contributing factors impacting PK/PD include molecular versus hydrodynamic size and physicochemical modifications that affect enzymatic stability (LME) and clearance pathways (SME vs. LME). Current understanding of the important properties of select small molecules and macromolecules in the context of predicting the onset/duration of action was reviewed with a focus on RNAi modes of action. Ophthalmology remains on the forefront of macromolecular therapeutic interventions. Due to the complexity of the therapeutic approaches, however, innovation must continue to advance for delivery of such molecules to the eye for the treatment of sight-threatening diseases.

REFERENCES

Acheampong AA, Shackleton M, Tang-Liu DD. 1995. Comparative ocular pharmacokinetics of brimonidine after a single dose application to the eyes of albino and pigmented rabbits. *Drug Metab Dispos.* 23:708–712.
Armstrong JK, Wenby RB, Meiselman HJ, Fisher TC. 2004. The hydrodynamic radii of macromolecules and their effect on red blood cell aggregation. *Biophys J.* 87:4259–4270.

Bainbridge JWB, Tan MH, Ali RR. 2006. Gene therapy progress and prospects: The eye. *Gene Ther.* 13:1191–1197.

Bartlet DW, Davis ME. 2006. Insights into the kinetics of siRNA-mediated gene silencing from live-cell and live-animal bioluminescent imaging. *Nucleic Acids Res.* 34:322–333.

Bishop PN. 2000. Structural macromolecules and supramolecular organisation of the vitreous gel. *Prog Retin Eye Res.* 19:323–344.

Bodor N, Buchwald P. 2005. Ophthalmic drug design based on the metabolic activity of the eye: Soft drugs and chemical delivery systems. *AAPS J.* 7:E820–E833.

Bretscher MS. 1984. Endocytosis: Relation to capping and cell locomotion. *Science.* 224:681–685.

Chastain JE. 2003. General considerations in ocular drug delivery, in *Ophthalmic Drug Delivery Systems*, edited by Mitra AK, 2nd ed., vol. 130, 65. New York: Marcel Dekker Inc.

Chen J, Aderman CM, Willett KL et al. 2009. Suppression of retinal neovascularization by EPO siRNA in a mouse model of proliferative retinopathy. *Invest Ophthalmol Vis Sci.* 50:1329–1335.

Chiang CH, Schoenwald RD. 1986. Ocular pharmacokinetic models of clonidine 3H hydrochloride. *J Pharm Biopharm.* 14:175–211.

Choi VW, McCarty DM, Samulski RJ. 2005. AAV hybrid serotypes: Improved vectors for gene delivery. *Curr Gene Ther.* 5:299–310.

Croxtall JD, van Hal PT, Choudhury Q, Gilroy DW, Flower RJ. 2002. Different glucocorticoids vary in their genomic and non-genomic mechanism of action in A549 cells. *Br J Pharmacol.* 135(2): 511–519.

Dalby B, Cates S, Harris A, Ohki EC, Tilkins ML. 2004. Advanced transfection with Lipofectamine 2000 reagent: Primary neurons, siRNA, and high-throughput applications. *Methods.* 33:95–103.

Dayhaw-Barker P. 2002. Retinal pigment epithelium melanin and ocular toxicity. *Int J Toxicology.* 21:451–454.

Dey S, Anand BS, Patel J, Mitra AK. 2003. Transporters/receptors in the anterior chamber: Pathways to explore ocular drug delivery strategies. *Expert Opin Biol Ther.* 3:23–44.

Duvvuri S, Rittenhouse KD, Mitra AK. 2005. Microdialysis assessment of drug delivery systems for vitreoretinal targets. *Adv Drug Deliv Rev.* 57:2080–2091.

Edelhauser HF, Rowe-Rendleman CL, Robinson MR et al. 2010. Ophthalmic drug delivery systems for the treatment of retinal diseases: Basic research to clinical applications. *Invest Ophthalmol Vis Sci.* 51:5403–5420.

Edman P, editor. 1993. *Biopharmacetics of Ocular Drug Delivery.* Boca Raton, FL: CRC Press.

Elbashir SM, Harborth J, Lendeckel W et al. 2001. Duplexes of 21-nucleotide RNAs mediate RNA interferences in cultured mammalian cells. *Nature.* 411:494–498.

Ghose AK, Viswanadhan VN, Wendoloski JJ. 1999. A knowledge-based approach in designing combinatorial or medicinal chemistry libraries for drug discovery. *J Comb Chem.* 1:55–68.

Helenius A, Mellman I, Wall D et al. 1983. Endosomes. *Trends Biochem Sci.* 8:245–250.

Hirano Y, Sakurai E, Matsubara A et al. 2010. Suppression of ICAM-1 in retinal and choroidal endothelial cells by plasmid small interfering RNAs *in vivo. Invest Opthalmol Vis Sci.* 51:508–515.

Hosoya VHL, Lee VH, Kim KJ. 2005. Roles of the conjunctiva in ocular drug delivery: A review of conjunctival transport mechanisms and their regulation. *Eur J Pharm Biopharm.* 60:227–240.

Hou Y, Xing L, Fu S et al. 2009. Downregulation of inducible co-stimulator (ICOS) by intravitreal injection of small interefering RNA (siRNA) plasmid suppresses ongoing experimental autoimmune uveoretinitis in rats. *Grafes Arch Clin Exp Ophthalmol.* 247:755–765.

Hussain A, Hirai S, Sieg J. 1980. Ocular absorption of propranolol in rabbits. *J of Pharm Sci.* 69:738–739.

Jackson AL, Linsley PS. 2010. Recognizing and avoiding siRNA off-target effects for target identification and therapeutic application. *Nat Rev Drug Discov.* 9:57–67.

Jenne N, Rauchenberger R, Hacker U, Kast T, Maniak M. 1998. Targeted gene disruption reveals a role for vacuolin B in the late endocytic pathway and exocytosis. *J Cell Sci.* 111:61–70.

Juliano R et al. 2009. Biological barriers to therapy with antisense and siRNA oligonucleotides. *Mol Pharm.* 6:686–695.

Kleinman ME, Yamada K, Takeda A et al. 2008. Sequence- and target-independent angiogenesis suppression by siRNA via TLR3. *Nature.* 452: 591–597.

Larsson BS. 1993. Interaction between chemicals and melanin. *Pigment Cell Res.* 6:127–133.

Lee BW, Chae HY, Tuyen TT, Kang D, Kim HA, Lee M, Ihm SH. 2009. A comparison of non-viral vectors for gene delivery to pancreatic beta-cells: Delivering a hypoxia-inducible vascular endothelial growth factor gene to rat islets. *Int J Mol Med.* 23(6):757–762.

Lee VHL. 1993. Precorneal, corneal and postcorneal factors, in *Ophthalmic Drug Delivery Systems*, edited by Mitra AK, 69. New York: Marcel Dekker Inc.

Lee YH, Kompella UB, Lee VHL. 1993. Systemic absorption pathways of topically applied adrenergic antagonists in the pigmented rabbit. *Exp Eye Res.* 57:341–349.

Le Goff MM, Bishop PN. 2008. Adult vitreous structure and postnatal changes. *Eye (Lond).* 22:1214–1222.

Lipinski CA, Lombardo F, Dominy BW, Feeney PJ. 2001. Experimental and computational approaches to estimate solubility and permeability in drug discovery and development settings. *Adv Drug Del Rev.* 46:3–26.

Matsuda T, Cepko CL. 2004. Electroporation and RNA interference in the rodent retina *in vivo* and *in vitro*. *Proc Natl Acad Sci USA.* 101:16–22.

Maurice DM, Mishma S. 1984. Ocular pharmacokinetics, in *Handbook of Experimental Pharmacology*, edited by Sears ML, vol. 69, 32. New York: Springer-Verlag.

McLaughlin CW, Peart D, Purves RD, Carré DA, Macknight AD, Civan MM. 1998. Effects of HCO3- on cell composition of rabbit ciliary epithelium: A new model for aqueous humor secretion. *Invest Ophthalmol Vis Sci.* 39:1631–1641.

Morrison JC, Freddo TF. Anatomy, microcirculation and ultrastructure of the ciliary body, in *The Glaucomas*, 2nd ed. p. 125–138. Maryland Heights, MO: C.V. Mosby.

Nguyen T, Menocal EM, Harborth J et al. 2008. RNAi therapeutics: An update on delivery. *Curr Opin Mol Ther.* 10:158–167.

Ohtori R, Sato H, Fukuda S, Ueda T et al. 1998. Pharmacokinetics of topical betaadrenergic antagonists in rabbit aqueous humor evaluated with the microdialysis method. *Exp Eye Res.* 66:487–494.

O'Neill SK, Veselits ML, Zhang M, Labno C et al. 2009. Endocytic sequestration of the B cell antigen receptor and toll-like receptor 9 in anergic cells. *Proc Natl Acad Sci USA.* 106: 6262–6267.

Revollo JR, Cidlowski JA. 2009. Mechanisms generating diversity in glucocorticoid receptor signaling. *Ann NY Acad Sci.* 1179:167–178.

Rittenhouse KD, Hirakawa B, Huang W, Basile AS, Johnson T, Schachar R. 2010. Dose-related gene silencing of RTP801 with the siRNA PF04523655 in Long Evans rat models of STZ induced diabetes and laser induced CNV. *Invest Ophthalmol Vis Sci.* Supplement 51:6447.

Rittenhouse KD, Kalabat D, Yang A, Vicini P, Johnson TR, Huang W, Hirakawa B, Basile AS, Schachar RA. 2011. Characterization of regional RTP801 gene expression within the retina and the concentration-effect relationship of PF-655, an RTP801-silencing siRNA, following intravitreous administration to diabetic rats. *Invest Ophthalmol Vis Sci.* Supplement 52:5641.

Rittenhouse KD, Peiffer R, Pollack G. 1998. Evaluation of microdialysis sampling of aqueous humor for *in vivo* models of ocular absorption and disposition. *J Pharm Biomed Anal.* 16:951–959.

Rittenhouse KD, Peiffer R, Pollack G. 1999. Microdialysis evaluation of the ocular pharmaco-kinetics of propranolol in the conscious rabbit. *Pharm Res.* 16:736–742.

Robinson MR, Lee SS, Kim H et al. 2007. A rabbit model for assessing the ocular barriers to the transscleral delivery of triamcinolone acetonide. *Exp Eye Res* 82:479–487.

SABiosciences Corporation et al. 2009. siRNA delivery methods into mammalian cells: Gene function study guide in stem cells. *Pathways.* 9:10–11.

Sato H, Uchida N, Fukuda S, Yuko K et al. 1996. Pharmacokinetics of norfloxacin and lome-floxacin in domestic rabbit aqueous humor analyzed by microdialysis. *J Jpn Ophthalmol Soc.* 100:513–519.

Scherphof GL. 1991. in *Targeted Drug Delivery*, edited by Juliano RL, 285–313. Berlin: Springer.

Schmitt C, Lotti VJ, Le Dourec JC. 1981. Penetration of five beta-adrenergic antagonists into the rabbit eye after ocular instillation. *Albrecht von Graefes Arch Klin Ophthalmol.* 217:167–174.

Schoenwald RD. 1993. Ocular pharmacokinetics/pharmacodynamics, in *Ophthalmic Drug Delivery Systems*, edited by Mitra AK. New York: Marcel Decker Inc.

Sen GL, Blau HM. 2006. A brief history of RNAi: The silence of the genes. *FASEB J.* 20:1293–1299.

Shah AR, Del Priore LV. 2009. Duration of action of intravitreal ranibizumab and bevacizumab in exudative AMD eyes based on macular volume measurements. *Br J Ophthalmol.* 93:1027–1032.

Shen J, Samul R, Silva RL et al. 2006. Suppression of ocular neovscularization with siRNA targeting VEGR-1. *Gene Ther.* 13:225–234.

Stay MS, Xu J, Randolph TW, Barocas VH. 2003. Computer simulation of convective and diffusive transport of controlled-release drugs in the vitreous humor. *Pharm Res.* 20:96–102.

Tomita H, Sugano E, Isago H, Tamai M. 2009. Channelrhodopsins provide a breakthrough insight into strategies for curing blindness. *J. Genet.* 88: 409–415.

Toris C. 2008. Aqueous humor dynamics I: Measurement in animal studies, in *The Eye's Aqueous Humor*, edited by Civan MM, Benos DJ, Simon SA. New York: Academic Press.

Van Adel BA, Kostic C, Deglon N, Ball AK, Arsenijevic Y. 2003. Delivery of ciliary neuro-trophic factor via lentiviral-mediated transfer protects axotomized retinal ganglion cells for an extended period of time. *Hum Gene Ther.* 14:103–115.

Vouri ML, Ali-Melkkila T, Kaila T, Lisalo E, Saari KM. 1993. Plasma and aqueous humor concentrations and systemic effects of topical betaxolol and timolol in man. *Acta Ophthalmol (Copenh).* 17:201–206.

Walther W, Stein U. 2000. Viral vectors for gene transfer: A review of their use in the treatment of human diseases. *Drugs.* 60:249–271.

Wang W, Sasai H, Chien D-S, Lee VHL. 1991. Lipophilicity influence on conjunctival drug penetration in the pigmented rabbit: A comparison with corneal penetration. *Curr Eye Res.* 10:571–579.

Whitehead KA, Langer R, Anderson DG. 2009. Knocking down barriers: Advances in siRNA delivery. *Nat Rev Drug Discov.* 8:129–138.

Wilson CG, Tan LE, Mains J. 2011. Principles of retinal drug delivery from within the vit-reous, in *Drug Product Development for the Back of the Eye (AAPS Advances in the Pharmaceutical Sciences Series)*, vol. 2, 125–158. New York: Springer.

15 Surface Modulation of Nanoparticles

Deepak Thassu and Kris Holt

CONTENTS

15.1 INTRODUCTION

Nanotechnology is one of the fastest-growing fields in engineered materials science and, focusing primarily on pharmaceutics, has been touted as the formulation salvation for many of the most promising new therapeutic agents. It seems that as we delve deeper into the workings of many of the tough-to-treat disease states, we find relatively small molecules or short lengths of protein that demonstrate excellent efficacy and specificity of treatment. Unfortunately, many of these new chemical and biological entities (NCEs and NBEs) exhibit solubilities similar to the glass and plastic vessels in which they were formed. There is little point in administering an oral dose of an extremely effective active pharmaceutical ingredient (API) if it passes through the gastrointestinal (GI) tract like a tomato seed, undissolved and unabsorbed. Typically, the new APIs are delivered in solution form with enormous amounts of solvents, solubilizers, and surfactants by oral or parenteral routes. However, the quantities of solubilizers and surfactants employed can lead to health issues such as sensitivity and impaired renal/hepatic function. The development of a nanoparticulate form of an API does not necessarily solve the solubility issues, but it does maximize the surface area from which slow dissolution can occur. In addition, development of amorphous nanoparticles from a crystalline active can drastically increase the intrinsic dissolution rate by eliminating the energy requirement needed to disrupt the crystalline lattice. Administration is also ineffective if the API cannot penetrate the physical barriers it faces, such as the skin or the cell membrane, or the chemical barriers such as the blood–brain barrier or the blood–ocular barrier. This is the role of surface modulation.

First, we present a quick definition of terms. Modulation is the varying of one or more properties—an alteration of what is present to what is desired. It is an engineering process undertaken to create the characteristics or chemistry desired. Modulation can occur during the formation of a nanoparticle to alter internal characteristics or after the particles have developed to alter the surfaces. Modifications are the individual acts or processes used to bring about the modulation desired. One can modify a surface with polymer to modulate the stability or specificity of the particles. Modulation is a result brought about by modifications.

While the nanoparticles are formed, modifications are made that alter the surface chemistry and physical properties of the particles, thus engendering stealth capability against the barriers they must pass or, conversely, attractiveness to the cells they must penetrate. Alternatively, surface modifications can create bioadhesiveness to retain the particles in place so that dissolution will present individual molecules to the barrier, slowly because of the low intrinsic dissolution rate but over an extended period of time.

It is worth noting that modulation is undertaken to engineer the particles toward a desired outcome. There are two main purposes for surface modulation: stability and selectivity. Both modifications can be applied simultaneously and a single modification may yield both attributes. In general, polymers and surfactants are added as modifiers to the surface of solid nanoparticles by adsorption and/or chemical bonding. Polymers especially have affinity for solid surfaces that interface with a liquid medium. Although some modulation is possible for semisolid particles and micelles, modifiers are generally chemically bound to the surface rather than adsorbed. It is also possible to modify the internal structure of solid amorphous nanoparticles through their coprecipitation with biodegradable polymers such as one of the poly-lactide-co-glycolides. Besides imparting specificity to certain tissues and barriers through traces of the polymer at the particle surface, biodegradation of the internal matrix after administration causes release of secondary particles that are orders of magnitude smaller than the parent. A 500-nm particle is already one-tenth the diameter of a human blood cell. Secondary breakup through biodegradation of the internal polymer matrix can release 5-nm particles having an aggregate 100 times as much surface area from which slow dissolution can occur.

15.2 SURFACE MODULATION FOR STABILITY

Surface modulation for stability is probably the most important after-development modification, especially for nanoparticles developed through solvent exchange methods. Without modulation, nanoparticles can continue to grow in size, flocc, and/or convert from amorphous to crystalline form. The driving force for stability alteration primarily stems from control of the zeta potential, the electrostatic charges that surround particles in solution. When the zeta potential is sufficiently high, whether positive or negative, the particles will repel each other and, thus, prevent aggregation. Addition of a water-soluble polymer solution to a nanosuspension formed by solvent exchange causes deposition to the polymer on the particle surfaces, normalizing the zeta potential over the range of particle sizes and compositions while thickening the media to attenuate particle movement. One can also modulate the particle surfaces so that they will repel each other during freezing to manufacture a lyophylate that will rapidly reconstitute to a stable nanosuspension.

In general, surface modulation is not undertaken for stability alone; it usually arises from the quest for more interesting properties or smaller particle size. As more polymer or surfactant is employed in an antisolvent (AS) to direct the formation of smaller particles or added to an oil/water emulsion premixture to effect smaller micelle production, stability will often be achieved without being sought for. When mixtures of surfactants and polymers are used to take advantage of their various individual properties, synergisms can occur that are far better than the expected addition of their individual actions and attributes.

15.3   SURFACE MODULATION FOR SPECIFICITY

This is the heart of surface and potentially the greatest area of future exploration toward tailored (personalized) therapy. There is extensive literature on enabling techniques to increase trans-barrier absorption through nanosizing and increasing particle hydrophobicity through coatings or through incorporation into hydrophobic carriers. High emphasis is placed on systemic delivery following GI absorption or parenteral administration, a shotgun dissemination of active in the hopes of attaining therapeutic concentration for the tissue or organ to be treated.

Logic dictates that the ideal therapy would affect only the tissue, organ, or invader that it is intended for. Under these conditions, the term "therapeutic level" ceases to have meaning and should be replaced with "therapeutic dose," that is, the amount of active required to effect treatment of the specific target without spillover into other tissues. This is an especially attractive treatment outcome for chemotherapeutics where the effect on nontargeted tissues and organs can lead to debilitating side effects.

The quest for early or better detection methods for various conditions and disease states has led to the discovery of a wealth of unique genetic, chemical, thermal, and pH markers that identify the condition from which they arise. The manufacture of nanoparticulate therapies on which surface specificity toward these unique markers has been engineered represents the magic bullet, a monotherapy so specific that it is essentially inert outside the target it was intended for. So far, antisense proteins are the only entities that approach the ideal. Ultimately, it is not outside the realm of possibility for prophylactic dosing. If surface modification renders a particle specific to a particular chemical marker and the molecule is inert to all other tissues and processes, then the particle has a theoretical infinite half-life until the chemical marker comes into being. In effect, one could create an inoculation against a particular cancer or the expression of an inherited disease state, a man-made immune system tailored for the individual patient.

15.4   MODULATION OF THE MATRIX

Modulation of the matrix presents a few additional difficulties, but also promises a few bonuses. For example, incorporation of biodegradable polymers within a nanoparticle developed by solvent exchange requires the simultaneous or near simultaneous precipitation of the polymer and API. An alternative would be to imbue the AS with nucleation seeds of some material having properties allowing for their steering or alteration.

Matrix nanoparticles are particularly interesting due to their potential for delivering chemotherapeutics to solid tumors. Localization of nanoparticulate chemotherapeutics within a solid tumor is initially targeted through extravasation through leaky vasculature and can be compounded with surface modulation to specifically seek these tissues. Once in place, degradation of the polymer matrix in situ breaks the nanoparticles into even smaller particles and thereby drastically increases the surface area of the poorly soluble active from which intrinsic dissolution occurs. This is a method of providing significant concentration only within the target tissue. Similar concentrated release could occur, with proper barrier penetration by the nanoparticles, for the eye or brain.

Inclusion of a seed or nucleation point within a nanoparticle opens another realm of possibilities for unique properties. If the seed is responsive to a magnetic field, it would be possible to steer the particles to the area of intended action. If the seed is metallic, it would respond dramatically to low-level microwave radiation and possibly to ultrasound. Within the realm of individualized therapy, any of these attributes could be used with a payload of amorphous API to a great effect.

15.5   NOVASPERSE AS A CASE STUDY

NovaSperse is a process developed by PharmaNova to generate nanoparticulate suspensions in a bottom-up approach with the ability to coprecipitate matrix modulators and achieve a desired target size while incorporating surface modulations toward project goals and particle activity. Stability is often the greatest concern, especially toward particle size growth and aggregation. Lyophilization is a quick fix for a nanosuspension produced by solvent displacement, provided that the solvent is not an appreciable percentage of the final mixture and that it has a significant vapor pressure at extremely low temperatures. The latter requirement allows for the removal of the solvent before initiating lyophilization, while the former ensures that the solvent will not interfere with the formation of a solid ice.

Initial development of nanoparticulate formulations derives from our NovaSperse feasibility screening process. Water-miscible organic solutions of an active are exposed to various concentrations of a number of surfactants, polymers, and salts to effect an organized, directed molecular aggregation of API from solution into spherical amorphous particles of limited size. Initial trials involve individual components, followed by the best examples combined with bile salts. Bile salts tend to be synergistic in combinations with other polymers and surfactants and form the basis for the NovaSperse IP. Further trials will then refine the AS so as to engineer the best particle size distribution with a desirable stability and concentration from the initial solvent displacement reaction. Secondary components can then be added to the reaction mixture to "kill" or "fix" the reaction, preventing addition of intermediate changes and thus imparting stability. This is also the point where we begin to modify the surface properties of the nanoparticles with additional polymers and/or surfactants to engender mucoadhesive, stealth, and targeting capabilities. We have also gone on to the addition of tertiary components, but this is usually reserved for products that are to be lyophilized. These tertiary components are generally bulking agents like sugars that are cake formers with the ability to hydrate rapidly and completely.

We have generated nanoparticulate chemotherapeutics that successfully and specifically targeted solid tumors. The modulations employed were stability and stealth to opsonization since the formulation was administered parenterally. We generated nanoparticulate chemotherapeutics with internal modulation in the form of coprecipitated biodegradable polymer, but this formulation is presently untested. We have also generated nanosuspensions for ocular instillation that demonstrated a ten fold increase in corneal penetration over that of a similar solution. The modulations employed in this case were for stability during lyophilization and mucoadhesiveness to hold the particles against the eye and inner eyelid, preventing their dilution and washout by tear production.

15.6 CONCLUSIONS

Surface modulation of nanoparticles for specific characteristics and attributes will define the majority of future research efforts. There are many methods of forming nanoparticles of poorly soluble actives, and most practitioners have devised various methods of stabilizing their formulations. Current and future research will be devoted toward crossing barriers, retaining position, and activating or enhancing release once in place. Modulation of surface and internal properties of nanoparticles will finally take some of the most promising new actives out of the laboratory and into patients, allowing healthcare practitioners more effective and individualized treatments.

16 Nanoparticles for the Treatment of Retinal Diseases

Amin Famili and Uday B. Kompella

CONTENTS

Drug-loaded nanoparticles prepared using synthetic polymers, natural polymers, proteins, lipids, and inorganic materials are currently being investigated for their therapeutic value in treating various diseases of the eye. These delivery systems are particularly useful for cellular delivery and targeted delivery, while allowing slow release of the therapeutic agent. Topical, periocular, intraocular, as well as systemic routes of administration are viable for nanoparticle delivery to the eye. Distribution and clearance of nanoparticles by each of these routes differs, resulting in distinct drug delivery kinetics. Through careful selection of nanoparticles based on their compatibility with the drug, uptake and retention at the disease target site, and drug release kinetics and safety, new ophthalmic nanomedicines can be developed.

16.1 BACKGROUND

Diseases affecting the posterior ocular tissues are a particular challenge from the perspective of delivery of therapeutic agents (Del Amo and Urtti 2008; Eljarrat-Binstock et al. 2010). Due to various anatomical and physiological factors that will be expounded herein, the posterior tissues are one of the most challenging targets for ocular drug delivery. Thus, there is a great need for the development of new techniques and technologies for efficient delivery of therapeutic agents to the posterior segment of the eye.

Back of the eye diseases, such as glaucoma, age-related macular degeneration, and diabetic retinopathy, which afflict the posterior segment tissues of the eye, are the major causes of severe vision loss and blindness worldwide. The prevalence and projected markets of these particular diseases are shown in Table 16.1. The vast prevalence of these diseases highlights the pressing need for effective therapeutic options.

Ophthalmic diseases present both unique challenges and opportunities to the drug delivery field. The challenges are vast and relate to the complex anatomy and physiology of the ocular tissues. However, ocular tissues also present unique opportunities for novel drug delivery system designs that are not possible in other areas of the body.

Unlike the majority of other pathologies, many ocular diseases are not adequately addressed by systemic therapies due to the limited ability of drugs to cross the blood–aqueous and blood–retinal barriers, leading to suboptimal bioavailability in the target tissues. The blood–aqueous barrier is maintained by uveal capillaries and the ciliary endothelium, whose zonulae occludentes, or tight junctions, prevent the diffusion of molecules between cells. This limits the delivery of molecules from systemic circulation into the anterior chamber, forcing them to be transported through the cells or else not all (Snell and Lemp 1998).

The blood–retinal barrier is maintained by various mechanisms, which together limit the entry of molecules into the retina and the vitreous body. The outer blood–retinal barrier in the retinal pigment epithelium (RPE) is maintained by the zonulae occludentes that limit paracellular transport. The inner blood–retinal barrier is maintained by the nonfenestrated endothelial cells of the retinal capillaries that also maintain tight intercellular junctions. These barriers protect the retinal tissues since the choroidal vasculature is leaky, allowing relatively easy access for even larger molecules into the choroidal extravascular space. While the outer blood–retinal barrier limits drug entry from the choroidal tissue into the retina, the inner blood–retinal barrier limits solute entry from the systemic route directly into the retina (Snell and Lemp 1998; Urtti 2006).

TABLE 16.1
Ocular Diseases Affecting the Retina, Their Prevalence, and Projected Market Value

Disease	U.S. Prevalence (Estimated, 2004)	U.S. Prevalence (Projected, 2020)	Global Market Value (Projected, 2014)
Glaucoma (open angle)	2.22 million (Friedman, Wolfs et al. 2004)	>3.00 million (Friedman, Wolfs et al. 2004)	$ 6.6 billion (Highsmith 2010)
Age-related macular degeneration	1.75 million (Friedman, O'Colmain et al. 2004)	>3.00 million (Friedman, O'Colmain et al. 2004)	$ 3.9 billion (Highsmith 2010)
Diabetic retinopathy	4.10 million (Kempen et al. 2004)	>6.00 million (Kempen et al. 2004)	

16.2 NANOPARTICLES

A variety of nanosystems with sizes ranging from 1 to 1000 nm, broadly referred to as nanoparticles in this chapter, can be employed for ophthalmic use. Based on the disease being treated and the drug of interest, a variety of materials can be chosen for preparing nanoparticles. Broadly, these materials can include synthetic polymers (e.g., poly(lactide-co-glycolide [PLGA]), natural polymers (e.g., chitosan), proteins (e.g., human serum albumin), lipids (e.g., dipalmitoylphosphatidylcholine), and some inorganic materials (e.g., gold, silver, and iron oxide). With each of these materials, a variety of architectures can be constructed for delivery systems. Some typical architectures for albumin, chitosan, polyamidoamine (PAMAM), PLGA, and liposome nanosystems are presented in Figure 16.1. Nanoparticles are selected on the basis of the disease features, cellular targets, drug compatibility, and desired duration of drug release. If there is a neovascular or inflammatory disorder, the barriers could be breached, allowing delivery of particle sizes that is otherwise not feasible. If the drug target is within the cell and the drug by itself has poor permeability, the cellular entry mechanisms should be built into the nanoparticle. The delivery system selected should allow efficient and high drug loading, while maintaining drug stability throughout the duration of drug release. Further, the drug should be released

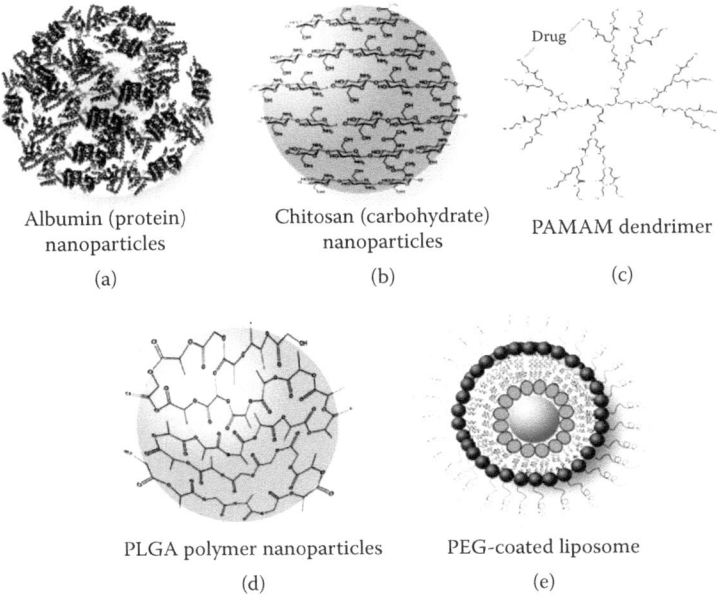

Albumin (protein) nanoparticles

(a)

Chitosan (carbohydrate) nanoparticles

(b)

Drug

PAMAM dendrimer

(c)

PLGA polymer nanoparticles

(d)

PEG-coated liposome

(e)

FIGURE 16.1 Nanoparticles currently under investigation for use as ocular drug delivery vehicles: (a) albumin nanoparticles, (b) chitosan nanoparticles, (c) polyamidoamine (PAMAM) dendrimers, (d) poly (lactic-co-glycolic) acid (PLGA) nanoparticles, and (e) polyethylene glycol (PEG)-coated liposomes. (With kind permission from Springer Science+Business Media: *Drug Product Development for the Back of the Eye (AAPS Advances in Pharmaceutical Science Series)*, "Nanotechnology and nanoparticles," 2011, 261–90, Durazo, S. A., and U. B. Kompella.)

in a controlled manner to meet the therapeutic goals. Last but not least, the chosen nanoparticle should be safe for human use.

Therapeutic agents with diverse physicochemical properties can be loaded into/ onto nanoparticles. Small-molecule drugs as well as large molecules including peptides, proteins, and nucleic acids are potential drug candidates for nanoparticle delivery systems. For poorly permeable molecules, nanoparticles might enhance cellular uptake and delivery across tissue barriers. For molecules readily degraded by enzymes (e.g., proteases and nucleases) within the body, nanoparticles might offer protection from degradation and enhance their stability. For molecules that require targeting for enhanced safety or efficacy, nanoparticles can be designed for selective tissue, cell, or organelle targeting. For all the above therapeutic agents, nanoparticles with reduced clearance from the target site or the system can be potentially designed.

16.3 ROUTES OF ADMINISTRATION

16.3.1 TOPICAL ADMINISTRATION

The most common approach to ocular therapeutic delivery has historically been through topical administration. This approach is an attractively simple solution that capitalizes on a unique aspect of the eye: namely, its external accessibility. However, the complex anatomy and physiology of the ocular tissues prevent this mechanism from achieving widespread applicability. The ocular tissues have evolutionarily developed intricate surface barriers for preventing the entry of foreign solutes into the eye, barriers that must be overcome to deliver drugs via this route.

The corneal tissue consists of a hydrophilic stroma surrounded by a hydrophobic epithelium and endothelium. The outer layers of the epithelium are connected by zonulae occludentes and maculae adherentes, which prevent free diffusion of molecules. Likewise, the endothelial cell layer is composed of cells with interdigitated plasma membranes and zonulae occludentes that present a virtually impermeable barrier. Penetration across the cornea is thus best achieved by small, lipophilic molecules that are able to penetrate the epithelial layers. However, their rate of absorption beyond this level is severely limited by the hydrophilic stroma. The molecules that do make it past this point can enter the anterior chamber, where they can easily access the iris and ciliary body (Urtti 2006).

Unfortunately, the corneal route is isolated to anterior segment drug delivery as bioavailability in posterior segments is limited to a very small fraction of that available to the anterior segment (Eljarrat-Binstock et al. 2010). Topically applied drugs can also enter and cross the conjunctival epithelium for subsequent transscleral delivery to the posterior segment of the eye. However, drug delivery to the posterior segment through topical administration requires high drug concentrations, doses, and dosing frequencies, limiting the utility of such a delivery mechanism for drug delivery to the back of the eye. From a user standpoint, they increase the cost of the treatment while also adding to the burden of the treatment by requiring frequent administration—a fact that may well reduce the success of the treatment since patient compliance is a major design challenge in drug product development. In fact, these two factors themselves, cost and compliance, were found in a recent survey to be two of the top four factors that ophthalmologists said would impact the likelihood that they would adopt a new drug delivery system (Weiner 2008a).

The other major consequences of this delivery technique are physiological concerns as related to local/systemic exposure and maintenance of the drug load within the therapeutic window. The former is directly related to the need for higher drug doses, which, in turn, raises the exposure both locally and systemically, increasing the likelihood of side effects. The latter relates to the combination of higher drug dose and frequency of administration, leading to pulsatile drug levels at the target tissue. Since all therapeutic compounds are only effective within a limited therapeutic window, this pulsatile behavior will alternately overshoot the desired window and then fall below it, limiting the effectiveness of the therapy.

While most efforts at developing a nanoparticle strategy for retinal drug delivery have not employed topical administration, some groups have. For example, Kompella et al. (2006) showed that uptake of negatively charged 20-nm nanoparticles in bovine corneal epithelium is between 1.1% and 1.6% 5 minutes after a topical eye drop application. Thus, despite a small size, some nanoparticles do not gain significant entry into the cornea following a short exposure time, which is the norm *in vivo* due to blinking and tear drainage. Interestingly, however, delivery of the above nanoparticles into the cornea could be elevated by about 4.5- and 8-fold following conjugation of a luteinizing hormone-releasing hormone (LHRH) receptor agonist (deslorelin) and transferrin on the surface of the nanoparticles. These ligand-conjugated nanoparticles, also known as functionalized nanoparticles, do not alter corneal epithelial barrier integrity, as evidenced by the lack of changes in the architecture and permeability of corneal epithelial tight junctions. The uptake and transport across conjunctiva was generally higher than that across the cornea for all nanoparticle groups assessed. Thus, functionalized nanoparticles can potentially enhance conjunctival uptake and transport of particles, followed by sustained transscleral drug delivery to the back of the eye. Indeed, particles sized in the range of 200 nm persist behind the conjunctiva *in vivo* for at least 2 months (Amrite and Kompella 2005). In another study, Eljarrat-Binstock et al. (2008) devised a system using polyacrylic hydrogels loaded with 20–45-nm fluorescent nanoparticles that were applied with iontophoresis at both the central cornea and the pars planar sclera. Nanoparticle penetration to the inner ocular tissues including the iris, ciliary body, choroid, and retina was suggested, with increasing fluorescence observed at 12 hours after administration. While this is a promising result given its ability to achieve high levels of inner ocular tissue penetration, its relative lack of sustained delivery characteristics may limit its applicability for retinal drug delivery.

16.3.2 Periocular Administration

As a result of the physiological barriers to transcorneal penetration and the resulting poor bioavailability of drugs at intraocular tissues, particularly posterior tissues, various alternative routes of administration have been investigated, as illustrated in Figure 16.2. The noncorneal route is one such alternative. In this route, drugs are absorbed across the conjunctiva and sclera and into the posterior tissues (Del Amo and Urtti 2008). Particularly for larger molecular radius drugs, this route can

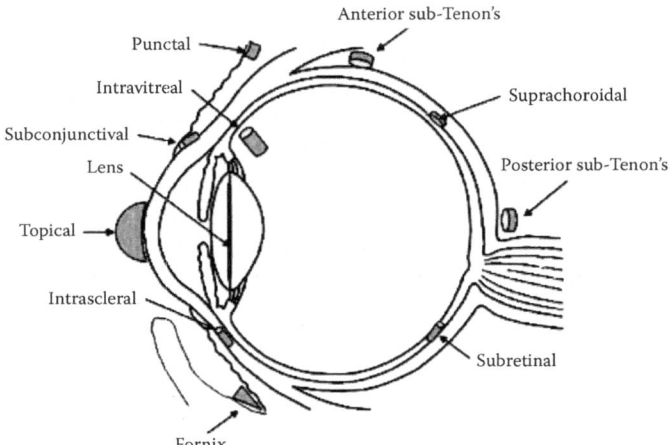

FIGURE 16.2 Potential drug delivery strategies for ophthalmic therapeutics. (From Weiner, A L, *Ocular Therapeutics: Eye on New Discoveries*, Elsevier Academic Press, New York, NY, 2008b.)

TABLE 16.2
Periocular Routes of Ophthalmic Drug Administration

Delivery Method	Procedural Invasiveness	Nanoparticles Investigated for Retinal Delivery?
Subconjunctival, sub-Tenon, and posterior juxtascleral	Low (Raghava et al. 2004)	Yes (Kompella et al. 2003; Amrite and Kompella 2005)
Retrobulbar	High (Raghava et al. 2004)	No
Peribulbar	Medium (Raghava et al. 2004)	No

provide significantly higher bioavailability in intraocular tissues, potentially providing a strategy for achieving therapeutic retinal drug levels (Ambati et al. 2000).

Other periocular delivery strategies have been investigated, as summarized in Table 16.2. These strategies are attractive in that they have the potential to provide localized drug administration while bypassing many of the barriers that limit the effectiveness of topical or systemic administration. As most of these strategies make use of transscleral absorption, they stand to benefit from the larger surface area of the sclera as compared to the cornea; the sclera accounts for 95% of the surface area of the globe (Geroski and Edelhauser 2000). Further, these strategies may advantageously make use of differences in scleral thickness at different locations to improve penetration. While anteriorly and posteriorly the sclera is relatively thick (means 0.53 and 0.9 mm, respectively), it is significantly thinner at the equator (mean 0.39 mm), indicating a target for increased transscleral penetration (Olsen et al. 1998).

An important consideration for periocular strategies is the need for these systems to provide sustained delivery of drugs. Due to their relative invasiveness, at least compared to topical administration and the potential complications that go along with it, the time between administration through these routes must be extended as long as possible. Further, any particle or device that is inserted here must be cleared by the body in some fashion so as to allow for repeated administrations. Injected solution forms of drugs and very small nanoparticles (e.g., 20 nm) are expected to be cleared rapidly from periocular spaces.

The periocular route has been investigated for nanoparticle administration to the retina, although to a much lesser degree than other drug delivery strategies. Kompella et al. (2003) have reported on the posterior subconjunctival administration of budesonide-containing poly(lactic acid) (PLA) nanoparticles for retinal delivery in a rat model. In a comparison between budesonide-loaded nanoparticles, budesonide-loaded microparticles, and budesonide solution, they found that while the microparticles exhibited more favorable release kinetics (particularly in the lack of a large initial burst), the nanoparticles were able to achieve similar initial drug absorption characteristics as solution, as measured by retinal, vitreous, and corneal drug levels. The nanoparticles were able to sustain tissue levels better than solution at the end of 1 week but not as well as the microparticles, which sustained drug levels during the 2-week study. While investigating the effect of nanoparticle size on periocular delivery, Amrite and Kompella (2005) found that subconjunctivally administered 20-nm nanoparticles were rapidly cleared from the periocular space while larger particles (200 and 2000 nm) were present out to 60 days at amounts close to their initial doses, indicating an ideal size range for sustained periocular delivery. In this study, the 20-nm particles disappeared from the periocular site with a half-life of about 5.5 hours. However, this disappearance did not result in significant entry of nanoparticles into the intraocular tissues, including the retina and choroid. Thus, since the previous budesonide study employed nanoparticles larger than 200 nm, it appears that inferior delivery compared to microparticles is a result of poor release kinetics as opposed to rapid clearance of the nanoparticles of budesonide. In other studies, periocularly administered PLGA microparticles of celecoxib were shown to sustain celecoxib delivery as well as antioxidant and antivascular leakage effects in the tissues of the posterior segment of the eye for 2 months (Ayalasomayajula and Kompella 2005; Amrite et al. 2008). In a follow-up study, the same group investigated the transscleral permeability and *in vivo* disposition of the various sizes of nanoparticles and found that while the 20-nm nanoparticles were transported across bovine sclera to a small degree after a 24-hour exposure, there was no significant transport across the sclera-choroid-RPE, largely due to the presence of a permeability barrier (Amrite et al. 2008). The 200-nm nanoparticles were not able to cross the sclera or sclera-choroid-RPE to any quantifiable extent. In the same study, *in vivo* disposition studies in a rat model indicated that periocularly administered 20-nm particles enter the organs of the reticuloendothelial system including the liver, spleen, and lymph nodes within 6 hours after dosing. Further, this study showed that elimination of circulatory clearance retains the 20-nm particles in the periocular region. Thus, circulatory clearance as well as permeability barriers restrict the entry of 20-nm particles into the retina following periocular administration. Particulate systems that can be retained in the periocular space are expected to slowly release the drug, which in turn crosses the sclera to reach the choroid-RPE and the retina.

Boddu et al. (2010) also investigated the utility of nanoparticles for subconjunctival administration, but from a thermosensitive gel. Their study encapsulated the corticosteroids dexamethasone, hydrocortisone acetate, and prednisolone acetate in PLGA nanoparticles, which were then suspended in PLGA-polyethylene glycol (PEG)-PLGA thermosensitive gels. Suspension in thermosensitive gels was found to diminish or eliminate the burst release effect commonly observed in nanoparticle systems and was also cited as a strategy to improve retention of nanoparticles at the site of injection.

16.3.3 INTRAVITREAL ADMINISTRATION

The final route of administration that has been heavily investigated recently is intravitreal injection. In this strategy, injections are made through the pars plana approximately 3.5–4 mm posterior to the surgical limbus. Much of the attractiveness of this strategy is derived from its direct injection of the drug to the intraocular space, thus circumventing the limitations of penetration. In this way, therapeutic doses can be readily achieved at the posterior tissues by injection of the drug alone (Arevalo et al. 2009), drug-eluting devices (e.g., Ozurdex® device, Haller et al. 2010), or drug-containing nanoparticles (Zhang et al. 2009). While this strategy has gained significant popularity recently, it also has many drawbacks that must be considered. Of particular note is the invasiveness of the procedure, which has been associated with various postoperative complications including infectious endophthalmitis (Ozkiriş and Erkiliç 2005; Jonas et al. 2008), retinal detachment (Jonas et al. 2008), traumatic cataract (Ozkiriş and Erkiliç 2005; Jonas et al. 2008), and elevated intraocular pressure (Ozkiriş and Erkiliç 2005). Therefore, as with the periocular strategies in Section 16.3.2, achieving a sustained release profile with significant duration is critical for the success of an intravitreal approach as repeated injections of excessive frequency will likely be rejected by ophthalmologists as well as patients.

Nanoparticles have been extensively applied to intravitreal applications due to their ease of delivery (Giordano et al. 1995; Sakurai et al. 2001; Merodio 2002; Bourges et al. 2003; de Kozak et al. 2004; Irache et al. 2005; Sakai et al. 2006; Xu et al. 2007; Kim et al. 2009; Zhang et al. 2009; Ryu et al. 2010; Jin et al. 2011). Unlike implants that contain a large dimension of polymer to control the release kinetics of the embedded drug or drug particles, nanoparticles can be readily suspended in a solution and injected through smaller needles into the vitreous cavity. The net benefit of this strategy, then, is reduced invasiveness of the procedure leading to less likelihood of negative side effects.

One consideration that must be given careful thought, however, is the altered fluid dynamic properties of nanoparticles versus larger implants or even microparticles. Despite the high viscosity of the vitreous humor, the small dimensions of nanoparticles allow them to more readily stay suspended in the vitreous. In contrast, the larger dimensions of microparticles will make it more likely for them to settle at the tissue interface of the vitreous cavity (Del Amo and Urtti 2008). The implications of this effect are twofold. Because nanoparticles are more likely to stay suspended in the vitreous humor, they are more likely to cause visual blurring when administered in large quantities because they will stay within the visual axis, whereas microparticles will settle out of the visual axis. On the other hand, the tendency of microparticles to settle at the tissue interface combined with the low diffusivity of particles in the highly viscous vitreous will result in higher local drug concentrations at the retinal tissues surrounding the microparticle depot (Stay et al. 2003), while nanoparticles are more likely to distribute the drug better across the retina.

This size-dependent nature of nanoparticle kinetics has been experimentally confirmed by Sakurai et al. (2001). In their experiments, they confirmed the presence of nanoparticles in the vitreous cavity for over a month compared to the control, in which fluorescence persisted for only 3 days. Further, they noted a strong correlation between particle diameter and elimination half-life from the vitreous cavity, with average half-lives of 5.4 days for 2-μm particles, 8.6 days for 200-nm particles, and 10.1 days for 50-nm particles. In addition to the more rapid clearance of the 2-μm particles, these particles were also the only ones that were not found in the retina. Although removal of larger particles by infiltrating macrophages is a possible explanation for the observed results, such pathways are not fully established.

Another factor that may affect nanoparticle localization in the retina after intravitreal injection is surface charge. While the vitreous is composed of 99% water, the various solutes present make it a more complex system. Of particular note is the presence of collagen and hyaluronic acid in relatively high concentrations (40–120 and 100–400 μg/mL, respectively) (Pitkänen et al. 2003). Due to the negative charge of vitreal hyaluronan and other present glycosaminoglycans, they have been found to interact with polymeric and liposomal gene delivery systems (Ruponen et al. 1999). This was confirmed experimentally for the case of intravitreally administered human serum albumin nanoparticles by Kim et al. (2009). They found that while anionic nanoparticles were able to freely diffuse to the retina, most of the cationic nanoparticles were bound and aggregated in the vitreous, leading to poor availability in the retina. As a result of their findings, they conclude that cationic charge is a more critical limiting factor than particle size in the vitreous mobility of nanoparticles.

The most commonly employed materials for the fabrication of nanoparticle-based drug delivery systems are the PLGA family of polymers, including the homopolymer PLA, owing to their bioerodibile behavior as well as their track record of Food and Drug Administration (FDA) approval for use in medical devices and delivery systems (Giordano et al. 1995; Jain 2000; Aukunuru et al. 2003; Bourges et al. 2003; Bejjani et al. 2005; Sakai et al. 2006; Gómez-Gaete et al. 2007; Xu et al. 2007; Zhang et al. 2009; Boddu et al. 2010). However, these material choices are not without their drawbacks. In particular, intravitreally administered PLGA particles have been implicated in some negative reactions. Giordano et al. (1995) noted a mild inflammatory response involving the vitreous, retina, and sometimes the choroid in the initial stages after injection at the site of deposition. However, this reaction did dissipate after 4 weeks, eventually becoming only a localized foreign body reaction.

Administration of PLA nanoparticles can also cause a mild inflammatory response, as demonstrated by Bourges et al. (2003). In their experimentation, however, they noted that this mild inflammatory response was observed in all three groups: those receiving the Nile red-loaded PLA nanoparticles, those receiving blank PLA nanoparticles, and those receiving injections of phosphate-buffered saline (PBS). They conclude that at least part of the inflammatory response is not so much a result of the nanoparticle system but the trauma of the intraocular penetration of the needle employed in intravitreal injections. This mild inflammatory response cannot be ignored, then, when designing a nanoparticle system that is to be administered through intravitreal injection, no matter the material employed in their fabrication.

16.4   NANOPARTICLES AND CELLULAR UPTAKE

One of the specific features that make nanoparticles such an attractive drug delivery mechanism is their uptake by specific cells. This benefit is derived from and relies on the cell efficiently internalizing and retaining the drug-loaded nanoparticle (Panyam and Labhasetwar 2003a). This uptake is a dynamic process (Panyam and Labhasetwar 2003b) and has been shown to be dependent on the size (Desai et al. 1997; Prabha 2002) and hydrophobic nature (Sahoo et al. 2002) of the nanoparticles.

In general, the mechanism of nanoparticle internalization can proceed through one of three major mechanisms, as demonstrated graphically in Figure 16.3. Note that these are best described as general categories with more specific mechanisms falling under each one. Phagocytosis involves cellular internalization of solid, usually larger particles (in the range of 0.1–10 µm) by use of the actin cytoskeleton to protrude a section of the plasma membrane and surround the particle (Pollard and Earnshaw 2008). Macropinocytosis is a bulk uptake mechanism in which the cell nonselectively ingests a volume of extracellular fluid, taking with it any solute or particle within that volume of fluid (Pollard and Earnshaw 2008). If a nanoparticle happens to be in this volume (generally between 50 and 1000 nm in diameter), it will be passively internalized. The final mechanism, receptor-mediated endocytosis, can be broken down into caveolae-mediated and clathrin-mediated endocytosis. In the former, which are especially abundant in endothelial cells, small invaginations of the plasma membrane mediate transcellular movement of serum proteins and nutrients from the bloodstream (Pollard and Earnshaw 2008). While these are generally static and stabilized through the protein caveolin, they can be internalized if triggered

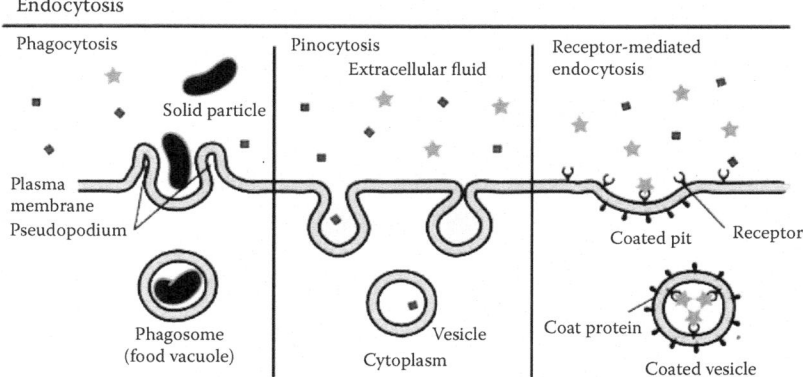

FIGURE 16.3   The major endocytotic mechanisms available for cellular uptake of particles. Nanoparticle internalization can potentially proceed through any of these mechanisms, depending on the nanoparticle size, surface charge, surface modification, and other characteristics.

by tyrosine phosphorylation. Clathrin-mediated endocytosis involves patches of the plasma membrane rich in ligand-receptor complexes, which can invaginate and pinch off to form clathrin-coated vesicles (Pollard and Earnshaw 2008). While a nonclathrin, noncaveolar endocytotic pathway is also recognized, it is not well characterized and likely has a specific role that does not relate to nanoparticle internalization.

In the treatment of retinal diseases, the focus is often on uptake by cells of the RPE. The RPE is a monolayer of cells located between the photoreceptor layer of the neurosensory retina and the choroid. Because of this critical location, it performs many functions associated with molecular transport between the vasculature of the choroid and the outer neural retina. Disruption of RPE function has been shown to result in retinal degeneration (D'Cruz et al. 2000; Gal et al. 2000) and has also been implicated (Dorey et al. 1989; Green and Enger 1993) in initiation and/or progression of age-related macular degeneration (Liang and Godley 2003).

Aukunuru and Kompella (2002) demonstrated that mass, surface area, and percent uptake of nanoparticles decline with an increase in size, with the uptake being as high as 18% for 20-nm particles in 3 hours. Based on such observations, many nanoparticle systems are designed with uptake by RPE cells as their goal (Bourges et al. 2003; Bejjani et al. 2005; Prow et al. 2008). In one such study, Bejjani et al. studied the *in vitro* kinetics of rhodamine-loaded PLA nanoparticle uptake, the toxicity of PLA and PLGA nanoparticles in RPE cells, and the efficacy of PLGA nanoparticles for gene transfer both *in vitro* and *in vivo* (Bejjani et al. 2005). The kinetics part of their work exhibited increasing cellular uptake of the nanoparticles with increasing concentration (up to 1 mg/mL) and increasing contact time duration (up to 6 hours). In their toxicity studies, neither PLA nor PLGA nanoparticles induced cellular toxicity at concentrations up to 4 mg/mL and contact times up to 48 hours. There was also no detectable interference with the metabolic or proliferative functions of the RPE cells. Most importantly, their *in vivo* experimentation, monitoring gene expression of red nuclear fluorescence, observed expression within 4 days after intraocular injection, with most of the expression localized within RPE nuclei. This protein expression remained apparent to day 14, while control eyes injected with blank nanoparticles or free plasmid showed no fluorescence.

While this study provides a promising outlook on the uptake of nanoparticles by RPE cells and the lack of cellular toxicity even at high nanoparticle concentrations, a limiting factor may be the relatively low number of cells that ultimately expressed the transfected genes. In this study, only 10–35% of cells expressed the genes, which may be a level too low to provide true therapeutic efficacy. Nevertheless, this continues to be an active area of research, and higher transfection efficiencies may be reasonably expected. One possible mechanism for this goal may involve the continued development of functionalized nanoparticles, as will be discussed in Section 16.6.

Paclitaxel-loaded albumin nanoparticles were approved by the FDA in 2005 for the treatment of cancers. These particles improved the safety and efficacy of paclitaxel therapy. Using a similar principle, Dr. Kompella's group prepared human serum albumin nanoparticles loaded with Cu/Zn-superoxide dismutase plasmid (Mo et al. 2007). Albumin nanoparticles exhibited internalization and high efficiency transfection (more than 80%) in human RPE cells. Investigation of intracellular routing mechanisms of these particles indicated that these particles entered

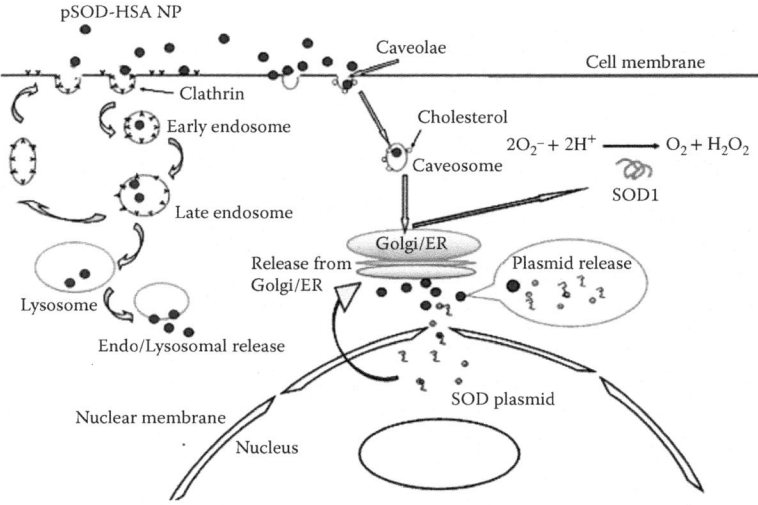

FIGURE 16.4 Proposed mechanisms for the endocytosis and activity of superoxide dismutase plasmid-human serum albumin nanoparticles (pSOD-HSA) for retinal gene delivery. (From Mo, Y. et al., 2007. *Molecular Vision* 13 (May)(January): 746–57. With permission.)

cells through caveolae- or clathrin-mediated pathways, protected the plasmid from degradation, allowed plasmid escape from lysosome entrapment, sustained plasmid release, allowed nuclear entry of human serum albumin and potentially plasmid, and enhanced gene expression and activity. The proposed mechanism of action is illustrated in Figure 16.4. Further, intravitreal delivery of albumin nanoparticles loaded with superoxide dismutase plasmids resulted in superior retinal gene expression.

16.5 NEOVASCULARIZATION AND NANOPARTICLES

A specific application that nanoparticles are particularly well suited in treating is neovascular disease. Neovascularization is of particular consequence when it disrupts retinal structure and function, leading to blindness through a number of different pathologies. These pathologies are of such significance that they account for the leading causes of blindness in infants, working-age adults, and the elderly, expressed as retinopathy of prematurity (Palmer et al. 1991), diabetic retinopathy (Klein and Klein 1997; Frank 2004; Kempen et al. 2004), and age-related macular degeneration (Resnikoff et al. 2004), respectively (Farjo and Ma 2010).

While treatment options that can inhibit neovascularization exist, with particular interest given to the anti-vascular endothelial growth factor (VEGF) agents pegaptanib (Gragoudas et al. 2005), bevacizumab (Bashshur et al. 2006; Spaide et al. 2006; Arevalo et al. 2009), and ranibizumab (Rosenfeld et al. 2006; Fung et al. 2007), these treatments require long-term, regular intravitreal injections to sustain therapeutic efficacy. An improved delivery system would significantly reduce the burden on patient and ophthalmologist and is of great interest. A nanoparticle-based delivery system is well-suited for treatment of neovascularization due to the characteristic

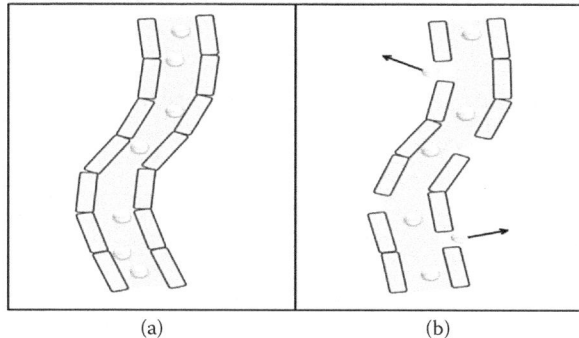

(a)   (b)

FIGURE 16.5 Enhanced penetration of nanoparticles in compromised vessels. (a) In healthy vessels, the intact endothelial layer prevents penetration of nanoparticles into the target tissue. (b) In neovascular disease, the hyperpermeable new vessels allow nanoparticles to exit through gaps in the endothelial layer, enhancing nanoparticle deposition at the site of neovascularization.

leakiness of newly formed blood vessels, allowing for passive diffusion or convection across the hyperpermeable new vessels, as demonstrated in Figure 16.5 (Yuan 1998; Cho et al. 2008). This effect coupled with compromised lymphatic clearance leads to the enhanced permeability and retention effect (Maeda et al. 2000; Cho et al. 2008), which results in nanoparticle accumulation at the site of neovascularization.

Various groups have investigated the particular application of nanoparticles for neovascular disease (Aukunuru et al. 2003; Kompella et al. 2003; Xu et al. 2007; Kim et al. 2009; Singh et al. 2009; Jin et al. 2011). In *in vitro* experimentation, Aukunuru et al. (2003) fabricated a PLGA nanoparticle formulation encapsulating a VEGF antisense oligonucleotide for inhibiting VEGF secretion. In a human RPE cell line, they showed that the cellular uptake of the oligonucleotide was significantly increased in the nanoparticle formulation versus when it was delivered in free solution. Further, they demonstrated that this strategy significantly reduced both $VEGF_{165}$ and $VEGF_{121}$ messenger RNA (mRNA) levels as well as VEGF protein secretion, while placebo nanoparticles and free solution oligonucleotide did not, indicating that the nanoparticle formulation enhances cellular uptake and, in turn, the effectiveness of the therapy.

For such a nanoparticle therapy to be commercialized, it will need to demonstrate similar successes *in vivo*, while also proving the absence of cellular toxicity. Another major consideration will be the amount of time that therapeutic levels can be maintained in the cellular targets. While Xu et al. (2007) were able to sustain therapeutic levels of dexamethasone through a nanoparticle formulation for 56 days, this is still only twice the 28 days between administrations of Lucentis. Clearly, much longer time frames of release are needed before such strategies will be adopted.

16.6 FUTURE DIRECTIONS

Surface modification of nanoparticles represents one of the next evolutionary steps for nanomedicine. While different types of surface modifications have been employed to different ends, one promising avenue available for drug delivery systems

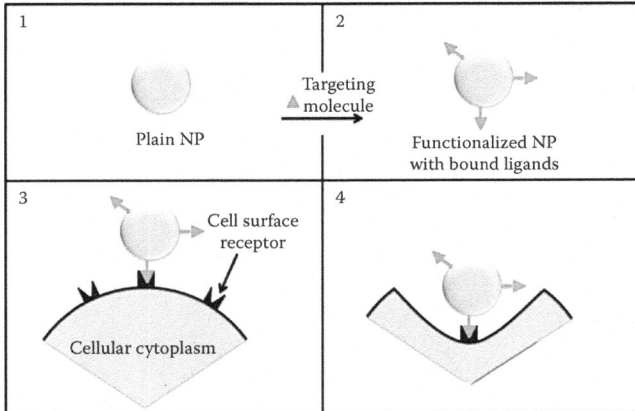

FIGURE 16.6 Graphical representation of the rationale behind nanoparticle surface modification. (1) A plain nanoparticle is prepared through conventional methods. (2) This nanoparticle is surface modified, for example, by binding a targeting molecule to its surface, to yield a functionalized nanoparticle. (3) After administration, this ligand is recognized by receptors on the surface of the target cell. (4) The nanoparticle, now bound to the surface of the cell, is internalized through receptor-mediated endocytosis. NP, nanoparticle.

is improvement in targeting specific regions through nanoparticle functionalization with specific targeting ligands, shown graphically in Figure 16.6 (Kreuter 2001; Hans and Lowman 2002).

In an exemplary embodiment this strategy, Singh et al. (2009) functionalized anti-VEGF intraceptor plasmid-loaded PLGA nanoparticles to facilitate receptor-mediated endocytosis in the retina after intravenous administration. Their surface modification involved a linear arginine-glycine-aspartic acid peptide (GRGDSPK), transferring, or a combination of the two. The peptide GRGDSPK was employed due to its known binding of integrin receptors, which are specifically overexpressed in age-related macular degeneration and proliferative diabetic retinopathy patients (Friedlander et al. 1996). A similar receptor-mediated endocytosis was the goal of transferrin functionalization. In their study, functionalized nanoparticles were found to be significantly more efficacious in reducing laser-induced choroidal neovascularization than vehicle and nonfunctionalized nanoparticle groups.

This promising new direction for nanomedicine capitalizes on knowledge of specific receptors and their upregulation in particular pathologies for improved therapeutic targeting and cellular uptake at the disease site. As demonstrated in the above example, this can allow therapies to be administered systemically while still targeting only the site where it is needed, potentially mitigating problems of side effects common with systemic administration. Further, functionalization of nanoparticles allows them to more readily employ receptor-mediated endocytosis pathways, increasing their cellular uptake, and, in turn, therapeutic efficacy. A well-designed functionalized nanoparticle-based therapeutic system, then, stands to be the next evolutionary step in the treatment of retinal diseases.

ACKNOWLEDGMENTS

This work was supported by NIH grants R01EY018940, R01EY017533, RC1EY020361, and R41 R41EY020097.

REFERENCES

Ambati, J, C S Canakis, J W Miller, E S Gragoudas, A Edwards, D J Weissgold, I Kim, F C Delori, and A P Adamis. 2000. "Diffusion of high molecular weight compounds through sclera." *Investigative Ophthalmology & Visual science* 41 (5) (April): 1181–5.

Amrite, A C, H F Edelhauser, S R Singh, and U B Kompella. 2008. "Effect of circulation on the disposition and ocular tissue distribution of 20 nm nanoparticles after periocular administration." *Molecular Vision* 14 (January): 150–60.

Amrite, A C, and U B Kompella. 2005. "Size-dependent disposition of nanoparticles and microparticles following subconjunctival administration." *The Journal of Pharmacy and Pharmacology* 57 (12) (December): 1555–63. doi:10.1211/jpp.57.12.0005.

Arevalo, J F, L Wu, J G Sanchez, M Maia, M J Saravia, C F Fernandez, and T Evans. 2009. "Intravitreal bevacizumab (Avastin) for proliferative diabetic retinopathy: 6-months follow-up." *Eye (London, England)* 23 (1) (January): 117–23. doi:10.1038/sj.eye.6702980.

Aukunuru, J V, S P Ayalasomayajula, and U B Kompella. 2003. "Nanoparticle formulation enhances the delivery and activity of a vascular endothelial growth factor antisense oligonucleotide in human retinal pigment epithelial cells." *The Journal of Pharmacy and Pharmacology* 55 (9) (September): 1199–206. doi:10.1211/0022357021701.

Aukunuru, J V, and U B Kompella. 2002. "*In vitro* delivery of nano- and micro-particles to human retinal pigment epithelial (ARPE-19) cells." *Drug Delivery Technology* 2 (50–57).

Ayalasomayajula, S P, and U B Kompella. 2005. "Subconjunctivally administered celecoxib-PLGA microparticles sustain retinal drug levels and alleviate diabetes-induced oxidative stress in a rat model." *European Journal of Pharmacology* 511 (2–3) (March): 191–8. doi:10.1016/j.ejphar.2005.02.019.

Bashshur, Z F, A Bazarbachi, A Schakal, Z A Haddad, C P El Haibi, and B N Noureddin. 2006. "Intravitreal bevacizumab for the management of choroidal neovascularization in age-related macular degeneration." *American Journal of Ophthalmology* 142 (1) (July): 1–9. doi:10.1016/j.ajo.2006.02.037.

Bejjani, R A, D BenEzra, H Cohen, J Rieger, C Andrieu, J-C Jeanny, G Gollomb, and F F Behar-Cohen. 2005. "Nanoparticles for gene delivery to retinal pigment epithelial cells." *Molecular Vision* 11 (February): 124–32.

Boddu, S H, J Jwala, R Vaishya, R Earla, P K Karla, D Pal, and A K Mitra. 2010. "Novel nanoparticulate gel formulations of steroids for the treatment of macular edema." *Journal of Ocular Pharmacology and Therapeutics: The Official Journal of the Association for Ocular Pharmacology and Therapeutics* 26 (1) (February): 37–48. doi:10.1089/jop.2009.0074.

Bourges, J-L, S E Gautier, F Delie, R A Bejjani, J-C Jeanny, R Gurny, D BenEzra, and F F Behar-Cohen. 2003. "Ocular drug delivery targeting the retina and retinal pigment epithelium using polylactide nanoparticles." *Investigative Ophthalmology & Visual Science* 44 (8) (August): 3562–9.

Cho, K, X Wang, S Nie, Z G Chen, and D M Shin. 2008. "Therapeutic nanoparticles for drug delivery in cancer." *Clinical Cancer Research: An Official Journal of the American Association for Cancer Research* 14 (5) (March 1): 1310–6. doi:10.1158/1078-0432. CCR-07-1441.

D'Cruz, P M, D Yasumura, J Weir, M T Matthes, H Abderrahim, M M LaVail, and D Vollrath. 2000. "Mutation of the receptor tyrosine kinase gene Mertk in the retinal dystrophic RCS rat." *Human Molecular Genetics* 9 (4) (March 1): 645–51.

de Kozak, Y, K Andrieux, H Villarroya, C Klein, B Thillaye-Goldenberg, M-C Naud, E Garcia, and P Couvreur. 2004. "Intraocular injection of tamoxifen-loaded nanoparticles: A new treatment of experimental autoimmune uveoretinitis." *European Journal of Immunology* 34 (12) (December): 3702–12. doi:10.1002/eji.200425022.

Del Amo, E M, and A Urtti. 2008. "Current and future ophthalmic drug delivery systems. A shift to the posterior segment." *Drug Discovery Today* 13 (3–4) (March): 135–43. doi:10.1016/j.drudis.2007.11.002.

Desai, M P, V Labhasetwar, E Walter, R J Levy, and G L Amidon. 1997. "The mechanism of uptake of biodegradable microparticles in Caco-2 cells is size dependent." *Pharmaceutical Research* 14 (11): 1568–73.

Dorey, C K, G Wu, D Ebenstein, A Garsd, and J J Weiter. 1989. "Cell loss in the aging retina. Relationship to lipofuscin accumulation and macular degeneration." *Investigative Ophthalmology & Visual Science* 30 (8) (August): 1691–9.

Durazo, S A, and U B Kompella. 2011. "Nanotechnology and nanoparticles." In *Drug Product Development for the Back of the Eye (AAPS Advances in Pharmaceutical Science Series)*, eds. U B Kompella and H F Edelhauser, 261–90. 2nd ed. New York, NY: Springer.

Eljarrat-Binstock, E, F Orucov, Y Aldouby, J Frucht-Pery, and A J Domb. 2008. "Charged nanoparticles delivery to the eye using hydrogel iontophoresis." *Journal of Controlled Release: Official Journal of the Controlled Release Society* 126 (2) (March): 156–61. doi:10.1016/j.jconrel.2007.11.016.

Eljarrat-Binstock, E, J Pe'er, and A J Domb. 2010. "New techniques for drug delivery to the posterior eye segment." *Pharmaceutical Research* 27 (4) (April): 530–43. doi:10.1007/s11095-009-0042-9.

Farjo, K M, and J-X Ma. 2010. "The potential of nanomedicine therapies to treat neovascular disease in the retina." *Journal of Angiogenesis Research* 2 (1) (January): 21. doi:10.1186/2040-2384-2-21.

Frank, R N. 2004. "Diabetic retinopathy." *The New England Journal of Medicine* 350 (1) (January 1): 48–58. doi:10.1056/NEJMra021678.

Friedlander, M, C L Theesfeld, M Sugita, M Fruttiger, M A Thomas, S Chang, and D A Cheresh. 1996. "Involvement of integrins alpha v beta 3 and alpha v beta 5 in ocular neovascular diseases." *Proceedings of the National Academy of Sciences of the United States of America* 93 (18) (September 3): 9764–9.

Friedman, D S, B J O'Colmain, B Muñoz, S C Tomany, C McCarty, P T de Jong, B Nemesure, P Mitchell, and J Kempen. 2004. "Prevalence of age-related macular degeneration in the United States." *Archives of Ophthalmology* 122 (4) (April): 564–72. doi:10.1001/archopht.122.4.564.

Friedman, D S, R C Wolfs, B J O'Colmain, B E Klein, H R Taylor, S West, M C Leske, P Mitchell, N Congdon, and J Kempen. 2004. "Prevalence of open-angle glaucoma among adults in the United States." *Archives of Ophthalmology* 122 (4) (April): 532–8. doi:10.1001/archopht.122.4.532.

Fung, A E, G A Lalwani, P J Rosenfeld, S R Dubovy, S Michels, W J Feuer, C A Puliafito, J L Davis, H W Flynn, and M Esquiabro. 2007. "An optical coherence tomography-guided, variable dosing regimen with intravitreal ranibizumab (Lucentis) for neovascular age-related macular degeneration." *American Journal of Ophthalmology* 143 (4) (April): 566–83. doi:10.1016/j.ajo.2007.01.028.

Gal, A, Y Li, D A Thompson, J Weir, U Orth, S G Jacobson, E Apfelstedt-Sylla, and D Vollrath. 2000. "Mutations in MERTK, the human orthologue of the RCS rat retinal dystrophy gene, cause retinitis pigmentosa." *Nature Genetics* 26 (3) (November): 270–1. doi:10.1038/81555.

Geroski, D H, and H F Edelhauser. 2000. "Drug delivery for posterior segment eye disease." *Investigative Ophthalmology & Visual Science* 41 (5) (April): 961–4.

Giordano, G G, P Chevez-Barrios, M F Refojo, and C A Garcia. 1995. "Biodegradation and tissue reaction to intravitreous biodegradable poly(D,L-lactic-co-glycolic)acid microspheres." *Current Eye Research* 14 (9) (September): 761–8. doi:10.3109/02713689508995797.

Gómez-Gaete, C, N Tsapis, M Besnard, A Bochot, and E Fattal. 2007. "Encapsulation of dexamethasone into biodegradable polymeric nanoparticles." *International Journal of Pharmaceutics* 331 (2) (March): 153–9. doi:10.1016/j.ijpharm.2006.11.028.

Gragoudas, E, A Adamis, E Cunninghamjr, M Feinsod, and D Guyer. 2005. "Pegaptanib for neovascular age-related macular degeneration." *American Journal of Ophthalmology* 139 (4) (April): 761–2. doi:10.1016/j.ajo.2005.02.003.

Green, W R, and C Enger. 1993. "Age-related macular degeneration histopathologic studies. The 1992 Lorenz E. Zimmerman Lecture." *Ophthalmology* 100 (10) (October): 1519–35.

Haller, J A, F Bandello, R Belfort, M S Blumenkranz, M Gillies, J Heier, A Loewenstein et al. 2010. "Randomized, sham-controlled trial of dexamethasone intravitreal implant in patients with macular edema due to retinal vein occlusion." *Ophthalmology* 117 (6) (June): 1134–46. e3. doi:10.1016/j.ophtha.2010.03.032.

Hans, M L, and A M Lowman. 2002. "Biodegradable nanoparticles for drug delivery and targeting." *Current Opinion in Solid State and Materials Science* 6 (4) (August): 319–27. doi:10.1016/S1359-0286(02)00117-1.

Highsmith, J. 2010. *Report PHM031C: Ophthalmic Therapeutic Drugs: Technologies and Global Markets.* Wellesley, MA: BCC Research, LLC.

Irache, J M, M Merodio, A Arnedo, M A Camapanero, M Mirshahi, and S Espuelas. 2005. "Albumin nanoparticles for the intravitreal delivery of anticytomegaloviral drugs." *Mini Reviews in Medicinal Chemistry* 5 (3): 293–305.

Jain, R A. 2000. "The manufacturing techniques of various drug loaded biodegradable poly(lactide-co-glycolide) (PLGA) devices." *Biomaterials* 21 (23) (December): 2475–90.

Jin, J, K K Zhou, K Park, Y Hu, X Xu, Z Zheng, P Tyagi, U B Kompella, and J-X Ma. 2011. "Anti-inflammatory and anti-angiogenic effects of nanoparticle-mediated delivery of a natural angiogenic inhibitor." *Investigative Ophthalmology & Visual Science* 52 (9) (February 25): 6230–7. doi:10.1167/iovs.10-6229.

Jonas, J B, U H Spandau, and F Schlichtenbrede. 2008. "Short-term complications of intravitreal injections of triamcinolone and bevacizumab." *Eye (London, England)* 22 (4) (April): 590–1. doi:10.1038/eye.2008.10.

Kempen, J H, B J O'Colmain, M C Leske, S M Haffner, R Klein, S E Moss, H R Taylor, and R F Hamman. 2004. "The prevalence of diabetic retinopathy among adults in the United States." *Archives of Ophthalmology* 122 (4) (April): 552–63. doi:10.1001/archopht.122.4.552.

Kim, H, S B Robinson, and K G Csaky. 2009. "Investigating the movement of intravitreal human serum albumin nanoparticles in the vitreous and retina." *Pharmaceutical Research* 26 (2) (February): 329–37. doi:10.1007/s11095-008-9745-6.

Klein, R, and B E Klein. 1997. "Diabetic eye disease." *Lancet* 350 (9072) (July 19): 197–204. doi:10.1016/S0140-6736(97)04195-0.

Kompella, U B., N Bandi, and S P Ayalasomayajula. 2003. "Subconjunctival nano- and microparticles sustain retinal delivery of budesonide, a corticosteroid capable of inhibiting VEGF expression." *Investigative Ophthalmology & Visual Science* 44 (3) (March): 1192–201. doi:10.1167/iovs.02-0791.

Kompella, U B, S Sundaram, S Raghava, and E R Escobar. 2006. "Luteinizing hormone-releasing hormone agonist and transferrin functionalizations enhance nanoparticle delivery in a novel bovine *ex vivo* eye model." *Micro* 12 (October): 1185–98.

Kreuter, J. 2001. "Nanoparticulate systems for brain delivery of drugs." *Advanced Drug Delivery Reviews* 47 (1) (March 23): 65–81.

Liang, F-Q, and B F Godley. 2003. "Oxidative stress-induced mitochondrial DNA damage in human retinal pigment epithelial cells: A possible mechanism for RPE aging and age-related macular degeneration." *Experimental Eye Research* 76 (4) (April 1): 397–403. doi:10.1016/S0014-4835(03)00023-X.

Maeda, H, J Wu, T Sawa, Y Matsumura, and K Hori. 2000. "Tumor vascular permeability and the EPR effect in macromolecular therapeutics: A review." *Journal of Controlled Release: Official Journal of the Controlled Release Society* 65 (1–2) (March 1): 271–84.

Merodio, M. 2002. "Ocular disposition and tolerance of ganciclovir-loaded albumin nanoparticles after intravitreal injection in rats." *Biomaterials* 23 (7) (April): 1587–94. doi:10.1016/S0142-9612(01)00284-8.

Mo, Y, M E Barnett, D Takemoto, H Davidson, and U B Kompella. 2007. "Human serum albumin nanoparticles for efficient delivery of Cu, Zn superoxide dismutase gene." *Molecular Vision* 13 (May) (January): 746–57.

Olsen, T W, S Y Aaberg, D H Geroski, and H F Edelhauser. 1998. "Human sclera: Thickness and surface area." *American Journal of Ophthalmology* 125 (2) (February): 237–41.

Ozkiriş, A, and K Erkiliç. 2005. "Complications of intravitreal injection of triamcinolone acetonide." *Canadian Journal of Ophthalmology. Journal Canadien D'ophtalmologie* 40 (1) (February): 63–8.

Palmer, E A, J T Flynn, R J Hardy, D L Phelps, C L Phillips, D B Schaffer, and B Tung. 1991. "Incidence and early course of retinopathy of prematurity. The Cryotherapy for Retinopathy of Prematurity Cooperative Group." *Ophthalmology* 98 (11) (November): 1628–40.

Panyam, J, and V Labhasetwar. 2003a. "Biodegradable nanoparticles for drug and gene delivery to cells and tissue." *Advanced Drug Delivery Reviews* 55 (3) (February 24): 329–47.

Panyam, J, and V Labhasetwar. 2003b. "Dynamics of endocytosis and exocytosis of poly(D,L-lactide-co-glycolide) nanoparticles in vascular smooth muscle cells." *Pharmaceutical Research* 20 (2) (February): 212–20.

Pitkänen, L, M Ruponen, J Nieminen, and A Urtti. 2003. "Vitreous is a barrier in nonviral gene transfer by cationic lipids and polymers." *Pharmaceutical Research* 20 (4) (April): 576–83.

Pollard, T D, and W C Earnshaw. 2008. "Endocytosis and the endosomal membrane system." In *Cell Biology*, 391–407. 2nd ed. Philadelphia, PA: Saunders Elsevier.

Prabha, S. 2002. "Size-dependency of nanoparticle-mediated gene transfection: Studies with fractionated nanoparticles." *International Journal of Pharmaceutics* 244 (1–2) (September 5): 105–15. doi:10.1016/S0378-5173(02)00315-0.

Prow, T W, I Bhutto, S Y Kim, R Grebe, C Merges, D S McLeod, K Uno et al. 2008. "Ocular nanoparticle toxicity and transfection of the retina and retinal pigment epithelium." *Nanomedicine: Nanotechnology, Biology, and Medicine* 4 (4) (December): 340–9. doi:10.1016/j.nano.2008.06.003.

Raghava, S, M Hammond, and U B Kompella. 2004. "Periocular routes for retinal drug delivery." *Expert Opinion on Drug Delivery* 1 (1) (November): 99–114. doi:10.1517/17425247.1.1.99.

Resnikoff, S, D Pascolini, D Etya'ale, I Kocur, R Pararajasegaram, G P Pokharel, and S P Mariotti. 2004. "Global data on visual impairment in the year 2002." *Bulletin of the World Health Organization* 82 (11) (November): 844–51. doi:/S0042-96862004001100009.

Rosenfeld, P J, R M Rich, and G A Lalwani. 2006. "Ranibizumab: Phase III clinical trial results." *Ophthalmology Clinics of North America* 19 (3) (September): 361–72. doi:10.1016/j.ohc.2006.05.009.

Ruponen, M, S Ylä-Herttuala, and A Urtti. 1999. "Interactions of polymeric and liposomal gene delivery systems with extracellular glycosaminoglycans: Physicochemical and transfection studies." *Biochimica et Biophysica Acta* 1415 (2) (January 8): 331–41.

Ryu, M, T Nakazawa, T Akagi, T Tanaka, R Watanabe, M Yasuda, N Himori et al. 2010. "Suppression of phagocytic cells in retinal disorders using amphiphilic poly(γ-glutamic acid) nanoparticles containing dexamethasone." *Journal of Controlled Release: Official Journal of the Controlled Release Society* (December). doi:10.1016/j.jconrel.2010.11.029.

Sahoo, S K, J Panyam, S Prabha, and V Labhasetwar. 2002. "Residual polyvinyl alcohol associated with poly (D,L-lactide-co-glycolide) nanoparticles affects their physical properties and cellular uptake." *Journal of Controlled Release: Official Journal of the Controlled Release Society* 82 (1) (July 18): 105–14.

Sakai, T, H Kohno, T Ishihara, M Higaki, S Saito, M Matsushima, Y Mizushima, and K Kitahara. 2006. "Treatment of experimental autoimmune uveoretinitis with poly(lactic acid) nanoparticles encapsulating betamethasone phosphate." *Experimental Eye Research* 82 (4) (April): 657–63. doi:10.1016/j.exer.2005.09.003.

Sakurai, E, H Ozeki, N Kunou, and Y Ogura. 2001. "Effect of particle size of polymeric nano-spheres on intravitreal kinetics." *Ophthalmic Research* 33 (1): 31–6.

Singh, S R, H E Grossniklaus, S J Kang, H F Edelhauser, B K Ambati, and U B Kompella. 2009. "Intravenous transferrin, RGD peptide and dual-targeted nanoparticles enhance anti-VEGF intraceptor gene delivery to laser-induced CNV." *Gene Therapy* 16 (5) (May): 645–59. doi:10.1038/gt.2008.185.

Snell, R S, and M A Lemp. 1998. *Clinical Anatomy of the Eye.* 2nd ed. Malden, MA: Wiley-Blackwell.

Spaide, R F, K Laud, H F Fine, J M Klancnik, C B Meyerle, L A Yannuzzi, J Sorenson, J Slakter, Y L Fisher, and M J Cooney. 2006. "Intravitreal bevacizumab treatment of choroidal neovascularization secondary to age-related macular degeneration." *Retina (Philadelphia, Pa.)* 26 (4) (April): 383–90. doi:10.1097/01.iae.0000238561.99283.0e.

Stay, M S, J Xu, T W Randolph, and V H Barocas. 2003. "Computer simulation of convective and diffusive transport of controlled-release drugs in the vitreous humor." *Pharmaceutical Research* 20 (1) (January): 96–102.

Urtti, A. 2006. "Challenges and obstacles of ocular pharmacokinetics and drug delivery." *Advanced Drug Delivery Reviews* 58 (11) (November): 1131–5. doi:10.1016/j.addr.2006.07.027.

Weiner, A L. 2008a. "Drug delivery system in ophthalmic applications." In *Ocular Therapeutics: Eye on New Discoveries*, eds. T Yorio, A F Clark, and M B Wax, 12–3. New York, NY: Academic Press.

Weiner, A L. 2008b. "Drug delivery systems in ophthalmic applications." In *Ocular Therapeutics: Eye on New Discoveries*, eds. T Yorio, A F. Clark, and M B. Wax, 7–43. 1st ed. New York, NY: Elsevier Academic Press.

Xu, J, Y Wang, Y Li, X Yang, P Zhang, H Hou, Y Shi, and C Song. 2007. "Inhibitory efficacy of intravitreal dexamethasone acetate-loaded PLGA nanoparticles on choroidal neovascularization in a laser-induced rat model." *Journal of Ocular Pharmacology and Therapeutics: The Official Journal of the Association for Ocular Pharmacology and Therapeutics* 23 (6) (December): 527–40. doi:10.1089/jop.2007.0002.

Yuan, F. 1998. "Transvascular drug delivery in solid tumors." *Seminars in Radiation Oncology* 8 (3) (July): 164–75. doi:10.1016/S1053-4296(98)80042-8.

Zhang, L, Y Li, C Zhang, Y Wang, and C Song. 2009. "Pharmacokinetics and tolerance study of intravitreal injection of dexamethasone-loaded nanoparticles in rabbits." *International Journal of Nanomedicine* 4 (January): 175–83.

17 Gene-Based Medicine for Ocular Diseases

Shannon M. Conley and Muna I. Naash

CONTENTS

The application of nanotechnology and nanoparticulate systems to gene delivery is predicated on the understanding that delivery of DNA alone is, at best, ineffective and, at worst, cytotoxic, and that therefore complexing nucleic acids with an additional agent is highly desirable. In theory, delivery of genetic material would be the ideal drug solution; a single dose or treatment leading to persistent, elevated gene expression would cause prolonged exposure of the cell/tissue to the therapeutic agent. This could be a replacement gene to combat a loss-of-function or recessive mutation, a knockdown therapy to target a gain-of-function allele, or a protective gene to retard or prevent degeneration for nonmonogenic disorders. In practice, however, achieving this ideal result has been challenging. In this chapter, we discuss the problems that have plagued attempts to develop therapeutically effective ocular gene delivery, the recent work that has been undertaken to combat those problems, and the application of gene-based medicine to the treatment of ocular diseases.

17.1 GENE-BASED MEDICINE

17.1.1 WHAT CONSTITUTES GENE-BASED MEDICINE?

Technically, gene-based medicine can be as simple as delivery of any therapeutic nucleic acid to a cell or tissue of interest. However, given the lack of efficacy and potential toxicity of naked DNA, delivering it alone is virtually useless. As a result, nucleic acids are almost always packaged with some outside agent prior to delivery. These agents are broadly classified as viral and nonviral. Much of the recent success in ocular gene therapy has arisen from the use of nonpathogenic adeno-associated viruses (AAVs) as a packaging agent for the DNA. Successful gene delivery by viruses relies on the inherent properties of the virus itself. Viruses are evolutionarily optimized to deliver their genetic cargo into the cell and moreover into the nucleus. Early use of retroviruses and adenoviruses led to many severe negative outcomes as a result of immune responses to the viral carrier or insertional mutagenesis (Cavazzana-Calvo et al. 2000; Raper et al. 2003). In contrast, AAV, a member of the Parvoviridae family, lacks many of these safety concerns, and no major adverse events have been reported in trials using AAV. AAV particles used for gene delivery are composed of a genome carrying the expression cassette of interest (with a maximum carrying capacity of ~4.8 kb) coated with a capsid comprising three proteins: VP1, VP2, and VP3. While wild-type (WT) AAV can replicate with the help of an outside virus, the recombinant AAVs used for gene delivery lack the *rep* genes necessary for this process and are therefore replication incompetent. There are many different serotypes and capsid variants of AAV, which contribute to tissue tropisms and are further discussed in Section 17.2.1.1.1. The AAV particle is ~20 nm in diameter and is icosahedral in shape (Grimm and Kay 2003). To give a sense of scale, dynamic light scattering estimates that plasmid DNA (pDNA) has a diameter of ~1200 nm (Read et al. 2010a).

Nonviral packaging agents are an increasingly popular option for delivery of nucleic acids. While some see development of nonviral methods to be a direct response to lingering concerns about the toxicity and other limitations of viral vectors, in reality, nonviral vectors are simply an additional alternative. As a result of many years of research, it has become evident that a one-size-fits-all approach will not be successful in the field of ocular gene delivery, and development of multiple clinically viable therapeutic options is both prudent and necessary. Nonviral packaging agents vary widely, but most aim to achieve compaction by neutralizing the negatively charged DNA in a manner vaguely reminiscent of native histone–chromatin interactions. Packaging options can be lipid based, such as liposomes and solid lipid nanoparticles (NPs); polymer based, such as chitosan or polylactide-co-glycolide (PLGA); or polypeptide based, such as polylysine. These compacting agents must be biocompatible and usually incorporate some cationic components.

The polylysine NPs that have been successfully used in the eye follow this pattern. They are composed of a single molecule of pDNA (up to 20 kb; Fink et al. 2006) compacted with a polyethylene glycolated (PEG) lysine 30-mer with a terminal cysteine (CK30) (Ziady et al. 2003; Farjo et al. 2006). The cationic polylysine enables significant compaction of the DNA, while the PEG confers colloidal stability under physiological conditions. These particles vary in shape depending on the lysine counterion present at the time of compaction and can be rod-like with a size of ~8–10 nm × 200 nm or ellipsoidal (~22 nm × 50 nm) (Farjo et al. 2006). Depending on the compaction agent and the compaction process, the size range for polypeptide-based NPs can vary considerably. While the polylysine particles are quite small, NPs compacted with PEG conjugated to a peptide from the glycosaminoglycan-binding domain of fibroblast growth factor (FGF) (termed POD) are spherical and ~130 nm in diameter (Read et al. 2010a). Similarly, PLGA NPs range from 200 to 400 nm (Singh et al. 2009).

Liposomes are typically composed of cationic lipids, which complex the negatively charged DNA. Often, some neutrally charged lipids are included as well, such as 1,2,-dioleoyl-3-phosphatidylethanolamine (DOPE), which can confer benefits on the particles in terms of transfection efficiency. Lipid-based vectors vary widely; evidence suggests that liposomes of ~200 nm are effectively taken up into the cell (Balazs and Godbey 2011). Liposomes usually self-assemble and are biodegradable and well tolerated (Naik et al. 2009). Solid lipid NPs are made a little differently from liposomes, and although they range from 50 to 1000 nm in size, research is ongoing to promote the preparation of homogeneous groups of particles for drug and gene delivery (Vitorino et al. 2011).

Technically, all of these systems (viral and nonviral) are nanoparticulate systems, although the typical connotation of nanoparticulate is limited to nonviral approaches. Effective gene delivery vehicles usually have a hydrodynamic diameter of 400 nm or less (Conley and Naash 2010). Many of the complexing agents mentioned here have been successfully used to deliver small-molecule drugs, and their application to gene delivery has been secondary. Finally, in addition to chemical or biological compacting agents, nanoparticulate delivery systems can also incorporate physical methods such as electroporation or iontophoresis to increase drug delivery.

17.1.2 IDEAL GENE DELIVERY CONSTRUCT

Simply put, the ideal gene delivery construct, like any drug, should be efficacious and safe. An efficacious gene delivery vector should be distributed evenly in and taken up efficiently and specifically by the target tissue, should be expressed at levels high enough to correct the disease phenotype but not so high as to cause overexpression-induced toxicity, and have an expression profile characterized by early-onset and long-lasting expression. Long-lasting expression has been particularly difficult to achieve with many nonviral delivery schemes in which episomal pDNA is quickly degraded or silenced, although newer approaches are overcoming this limitation. Several criteria must be met for a vector to be considered safe. Clearly, there should be no significant immune response, either systemic or local, and ideally, the vector should have low systemic bioavailability (i.e., it should stay where it is delivered). Furthermore, it should exhibit no ectopic expression. Finally, the vector should not cause insertional mutagenesis.

In addition to safety and efficacy, practical concerns for ocular gene therapy must also be considered. Surface instillation may be the easiest and least invasive route of delivery, but does not lead to significant intraocular expression in the posterior segment. Subretinal injection can precisely target gene delivery to the photoreceptors and retinal pigment epithelium (RPE), but leads to a temporary retinal detachment and possible long-term sequelae.

17.2 GENE DELIVERY PROCESS

One of the primary differences between the delivery of genes to the retina and the delivery of other small-molecule drugs is that successful transmission of genetic material to the nucleus is an absolute requirement for treatment efficacy. Thus, the critical features for gene delivery (in contrast to other drugs) are those things that happen at the cellular level. Concerns about intraocular dosing, route of injection, movement through the extracellular space, and so on are likely to be similar for delivery of genes and other drugs, especially given that many of the compounds used to complex and deliver genetic material are also used to deliver other non-nucleic acid-based drugs. As these issues have been covered extensively in other chapters, they will not be belabored here, and we will begin our discussion at the cell surface. There are four steps required for proper gene expression, a prerequisite for efficacious therapy. They are (1) binding and uptake into the cell, (2) trafficking to the nucleus, (3) permeation of the nuclear membrane, and (4) uncoating and expression in the nucleus. Here, we consider the mechanisms used by different types of nanoparticulate systems to accomplish these steps and possible ways to experimentally intervene to optimize them.

17.2.1 BINDING AND UPTAKE INTO THE CELL

17.2.1.1 Receptor-Mediated Uptake

17.2.1.1.1 AAV

AAV internalization depends on receptor binding to the viral capsid, of which over 120 variants exist. AAV2 is the most widely studied serotype, and much of the available information comes from studies on AAV2 and AAV5. As capsid variations are thought to be the primary determinants of tissue tropism, AAV exhibits significant variability in cell-type specificity. AAV capsid biology is a field unto itself and significant effort has gone into optimizing capsid content to improve or more specifically target viral uptake to the tissue of interest. Examples of ongoing efforts to optimize transfectivity include screening AAV variants that demonstrate transfectivity in multiple species to identify capsid variants that safely infect human tissue without being blocked by the neutralizing antibodies commonly found in human hosts (Grimm and

Kay 2003; Wu, Asokan et al. 2006); by genetically modifying well-known capsid variants to target them more specifically, for example, by incorporating a known receptor ligand into the viral capsid (Girod et al. 1999); or by directed evolution, a process that exposes a diverse library of capsid molecules and WT (i.e., replication competent) virus to selection pressure in order to optimize a particular feature such as high-affinity receptor binding (Vandenberghe et al. 2009). These issues are widely and competently reviewed elsewhere, and the reader is warned that the issue of tissue tropism is both complex and critical for efficient transduction. To add further layers of complication, tropism is often species specific, with different serotypes exhibiting differing degrees of transduction efficiency in rodents, large animals, and primate model systems, making it often difficult to predict which variants will be most efficient in humans (Vandenberghe et al. 2009).

The receptors for some AAV serotypes have been identified. For example, AAV1, AAV5, and AAV6 are known to bind to N-linked sialic acid (Kaludov et al. 2001; Wu, Miller et al. 2006), while AAV2, AAV3, and AAV2.5 (a modified variant) exhibit a preference for heparan sulfate proteoglycan (Summerford and Samulski 1998; Handa et al. 2000; Rabinowitz et al. 2002), and AAV8 and AAV9 bind laminin receptor (Akache et al. 2006). In some cases, a co-receptor is involved in binding: $\alpha V\beta 5$ integrin and FGF receptor 1 (Qiu and Brown 1999a; Summerford et al. 1999) serve as co-receptors for AAV2 and platelet-derived growth factor receptor serves as a co-receptor for AAV5 (Di Pasquale et al. 2003).

After receptor binding, the virus must be taken into the cell. The endocytic process for AAV2 and AAV5 has been characterized, and in spite of the difference in receptor, both vectors are taken into the cell in clathrin-coated pits (Ding et al. 2005), although this may vary by tissue. In the case of AAV2, the uptake process relies on a signaling cascade initiated by the receptor involving the small GTPase, Rac1 (Sanlioglu et al. 2000). Reports vary on whether endocytosis is a rate-limiting step for AAV gene delivery, but it is clear that expression of a cell surface receptor capable of binding the virus is critical (Ding et al. 2005). AAV2 is capable of driving significant gene expression in the retina, particularly in photoreceptors and RPE cells, and all three current ocular AAV clinical trials utilize AAV2-based vectors (Bainbridge et al. 2008; Cideciyan et al. 2008; Maguire et al. 2008; Cideciyan et al. 2009). Subsequent work has demonstrated, however, that other serotypes may drive higher expression in photoreceptors and RPE (AAV5) or specifically in RPE (AAV1 and AAV6) (for review, see Grimm and Kay 2003). Two newer AAV capsid variants, AAVrh8 and AAVrh10, can also efficiently transduce inner retinal cells, including ganglion, horizontal, and amacrine cells (Giove et al. 2010), while AAV6, AAV8, and AAV9 can transduce corneal cells (Sharma et al. 2010), and an AAV6 variant called ShH10 specifically targets Müller cells (Klimczak et al. 2009).

17.2.1.1.2 Polypeptide NPs

One of the most well-characterized ocular nonviral gene delivery systems is polypeptide NPs composed of PEG-conjugated lysine 30-mers (CK30PEG10K) and DNA. These CK30PEG NPs have been effective in the eye (see discussion in Section 17.3.1.2), lung, and brain and have a well-described receptor-mediated uptake pathway. CK30PEG NPs are known to bind to the cell surface receptor nucleolin (Chen, Kube et al. 2008).

Nucleolin is a cell surface receptor known to shuttle back and forth to the nucleus/ nucleolus, and the CK30PEG NPs have been found to bind to it on the cell surface with a K_d of ~25 nM (Chen, Kube et al. 2008). Lack of colocalization with markers of clathrin- and caveolin-mediated endocytosis suggests that CK30PEG NP internalization occurs by a different route (Chen, Kube et al. 2008; Chen et al. 2010). Recent studies have demonstrated that nucleolin accumulates on the cell surface in lipid rafts, dynamic aggregations of cholesterol and glycosphingolipids, and that lipid raft-mediated endocytosis is the mechanism by which nucleolin and CK30PEG NPs are taken into the cell, completely bypassing traditional endocytic machinery (Chen et al. 2010). This process is quite efficient, and the tissues that take up CK30PEG NPs efficiently correlate with those expressing cell surface nucleolin (Conley and Naash 2010). In the eye, we have demonstrated that most retinal cell types express nucleolin and have shown that retinal cells and cells of the anterior segment (including cornea, lens, and trabecular meshwork) readily take up CK30PEG NPs (see further discussion in Section 17.3.1.2) (Farjo et al. 2006).

17.2.1.1.3 Liposomes and Solid Lipid NPs

Untargeted liposomes have been demonstrated to be taken up into cells by both clathrin-dependent and caveolin-dependent endocytosis; however, the receptor responsible has not been identified (Pichon et al. 2010). Receptor targeting has been used to enhance liposome-mediated transfection and to increase cellular specificity; as examples, transferrin and epidermal growth factor (EGF) have been included on the surface of the liposome to target them to specific receptors (Zhai et al. 2010; Bunuales et al. 2011). In the eye, a peptide ligand for vascular endothelial growth factor (VEGF) receptor 2 has been incorporated into drug delivery liposomes for the treatment of choroidal neovascularization (Li, Zhang et al. 2010) but has not yet been tested on gene delivery liposomes. Significant research on liposome targeting has been done in the cancer field. These liposomes commonly employ a subtype of receptor targeting and are termed immunoliposomes due to the incorporation of monoclonal antibodies on their surface to aid targeting and uptake (Park et al. 2001).

Solid lipid NPs have been used extensively to deliver drugs, but their use for gene delivery in the eye has been limited to *in vitro* studies in transformed human cultured RPE cells (ARPE-19) cells. In *in vitro* experiments, clathrin-mediated endocytosis was implicated in the uptake of solid lipid NPs, but no specific receptor has been identified (del Pozo-Rodriguez et al. 2008).

17.2.1.2 Nonreceptor-Mediated Uptake

17.2.1.2.1 Cell-Penetrating Peptides

Cell-penetrating peptides (CPPs) are a class of peptides that pass cargo through the cell membrane without the aid of a receptor. Although these peptides vary in composition, they are often alpha-helical and may contain either arginine or lysine motifs or a hydrophobic core (Rhee and Davis 2006). The cell entry mechanism seems to vary from peptide to peptide, but experiments demonstrating equal uptake of l- and d-peptide isomers indicate that the process is not receptor-mediated (Futaki et al. 2001) and does not appear to occur by any standard endocytic routes. Some evidence suggests that penetration may occur by a specialized form of endocytosis called fluid-phase macropinocytosis, which may also be dependent on lipid rafts (Wadia et al. 2004). Other studies have proposed a mechanism of entry for highly basic CPPs such as the TAT peptide (which originated on the HIV-1 virus), postulating that interactions between the charged arginine/lysine residues and the phosphate groups on the proximal leaflet of the lipid bilayer initiate the formation of a pore. As the TAT peptide concentration increases, positively charged arginine and lysine side chains are attracted to phosphate groups on the distal side of the plasma membrane and facilitate movement of the peptide through the pore (Herce and Garcia 2007). One well-studied CPP is C105Y, a synthetic peptide whose sequence originates in the α-1-antitrypsin gene. Consistent with other CPPs (Wadia et al. 2004), internalization of C105Y is not energy dependent, in contrast to clathrin-mediated endocytosis; however, passage of the peptide into the cell was shown to be energy-dependent, suggesting that nuclear import of CPPs may rely on traditional methods (described in Section 17.2.3). Peptides such as C105Y have been attached to DNA NPs to facilitate receptor-independent DNA uptake and gene expression (Ziady et al. 1999). For example, in the eye, a CPP called POD (originally from the glycosaminoglycan-binding domain of the FGF receptor) has been used as both a compacting peptide and a CPP for intraocular NP delivery (Johnson et al. 2008; Read et al. 2010a). NPs compacted with this peptide are capable of driving expression of both a reporter gene and glial-derived neurotrophic factor (GDNF) in multiple ocular tissues, including photoreceptors, ganglion cells, and the RPE (Read et al. 2010a, b).

17.2.2 INTRACELLULAR TRAFFICKING

17.2.2.1 Trafficking of AAVs after Clathrin-Mediated Endocytosis

Although AAVs can be taken up by multiple pathways, those virions taken up by clathrin- or caveolin-mediated endocytosis typically converge at the early endosome stage. Again, the processing steps for AAV vary significantly across serotypes and across cell types. However, some general trends regarding the intracellular trafficking of AAV seem to be emerging. The first is that endosomal processing is a critical step for driving subsequent effective gene expression. AAV2 which bypasses endosomal processing by being injected directly into the cytoplasm of HeLa cells, fails to accumulate in the nucleus and fails to express its transgene (Ding et al. 2005), suggesting that endosomal "priming" of the capsid is important for nuclear transport and/or viral uncoating. This process may be dependent on acidification of the virus in the late endosome/lysosome; in some cases where acidification is blocked, AAV transduction decreases (Douar et al. 2001; Hansen et al. 2001). It is also becoming evident that in some cases, inhibition of the proteasome promotes intracellular trafficking or nuclear import and that this process is enhanced in particles that have undergone prior acidification (Douar et al. 2001). Similarly, it has been demonstrated that EGF receptor-mediated tyrosine phosphorylation of viral capsid proteins leads to ubiquitination and degradation of viral particles (Zhong et al. 2008; Markusic et al. 2010). A new experimental response to this pathway is to introduce tyrosine to phenylalanine substitutions in amino acid residues on the capsid surface in order to minimize capsid phosphorylation and degradation. This approach has been successfully used to enhance AAV-mediated gene expression in both retina and liver

(Markusic et al. 2010; Petrs-Silva et al. 2011). The precise vesicular pathways through which AAVs are processed are extremely variable, and the reader is referred to Ding et al. (2005) for a more extensive review of this issue. It is not clear precisely at which stage AAVs exit the endosomal processing pathway, but it is reasonably clear that this is necessary for subsequent nuclear translocation. Nuclear translocation of AAV is quite slow and inefficient and may be considered a rate-limiting step in AAV transduction (Ding et al. 2005). Although it is not precisely clear how AAV traffics after being released from the endosome/lysosome, translocation of AAV2 has been shown to be inhibited by blocking both tubulin and actin filaments (Sanlioglu et al. 2000), suggesting that they play a role in the process.

17.2.2.2 Trafficking after Nucleolin-Mediated Uptake

Raft-mediated endocytosis is a fairly generic name referring to endocytic processes that are cholesterol dependent. In fact, caveolin-mediated endocytosis falls into this category, as does the nucleolin-mediated process by which CK30PEG NPs enter the cell; however, apart from this, nucleolin-mediated trafficking is distinct from other forms of endocytosis. Nucleolin rapidly traffics from the cell surface to the nucleus and vice versa and is found in the cytoplasm and enriched in the nucleolus in addition to its localization to the cell surface (Chen, Kube et al. 2008). It does not have a membrane targeting sequence, and it has been demonstrated that its plasma-membrane, lipid-raft localization is due to interactions with the raft protein flotillin (Chen et al. 2010). After NP binding, nucleolin-mediated NP trafficking proceeds through dynein-mediated, microtubule-dependent transport toward the nucleus through a process independent of actin filaments (Chen et al. 2010). For nucleolin-mediated NP trafficking, this process is very rapid; NPs accumulate in the cytoplasm within 15 minutes of being delivered to the extracellular space and accumulate in the nucleolus within 1 hour (Chen, Kube et al. 2008). Other NPs may take advantage of the microtubule network, but there is no evidence that other NPs are trafficked by nucleolin. For example, data from COS-7 cells exposed to DNA NPs compacted with polyethyleneimine (PEI) demonstrated that they were actively transported through motor proteins along the microtubule system and accumulated rapidly in the peri-nuclear region (Suh et al. 2003).

17.2.2.3 Intracellular Trafficking of Liposomes

As with AAVs, liposomes enter the endosomal processing pathway after clathrin- or caveolin-mediated endocytosis. However, in contrast to AAVs, which are released from endosomes/lysosomes through efficient, although poorly understood, mechanisms, release of liposomes and solid lipid NPs from the endosomal pathway has been a major barrier to their use for clinically relevant ocular gene delivery. In fact, one of the reports on the use of solid lipid NPs to deliver genes to ocular tissues (ARPE-19 cells) reported limited transfection, attributed to accumulation, and subsequent degradation of the vector in lysosomes (del Pozo-Rodriguez et al. 2008). Researchers have experimented with incorporation of peptides modeled on viral fusion peptides to promote endosomal escape, while others have utilized a lytic peptide called melittin. Unfortunately, these often destabilize the plasma membrane as well as endosomal vesicular membranes, causing cytotoxicity (Pichon et al. 2010). Some common lipid components of liposomes such as DOPE have fusogenic activity, and incorporation during liposome formulation can improve endosomal escape. After release from endosomes, however, pDNA is exposed directly to the cytosol. This leads to rapid degradation of the pDNA by DNases; pDNA in the cytosol has an estimated half-life of only 50–90 minutes (Lechardeur et al. 1999), thus further reducing the efficiency of downstream gene expression. Consistent with this, we have demonstrated that transfection of primary ocular cells (trabecular meshwork) with lipid-based agents is enhanced by treatment with DNase inhibitors (Hoffman et al. 2005). There is evidence to suggest that nonspecific binding of plasmids to intracellular actin or keratin inhibits trafficking toward the nucleus, and indeed, pDNA microinjected into the cytosol has been observed to remain near the site of injection, suggesting limited intracellular diffusion (Dowty et al. 1995). On the other hand, there is some evidence that pDNA can be actively transported along microtubules (Vaughan and Dean 2006). Efficient intracellular trafficking without degradation is a major hurdle for lipid-based gene delivery.

17.2.3 NUCLEAR IMPORT

In contrast to many other non-nucleic acid drugs that may have extracellular or cytoplasmic targets, DNA-based compounds must be transported into the nucleus to have any beneficial effect. However, nuclear import has long been recognized as a significant gene transfer barrier. Over 30 years ago, studies demonstrated that gene expression after injection of pDNA into the cytosol was undetectable, while 50–100% of cells expressed the reporter gene (thymidine kinase) after direct nuclear injection (Capecchi 1980). All transport (passive and active) into the nucleus occurs through nuclear pore complexes (NPCs) embedded in the nuclear envelope (Wente 2000). While small molecules (up to 25 nm or 45 kDa) may be able to diffuse passively through nuclear pores, most larger molecules and complexes are actively transported through the NPC through the interactions between nuclear localization sequences (NLSs) and one of the nuclear pore components (importin-α) (Wagstaff and Jans 2007). Based on these two mechanisms (passive diffusion of small molecules and active transport of complexes with a NLS), the ideal gene therapy vector will be quite small or able to assemble with components of a nuclear targeting complex.

17.2.3.1 Nuclear Import of Viruses

The exact process by which AAVs are taken into the nucleus is not known; they are of a size that might permit diffusion across the nuclear pore, but this has not been clearly established. Some studies have suggested that nuclear import of AAVs might not depend on classical nuclear receptor/nuclear import machinery, while others have demonstrated that AAVs can associate with the protein nucleolin (Qiu and Brown 1999b) in the cytoplasm, facilitating nuclear uptake; still others have suggested that the capsid protein VP2 may contain a nontraditional nuclear localization signal (Hoque et al. 1999) facilitating nuclear entry.

17.2.3.2   Nuclear Import of Nucleolin-Bound NPs

A critical, often rate-limiting step for nonviral gene expression is transport through the nuclear membrane, and vectors that complex with nuclear-targeted proteins have a much better chance of penetrating the nuclear envelope than vectors lacking this association. Recent experiments have demonstrated that nucleolin carrying CK30PEG NPs in the cytoplasm binds to a member of the glucocorticoid nuclear receptor family (GCR) (Chen et al. 2010). Nucleolin binds to the DNA-binding domain of the GCR (Schulz et al. 2001), at which time the receptor assembles with other proteins into a nuclear import complex (Heitzer et al. 2007). Interestingly, the DNA-binding domain of GCR is ordinarily blocked in the absence of ligand (cortisone). Consistent with the idea that increasing assembly of the nuclear import complex would increase NP transfection efficiency, incubation of tissue culture cells with cortisone increased both nucleolin–GCR interactions and CK30PEG NP-based transfectivity and gene expression (Chen et al. 2010).

17.2.3.3   Nuclear Import of Lipid-Bound or Naked Plasmid DNA

In addition to endosomal escape, another chronic problem for nonviral vectors has been nuclear import. While CK30PEG NPs can enter the nucleus as a result of their interaction with nucleolin, and also may diffuse through nuclear pores (due to their small size), neither of these benefits typically apply to liposomes or solid lipid NPs. In many instances, nuclear import must rely on cell division and the accompanying breakdown of the nuclear envelope, making them quite inefficient for delivery to postmitotic cells. For example, solid lipid NPs were twice as effective at transfecting HEK293 (human embryonic kidney) cells than ocular ARPE-19 cells, a difference the authors attributed to the slower rate of division of ARPE-19 cells versus HEK293 cells (del Pozo-Rodriguez et al. 2008). Similarly, primary trabecular meshwork cells took up liposome-complexed DNA into almost 100% of cells, but DNA was only detected in the nuclei of ~4% of those cells and gene expression was correspondingly low (Hoffman et al. 2005).

Passive diffusion of pDNA through nuclear pores does not occur as a result of large size, although small oligonucleotides (<250 bp) can rapidly penetrate the nuclear envelope (Lukacs et al. 2000). There is some evidence suggesting that pDNA can be actively transported into the nucleus, but the process is still quite inefficient and relies on specific DNA sequences such as the SV40 origin of replication and early promoter regions (Dean et al. 1999). As a result of these limitations, significant energy has gone into increasing the nuclear import of plasmid and lipid complexed DNA. One approach that has been explored with some success is incorporating an NLS peptide. This can be accomplished by condensing the DNA with an NLS peptide, by binding the NLS peptide to coating agent, or by binding the NLS directly to the DNA. In addition, some have tried to increase nuclear import by complexing the DNA with adenoviral or retroviral proteins that have NLSs. Finally, pDNA can be complexed with mammalian proteins, such as transcription factors, that traffic to the nucleus. For more information on these options, the reader is referred to Wagstaff and Jans (2007).

17.2.4 GENE EXPRESSION

Once the vector enters the nucleus, the last significant step is generating gene expression.

17.2.4.1 Genome Conversion, Virus Uncoating, and Expression

AAV uncoating (i.e., separation of the capsid from the genome) occurs in the nucleus, and evidence suggests that uncoated AAVs (specifically AAV2) may remain stably sequestered in the nucleolus for an extended period of time prior to being mobilized to the nucleoplasm for gene expression (Johnson and Samulski 2009). Evidence suggests that this process may be accelerated by nucleolar disruption during mitosis, by delivery of a genotoxic agent (such as hydroxyurea), or by coinfection, a natural requirement for the replication of WT AAV (Johnson and Samulski 2009). AAVs are classified as dependoviruses, meaning that for successful replication, they require the presence of an additional virus (such as adenovirus or herpes simplex virus). Different AAV serotypes have differential requirements for helper viruses, and although helper viruses are not specifically required for gene expression, there is some suggestion that AAV serotypes with a lesser requirement for helper viruses (such as AAV5) may be able to drive higher levels of gene expression (Grimm and Kay 2003). After being uncoated, single-stranded AAV DNA is converted to double-stranded DNA by the nuclear machinery, after which transcription can be initiated. Insertional mutagenesis has historically been a serious concern with viral vectors, particularly with retroviruses (Cavazzana-Calvo et al. 2000), but although WT AAV integrates into the genome, replacement of the *rep* and *cap* open reading frames with the therapeutic expression cassette in recombinant AAV eliminates targeted integration (Flotte 2005). Therefore, the majority of recombinant AAV2 vectors used for gene therapy remain episomal, although very small amounts of random integration can be detected. Although potentially problematic in rapidly dividing tissues, stably episomal vectors are not necessarily a disadvantage in the primarily postmitotic eye, provided they can continue to be expressed long-term.

17.2.4.2 Expression of Nonviral Vectors

Once in the nucleus, pDNA (from liposomes, NPs, or other vectors) remains episomal and can be transcribed. Several factors can contribute to the attenuated duration of gene expression often associated with nonviral DNA delivery. These can include (1) loss of vector during mitosis; (2) gene silencing as a result of the promoter (e.g., the cytomegalovirus CMV promoter is silenced within a few days of gene delivery); (3) gene silencing due to the presence of bacterial elements in the plasmid backbone, possibly a result of heterochromatin spreading; (4) silencing due to vector degradation or instability; (5) silencing due to vector content such as CpG islands; and (6) silencing due to improper nuclear localization or conformation within the nucleus. Significant research into vector optimization has been ongoing to help overcome these difficulties, and it is of particular importance in the eye since most ocular genetic diseases are chronic in nature (requiring long-term treatment), the cells of the retina are postmitotic and so gene expression from episomal vectors could be maintained if vectors were not silenced, and the invasiveness of dosing makes minimizing treatment frequency important.

17.2.4.2.1 Promoting Vector Integration

Two major systems have been incorporated into nonviral vectors to promote genome integration and combat vector loss due to cell division or episomal silencing. The first is called the sleeping beauty transposon–transposase system and involves delivery of a plasmid containing two expression cassettes—one encoding a transposase enzyme, and the other carrying the therapeutic gene flanked by transposon elements (Conley and Naash 2010). Transposons (and the DNA they flank) are able to migrate from one chromosomal locus to another (or in this case from the vector to the genome) with the help of transposase enzyme. Although the integration is not site specific, the transposons show a preference for microsatellite repeats and are less likely to integrate into actively transcribed genes than viral vectors (Liu et al. 2005). Integration does not occur in every transfected cell, but significant research is going into optimizing the transposase enzyme for maximum insertional activity (Baus et al. 2005). This technology has not yet been utilized in the eye but has been used to drive persistent gene expression in various tissues, including the liver and the skin (Yant et al. 2000; Ortiz-Urda et al. 2003). The second system is the ΦC31 integrase. As with the sleeping beauty system, the ΦC31 system relies on codelivery of a bacterial integrase gene along with the therapeutic gene that is flanked by *AttB* sites (Chalberg et al. 2005). In the native bacteriophage from which this system originates, *AttB* sites recombine with *AttP* sites. In the mammalian system, the *AttB* sites recombine with a limited number of pseudo *AttP* sites found in the mammalian genome. The mammalian pseudo *AttP* sites are not reported to be near cancer-causing genes, theoretically making ΦC31-mediated integrations unlikely to cause cancer-associated insertional mutagenesis (Ehrhardt et al. 2005). This technology has been successfully tested in the rat retina; gene expression from a reporter gene vector codelivered with an integrase vector lasted for the duration of the study period (4.5 months), while expression decreased to baseline by ~1 month in animals that received only the reporter vector (Chalberg et al. 2005). Unfortunately, significant chromosomal rearrangement has been observed after transduction with ΦC31-integrase-containing vectors, suggesting that the process may need to be further optimized prior to widespread use (Ehrhardt et al. 2006).

17.2.4.2.2 Silencing due to Promoter

Promoter choice is critical for effective gene therapy. Promoters vary significantly in their tissue specificity (or lack thereof), levels of expression, and duration of expression. In addition, multiple enhancer and regulatory elements can be combined with the promoter to alter those factors, and many of these elements have been incorporated into gene delivery vectors. Promoter selection will vary from application to application, and selection must be based on the requirements of a given disease target or model. Many ocular cell-type-specific promoters have been characterized and used in ocular gene therapy trials. Some of the more-often used include (1) the mouse opsin (MOP) promoter, which drives high levels of gene expression in rods and some expression in cones (Quiambao et al. 1997; Ali et al. 2000; Cai et al. 2010); (2) the rhodopsin kinase promoter, which drives gene expression in rods and cones (Beltran et al. 2010; Boye et al. 2010); (3) the interphotoreceptor retinoid-binding protein (IRBP) promoter, which drives gene expression in rods and cones (Cai et al. 2009); (4) the vitelliform macular dystrophy 2 promoter, which drives gene expression in RPE (Esumi et al. 2004; Kachi et al. 2006); (5) the β-3 nicotinic receptor promoter, which drives expression in ganglion cells (Skowronska-Krawczyk et al. 2005); and (6) the RPE65 promoter to drive RPE expression (Bainbridge et al. 2008). Ocular gene therapy studies have also effectively used ubiquitous promoters including the chicken β-actin (CBA) (Maguire et al. 2008; Cai et al. 2009) and the CAG (the CBA promoter and the CMV immediate early enhancer) promoters (Cideciyan et al. 2008; Cideciyan et al. 2009).

Generally speaking, tissue-specific promoters are favored as they lead to less ectopic expression and fewer safety concerns. However, for intraocular treatment, ectopic expression is often less of a concern than with other types of delivery since the choice of injection site can limit access of the genetic material to a particular tissue of interest and systemic bioavailability of genetic material after intraocular delivery is usually low. Furthermore, in some cases, if the ultimate effect of the transgenic protein is extracellular (i.e., delivery of a gene coding for a secreted neurotrophic factor), therapeutic benefits can occur after expression in multiple cell types. Finally, promoters that are ubiquitously expressed may have the appearance of cell-type specificity if the transgene they are driving is cell-type specific. For example, we have demonstrated that the non-cell-type-specific CBA promoter drives expression of the outer segment protein retinal degeneration slow (RDS) only in photoreceptors after subretinal injection (Cai et al. 2009), even though we have observed that it drives eGFP (enhanced green fluorescent protein) expression in multiple ocular cell types (unpublished data). This difference is due to the highly specialized nature of RDS; any cells without outer segments that might ectopically express RDS likely degrade it rapidly as it would not be processed properly. A similar phenomenon has been observed when the CBA promoter is used to drive RPE65 expression (from an AAV vector) in *rd12* mutant mice; in treated mice, transgene-driven RPE65 expression is localized exclusively to the RPE layer (Pang et al. 2006). The choice of promoter will affect both the efficacy and the safety profile of the therapy and should be considered carefully. In addition, the size limits for AAV-based vectors (~4.8 kb) mean that promoter size may also be a selection criterion in some cases.

17.2.4.2.3 Nuclear Targeting

Significant improvements in the persistence of episomal vectors can be achieved by incorporating sequences that promote localization of the vector to transcriptionally active regions of the nucleus. RNA polymerase and transcription complexes have been shown to be localized to the nuclear cytoskeleton and are thought to have a relatively fixed location in the nucleus in contrast to the chromatin that moves around them (Cook 1999). Associated with this nuclear cytoskeleton are DNA-binding scaffolding proteins, which are known to bind to specific DNA elements called scaffold/matrix attachment regions or S/MARs (Jackson et al. 2006; Harraghy et al. 2008). These regions have been identified in multiple genes, and the one found in the human β-interferon gene is one of the most well characterized (Bode et al. 1992; Jenke, Scinteie et al. 2004; Jenke, Stehle et al. 2004). This region can be incorporated into gene therapy vectors to promote gene transcription. In addition to promoting localization to transcriptionally active regions of the nucleus, inclusion of S/MAR regions is thought to promote gene transcription by reducing superhelical stress and by promoting association with modified histones characteristic of active chromatin. S/MARs have been used to prolong gene expression in multiple tissues including the liver and the brain. Notably, CK30PEG NPs carrying a luciferase reporter gene flanked by S/MARs drove gene expression in the rat striatum for up to 1 year, suggesting that S/MARs may be effective for transduction of neural tissue (Kaytor et al. 2009). Interestingly, there is an S/MAR motif in the 5′ region of the tyrosinase gene (part of the melanin synthesis pathway), which is highly expressed in the RPE during development. When included in the expression cassette, this S/MAR has been shown to promote transgene expression (in transgenic mouse lines) (Porter et al. 1999). Studies on incorporating S/MARs into vectors being used in the eye are ongoing, and data on the use of S/MARs in the RPE are particularly promising.

17.2.4.2.4 *Inhibiting Heterochromatin Spreading*

DNA found in heterochromatin is transcriptionally silent. Heterochromatin and euchromatin are differentiated by the pattern of histone modifications they feature, and alterations in these modifications can silence or activate regions of the genome (Jenuwein and Allis 2001; Richards and Elgin 2002). Heterochromatin spreading is a well-documented phenomenon wherein histone-modifying enzymes can suppress an entire genomic locus starting from one isolated region of heterochromatin (Richards and Elgin 2002). It has been demonstrated that some bacterial plasmid backbone regions are associated with heterochromatin-like histone modifications (Suzuki et al. 2006; Riu et al. 2007) and that heterochromatin spreading can cause silencing of episomal transgenes (Riu et al. 2007). Solutions to this problem typically take one of two forms. The first solution is to introduce insulating elements into the vector across which heterochromatin will not spread. For example, inclusion of the chicken cHS4 insulating region or S/MAR regions (which have insulating properties in addition to their other benefits) have been shown to limit heterochromatin spreading and promote prolonged gene expression (Noma et al. 2001; Jackson et al. 2006; Chen, Riu et al. 2008). The second solution is to remove the problematic bacterial backbone element. This can be done by delivering a linearized expression cassette rather than pDNA (Chen et al. 2004) or by delivering DNA minicircles containing only the expression cassette (Chen et al. 2003, 2005). Minicircles are ingeniously produced from a larger vector that contains the therapeutic expression cassette flanked by I-SceI bacterial endonuclease sites as well as an expression cassette containing the endonuclease. Recombination in the bacteria allows production of a large quantity of DNA minicircles containing only the expression cassette, which can then be purified and delivered, while the remaining vector DNA elements are degraded (Chen et al. 2003, 2005). Minicircles have been used to prolong DNA expression in liver, skeletal muscle, and skin (Chen et al. 2003, 2005; Stenler et al. 2009; Yoon et al. 2009).

17.2.4.2.5 *CpG Methylation*

The role of CpG methylation in gene silencing is controversial and may well be vector and tissue specific. Experiments using liposomes to deliver transgenes to C3A human hepatoblastoma cells and experiments using lentiviral vectors to deliver transgenes to murine embryonic stem cells support the idea that methylation of CpGs in the vector or promoter may contribute to gene silencing (Hong et al. 2001; He et al. 2005).

Similarly, levels of expression were higher when CpG-depleted vectors were delivered to the lung by PEI NPs rather than standard vectors (Hyde et al. 2008). In contrast, other groups have shown that methylation status and CpG depletion had no effect on gene expression after hydrodynamic delivery to the liver (Chen, Riu et al. 2008). Preliminary data from our collaborators using CpG-depleted vectors to deliver genes through CK30PEG NPs to the lung and the brain support the idea that the role of methylation in gene silencing may be tissue dependent (Mark Cooper, personal communication).

17.3 SUCCESSFUL OCULAR GENE DELIVERY

The majority of ocular gene therapy thus far has focused on the retina, although corneal gene transfer is also undergoing rapid development (Jani et al. 2007; Johnson et al. 2008; Klausner et al. 2010; Sharma et al. 2010). While many different types of vectors have been tested *in vitro* and *in vivo* for their ability to drive reporter gene expression, few have been able to promote rescue in animal models.

17.3.1 GENE REPLACEMENT THERAPY

17.3.1.1 Gene Replacement Using AAV

AAVs have been used experimentally to treat many animal models of ocular disease and are the most advanced ocular gene delivery vehicles in terms of clinical use. Animal models tested with various AAV serotypes and vectors include the *rd12* model of Leber's congenital amaurosis (LCA, RPE65 mutation) (Pang et al. 2006; Li, Li et al. 2010), the *rds* model of autosomal dominant retinitis pigmentosa (ADRP) (Ali et al. 2000), the whirlin knockout model of Usher's syndrome type IID (Zou et al. 2011), the guanylate cyclase-1 knockout (a model for LCA) (Boye et al. 2010), the RPGRIP1 knockout (a model for LCA) (Pawlyk et al. 2010), and the RS1 knockout (a model for X-linked juvenile retinoschisis) (Park et al. 2009), among many others.

Currently, multiple clinical trials are under way for the treatment of LCA with AAV2-based therapies. LCA is a severe, early-onset form of retinal degeneration associated with mutations in the *Rpe65* gene. The first major milestone was reported in 2001, when a multiuniversity team of researchers reported long-lasting improvement in vision in blind Briard dogs (Acland et al. 2001). As a result of a naturally occurring mutation, these dogs are a well-studied LCA model, and one of the team's successfully treated dogs, named Lancelot, has become something of a mascot for ocular gene therapy. As a result of successful preliminary work, three independent phase 1 clinical trials using AAV2 for the treatment of LCA were undertaken. All three groups delivered the human RPE65 cDNA (under the control of the CBA promoter, Maguire et al. 2008; the human RPE65 promoter, Bainbridge et al. 2008; or the CAG promoter, Jacobson et al. 2006, Cideciyan et al. 2008). As the trials were initially designed as safety and tolerability (not efficacy) studies, there were only a small number of patients in each group. Furthermore, each trial had different outcome measures and so it has been difficult to determine which treatment provided the best improvement in vision. Most importantly, in all these initial trials, the treatments were well tolerated and resulted in no significant accumulation of antibodies against the vector or other signs of immune response, and no serious adverse events were reported (Bainbridge et al. 2008; Cideciyan et al. 2008, 2009; Hauswirth et al. 2008; Maguire et al. 2008). All three trials reported encouraging early results (Bainbridge et al. 2008; Cideciyan et al. 2008; Maguire et al. 2008), and two of the groups have published follow-up data for up to 1.5 years posttreatment (Cideciyan et al. 2009; Simonelli et al. 2010). At 1 year (Cideciyan et al. 2009) and 1.5 years posttreatment (Simonelli et al. 2010), authors reported that participants in the two separate studies continued to be healthy and that AAV2 antibody titers remained lower than the population mean and were not higher than at baseline. Cideciyan et al. (2008, 2009) had previously reported that at 3-months posttreatment, a significant increase in light sensitivity was detected in treated eyes, and they further reported that this improvement persisted to 1 year. In a follow-up to the initial Maguire et al. (2008) study, researchers reported that improvements in pupillary light response (first observed at 1 month after treatment) persisted through 1.5 years (the latest time point examined). Similarly, they reported that subjective measures of visual function, including visual acuity and mobility in an obstacle course, continued to improve through the extended study period. These promising initial results have led to the implementation of several expanded trials in the United States (http://clinicaltrials .gov) as well as an additional trial in Israel (Banin et al. 2010).

In addition to clinical trials with AAVs for LCA, animal studies are still ongoing. Most of these studies are focusing on other AAV serotypes, particularly AAV5. Studies in *rd12* mice (an RPE65 mutant model) have recently shown that cone photoreceptors could be preserved and opsin expression reinitiated after late (postnatal day 90) treatment with AAV5 RPE65 (Li, Li et al. 2010). These results are particularly exciting given that it is often difficult to intervene in patients before disease onset, so treatments effective after degeneration has begun are desirable.

17.3.1.2 Gene Replacement Using CK30PEG NPs

Currently, CK30PEG NPs are still undergoing preclinical testing for ocular gene delivery, although they have been successfully used in a phase 1/2 clinical trial to deliver the cystic fibrosis (CF) transmembrane receptor to the airways of CF patients (Konstan et al. 2004), and another CF trial is in the application stages. Initial ocular studies using an eGFP reporter gene under the control of the CMV promoter used two different NP formulations, rod-shaped and ellipsoid-shaped NPs (as described in Section 17.1). These NPs were delivered to the subretinal or intravitreal space of the mouse eye and gene expression was examined (Farjo et al. 2006). Intravitreal injection led to high levels of gene expression in the lens with modest expression in the retina, cornea, and trabecular meshwork. Very little expression was detected in the RPE or choroid/sclera. Subretinal injection, in contrast, led to significant retinal and RPE gene expression as well as expression in the optic nerve head. In these studies, acetate (rod-shaped) compacted NPs were more efficient than ellipsoidal NPs (Farjo et al. 2006), and those were chosen for subsequent work.

In later work, we have tested the ability of CK30PEG NPs to drive long-term gene expression in the eye (CMV-driven expression was shut down by postinjection day 7 [PI-7]) and to rescue the ADRP phenotype of the RDS mouse model ($rds^{+/-}$). The $rds^{+/-}$ mouse exhibits haploinsufficiency-based early onset, slow rod degeneration, followed by late-onset cone degeneration, and exhibits ADRP symptoms similar to those seen in patients with loss-of-function mutations in RDS (Cheng et al. 1997; Farjo and Naash 2006). CK30PEG rod-shaped NPs carried the therapeutic gene (normal mouse peripherin/RDS, called NMP) under the control of one of three different promoters chosen to drive longer-term gene expression (Cai et al. 2008, 2009). The first two promoters were photoreceptor specific (MOP and IRBP) (Liou et al. 1991; Yokoyama et al. 1992; Flannery et al. 1997; Quiambao et al. 1997) while the final promoter chosen was the ubiquitously expressed CBA promoter. NPs or controls were delivered to the subretinal space of $rds^{+/-}$ mice at postnatal day 5. All three promoters drove high levels of gene expression (message and protein), which peaked between PI-2 and PI-7 and stabilized at levels approximately twofold higher than in control eyes for the duration of the study (4 months) (Cai et al. 2009, 2010). In all the NP-injected mice, RDS transgene expression was properly localized to the outer segment and gene expression was detected in cells outside of the retinal detachment injection bleb. This observation is significant as limited retinal distribution of gene expression has been a problem with some other vectors (Conley and Naash 2010; Herzog et al. 2010; Roy et al. 2010). Delivery of all three NPs led to significant improvement in the ADRP phenotype as measured structurally, functionally, and biochemically (Cai et al. 2009, 2010). Consistent with the persistence of elevated gene expression levels, phenotypic improvements also persisted through the duration of the study. Encouragingly, our ongoing studies have demonstrated that when injected in this murine model, IRBP-NMP NPs are capable of driving gene expression and promoting phenotypic improvement for up to 15 months, which is the longest time point examined (unpublished data).

Although gene expression was detected throughout the retina, structural improvement was more pronounced on the injected side of the eye, and even when gene expression was driven by the rod-dominant MOP promoter, improvements in cone function were significantly more pronounced than improvements in rod function. We hypothesize that this pattern may be due to the differential requirement of rods versus cones for RDS. Rods have a higher demand for RDS than cones, and thus, the rescue of rods may require higher expression levels than the rescue of cones. Additionally, rods begin to degenerate early in $rds^{+/-}$ eyes, while cones degenerate later and so the timing of treatment delivery may be critical. In support of this hypothesis, in studies wherein we delivered MOP-NMP NPs at P21 (instead of P5), NP-driven gene expression levels were similar to those seen after early injection, but rescue was significantly less pronounced (Cai et al. 2010), likely due to ongoing photoreceptor degeneration in the older $rds^{+/-}$ mouse.

CK30PEG NPs are also safe and well tolerated in the eye. In our initial reporter gene studies, NP treatment did not lead to functional deficits (Farjo et al. 2006), supporting the idea that the NPs were fairly nontoxic. To evaluate whether NPs induce an immune response after ocular delivery, P22 WT animals were subretinally injected with varying doses of eGFP CK30PEG NPs and immune reactivity was evaluated at 1, 2, 4, or 7 days PI (Ding et al. 2009). No infiltration of polymorphonuclear neutrophils, lymphocytes, or macrophages was detected, and no NP-related elevation in cytokines was detected. We observed a transient increase in KC (murine IL-8) mRNA levels and in monocyte chemotactice protein-1 mRNA and protein levels on PI-1, but levels were resolved by PI-2 and were detected in both

NP- and vehicle-treated eyes (Ding et al. 2009). Since we had previously observed that $rds^{+/-}$ eyes were more sensitive to subretinal injection–related damage than WT eyes (Nour et al. 2003; Cai et al. 2009), we undertook similar toxicity testing in $rds^{+/-}$ mice injected with MOP-NMP CK30PEG NPs (Cai et al. 2010). As with WT mice, no significant immune response was detected in NP-treated $rds^{+/-}$ animals. These data support the idea that NPs are safe and well tolerated after subretinal injection. Currently, studies are under way to optimize these NPs for use in other disease models and for intravitreal delivery. Although ongoing clinical trials are utilizing subretinal injection, this process is quite invasive, and delivery by intravitreal injection (which does not induce retinal detachment) would be preferred.

17.3.2 OTHER STRATEGIES

In addition to traditional gene replacement, ocular research on other types of genetic therapies is exploding. Approaches include delivery of neurotrophic factors to retard retinal degeneration (LaVail et al. 1998; Cayouette et al. 1999; Liang et al. 2001; Campochiaro et al. 2006), delivery of antiangiogenic factors to inhibit retinal and corneal neovascularization (Pechan et al. 2009; Zhou et al. 2010; Maclachlan et al. 2011), and delivery of endoplasmic reticulum (ER) chaperones to promote folding and trafficking of mutant photoreceptor proteins (Gorbatyuk et al. 2010).

17.3.2.1 Gene Delivery of VEGF Inhibitors

Genetic inhibition of neovascularization (wet age-related macular degeneration, [AMD]) is following on the heels of the success of small-molecule drugs for the same purpose. A new clinical trial testing the safety of utilizing AAV2 to deliver a gene coding for a soluble portion of the VEGF receptor (termed sFLT-01) is now recruiting patients (clinicaltrials.gov, NCT01024998). This treatment is designed to suppress VEGF levels and reduce choroidal neovascularization in patients with wet AMD. The approach was effective at suppressing angiogenesis in a mouse disease model (Pechan et al. 2009). In cynomolgus monkeys intravitreally treated with AAV2-sFLT-01 (under the control of the CBA promoter), sFLT-01 gene expression was detected for up to 1 year in the aqueous humor and vitreous, although the levels were quite variable (Maclachlan et al. 2011). Toxicological review of the animals revealed that 14/18 AAV-treated animals developed vitreal inflammation between 1 month and 5 months postinjection (Maclachlan et al. 2011). This inflammation had resolved in the majority of animals by 12 months postinjection and did not correlate with measured sFLT-01 levels. In the majority of animals, the inflammation was graded as trace to mild with only a few animals showing moderate to marked inflammation and no animals with severe inflammation. This inflammation (measured in live animals) correlated with histological markers of inflammation in the animals that were collected mid-study. No sFLT-01 protein or antibodies against sFLT-01 were detected in the serum of treated animals; however, very low levels of sFLT-01 vector DNA were found in the serum and other tissues. In addition, dose- and time-dependent increases in AAV2 antibody titer were observed (Maclachlan et al. 2011). Two animals had preexisting serum AAV2 titers, and one of them exhibited very low expression of sFLT-01, suggesting that in common with prior ideas, the presence of neutralizing antibodies may impact the efficacy of AAV treatment, although this was shown not to be the case in dogs treated with AAV2-RPE65 (Annear et al. 2011). These and other encouraging preliminary data supported the initiation of a phase 1 safety and tolerability study in patients with neovascular AMD, and the results are eagerly anticipated.

17.3.2.2 Delivery of Protective Neurotrophic Factors

Although gene replacement for monogenic inherited diseases has been the primary strategy for gene therapy researchers for some time, recently, there has been an increase in studies designed to deliver protective neurotrophic factors. The utility of this approach is evident; one of the significant limitations of gene replacement therapy is that a different therapy has to be developed for each different mutated gene, and in the case of genes with multiple gain-of-function mutations, a different therapy may be needed for each disease allele. In combination with the relatively low number of patients expressing each individual allele, this strategy is not ideal from a drug development standpoint. In addition to being an allele/gene-independent approach, neurotrophic factors can be used to protect against degenerative retinal diseases regardless of their genetic component. For example, PEDF (pigment epithelial derived factor) has been used to delay degeneration in at least two models of monogenic retinal degeneration, the *rds* mouse and the *rd* mouse (Cayouette et al. 1999), but has also been used to retard glaucomatous retinal degeneration (see discussion later). Each neurotrophic factor has a different protective and safety profile. Overexpression of neurotrophic factors can be a concern, and some, such as CNTF (ciliary neurotrophic factor), have generated concerns about toxic reductions in retinal function in spite of having a protective effect against cell death (Liang et al. 2001; Schlichtenbrede et al. 2003; Buch et al. 2006; Pease et al. 2009).

Gene delivery of neurotrophic factors has been tested for efficacy in too many models of retinal degeneration to describe them all; here we mention some examples of their use for the treatment of glaucomatous optic neuropathy and retinal ganglion cell (RGC) degeneration. CNTF acts through the JAK/STAT pathway, and delivery of AAV-CNTF led to a reduction in RGC loss in the rat laser-induced glaucoma model (Pease et al. 2009). BDNF (brain derived neurotrophic factor) is synthesized by RGCs and astrocytes, but is also sent to the retina by retrograde transport from the brain, a pathway that may be blocked in glaucomatous eyes. The effect of BDNF delivery on ganglion cell death/optic nerve atrophy is controversial. BDNF delivered by adenovirus transiently improved RGC survival (for 10 days) after optic nerve transaction (Di Polo et al. 1998). In contrast, single injections of AAV-BDNF to the mouse laser-induced glaucoma model had no neuroprotective effect (Pease et al. 2009). Some benefit was reported after multiple injections (Ko et al. 2000), but only when accompanied by a free-radical scavenger. PEDF is thought to have antiangiogenic, neurotrophic, and anti-inflammatory effects, and PEDF levels are significantly reduced in glaucomatous human eyes and in the eyes of the DBA/2J mouse glaucoma model (Zhou et al. 2009). PEDF has been used to protect RGCs *in vitro* (Pang et al. 2007), and AAV-PEDF delivery to DBA/2J mice resulted in improvements in visual acuity lasting out to 11 months (latest time point examined) and a reduction in RGC loss

(Zhou et al. 2009). As all these studies suggest, there is significant room for expansion in this area of research, and it is likely to remain a primary area of focus for future ocular gene therapy.

17.4   CONCLUDING REMARKS

From a clinical development standpoint, the amount of time it takes to get a treatment ready for clinical use is so lengthy that at some point, a vector must be chosen and preclinical studies undertaken. For ocular gene therapy, the earliest vector that generated safe and efficient gene expression was AAV2, and thus, studies using it are the most advanced, in spite of more recent data suggesting that other serotypes or engineered vectors may be more optimal. The relatively modest disease improvement seen thus far with ocular genetic therapies makes the ongoing development of additional vectors and delivery methods critical. For every limiting step in the gene transduction and expression process, thoughtful researchers have tested ways to overcome the barrier, and incorporation of one or many of these novel features, particularly in nonviral nanoparticulate delivery systems, can significantly improve vector efficacy. Rational testing strategies and caution should be employed, however, as the number of different options to test can become overwhelming. It is important to remember that in addition to scientific concerns like treatment efficacy and safety, any clinically useful genetic therapy must be able to get FDA approval, needs to be stable, and ideally should be easy and cheap to produce. Simple delivery strategies may not always work, but the simplest one that is efficient is likely to be the best. In the face of these practical limitations, the other thing that the past 30 years of research have taught us is that generating persistent targeted gene expression is complicated, disease specific, and target-tissue specific.

Finally, there are an enormous number of variables influencing the efficacy of ocular gene delivery, and transition from animal models to humans continues to be challenging. In spite of the exciting results from the current round of clinical trials, none of the current (or upcoming) treatments provide true cures for retinal degenerative diseases, suggesting that significant work remains. As vectors and delivery strategies improve, a lingering challenge for curative ocular gene delivery is the variability in onset and severity of many retinal diseases. Genetic treatments have the highest probability of being effective if they can be delivered prior to the onset of degeneration, highlighting the idea that it may be easier to develop a preventative cure than a regenerative one. However, this underscores the difficulties that may arise from treating human patients in whom the onset of degeneration may precede presentation of the phenotype and clinical diagnosis. In addition, incomplete disease penetrance coupled with the relative invasiveness of intraocular treatment make treatment of asymptomatic individuals less than ideal. In spite of these limitations, the field of ocular gene therapy is quite highly developed and many groups are building on all the positive preliminary results to rapidly develop and test exciting new therapeutic approaches, thus ensuring that the next few years will generate exciting results and significant progress for the genetic treatment of retinal and other ocular diseases.

REFERENCES

Acland GM, Aguirre GD, Ray J et al. 2001. Gene therapy restores vision in a canine model of childhood blindness. *Nat Genet* 28:92–95.

Akache B, Grimm D, Pandey K et al. 2006. The 37/67-kilodalton laminin receptor is a receptor for adeno-associated virus serotypes 8, 2, 3, and 9. *J Virol* 80:9831–9836.

Ali RR, Sarra GM, Stephens C et al. 2000. Restoration of photoreceptor ultrastructure and function in retinal degeneration slow mice by gene therapy. *Nat Genet* 25:306–310.

Annear MJ, Bartoe JT, Barker SE et al. 2011. Gene therapy in the second eye of RPE65-deficient dogs improves retinal function. *Gene Ther* 18:53–61.

Bainbridge JW, Smith AJ, Barker SS et al. 2008. Effect of gene therapy on visual function in Leber's congenital amaurosis. *N Engl J Med* 358:2231–2239.

Balazs D, Godbey W. 2011. Liposomes for use in gene delivery. *J Drug Deliv* 2011:Article ID 326497.

Banin E, Bandah-Rosenfeld D, Obolensky A et al. 2010. Molecular anthropology meets genetic medicine to treat blindness in the North African Jewish population: Human gene therapy initiated in israel. *Hum Gene Ther* 21:1749–1757.

Baus J, Liu L, Heggestad AD et al. 2005. Hyperactive transposase mutants of the Sleeping Beauty transposon. *Mol Ther* 12:1148–1156.

Beltran WA, Boye SL, Boye SE et al. 2010. rAAV2/5 gene-targeting to rods: Dose-dependent efficiency and complications associated with different promoters. *Gene Ther* 17:1162–1174.

Bode J, Kohwi Y, Dickinson L et al. 1992. Biological significance of unwinding capability of nuclear matrix-associating DNAs. *Science* 255:195–197.

Boye SE, Boye SL, Pang J et al. 2010. Functional and behavioral restoration of vision by gene therapy in the guanylate cyclase-1 (GC1) knockout mouse. *PLoS One* 5:e11306.

Buch PK, MacLaren RE, Duran Y et al. 2006. In contrast to AAV-mediated Cntf expression, AAV-mediated Gdnf expression enhances gene replacement therapy in rodent models of retinal degeneration. *Mol Ther* 14:700–709.

Bunuales M, Duzgunes N, Zalba S et al. 2011. Efficient gene delivery by EGF-lipoplexes *in vitro* and *in vivo*. *Nanomedicine (Lond)* 6:89–98.

Cai X, Conley S, Naash M. 2008. Nanoparticle applications in ocular gene therapy. *Vision Res* 48:319–324.

Cai X, Conley SM, Nash Z et al. 2010. Gene delivery to mitotic and postmitotic photoreceptors via compacted DNA nanoparticles results in improved phenotype in a mouse model of retinitis pigmentosa. *FASEB J* 24:1178–1191.

Cai X, Nash Z, Conley SM et al. 2009. A partial structural and functional rescue of a retinitis pigmentosa model with compacted DNA nanoparticles. *PLoS One* 4:e5290.

Campochiaro PA, Nguyen QD, Shah SM et al. 2006. Adenoviral vector-delivered pigment epithelium-derived factor for neovascular age-related macular degeneration: Results of a phase I clinical trial. *Hum Gene Ther* 17:167–176.

Capecchi MR. 1980. High efficiency transformation by direct microinjection of DNA into cultured mammalian cells. *Cell* 22:479–488.

Cavazzana-Calvo M, Hacein-Bey S, de Saint Basile G et al. 2000. Gene therapy of human severe combined immunodeficiency (SCID)-X1 disease. *Science* 288:669–672.

Cayouette M, Smith SB, Becerra SP et al. 1999. Pigment epithelium-derived factor delays the death of photoreceptors in mouse models of inherited retinal degenerations. *Neurobiol Dis* 6:523–532.

Chalberg TW, Genise HL, Vollrath D et al. 2005. phiC31 integrase confers genomic integration and long-term transgene expression in rat retina. *Invest Ophthalmol Vis Sci* 46:2140–2146.

Chen ZY, He CY, Ehrhardt A et al. 2003. Minicircle DNA vectors devoid of bacterial DNA result in persistent and high-level transgene expression *in vivo*. *Mol Ther* 8:495–500.

Chen ZY, He CY, Kay MA. 2005. Improved production and purification of minicircle DNA vector free of plasmid bacterial sequences and capable of persistent transgene expression *in vivo*. *Hum Gene Ther* 16:126–131.

Chen ZY, He CY, Meuse L et al. 2004. Silencing of episomal transgene expression by plasmid bacterial DNA elements *in vivo*. *Gene Ther* 11:856–864.

Chen X, Kube DM, Cooper MJ et al. 2008. Cell surface nucleolin serves as receptor for DNA nanoparticles composed of pegylated polylysine and DNA. *Mol Ther* 16:333–342.

Chen ZY, Riu E, He CY et al. 2008. Silencing of episomal transgene expression in liver by plasmid bacterial backbone DNA is independent of CpG methylation. *Mol Ther* 16:548–556.

Chen X, Shank S, Davis PB et al. 2010. Nucleolin-mediated cellular trafficking of DNA nanoparticle is lipid raft and microtubule dependent and can be modulated by glucocorticoid. *Mol Ther* 19:93–102.

Cheng T, Peachey NS, Li S et al. 1997. The effect of peripherin/rds haploinsufficiency on rod and cone photoreceptors. *J Neurosci* 17:8118–8128.

Cideciyan AV, Aleman TS, Boye SL et al. 2008. Human gene therapy for RPE65 isomerase deficiency activates the retinoid cycle of vision but with slow rod kinetics. *Proc Natl Acad Sci U S A* 105:15112–15117.

Cideciyan AV, Hauswirth WW, Aleman TS et al. 2009. Human RPE65 gene therapy for Leber congenital amaurosis: Persistence of early visual improvements and safety at 1 year. *Hum Gene Ther* 20:999–1004.

Conley SM, Naash MI. 2010. Nanoparticles for retinal gene therapy. *Prog Retin Eye Res* 29:376–397.

Cook PR. 1999. The organization of replication and transcription. *Science* 284:1790–1795.

Dean DA, Dean BS, Muller S et al. 1999. Sequence requirements for plasmid nuclear import. *Exp Cell Res* 253:713–722.

del Pozo-Rodriguez A, Delgado D, Solinis MA et al. 2008. Solid lipid nanoparticles for retinal gene therapy: Transfection and intracellular trafficking in RPE cells. *Int J Pharm* 360:177–183.

Di Pasquale G, Davidson BL, Stein CS et al. 2003. Identification of PDGFR as a receptor for AAV-5 transduction. *Nat Med* 9:1306–1312.

Di Polo A, Aigner LJ, Dunn RJ et al. 1998. Prolonged delivery of brain-derived neurotrophic factor by adenovirus-infected Muller cells temporarily rescues injured retinal ganglion cells. *Proc Natl Acad Sci U S A* 95:3978–3983.

Ding W, Zhang L, Yan Z et al. 2005. Intracellular trafficking of adeno-associated viral vectors. *Gene Ther* 12:873–880.

Ding XQ, Quiambao AB, Fitzgerald JB et al. 2009. Ocular delivery of compacted DNA-nanoparticles does not elicit toxicity in the mouse retina. *PLoS One* 4:e7410.

Douar AM, Poulard K, Stockholm D et al. 2001. Intracellular trafficking of adeno-associated virus vectors: Routing to the late endosomal compartment and proteasome degradation. *J Virol* 75:1824–1833.

Dowty ME, Williams P, Zhang G et al. 1995. Plasmid DNA entry into postmitotic nuclei of primary rat myotubes. *Proc Natl Acad Sci U S A* 92:4572–4576.

Ehrhardt A, Engler JA, Xu H et al. 2006. Molecular analysis of chromosomal rearrangements in mammalian cells after phiC31-mediated integration. *Hum Gene Ther* 17:1077–1094.

Ehrhardt A, Xu H, Huang Z et al. 2005. A direct comparison of two nonviral gene therapy vectors for somatic integration: *In vivo* evaluation of the bacteriophage integrase phiC31 and the Sleeping Beauty transposase. *Mol Ther* 11:695–706.

Esumi N, Oshima Y, Li Y et al. 2004. Analysis of the VMD2 promoter and implication of E-box binding factors in its regulation. *J Biol Chem* 279:19064–19073.

Farjo R, Naash MI. 2006. The role of Rds in outer segment morphogenesis and human retinal disease. *Ophthalmic Genet* 27:117–122.

Farjo R, Skaggs J, Quiambao AB et al. 2006. Efficient non-viral ocular gene transfer with compacted DNA nanoparticles. *PLoS One* 1:e38.

Fink TL, Klepcyk PJ, Oette SM et al. 2006. Plasmid size up to 20 kbp does not limit effective *in vivo* lung gene transfer using compacted DNA nanoparticles. *Gene Ther* 13:1048–1051.

Flannery JG, Zolotukhin S, Vaquero MI et al. 1997. Efficient photoreceptor-targeted gene expression *in vivo* by recombinant adeno-associated virus. *Proc Natl Acad Sci U S A* 94:6916–6921.

Flotte TR. 2005. Recent developments in recombinant AAV-mediated gene therapy for lung diseases. *Curr Gene Ther* 5:361–366.

Futaki S, Suzuki T, Ohashi W et al. 2001. Arginine-rich peptides: An abundant source of membrane-permeable peptides having potential as carriers for intracellular protein delivery. *J Biol Chem* 276:5836–5840.

Giove TJ, Sena-Esteves M, Eldred WD. 2010. Transduction of the inner mouse retina using AAVrh8 and AAVrh10 via intravitreal injection. *Exp Eye Res* 91:652–659.

Girod A, Ried M, Wobus C et al. 1999. Genetic capsid modifications allow efficient re-targeting of adeno-associated virus type 2. *Nat Med* 5:1052–1056.

Gorbatyuk MS, Knox T, LaVail MM et al. 2010. Restoration of visual function in P23H rhodopsin transgenic rats by gene delivery of BiP/Grp78. *Proc Natl Acad Sci U S A* 107:5961–5966.

Grimm D, Kay MA. 2003. From virus evolution to vector revolution: Use of naturally occurring serotypes of adeno-associated virus (AAV) as novel vectors for human gene therapy. *Curr Gene Ther* 3:281–304.

Handa A, Muramatsu S, Qiu J et al. 2000. Adeno-associated virus (AAV)-3-based vectors transduce haematopoietic cells not susceptible to transduction with AAV-2-based vectors. *J Gen Virol* 81:2077–2084.

Hansen J, Qing K, Srivastava A. 2001. Adeno-associated virus type 2-mediated gene transfer: Altered endocytic processing enhances transduction efficiency in murine fibroblasts. *J Virol* 75:4080–4090.

Harraghy N, Gaussin A, Mermod N. 2008. Sustained transgene expression using MAR elements. *Curr Gene Ther* 8:353–366.

Hauswirth WW, Aleman TS, Kaushal S et al. 2008. Treatment of Leber congenital amaurosis due to RPE65 mutations by ocular subretinal injection of adeno-associated virus gene vector: Short-term results of a phase I trial. *Hum Gene Ther* 19:979–990.

He J, Yang Q, Chang LJ. 2005. Dynamic DNA methylation and histone modifications contribute to lentiviral transgene silencing in murine embryonic carcinoma cells. *J Virol* 79:13497–13508.

Heitzer MD, Wolf IM, Sanchez ER et al. 2007. Glucocorticoid receptor physiology. *Rev Endocr Metab Disord* 8:321–330.

Herce HD, Garcia AE. 2007. Molecular dynamics simulations suggest a mechanism for translocation of the HIV-1 TAT peptide across lipid membranes. *Proc Natl Acad Sci U S A* 104:20805–20810.

Herzog RW, Cao O, Srivastava A. 2010. Two decades of clinical gene therapy—Success is finally mounting. *Discov Med* 9:105–111.

Hoffman EA, Conley SM, Stamer WD et al. 2005. Barriers to productive transfection of trabecular meshwork cells. *Mol Vis* 11:869–875.

Hong K, Sherley J, Lauffenburger DA. 2001. Methylation of episomal plasmids as a barrier to transient gene expression via a synthetic delivery vector. *Biomol Eng* 18:185–192.

Hoque M, Ishizu K, Matsumoto A et al. 1999. Nuclear transport of the major capsid protein is essential for adeno-associated virus capsid formation. *J Virol* 73:7912–7915.

Hyde SC, Pringle IA, Abdullah S et al. 2008. CpG-free plasmids confer reduced inflammation and sustained pulmonary gene expression. *Nat Biotechnol* 26:549–551.

Jackson DA, Juranek S, Lipps HJ. 2006. Designing nonviral vectors for efficient gene transfer and long-term gene expression. *Mol Ther* 14:613–626.

Jacobson SG, Boye SL, Aleman TS et al. 2006. Safety in nonhuman primates of ocular AAV2-RPE65, a candidate treatment for blindness in Leber congenital amaurosis. *Hum Gene Ther* 17:845–858.

Jani PD, Singh N, Jenkins C et al. 2007. Nanoparticles sustain expression of Flt intraceptors in the cornea and inhibit injury-induced corneal angiogenesis. *Invest Ophthalmol Vis Sci* 48:2030–2036.

Jenke AC, Scinteie MF, Stehle IM et al. 2004. Expression of a transgene encoded on a nonviral episomal vector is not subject to epigenetic silencing by cytosine methylation. *Mol Biol Rep* 31:85–90.

Jenke AC, Stehle IM, Herrmann F et al. 2004. Nuclear scaffold/matrix attached region modules linked to a transcription unit are sufficient for replication and maintenance of a mammalian episome. *Proc Natl Acad Sci U S A* 101:11322–11327.

Jenuwein T, Allis CD. 2001. Translating the histone code. *Science* 293:1074–1080.

Johnson JS, Samulski RJ. 2009. Enhancement of adeno-associated virus infection by mobilizing capsids into and out of the nucleolus. *J Virol* 83:2632–2644.

Johnson LN, Cashman SM, Kumar-Singh R. 2008. Cell-penetrating peptide for enhanced delivery of nucleic acids and drugs to ocular tissues including retina and cornea. *Mol Ther* 16:107–114.

Kachi S, Esumi N, Zack DJ et al. 2006. Sustained expression after nonviral ocular gene transfer using mammalian promoters. *Gene Ther* 13:798–804.

Kaludov N, Brown KE, Walters RW et al. 2001. Adeno-associated virus serotype 4 (AAV4) and AAV5 both require sialic acid binding for hemagglutination and efficient transduction but differ in sialic acid linkage specificity. *J Virol* 75:6884–6893.

Kaytor MD, Weatherspoon MR, Green KJ et al. 2009. *In vivo* delivery of nucleic acid to the brain using DNA nanoparticles. *Mol Ther* 17:S519.

Klausner EA, Zhang Z, Chapman RL et al. 2010. Ultrapure chitosan oligomers as carriers for corneal gene transfer. *Biomaterials* 31:1814–1820.

Klimczak RR, Koerber JT, Dalkara D et al. 2009. A novel adeno-associated viral variant for efficient and selective intravitreal transduction of rat Muller cells. *PLoS One* 4:e7467.

Ko ML, Hu DN, Ritch R et al. 2000. The combined effect of brain-derived neurotrophic factor and a free radical scavenger in experimental glaucoma. *Invest Ophthalmol Vis Sci* 41:2967–2971.

Konstan MW, Davis PB, Wagener JS et al. 2004. Compacted DNA nanoparticles administered to the nasal mucosa of cystic fibrosis subjects are safe and demonstrate partial to complete cystic fibrosis transmembrane regulator reconstitution. *Hum Gene Ther* 15:1255–1269.

LaVail MM, Yasumura D, Matthes MT et al. 1998. Protection of mouse photoreceptors by survival factors in retinal degenerations. *Invest Ophthalmol Vis Sci* 39:592–602.

Lechardeur D, Sohn KJ, Haardt M et al. 1999. Metabolic instability of plasmid DNA in the cytosol: A potential barrier to gene transfer. *Gene Ther* 6:482–497.

Li T, Zhang M, Han Y et al. 2010. Targeting therapy of choroidal neovascularization by use of polypeptide- and PEDF-loaded immunoliposomes under ultrasound exposure. *J Huazhong Univ Sci Technolog Med Sci* 30:798–803.

Li X, Li W, Dai X et al. 2010. Gene therapy rescues cone structure and function in the 3-month-old rd12 mouse: A model for midcourse RPE65 leber congenital amaurosis. *Invest Ophthalmol Vis Sci* 52:7–15.

Liang FQ, Aleman TS, Dejneka NS et al. 2001. Long-term protection of retinal structure but not function using RAAV.CNTF in animal models of retinitis pigmentosa. *Mol Ther* 4:461–472.

Liou GI, Matragoon S, Yang J et al. 1991. Retina-specific expression from the IRBP promoter in transgenic mice is conferred by 212 bp of the 5'-flanking region. *Biochem Biophys Res Commun* 181:159–165.

Liu G, Geurts AM, Yae K et al. 2005. Target-site preferences of Sleeping Beauty transposons. *J Mol Biol* 346:161–173.

Lukacs GL, Haggie P, Seksek O et al. 2000. Size-dependent DNA mobility in cytoplasm and nucleus. *J Biol Chem* 275:1625–1629.

Maclachlan TK, Lukason M, Collins M et al. 2011. Preclinical safety evaluation of AAV2-sFLT01: A gene therapy for age-related macular degeneration. *Mol Ther* 19:326–334.

Maguire AM, Simonelli F, Pierce EA et al. 2008. Safety and efficacy of gene transfer for Leber's congenital amaurosis. *N Engl J Med* 358:2240–2248.

Markusic DM, Herzog RW, Aslanidi GV et al. 2010. High-efficiency transduction and correction of murine hemophilia B using AAV2 vectors devoid of multiple surface-exposed tyrosines. *Mol Ther* 18:2048–2056.

Naik R, Mukhopadhyay A, Ganguli M. 2009. Gene delivery to the retina: focus on non-viral approaches. *Drug Discov Today* 14:306–315.

Noma K, Allis CD, Grewal SI. 2001. Transitions in distinct histone H3 methylation patterns at the heterochromatin domain boundaries. *Science* 293:1150–1155.

Nour M, Quiambao AB, Peterson WM et al. 2003. P2Y(2) receptor agonist INS37217 enhances functional recovery after detachment caused by subretinal injection in normal and rds mice. *Invest Ophthalmol Vis Sci* 44:4505–4514.

Ortiz-Urda S, Lin Q, Yant SR et al. 2003. Sustainable correction of junctional epidermolysis bullosa via transposon-mediated nonviral gene transfer. *Gene Ther* 10:1099–1104.

Pang IH, Zeng H, Fleenor DL et al. 2007. Pigment epithelium-derived factor protects retinal ganglion cells. *BMC Neurosci* 8:11.

Pang JJ, Chang B, Kumar A et al. 2006. Gene therapy restores vision-dependent behavior as well as retinal structure and function in a mouse model of RPE65 Leber congenital amaurosis. *Mol Ther* 13:565–572.

Park JW, Kirpotin DB, Hong K et al. 2001. Tumor targeting using anti-her2 immunoliposomes. *J Control Release* 74:95–113.

Park TK, Wu Z, Kjellstrom S et al. 2009. Intravitreal delivery of AAV8 retinoschisin results in cell type-specific gene expression and retinal rescue in the Rs1-KO mouse. *Gene Ther* 16:916–926.

Pawlyk BS, Bulgakov OV, Liu X et al. 2010. Replacement gene therapy with a human RPGRIP1 sequence slows photoreceptor degeneration in a murine model of Leber congenital amaurosis. *Hum Gene Ther* 21:993–1004.

Pease ME, Zack DJ, Berlinicke C et al. 2009. Effect of CNTF on retinal ganglion cell survival in experimental glaucoma. *Invest Ophthalmol Vis Sci* 50:2194–2200.

Pechan P, Rubin H, Lukason M et al. 2009. Novel anti-VEGF chimeric molecules delivered by AAV vectors for inhibition of retinal neovascularization. *Gene Ther* 16:10–16.

Petrs-Silva H, Dinculescu A, Li Q et al. 2011. Novel properties of tyrosine-mutant AAV2 vectors in the mouse retina. *Mol Ther* 19:293–301.

Pichon C, Billiet L, Midoux P. 2010. Chemical vectors for gene delivery: Uptake and intracellular trafficking. *Curr Opin Biotechnol* 21:640–645.

Porter SD, Hu J, Gilks CB. 1999. Distal upstream tyrosinase S/MAR-containing sequence has regulatory properties specific to subsets of melanocytes. *Dev Genet* 25:40–48.

Qiu J, Brown KE. 1999a. Integrin alphaVbeta5 is not involved in adeno-associated virus type 2 (AAV2) infection. *Virology* 264:436–440.

Qiu J, Brown KE. 1999b. A 110-kDa nuclear shuttle protein, nucleolin, specifically binds to adeno-associated virus type 2 (AAV-2) capsid. *Virology* 257:373–382.

Quiambao AB, Peachey NS, Mangini NJ et al. 1997. A 221-bp fragment of the mouse opsin promoter directs expression specifically to the rod photoreceptors of transgenic mice. *Vis Neurosci* 14:617–625.

Rabinowitz JE, Rolling F, Li C et al. 2002. Cross-packaging of a single adeno-associated virus (AAV) type 2 vector genome into multiple AAV serotypes enables transduction with broad specificity. *J Virol* 76:791–801.

Raper SE, Chirmule N, Lee FS et al. 2003. Fatal systemic inflammatory response syndrome in a ornithine transcarbamylase deficient patient following adenoviral gene transfer. *Mol Genet Metab* 80:148–158.

Read SP, Cashman SM, Kumar-Singh R. 2010a. A poly(ethylene) glycolylated peptide for ocular delivery compacts DNA into nanoparticles for gene delivery to post-mitotic tissues *in vivo*. *J Gene Med* 12:86–96.

Read SP, Cashman SM, Kumar-Singh R. 2010b. POD nanoparticles expressing GDNF provide structural and functional rescue of light-induced retinal degeneration in an adult mouse. *Mol Ther* 18:1917–1926.

Rhee M, Davis P. 2006. Mechanism of uptake of C105Y, a novel cell-penetrating peptide. *J Biol Chem* 281:1233–1240.

Richards EJ, Elgin SC. 2002. Epigenetic codes for heterochromatin formation and silencing: Rounding up the usual suspects. *Cell* 108:489–500.

Riu E, Chen ZY, Xu H et al. 2007. Histone modifications are associated with the persistence or silencing of vector-mediated transgene expression *in vivo*. *Mol Ther* 15:1348–1355.

Roy K, Stein L, Kaushal S. 2010. Ocular gene therapy: An evaluation of recombinant adeno-associated virus-mediated gene therapy interventions for the treatment of ocular disease. *Hum Gene Ther* 21:915–927.

Sanlioglu S, Benson PK, Yang J et al. 2000. Endocytosis and nuclear trafficking of adeno-associated virus type 2 are controlled by rac1 and phosphatidylinositol-3 kinase activation. *J Virol* 74:9184–9196.

Schlichtenbrede FC, MacNeil A, Bainbridge JW et al. 2003. Intraocular gene delivery of ciliary neurotrophic factor results in significant loss of retinal function in normal mice and in the Prph2Rd2/Rd2 model of retinal degeneration. *Gene Ther* 10:523–527.

Schulz M, Schneider S, Lottspeich F et al. 2001. Identification of nucleolin as a glucocorticoid receptor interacting protein. *Biochem Biophys Res Commun* 280:476–480.

Sharma A, Tovey JC, Ghosh A et al. 2010. AAV serotype influences gene transfer in corneal stroma *in vivo*. *Exp Eye Res* 91:440–448.

Simonelli F, Maguire AM, Testa F et al. 2010. Gene therapy for Leber's congenital amaurosis is safe and effective through 1.5 years after vector administration. *Mol Ther* 18:643–650.

Singh SR, Grossniklaus HE, Kang SJ et al. 2009. Intravenous transferrin, RGD peptide and dual-targeted nanoparticles enhance anti-VEGF intraceptor gene delivery to laser-induced CNV. *Gene Ther* 16:645–659.

Skowronska-Krawczyk D, Matter-Sadzinski L, Ballivet M et al. 2005. The basic domain of ATH5 mediates neuron-specific promoter activity during retina development. *Mol Cell Biol* 25:10029–10039.

Stenler S, Andersson A, Simonson OE et al. 2009. Gene transfer to mouse heart and skeletal muscles using a minicircle expressing human vascular endothelial growth factor. *J Cardiovasc Pharmacol* 53:18–23.

Suh J, Wirtz D, Hanes J. 2003. Efficient active transport of gene nanocarriers to the cell nucleus. *Proc Natl Acad Sci U S A* 100:3878–3882.

Summerford C, Bartlett JS, Samulski RJ. 1999. AlphaVbeta5 integrin: A co-receptor for adeno-associated virus type 2 infection. *Nat Med* 5:78–82.

Summerford C, Samulski RJ. 1998. Membrane-associated heparan sulfate proteoglycan is a receptor for adeno-associated virus type 2 virions. *J Virol* 72:1438–1445.

Suzuki M, Kasai K, Saeki Y. 2006. Plasmid DNA sequences present in conventional herpes simplex virus amplicon vectors cause rapid transgene silencing by forming inactive chromatin. *J Virol* 80:3293–3300.

Vandenberghe LH, Wilson JM, Gao G. 2009. Tailoring the AAV vector capsid for gene therapy. *Gene Ther* 16:311–319.

Vaughan EE, Dean DA. 2006. Intracellular trafficking of plasmids during transfection is mediated by microtubules. *Mol Ther* 13:422–428.

Vitorino C, Carvalho FA, Almeida AJ et al. 2011. The size of solid lipid nanoparticles: An interpretation from experimental design. *Colloids Surf B Biointerfaces* 84:117–130.

Wadia JS, Stan RV, Dowdy SF. 2004. Transducible TAT-HA fusogenic peptide enhances escape of TAT-fusion proteins after lipid raft macropinocytosis. *Nat Med* 10:310–315.

Wagstaff KM, Jans DA. 2007. Nucleocytoplasmic transport of DNA: Enhancing non-viral gene transfer. *Biochem J* 406:185–202.

Wente SR. 2000. Gatekeepers of the nucleus. *Science* 288:1374–1377.

Wu Z, Asokan A, Samulski RJ. 2006. Adeno-associated virus serotypes: Vector toolkit for human gene therapy. *Mol Ther* 14:316–327.

Wu Z, Miller E, Agbandje-McKenna M et al. 2006. Alpha2,3 and alpha2,6 N-linked sialic acids facilitate efficient binding and transduction by adeno-associated virus types 1 and 6. *J Virol* 80:9093–9103.

Yant SR, Meuse L, Chiu W et al. 2000. Somatic integration and long-term transgene expression in normal and haemophilic mice using a DNA transposon system. *Nat Genet* 25:35–41.

Yokoyama T, Liou GI, Caldwell RB et al. 1992. Photoreceptor-specific activity of the human interphotoreceptor retinoid-binding protein (IRBP) promoter in transgenic mice. *Exp Eye Res* 55:225–233.

Yoon CS, Jung HS, Kwon MJ et al. 2009. Sonoporation of the minicircle-VEGF(165) for wound healing of diabetic mice. *Pharm Res* 26:794–801.

Zhai G, Wu J, Yu B et al. 2010. A transferrin receptor-targeted liposomal formulation for docetaxel. *J Nanosci Nanotechnol* 10:5129–5136.

Zhong L, Li B, Jayandharan G et al. 2008. Tyrosine-phosphorylation of AAV2 vectors and its consequences on viral intracellular trafficking and transgene expression. *Virology* 381:194–202.

Zhou SY, Xie ZL, Xiao O et al. 2010. Inhibition of mouse alkali burn induced-corneal neo-vascularization by recombinant adenovirus encoding human vasohibin-1. *Mol Vis* 16:1389–1398.

Zhou X, Li F, Kong L et al. 2009. Anti-inflammatory effect of pigment epithelium-derived factor in DBA/2J mice. *Mol Vis* 15:438–450.

Ziady AG, Ferkol T, Dawson DV et al. 1999. Chain length of the polylysine in receptor-targeted gene transfer complexes affects duration of reporter gene expression both *in vitro* and *in vivo*. *J Biol Chem* 274:4908–4916.

Ziady AG, Gedeon CR, Miller T et al. 2003. Transfection of airway epithelium by stable PEGylated poly-L-lysine DNA nanoparticles *in vivo*. *Mol Ther* 8:936–947.

Zou J, Luo L, Shen Z et al. 2011. Whirlin replacement restores the formation of the USH2 protein complex in whirlin knockout photoreceptors. *Invest Ophthalmol Vis Sci* 52:2343–2351.

18 Stealth-Type Polymeric Nanoparticles with Encapsulated Betamethasone Phosphate for Treatment of Intraocular Inflammation

Tsutomu Sakai, Tsutomu Ishihara, and Megumu Higaki

CONTENTS

18.1 BACKGROUND

A recent focus of nanotechnology in medicine has been on drug delivery systems (DDS) using therapeutic colloidal nanocarriers, including polymeric nanoparticles (NPs), liposomes, and micelles. Polymeric NPs have some distinct advantages over liposomes. For example, it is possible for the drug release profile of polymeric NPs to be modulated, and these NPs are more stable in biological fluids. Among the polymeric NPs for controlled drug delivery, biodegradable and biocompatible poly (D, L-lactic acid)/poly (D, L-lactic/glycolic acid)/(PLA/PLGA)-based NPs have been investigated as carriers for therapeutic bioactive molecules. PLA/PLGA has been studied for many years and is approved by the U.S. Food and Drug Administration for human therapy. Also, poly(ethylene glycol) (PEG) is used for surface modification of NPs to reduce opsonization and prevent interactions with the mononuclear phagocyte system (MPS) (Bazile et al. 1995). The main advantages of PEG NPs (stealth NPs) compared to other long-circulating systems are their shell stability and their ability to control the release of the encapsulated compound. In this regard, PEG-PLA/PLGA helps to stabilize the inner core, reduce droplet size, and encapsulate drugs. Thus, stealth NPs could markedly improve the pharmacological properties of drugs.

Glucocorticoids are effective in the treatment of uveitis (Wakefield et al. 1986; Sasamoto et al. 1990), but their systemic application is limited because of a high incidence of serious adverse effects, particularly in long-term treatment (Dukes 1996). Since intravenously administered glucocorticoids distribute throughout the body and rapidly disappear, a relatively high dose is needed to achieve an effective concentration at an inflamed target site. Moreover, the diverse physiological activities of glucocorticoids in many different tissues increase the risk of adverse effects. These problems require the development of a delivery system for glucocorticoids that both enhances localization to the target site and sustains drug release (Torchilin 2005; YihM and Al-Fandi 2006; Peer et al. 2007).

We have engineered betamethasone phosphate (BP) encapsulated in biocompatible and biodegradable blended NPs of PLA homopolymers and PEG-block-PLA copolymers (stealth nanosteroids) and examined the therapeutic activity of stealth nanosteroids in experimental arthritis and asthma models (Ishihara et al. 2009a; Matsuo et al. 2009). In mice with collagen-induced arthritis (CIA), a single intravenous injection of stealth nanosteroids resulted in complete remission of the inflammatory response. The strong therapeutic benefit may have been due to prolonged blood circulation and targeting to the inflamed joint in addition to its sustained release *in situ*.

In the present study, we determined the distribution of stealth NPs in the inflamed uvea and retina, and examined the anti-inflammatory activity of stealth nanosteroids in rats with experimental autoimmune uveoretinitis (EAU).

18.2 PRODUCTION AND USE OF NANOPARTICLES

18.2.1 PREPARATION OF NANOPARTICLES

NPs were prepared using the oil-in-water solvent diffusion method, as reported previously (Higaki et al. 2005; Ishihara et al. 2009c). The stealth nanosteroids (diameter of approximately 120 nm) were composed of the PLA homopolymer and a block copolymer of PEG and PLA (PEG content in the polymer blend: 10 wt%) (Ishihara et al. 2009b).

Vehicle-only NPs without BP were also prepared as controls. Conventional PLA nanosteroids (nonstealth nanosteroids) formed from PLA homopolymers alone without PEG-PLA copolymers were also prepared by the addition of an acetone solution of 50 mg PLA, 7.5 mg DEA, 68 μL of 1 M zinc chloride, and 28 μL of 350 mg/mL BP to 0.5% Pluronic F68 (Ishihara et al. 2005).

18.2.2 Biodistribution of Nanoparticles in EAU Rats

Accumulation of stealth NPs labeled with Cy7 in inflamed eyes was assessed by *in vivo* fluorescence imaging. A 500 μL aliquot of stealth or nonstealth NPs containing Cy7-dodecylamine or free Cy7-dodecylamine was injected into the tail vein of EAU rats on day 13 after immunization. EAU was induced by *S*-antigen peptide in Lewis rats. After 24 hours, the rats were anesthetized with sodium pentobarbital and placed into a whole-body animal *in vivo* imaging system (Optix, GE Healthcare) equipped with band-pass excitation at 750 nm and long-pass emission filters at 770 nm to obtain near-infrared fluorescence images of the eyes.

A proportion of the rats with EAU were injected into the tail vein with stealth or nonstealth nanosteroids on day 13 after immunization. After 24 hours, the rats were sacrificed by cervical dislocation and the inflamed uveas and retinas were excised and washed quickly with cold water. The concentration of BP was quantified using a time-resolved fluoroimmunoassay (TR-FIA) kit (Higaki et al. 2005; Ishihara et al. 2005, 2009b,c). The detection limit was 0.01 μg/mL.

18.2.3 Drug Treatment

An intravenous injection of 0.5 mL of stealth nanosteroids (containing 100 μg BP), nonstealth nanosteroids (containing 100 μg BP), or saline was given to rats with EAU via the tail vein on day 13 after immunization. The treatment was carried out when the clinical score reached a mean of 2, which is about half the maximum score in the course of the study. The rats continued to be examined and the clinical score was recorded until day 33 after immunization.

18.2.4 Histopathology

Rats in each group were sacrificed with sodium pentobarbital (i.v.) 7 days after the intravenous injection described previously. Eyes were enucleated and immersion fixed for 10 minutes in 4% paraformaldehyde in a 0.1 N sodium cacodylate buffer (pH 7.4). After the corneas and lens were removed, the eyecups were cut into half. For high-resolution transmitted light microscopy, half of the eyecups were fixed in 1% glutaraldehyde and 1% paraformaldehyde in a 0.086 M sodium phosphate buffer (pH 7.3) overnight at 4°C, fixed in 2% phosphate-buffered osmium tetroxide for 1 hour, and then embedded in epoxy resin. Sections of 1 μm were prepared and stained with toluidine blue. The histological severity of EAU was graded in five histological sections from each animal in a blinded fashion using a previously described semiquantitative system (Verwaerde et al. 2003).

18.3 RESULTS USING NANOPARTICLES

18.3.1 BIODISTRIBUTION OF NANOPARTICLES IN EAU RATS

In vivo imaging showed accumulation of Cy7-labeled stealth NPs in inflamed eyes of rats with EAU at 24 hours after administration (Figure 18.1a), whereas nonstealth Cy7 was not found in the eyes (Figure 18.1b). The eyes were excised and the accumulation of Cy7-labeled stealth NPs in the back of the eye was evaluated (Figure 18.1c). The NPs showed significant accumulation in inflamed eyes compared with nonstealth Cy7 (counts: 2317 ± 475 vs. 1747 ± 70, $p < 0.05$; Figure 18.1d). Meanwhile, the average BP concentrations in inflamed eyes of rats with EAU were 0.13 ± 0.03 and 0.05 ± 0.01 mg/kg at 24 hours and 72 hours, respectively, after treatment with stealth nanosteroids (Figure 18.2). The BP levels at 24 hours and 72 hours after treatment with nonstealth nanosteroids were 0.02 ± 0.01 mg/kg and below the detection limit, respectively.

18.3.2 THERAPEUTIC EFFECTS OF STEALTH NANOSTEROIDS

The clinical scores for rats with EAU (Figure 18.3) reached a maximum on day 15 after immunization (3.29 ± 0.78), after which the inflammation gradually resolved. Rats with EAU treated with stealth nanosteroids showed reduced scores on day 15–33 after immunization compared with rats treated with nonstealth nanosteroids ($p < 0.01$ or $p < 0.05$). Significant differences between treatments with stealth and nonstealth nanosteroids were observed both in the early (day 15: 1.33 ± 0.29 vs.

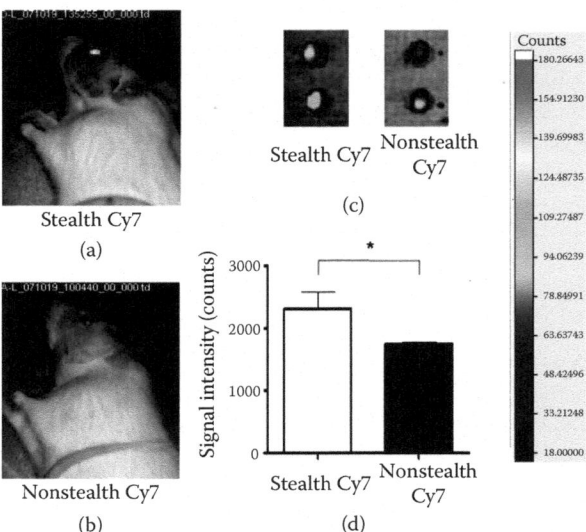

FIGURE 18.1 (See color insert.) Accumulation of nanoparticles containing Cy7-dodecylamine in inflamed eyes of rats with EAU. (a) Stealth and (b) nonstealth nanoparticles containing Cy7-dodecylamine were intravenously administered to Lewis rats, and the eyes were observed with an Optix *in vivo* fluorescence imaging system. Representative data are shown. (c) Fluorescence images of excised eyes. (d) Counts of the signal intensity of each eye. *$p < 0.05$ (stealth vs. nonstealth Cy7).

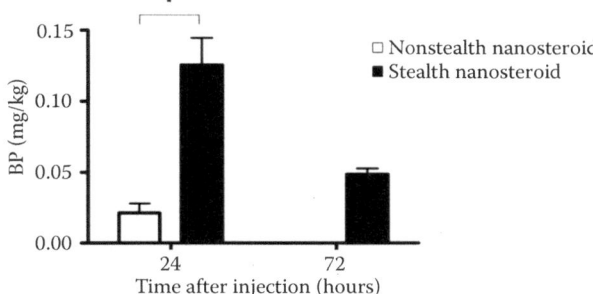

FIGURE 18.2 Betamethasone phosphate (BP) concentration in the eye. EAU rats were intravenously administered nonstealth or stealth nanosteroids through the tail vein. The mean ± SD of the BP concentration in four rats in each group is shown at 24 hours and 72 hours after administration. The BP level in rats treated with stealth nanosteroids was significantly higher than that in rats treated with nonstealth nanosteroids. *$p < 0.05$ (stealth vs. nonstealth nanosteroids).

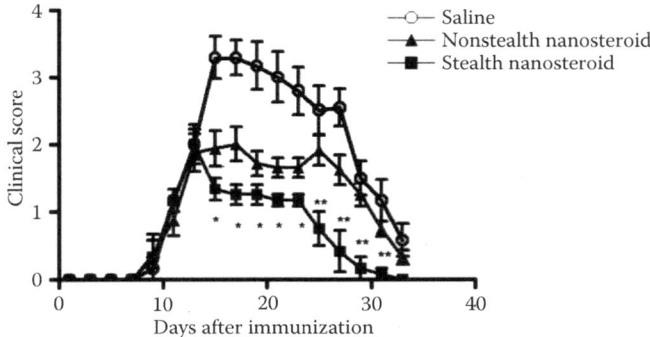

FIGURE 18.3 Clinical scores of rats with EAU. The clinical scores of rats treated with 100 μg of stealth nanosteroids were significantly lower than those of rats treated with 100 μg of nonstealth nanosteroids on days 2–20 ($p < 0.05$). o, saline; ▲, nonstealth nanosteroids (100 μg); ■, stealth nanosteroids (100 μg) ($n = 6$ in each group). The results are shown as the mean ± SD. *$p < 0.05$, **$p < 0.01$ (stealth vs. nonstealth nanosteroids).

1.94 ± 0.73, $p < 0.05$) and in the late (day 27: 0.42 ± 0.52 vs. 1.63 ± 0.54, $p < 0.01$) stages of the disease progression. The scores for rats with EAU treated with 100 μg of nonstealth nanosteroids were lower than those for saline-treated rats (Figure 18.3).

18.3.3 Histopathology

Representative histopathologic features of the retinas of rats with EAU 7 days after treatment are shown in Figure 18.4. Disruption in the inner segment (IS) and outer segment (OS) of all surviving photoreceptors was observed in all areas in the saline-treated rats (Figure 18.4a). Rats treated with stealth nanosteroids displayed marked preservation of structural integrity (Figure 18.4c), whereas those treated with non-stealth nanosteroids showed mild infiltration of inflammatory cells and disruption

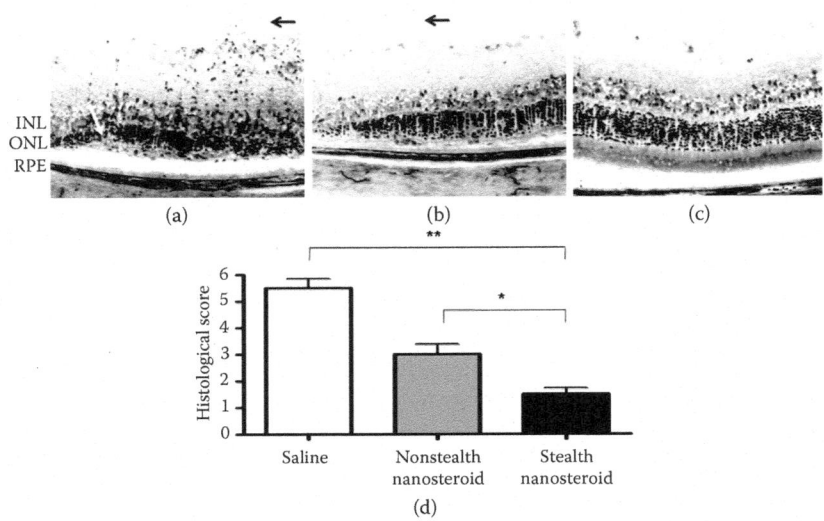

INL
ONL
RPE

(a)   (b)   (c)

Histological score

Saline   Nonstealth   Stealth
              nanosteroid   nanosteroid

(d)

FIGURE 18.4 Histopathologic features and scores of rats with EAU. (a) Representative photographs taken 7 days after the treatment with saline, (b) 100 µg of nonstealth nanosteroids, and (c) 100 µg of stealth nanosteroid are shown. Note the disruption of the inner and outer segments in all areas for saline-treated rats (a) and the preservation of structural integrity with stealth nanosteroids (c). White arrows: retinal folds and small granuloma formation. Black arrows: inflammatory cellular infiltrates in the vitreous. INL, inner nuclear layer; ONL, outer nuclear layer; RPE, retinal pigment epithelium; Ch, choroid (original magnification, ×100). (d) The severity of EAU in rats treated with saline, nonstealth nanosteroids, or stealth nanosteroids was graded 7 days after treatment. The scores for rats treated with stealth nanosteroids (1.5 ± 0.6) were significantly lower than for rats treated with saline (5.5 ± 0.8, $p < 0.01$**) and nonstealth nanosteroids (3.0 ± 0.9, $p < 0.05$*). Data are shown as the mean ± SD ($n = 6$ in each group).

in the IS and OS (Figure 18.4b). Histological scores for rats treated with stealth nanosteroids were significantly lower than those for rats treated with nonstealth nanosteroids (1.50 ± 0.55 vs. 3.00 ± 0.89, $p < 0.05$; Figure 18.4d).

18.4 SUMMARY AND FUTURE CHALLENGES

Systemic administration of stealth nanosteroids resulted in higher anti-inflammatory activity than nonstealth nanosteroids in EAU rats. A 40% decrease in intraocular inflammation was obtained within a few days and maintained for 20 days following a single injection of stealth nanosteroids, while the same dose of nonstealth nanosteroids gave a significantly weaker response. Inflammation-dependent accumulation of stealth NPs in EAU rats was demonstrated using *in vivo* imaging, suggesting that stealth NPs escape from hepatic uptake and have a prolonged blood half-life (Ishihara et al. 2009a,b). Thus, the strong therapeutic benefit of stealth nanosteroids in EAU rats may be due to targeting to inflamed eyes, in addition to sustained release *in situ* and prolonged blood circulation.

Stealth NPs were specifically designed to enable systemic delivery of steroids (Ishihara et al. 2009a,b; Matsuo et al. 2009).

To avoid entrapment by the reticuloendothelial system, the outer layer of the NPs is PEGylated, which is essential for the accumulation of NPs at the inflammatory site. This suggests that a long half-life in the circulation is critical for the overall efficacy of targeting of stealth NPs *in vivo*. In fact, we have previously shown that the half-life of stealth nanosteroids is markedly longer than that of nonstealth nanosteroids (Ishihara et al. 2009b). To facilitate accumulation at an inflammatory site via an enhanced permeability and retention (EPR) effect (Maeda et al. 2000) and to suppress renal excretion and reticuloendothelial uptake, the diameter of the NPs was set at approximately 120 nm. With these properties, the stealth NPs can optimally deliver steroids to inflammatory sites following systemic administration *in vivo* (Ishihara et al. 2009a,b; Matsuo et al. 2009).

Preferential accumulation and longer residence of BP in inflammatory sites appear to enhance its therapeutic effects, while rapid clearance of BP in conventional therapeutic strategies seems to virtually abolish these effects in rapid order. In the present study, BP delivered by stealth NPs preferentially accumulated in target sites of EAU rats after 24 hours, compared to delivery using nonstealth NPs. In addition, BP delivered by stealth NPs remained at these sites at 72 hours after administration, at which time BP delivered by nonstealth NPs was not detectable. This suggests that the stealth nanosteroids had a considerably longer residence times. Taken together, these results suggest that the BP delivered by stealth NPs may be useful for the treatment of inflammatory disorders such as chronic uveitis. At sites of inflammation, activated phagocytes may phagocytose extravasated PEG-coated vesicles. The phagocytic process is probably enhanced as a result of the combined effect of an elevated concentration of phospholipase A2 at these sites and an increased opsonization.

In future directions, applications of stealth NPs for intraocular inflammation present additional challenges and opportunities for clinical interventions. In particular, immunosuppressive drugs are attractive candidates as drugs to encapsulate with stealth NPs for the treatment of intraocular inflammation associated with autoimmune disease. We have shown the effectiveness of PEG-PLA/PLA NPs containing FK506 for the treatment of inflammatory diseases (Higaki 2009). In CIA mice, a decrease in paw inflammation was obtained in 3 days and maintained for 12 days with a single injection of stealth nanoFK506, while the same dose of free FK506 showed a significantly weaker response. Thus, NP-based novel therapies with immunosuppressive drugs may develop as the treatment of choice for autoimmune-related ocular disease.

REFERENCES

Bazile D, Prudhomme C, Bassoullet MT, et al. Stealth Me.PEG-PLA nanoparticles avoid uptake by the mononuclear phagocytes system. *J Pharm Sci.* 1995;84:493–498.

Dukes MNG. Corticotrophins and corticosteroids. In Dukes MNG (Ed.), *Meyler's Side Effects of Drugs.* Amsterdam: Elsevier; 1996. pp. 1189–1209.

Higaki M. Recent development of nanomedicine for the treatment of inflammatory diseases. *Inflamm Regen.* 2009;29:112–117.

Higaki M, Ishihara T, Izumo N, Takatsu M, Mizushima Y. Treatment of experimental arthritis with PLGA nanoparticles encapsulating betamethasone sodium phosphate. *Ann Rheum Dis*. 2005;64:1132–1136.

Ishihara T, Izumo N, Higaki M, et al. Role of zinc in formulation of PLGA/PLA nanoparticles encapsulating betamethasone phosphate and its release profile. *J Control Rel*. 2005;105:68–76.

Ishihara T, Kubota T, Choi T, Higaki M. Treatment of experimental arthritis with stealth-type polymeric nanoparticles encapsulating betamethasone phosphate. *J Pharmacol Exp Ther*. 2009a;329:412–417.

Ishihara T, Kubota T, Choi T, et al. Polymeric nanoparticles encapsulating betamethasone phosphate with different release profiles and stealthiness. *Int J Pharm*. 2009b;375: 148–154.

Ishihara T, Takahashi M, Higaki M, Mizushima Y. Efficient encapsulation of a water-soluble corticosteroid in biodegradable nanoparticles. *Int J Pharm*. 2009c;365:200–205.

Maeda H, Wu J, Sawa T, Matsumura Y, Hori K. Tumor vascular permeability and the EPR effect in macromolecular therapeutics: A review. *J Control Rel*. 2000;65:271–284.

Matsuo Y, Ishihara T, Ishizaki J, et al. Effect of betamethasone phosphate loaded polymeric nanoparticles on a murine asthma model. *Cell Immunol*. 2009;260:33–38.

Peer D, Karp JM, Hong S, et al. Nanocarriers as an emerging platform for cancer therapy. *Nat Nanotechnol*. 2007;2:751–760.

Sasamoto Y, Ohno S, Matsuda H. Studies on corticosteroid therapy in Vogt-Koyanagi-Harada disease. *Ophthalmologica*. 1990;201:162–167.

Torchilin VP. Recent advances with liposomes as pharmaceutical carriers. *Nat Rev Drug Discov*. 2005;4:145–160.

Verwaerde C, Naud MC, Delanoye A, et al. Ocular transfer of retinal glial cells transduced *ex vivo* with adenovirus expressing viral IL-10 or CTLA4-Ig inhibits experimental auto-immune uveoretinitis. *Gene Ther*. 2003;10:1970–1981.

Wakefield D, McCluskey P, Penny R. Intravenous pulse methylprednisolone therapy in severe inflammatory eye disease. *Arch Ophthalmol*. 1986;104:847–851.

Yih TC, Al-Fandi M. Engineered nanoparticles as precise drug delivery systems. *J Cell Biochem*. 2006;97:1184–1190.

Section V

Technology and Materials Development

19 Nanotechnology-Guided Imaging of Retinal Vascular Disease

Joshua R. Trantum, John S. Penn, and Ashwath Jayagopal

CONTENTS

19.1 INTRODUCTION

Therapeutic strategies based on nanotechnology, the control of matter on a 1–100 nm scale, have great potential for improving clinical management of ocular disease. The nanoscale regime is the scale of biological machinery and essential processes in living systems, and designing devices on this scale is enabling unique applications in medicine. From the emerging field of nanomedicine, new clinical approaches have been developed, which take advantage of unique properties achieved on the nanoscale, including controlled release of therapies [1], site-specific drug delivery to diseased tissues [2], and improved stability of therapies *in vivo* [3]. These features can reduce dosage required for the intended therapeutic response, as well as off-target effects, which contribute to toxicity. Engineering on the nanoscale can influence the manner and efficiency in which organic or inorganic materials interact with biological systems [4]. For this reason, significant efforts in the field of nanomedicine have also been directed toward the development of nanoscale imaging agents. Similar to applications involving nanoengineering of therapies, imaging agents manipulated on the nanoscale can be designed to home specifically to dysfunctional tissues to aid in the early detection of disease [5]. Furthermore, the physicochemical attributes of the imaging agent can be altered to improve the contrast properties of the imaging agent for higher sensitivity detection of disease in medical imaging procedures, or to enable multimodal imaging applications for enhanced disease characterization [6,7]. A number of other unique and desirable features of nanoscale imaging agents contribute to their significant clinical relevance leading into the future, such as prolonged circulation times [8].

A specific arena in which nanoscale contrast agents are poised to improve disease detection and management involves vascular diseases of the retina. Retinal vascular diseases such as diabetic retinopathy and neovascular age-related macular degeneration (AMD) are the leading causes of blindness, and efforts to improve diagnostic and prognostic ocular imaging techniques are needed to improve the efficacy of clinical management. Nanotechnology can play a significant role in improving imaging strategies for retinal vascular disease by expanding the clinician's toolkit for interrogating tissue properties, through emerging applications such as nanotechnology-guided imaging of specific biomolecular mediators of disease, and identifying surrogate biomarkers of therapeutic response and disease susceptibility [9–11]. A primary means of accomplishing these goals involve the development of nanoscale contrast agents, which enhance medical images in various modalities to better highlight specific areas of interest.

In this chapter, an overview of contrast agents applicable to retinal vascular imaging is presented. Next, the utility of nanotechnology-guided approaches for imaging retinal vascular diseases is discussed, using recent examples from the literature and emerging applications. Special emphasis is placed on describing the utility of nanoscale imaging agents for the assessment of therapeutic response, detecting retinal disease in subclinical stages, and elucidating cellular and molecular participants in retinal vascular disease to facilitate the development of targeted therapies.

19.2   OVERVIEW OF NANOSCALE IMAGING AGENTS

The retinal vasculature is readily accessible to a number of imaging modalities, which is ideal for the development of diverse nanoscale contrast agents for probing different tissue properties. The purpose of a contrast agent as an adjuvant to medical imaging is to improve the visibility of the intended imaging target using a medical imaging instrument. The target can be a specific cell or tissue type, a specific environmental parameter such as pH or oxygen tension, or even a specific protein or nucleic acid sequence [12].

The anatomy of a nanoscale contrast agent can be described as a core, which exhibits intrinsic contrast in medical imaging procedures, with concentric functional layers radiating outward from the core surface in order to impart important properties upon the biodistribution or contrast properties of the nanoparticulate contrast agent [13,14]. The core of an optical contrast agent is typically an organic dye or metallic compound, although for other imaging modalities, such as ultrasound, a perfluorocarbon gas-filled microbubble is used [15]. The core can be modified to improve intrinsic contrast; for example, multiple dyes may be used within a core in order to enhance the brightness of the contrast agent [16,17], or the metallic core may be alloyed with an additional metallic shell to increase photostability or alter spectral emission wavelength [18].

Other coating layers of the nanoparticle are associated with stabilizing the contrast agent *in vivo* to ensure long circulation times without degradation using lipids or polymers such as polyethylene glycol (PEG) to protect the core from immune-mediated clearance or other degradative mechanisms [3,19]. In addition, outer layers can be engineered to confer homing capabilities upon the contrast agent to enable accumulation and retention within target tissues for detection by medical imaging [20]. In this layer, high-affinity biological ligands are immobilized to enable binding of the contrast agent to the intended target. Ligands commonly used for this purpose include antibodies [21], peptides [9,22,23], aptamers [24], sugars [25], and small molecules [26]. Nanoscale contrast agents based on these principles have improved the specificity and sensitivity of medical imaging and have enabled new insights into the molecular complexity of disease.

On account of the optical transparency of the eye, contrast agents used for ocular vascular imaging primarily involve those with optical contrast properties for light-based imaging modalities, such as optical coherence tomography (OCT) and ophthalmoscopy. However, emerging applications in nanomedicine involving contrast agent development for MRI, ultrasound, and other modalities also hold substantial clinical relevance and may be useful for uncovering novel insights from diseased retinal vasculature.

19.2.1 Optical Nanoscale Contrast Agents for Fluorescence Imaging

Instrumentation for fluorescence-based imaging of the retina is readily available in the ophthalmology clinic, and is used extensively in diagnosis and monitoring of retinal vascular diseases. In angiographic procedures intended to diagnose vascular disease, organic dyes such as fluorescein or indocyanine green are used as blood pooling agents to highlight the vasculature for subsequent morphological analysis by the clinician. Typically, the angiogram provides valuable diagnostic information in the form of morphological data, such as retinal vessel caliber or the presence of hemorrhage and ocular neovascularization. However, using similar retinal imaging instrumentation (e.g., fundoscopy [27], scanning laser ophthalmoscopy [28]), contrast agents can be engineered on the nanoscale to enhance the detail derived from an angiogram beyond that achieved by organic dyes.

Several optical contrast agents hold considerable promise in enhancing the information obtainable from an angiogram, by directly addressing limitations of the dye-based approaches used currently in the clinic. Advantages of nanoscale contrast agents compared to currently employed contrast agents in the clinical setting include increased photostability, increased photointensity, amenability to surface conjugation to homing ligands, and improved *in vivo* stability. As a key example of these developments, Quantum dots (QD), or semiconducting nanocrystals, feature enhanced brightness attributable to their high quantum efficiency, and do not lose fluorescence as rapidly as organic dyes with continued illumination (i.e., photobleach) [9,23,29–31]. Furthermore, their size can be precisely adjusted within single nanometers to alter their bandgap energy, which allows tuning of QD emission wavelengths from violet to infrared spectra. Through several advances in nanoparticulate surface engineering, QD are amenable to surface conjugation of bioaffinity ligands and stealth polymers such as PEG to prolong their *in vivo* circulation lifetimes upon systemic injection (Figure 19.1). To date, QD have been used extensively *in vivo* in

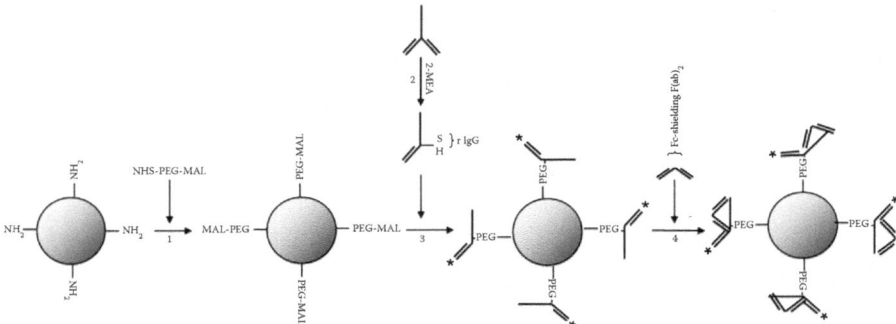

FIGURE 19.1 Scheme for the synthesis of antibody-targeted quantum dots (QD) for multi-spectral imaging for retinal vascular biomarkers. The process allows for specific hinge region attachment of IgG fragments upon the QD surface for enhanced affinity. Polyethylene glycol (PEG) spacers reduce nonspecific binding of QD and premature clearance from the bloodstream. (Reproduced with permission from Jayagopal, A. et al., *Bioconjugate Chemistry*, 18(5), 1424–1433, 2007. Copyright [2007] American Chemical Society.)

preclinical studies for imaging tumor vasculature and multiple organs via vascular perfusion with QD [30,32–34], as well as specific cells and biomolecules within retinal vasculature [21]. However, their heavy metal composition may limit their eventual clinical utility, and QD toxicity issues have yet to be completely resolved [35].

Other nanoparticulate platforms show considerable promise for improving optical imaging of retinal vasculature in the clinic. Lipid-based nanocarrier formulations such as liposomes and micelles [21,36,37], as well as polymeric nanoparticles [38], have been evaluated for ocular drug delivery applications for over 20 years, with the goal of improving stability and efficacy of encapsulated therapeutics. However, using optical imaging instrumentation, these nanoparticles could also be co-loaded with an optical contrast agent to permit image-guided drug delivery applications, and to confirm successful delivery of therapy to the diseased site in the clinic. Several such multimodal applications have been described for vascular imaging [6], although applications specific for retinal vasculature have not been reported. Like QD, these approaches allow coating of the contrast agent core with concentric layers for *in vivo* stability and site-specific targeting of diseased tissue from the bloodstream. Furthermore, these nanocarriers can be synthesized on a pharmaceutical grade scale based on FDA-approved, biocompatible lipids and polymers (e.g., poly-lactic and glycolic acids, PEG) [39]. Therefore, several optical imaging approaches utilizing nanotechnology are poised for incorporation into retinal vascular imaging procedures.

19.2.2 Optical Nanoscale Contrast Agents for OCT

OCT has emerged as a primary imaging modality for the diagnosis of retinal diseases including macular edema and AMD [40–42]. In this laser-guided technique, backscattered light waves are converted into high-resolution, cross-sectional images of the retina, enabling what is commonly referred to as *in vivo* optical biopsies. OCT is an established technology for imaging retinal tissue morphology, and applications for imaging specific molecules in tissue based on endogenous contrast properties are emerging [43–46]. However, investigation into exogenous contrast enhancement techniques for improving signal-to-noise ratios of OCT imaging and/or permitting imaging of specific cells or biomolecules is only in early stages. Nevertheless, several reports indicate that nanoscale contrast agents show promise for enhancing the detail obtained from OCT imaging in the retina.

Several nanoscale contrast agents have promise for improving retinal OCT imaging. A primary method of generating contrast in OCT images involves scattering, or the altering of the index of refraction of light within the region of interest. This alters the intensity of backscattered light entering the detector and can be used to detect specific abnormalities in tissue. For this purpose, a number of nanoparticle-based scattering OCT contrast agents have been developed on the basis of cores consisting of iron oxide, gold, melanin, and perfluorocarbon gas-filled microbubbles [47]. These agents can be coated with proteins, including antibodies, for targeting specific cell surface molecules for targeted imaging. These agents have an established history for use as intravascular contrast agents, and it is likely that they will be effective as OCT contrast agents for imaging retinal vascular abnormalities. While these agents can be synthesized on the microscale or nanoscale, the nanoscale regime permits extravasation from blood vessels into tissue, enhancing diagnostic value of these agents [45]. Furthermore, the nanoscale would likely permit enhanced cell–nanoparticle interactions for *in vivo* targeting purposes. However, as predicted by Rayleigh scattering principles, nanoparticles are generally not ideal scattering contrast agents due to their small diameter available for light scattering. Recently, there has been substantial interest in developing gold nanoparticles in various geometries as OCT contrast agents, due to their surface plasmon resonance (SPR) properties, which enable large optical cross sections achieved by smaller particle sizes [45,48,49]. Gold SPR agents on the nanoscale, typically below 100 nm, are capable of absorbing incident laser light and altering backscattered light spectra to enable much greater OCT contrast than that achievable with similarly sized nanoparticles. Geometries studied to date for use in biomedical imaging applications include gold-coated silica nanoshells [50], as well as gold nanospheres [51], and nanorods [52], although specific OCT vascular imaging applications are in their infancy. Gold-based contrast agents are promising candidates for incorporating into clinical applications, due to their biocompatibility, size and shape tunability (from the nano to microscale regime), and amenability to surface conjugation of targeting ligands, especially via facile thiol-based chemistries [52].

In addition, to scattering and absorbing contrast agents for OCT, other nanoscale contrast agents synthesized on the nanoscale can be used for the derivation of specific properties from the tissue. An emerging technique called magnetomotive technique incorporates superparamagnetic iron oxide nanoparticles under the influence of an external magnetic field to alter OCT signal by altering tissue scattering. Alternatively, a fluorescent dye may be coupled to the iron oxide or a ferromagnetic fluid encapsulated within a carrier shell to enable OCT detection based on dynamic blinking fluorescence induced by magnetic field application. Such techniques can be used to distinguish moving signals from stationary signal for further contrast enhancement or to identify single cells and/or biomolecules in the living tissue [53,54]. Given the importance of OCT in retinal imaging, the development of nanoscale OCT contrast agents and complementary imaging techniques such as those described earlier, which enhance gold standard OCT techniques will have a great impact on patients who undergo screenings, and clinicians will be able to maximize the detail of information achieved by OCT screening.

19.2.3 ALTERNATIVE NANOSCALE CONTRAST AGENTS

Emerging imaging modalities for diagnosing and monitoring retinal vascular diseases have been accompanied by developments in contrast agents for these technologies, and a limited number of them benefit from nanoscale engineering. Of these, magnetic resonance imaging is being developed as a powerful, noninvasive modality for imaging-specific metrics related to retinal vascular diseases, including blood–retinal barrier (BRB) stability, retinal oxygenation levels, and retinal ion demand [55]. Several nanoscale contrast agents have been developed, which may be useful in expanding the application of MRI for retinal disease diagnosis. Paramagnetic iron oxide nanoparticles are established T_2 contrast agents, which can be surface functionalized for targeting cell surface receptors, and cells such as macrophages endocytose them [56–58]. Alternatively, by altering the diameter of iron oxide nanoparticles, they conceivably could be useful for probing retinal vessel permeability using quantitative analysis. Recently, the first biocompatible T_1 nanoparticulate contrast agents were synthesized, based on manganese oxide [59]. This is a significant development, as T_1-weighted imaging techniques are highly desirable for enhancing "whitening" images (positive enhancement) to facilitate disease detection, whereas T_2-weighted techniques typically darken images (negative enhancement), which can lead to difficulties in diagnosis as dark image voxels could be confused with blood or calcification. Furthermore, manganese in ionic form is thought to compete for Ca^{++} in the body, potentially contributing to neurotoxicity *in vivo* [60]. These MnO nanoparticles will likely permit clinical application harnessing manganese's T_1 relaxivity properties without these adverse effects. While MRI is not typically a standard imaging modality in ophthalmic diagnosis for diseases that are not malignant, the unique types of information yielded by MRI in conjunction with nanoscale contrast agents may advance their applications in the field. Furthermore, it is important to note here that MRI contrast agents can often be incorporated into multimodal imaging agents using nanoscale engineering strategies [61–63]. For example, iron oxide nanoparticles can be loaded into liposomes with gadolinium to enable concurrent T_1- and T_2-weighted MRI imaging in the same patient, or for simultaneous MRI and optical imaging, QD may be combined with iron oxide within lipid vehicles or polymeric nanoparticles. Therefore, a key factor driving investigation into developing nanoscale imaging agents is the powerful applications achieved by multimodal imaging.

Other emerging modalities applicable to retinal vascular disease, which may benefit from nanoscale contrast agent development are based on acoustic contrast. Ultrasound has typically been utilized in other areas of the body to assess hemodynamic parameters, but has also been utilized for molecular imaging. In this arena, surface-functionalized nanoscale gas-filled bubbles may be useful as ultrasound contrast agents for the detection of specific cell surface receptors within blood vessels, as an adjuvant to conventional ultrasound-mediated imaging for quantification of blood flow [64]. Furthermore, efforts to expand upon the utility of acoustic contrast agents through advances in imaging modalities are underway. Photoacoustic imaging is based upon the detection of acoustic waves triggered by the absorption of pulsed light within tissue, and is useful for detecting clinically relevant absorbing compounds in the eye, such as melanin and hemoglobin. Functionalized nanobubble contrast agents loaded with absorbing inks are being developed as a means of providing contrast using this approach [65]. In the prospective clinical application using photoacoustic imaging, areas with accumulated nanobubbles illuminated by laser would be selectively excited to produce acoustic waves, which can be imaged, thereby minimizing background contrast and enabling detection of a particular cell or biomolecule within tissue. Therefore, acoustic contrast agents based on nanotechnology may expand the overall utility of these imaging modalities and related multimodal approaches in ophthalmic diagnosis.

19.3 APPLICATIONS OF NANOTECHNOLOGY TOWARD IMAGING OF RETINAL VASCULAR DISEASES

Reports describing the use of nanoscale imaging agents to evaluate retinal vascular diseases are limited, but promising, and support the overall utility of incorporating nanomedicine-based approaches in this field. Significant progress in developing nanoscale contrast agents has been demonstrated for applications in imaging diabetic eye diseases, diseases of the macula, and ocular inflammatory diseases such as uveitis. Highlights of recent research in nanoscale contrast agent development and characterization and opportunities for advancement in prominent retinal vascular diseases are discussed.

19.3.1 DIABETIC EYE DISEASE

Early detection of diabetic retinopathy and other diabetic complications in the eye significantly reduces the risk of blinding complications in patients by enabling timely therapeutic interventions and preventative measures [66]. Both OCT and angiography are useful diagnostic tools for diagnosing and staging this disease [67]. At early and late stages of the disease, nanoscale contrast agents can enhance the diagnostic utility of these imaging techniques for improved clinical management.

In the early stages of the disease, breakdown of the BRB, which is maintained by retinal capillary endothelial cell tight junctions and pericytes, is a major contributor to vision loss due to diabetes. Specifically, breakdown of the inner BRB leads to diabetic macular edema, which is a leading cause of blindness in diabetic patients, and loss of barrier integrity precedes a number of complications that cause blindness in diabetic retinopathy [68–70]. Techniques based on fluorescent dyes and macromolecules such as dextran can be utilized to probe BRB integrity in diabetes [71]. Specifically, by injecting molecules with varying molecular weights, the permissibility of the BRB can be quantified. As the molecular sieving properties of the BRB have been extensively characterized, BRB permeability is a very useful imaging biomarker [68]. Toward this goal, gold nanoparticles are precisely size-tunable contrast agents for OCT, which pass through the BRB when sized at 20 nm, but do not normally penetrate the BRB when sized at >100 nm [72], and due to their high bio-compatibility, they could be readily incorporated into imaging strategies for probing barrier integrity. Fluorescence imaging techniques could use fluorescent dye-loaded nanoparticles or proteins such as human serum albumin to similarly measure BRB function *in vivo*, analogous to *ex vivo* Evans Blue fluorophotometric techniques [73]. In addition, MRI-based contrast agents based on gadolinium were demonstrated to be useful for permeability quantification using dynamic contrast-enhanced MRI [74], presenting an opportunity to apply appropriate nanoscale contrast agents to imaging disease, such as liposome-based contrast agents [62]. Other areas of investigation involving BRB function include targeting of tight junction molecules such as occluding and claudins to monitor *in vivo* expression, provided that targeting strategies do not themselves compromise barrier function.

Endothelial dysfunction and subsequent retinal capillary dropout in diabetic retinopathy is another critical event in disease progression, which can be imaged at the molecular level using nanotechnology, although a few attempts have been made to perform molecular imaging of retinal vessels. The authors used QD to perform multispectral molecular fluorescence imaging of endothelial surface molecules in a streptoxotocin (STZ)-induced rat model of diabetes (Figure 19.2) [21]. Specifically, QD are

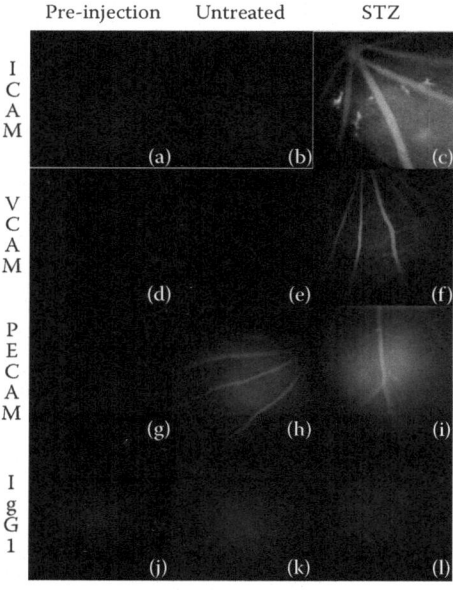

FIGURE 19.2 (**See color insert.**) QD-mediated imaging of vascular biomarkers in STZ-induced diabetic rat retinas. STZ treatment induced enhanced expression of ICAM-1 and VCAM-1 as evidenced by increased QD binding in retinal vessels. Minimal binding was observed for ICAM- and VCAM-targeted QD probes in age-matched controls, but binding of QD probes targeted toward PECAM, a constitutively expressed endothelial cell surface, was similar in both experimental groups. A nonspecific IgG-functionalized QD did not accumulate in retinas of either group. (Reproduced with permission from Jayagopal, A. et al., *Bioconjugate Chemistry*, 18(5), 1424–1433, 2007. Copyright [2007] American Chemical Society.)

surface functionalized with PEG to prolong QD circulation lifetimes via steric hindrance from the hydrophilic polymer, which reduces immunorecognition and premature clearance by reticuloendothelial tissues such as the liver and spleen. Furthermore, QD are coated with antibodies directed against specific biomarkers of endothelial inflammatory responses, such as intercellular adhesion molecule (ICAM-1) and vascular cell adhesion molecule (VCAM-1), both known to be upregulated in the STZ model [75]. QD can be tuned to emit fluorescence within visible or infrared spectra, and only require one excitation wavelength irrespective of emission signatures. Therefore, QD enable multiplexed, simultaneous imaging of biomolecules *in vivo*, which is valuable for preclinical longitudinal imaging studies, and cannot be practically performed using dyes. However, the cadmium content of QD precludes their usage for clinical applications. To address this, QD featuring tolerable metallic composition are under development [76]. The perceived clinical application using this technology would be early detection of capillary insult, which would facilitate more timely therapeutic intervention to minimize disease progression, or possibly identify early on areas in need of photocoagulation therapies. Alternatively, this type of molecular imaging could be valuable in assessing treatment response in patients, and identify those refractory to conventional therapies early in the time course of clinical management. These applications can take advantage of preexisting instrumentation, such as fundus cameras and scanning laser ophthalmoscopes in the clinic.

A consequence of endothelial dysfunction in diabetic retinopathy is leukostasis, in which endothelial cell surface inflammatory molecules provoke the cascade on rolling, arrest, and extravasation of leukocytes on and across retinal vessel walls [77]. Leukocytes have been implicated in capillary degeneration and increased vessel permeability in diabetes, and techniques to visualize leukocyte recruitment and accumulation within retinal tissue would be valuable in the clinic. Early efforts to study recruitment were centered upon a DNA-intercalating dyes called acridine orange (AO) to perform quantitative fluorographic imaging of leukocytes in diabetic retinas [78,79]. AO can be intravenously injected and rapidly penetrates leukocyte cell membranes, and fluorescence is enhanced upon nucleic acid binding within the cell. However, this dye is promiscuous and labels all cells with a nucleus, and it can be challenging to distinguish arrested leukocytes from the equally fluorescing endothelium. Furthermore, AO is not a high-quantum-efficiency dye and fades quickly. Therefore, the authors developed antibody-functionalized QD probes targeting circulating leukocyte subpopulations to study their role in retinal vascular diseases such as diabetic retinopathy and uveitis using established animal models (Figure 19.3) [21]. Similar to multispectral imaging of endothelial surface biomarkers, QD were also shown to be useful for monitoring leukocyte subtypes such as neutrophils in retinal inflammatory processes, and were much brighter and photostable over time than AO (Figure 19.4), permitting long-term imaging of leukocytes *in vivo*. In addition, QD-antibody probes could be injected systemically for labeling of leukocytes *in situ*; there was no need to prelabel whole blood *ex vivo*, simplifying experimental procedures.

A major vision-threatening complication of diabetes is proliferative diabetic retinopathy, characterized by aberrant growth of blood vessels (angiogenesis) above the inner limiting membrane into the vitreous. Early detection of this disease component

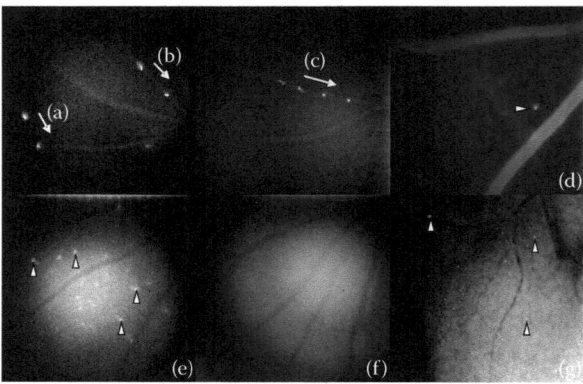

FIGURE 19.3 Detection of *in vivo*-labeled circulating leukocytes in rat models of uveitis and diabetes. QD-targeting RP-1, a marker of rat neutrophils, was used to image normal circulation (a), rolling (b), and arrest (d, e). In normal animals, no arrest was observed (f). Cells were also detectable in retinal flat mounts (g). (a–d, g): STZ model, (f) endotoxin-induced uveitis model, (e) age-matched control for (f). (Reproduced with permission from Jayagopal, A. et al., *Bioconjugate Chemistry*, 18(5), 1424–1433, 2007. Copyright [2007] American Chemical Society.)

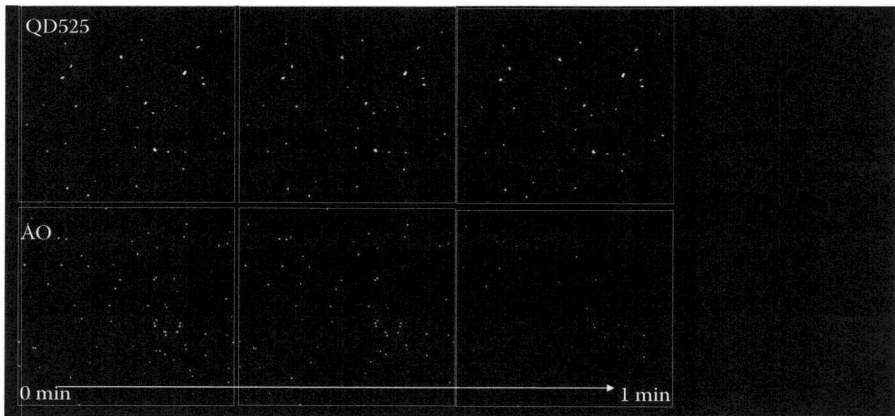

FIGURE 19.4 Optical properties for QD- (top row) and AO-labeled (bottom row) neutrophils *ex vivo*. RP-1-targeted QD were used to label isolated neutrophils using previously published methods [21], and fluorescence intensity was compared during continuous illumination under a fluorescence microscope. QD remain bright and photostable over time, and lose only 4% of initial fluorescence over 30 minutes of illumination, whereas AO-labeled neutrophils are initially dim compared to QD and fade quickly.

is critical to preserving vision in the patient, and several imaging biomarkers may be targeted by nanoscale contrast agents for this purpose, such as oxygen delivery and production of vascular endothelial growth factor (VEGF) in tissue. Local tissue ischemia is thought to be caused in diabetic retinopathy by closure of blood vessels secondary to endothelial cell and pericyte death, and leads to angiogenesis, a process stimulated by VEGF. Therefore, clinical strategies for imaging ischemia and VEGF in retinal vasculature *in vivo* would greatly aid treatment. Recently, a nanoparticle featuring boron difluoride coupled with a poly (lactic acid) polymer was shown to exhibit interesting oxygen-sensitive properties [80–83]. Specifically, these nanoparticles are capable of emitting long-lived, green phosphorescence (i.e., "glow in the dark" emission), the intensity of which is responsive to oxygen concentrations. Although these probes were used to interrogate tumor hypoxia, with imaging instrumentation similar to that used for fluorescence imaging (e.g., photomultiplier tubes or CCD cameras), the probes may be useful for imaging hypoxia in retinal tissue noninvasively.

Applications in oncology have also demonstrated that VEGF can be measured using optical molecular imaging techniques [84]. In a work by Chang et al., VEGF was monitored in tumors using a hyperspectral imaging system, by injecting dye-labeled anti-VEGF antibodies capable of binding to soluble and extracellular matrix-bound VEGF. Given the importance of VEGF in diabetic retinopathy initiation and progression [85,86], clinically relevant techniques to probe concentration and localization within the retina could be powerful diagnostic tools. Based on the findings of this work, it is conceivable that anti-VEGF antibodies could be conjugated to nanoparticles such as QD or gold nanoparticles (the latter being biocompatible) and injected into the eye to visualize local spikes in VEGF production, which could be a harbinger for edema and/or angiogenesis in diabetic patients, and permit timely treatments. Several other molecularly targeted imaging agents are being developed for the detection of angiogenesis, including integrin and extracellular matrix binding agents, using MRI and/or optical imaging strategies [63,87–90]. Therefore, it is likely that a nanoscale toolbox for probing the multiple cellular and molecular components of angiogenesis in diabetic retinopathy can be adapted for clinical applications.

19.3.2 MACULAR DISEASE

The technologies described earlier for imaging diabetic retinopathy are also relevant for imaging multiple biological processes in diseases of the macula. For example, AMD, a leading cause of vision loss in the elderly, exhibits significant inflammatory components, barrier integrity breakdown (i.e., in the outer BRB consisting of a retinal pigment epithelial monolayer), and angiogenesis. Therefore, many of the nanoscale imaging agents described earlier can be used to probe diseases in the macula. However, several cellular and molecular components of AMD necessitate development of unique nanoscale contrast agents to image disease-specific processes. First, AMD is an inflammatory disease in which the macrophage cell type is extensively involved, in both initiation and progression [91–93]. Imaging macrophages in the retina and choroid may have useful diagnostic value when managing AMD treatment and assessing treatment response. To this end, several classes of imaging contrast agents have been developed for imaging macrophages in the atherosclerotic plaque [6,9,94,95]. These probes include MRI contrast agents, QD, gold nanoparticles, perfluorocarbon bubbles, and dye-loaded nanoparticles. Generally, macrophages are capable of endocytosing a broad size range of nanoparticles, and surface ligands such as dextrans or cell-penetrating peptides can greatly enhance the efficiency of this process. The capability of imaging macrophage content in the AMD patient will certainly expand detection and monitoring of AMD.

Drusen, deposits of cellular waste underneath the retina, are also a critical component of macular disease diagnosis that warrants imaging agent development. Drusen normally appear in the eye as a consequence of aging, but excessive deposition of drusen warrants evaluation for AMD, and therefore it is important to monitor drusen in the at-risk population. Current techniques such as fundoscopy, OCT, ultrasound, and scanning laser ophthalmoscopy are all capable of imaging drusen deposits [96], but this task can be complicated if drusen are obstructed by other structures or buried deep within the optic nerve [97]. Over the past decade, many of the molecular components of drusen were identified [98,99] and include a variety of lipids and proteins, opening up a significant opportunity for the development of molecularly targeted contrast agents for enhancing drusen detection and enabling complex characterization studies. Given the amenability of many nanoscale contrast agents to imaging using OCT and other optical instrumentation, the use of nanoscale agents as adjuvants to standard morphological imaging of drusen would be an important clinical advance.

While applications of nanotechnology toward retinal vascular imaging is in its infancy, the potential of nanotechnology for improving our understanding of diseases and advancing diagnostic capabilities is very significant. Other areas of investigation by the authors include leukocyte and endothelial targeting agents for monitoring uveitis, uveal melanoma, and the investigation of retinal vasculature as a sentinel for diseases in which the eye is not a primary component. Specifically, the authors are monitoring retinal molecular expression to monitor atherosclerosis and stroke susceptibility, neurodegenerative diseases, and circulating tumor cells in cancer. The optical accessibility and transparency of the eye make eye-based diagnostic procedures attractive compared to many other diagnostic procedures for these diseases, which include x-ray angiography, CT, and blood sampling.

19.4   CONSIDERATIONS FOR CLINICAL TRANSLATION OF NANOSCALE IMAGING AGENTS

As is characteristic for technology development in medicine, nanotechnology faces many barriers to clinical adaptation. The safety profile of nanoscale agents is under great scrutiny. As these devices are on the scale of biology, they are capable of adverse, off-target effects due to their ability to biodistribute in many tissues not typically accessible to exogenously administered substances, including the central nervous system and lungs (e.g., carbon nanotubes). It is thought that advances in imparting targeting capabilities upon nanoparticles and employing polymer coatings to reduce nonspecific tissue accumulation will minimize these risks. Many nanomaterials have only been partially characterized and the full extent of their effects on the human body has yet to be confirmed. Therefore, nanotechnology applications based on fewer materials that have a longer history of investigation, such as gold nanoparticles, are more attractive candidates for clinical approval and will likely face fewer regulatory reviews. In addition, evaluating a contrast agent for local intraocular injection, as opposed to systemic administration, is likely to be viewed upon favorably when evaluating it for clinical applications. As a closed, privileged tissue with several local injection routes, the eye is an ideal site for developing nanoscale contrast agents.

A second practicality that must be considered for nanoscale imaging agents involves the challenges associated with synthesis on pharmaceutical scales. Many of the nanotechnologies discussed here have only been synthesized in the laboratory, and it may not be possible to scale up production for distribution to the population. Advances in production capabilities necessarily lag behind invention of the nanoscale device itself and require great investments. The production process must also be cost-effective, which can be difficult since nanomaterials are often difficult to procure at the necessary levels for production on a large scale.

Despite the challenges, the advantages in diagnostic capabilities made possible only by nanotechnology will be the forces that drive its continued development and translation in the coming decades. Advances in nanotechnology-based therapies will certainly be a major factor in concurrent growth of nanoscale imaging agents based on similar materials. As of this writing, a number of professional societies for eye research have recently developed subsections promoting multidisciplinary applications of nanotechnology for imaging and therapy. The prospect of having material scientists, engineers, clinicians, and biologists cultivating designs and applications for nanomedicine in ophthalmology is exciting and will accelerate discovery in this field.

REFERENCES

1. L. Zhang, et al., Nanoparticles in medicine: Therapeutic applications and developments. *Clinical Pharmacology and Therapeutics*, 2008, 83(5): 761–69.
2. A.M. Chacko, et al., Targeted nanocarriers for imaging and therapy of vascular inflammation. *Current Opinion in Colloid & Interface Science*, 2011, 16(3): 215–27.
3. M.C. Woodle, Controlling liposome blood clearance by surface-grafted polymers. *Advanced Drug Delivery Reviews*, 1998, 32(1–2): 139–52.
4. M.M. van Schooneveld, et al., Improved biocompatibility and pharmacokinetics of silica nanoparticles by means of a lipid coating: a multimodality investigation. *Nano Letters*, 2008, 8(8): 2517–25.
5. D.P. Cormode, et al., A versatile and tunable coating strategy allows control of nanocrystal delivery to cell types in the liver. *Bioconjugate Chemistry*, 2011, 22(3): 353–61.
6. D.P. Cormode, et al., Nanotechnology in medical imaging: Probe design and applications. *Arteriosclerosis, Thrombosis, and Vascular Biology*, 2009, 29(7): 992–1000.
7. S.D. Caruthers, et al., Nanotechnological applications in medicine. *Current Opinion in Biotechnology*, 2007, 18(1): 26–30.
8. H.S. Choi, and J.V. Frangioni, Nanoparticles for biomedical imaging: Fundamentals of clinical translation. *Molecular Imaging*, 2010, 9(6): 291–310.
9. A. Jayagopal, et al., Quantum dot mediated imaging of atherosclerosis. *Nanotechnology*, 2009, 20(16): 165102.
10. R.A. Smith, et al., Molecular imaging metrics to evaluate response to preclinical therapeutic regimens. *Frontiers in Bioscience : A Journal and Virtual Library*, 2011, 1: 393–410.
11. C. Shah, et al., Imaging biomarkers predict response to anti-HER2 (ErbB2) therapy in preclinical models of breast cancer. *Clinical Cancer Research: An Official Journal of the American Association for Cancer Research*, 2009, 15(14): 4712–21.

12. M.A. Hahn, et al., Nanoparticles as contrast agents for *in-vivo* bioimaging: Current status and future perspectives. *Analytical and Bioanalytical Chemistry*, 2011, 399(1): 3–27.

13. D. Thassu, et al., *Nanoparticulate drug Delivery Systems*, Informa Healthcare, New York, 2007.

14. S.A. Wickline, and G.M. Lanza, Nanotechnology for molecular imaging and targeted therapy. *Circulation*, 2003, 107(8): 1092–95.

15. G.M. Lanza, and S.A. Wickline, Targeted ultrasonic contrast agents for molecular imaging and therapy. *Current Problems in Cardiology*, 2003, 28(12): 625–53.

16. A. Jayagopal, et al., Functionalized solid lipid nanoparticles for transendothelial delivery. *IEEE Transactions on Nanobioscience*, 2008, 7(1): 28–34.

17. A. Auger, et al., A comparative study of non-covalent encapsulation methods for organic dyes into silica nanoparticles. *Nanoscale Research Letters*, 2011, 6(1): 328.

18. S.J. Rosenthal, et al., Biocompatible quantum dots for biological applications. *Chemistry & Biology*, 2011, 18(1): 10–24.

19. W. Gao, et al., Poly(ethylene glycol) with observable shedding. *Angewandte Chemie*, 2010, 49(37): 6567–71.

20. C. Garnacho, et al., Differential intra-endothelial delivery of polymer nanocarriers targeted to distinct PECAM-1 epitopes. *Journal of Controlled Release: Official Journal of the Controlled Release Society*, 2008, 130(3): 226–33.

21. A. Jayagopal, et al., Surface engineering of quantum dots for *in vivo* vascular imaging. *Bioconjugate Chemistry*, 2007, 18(5): 1424–33.

22. P.M. Winter, et al., Endothelial alpha(v)beta3 integrin-targeted fumagillin nanoparticles inhibit angiogenesis in atherosclerosis. *Arteriosclerosis Thrombosis, and Vascular Biology*, 2006, 26(9): 2103–09.

23. A. Jayagopal, et al., Insights into atherosclerosis using nanotechnology. *Current Atherosclerosis Reports*, 2010, 12(3): 209–15.

24. D.J. Javier, et al., Aptamer-targeted gold nanoparticles as molecular-specific contrast agents for reflectance imaging. *Bioconjugate Chemistry*, 2008, 19(6): 1309–12.

25. F.S. Villanueva, et al., Myocardial ischemic memory imaging with molecular echocardiography. *Circulation*, 2007, 115(3): 345–52.

26. G. Zheng, et al., Rerouting lipoprotein nanoparticles to selected alternate receptors for the targeted delivery of cancer diagnostic and therapeutic agents. *Proceedings of the National Academy of Sciences of the United States of America*, 2005, 102(49): 17757–62.

27. G. Liew, et al., Retinal vascular imaging: A new tool in microvascular disease research. *Circulation. Cardiovascular Imaging*, 2008, 1(2): 156–61.

28. M.A. Mainster, et al., Scanning laser ophthalmoscopy. Clinical applications. *Ophthalmology*, 1982, 89(7): 852–57.

29. S.K. Chakraborty, et al., Cholera toxin B conjugated quantum dots for live cell labeling. *Nano Letters*, 2007, 7(9): 2618–26.

30. B. Ballou, et al., Sentinel lymph node imaging using quantum dots in mouse tumor models. *Bioconjugate Chemistry*, 2007, 18(2): 389–96.

31. J.D. Smith, et al., The use of quantum dots for analysis of chick CAM vasculature. *Microvascular Research*, 2007, 73(2): 75–83.

32. A.M. Smith, et al., Quantum dot nanocrystals for *in vivo* molecular and cellular imaging. *Photochemistry and Photobiology*, 2004, 80(3): 377–85.

33. W.C. Chan, and S. Nie, Quantum dot bioconjugates for ultrasensitive nonisotopic detection *Science*, 1998, 281(5385): 2016–18.

34. B. Ballou, et al., Noninvasive imaging of quantum dots in mice. *Bioconjugate Chemistry*, 2004, 15(1): 79–86.

35. Y. Su, et al., *In vivo* distribution, pharmacokinetics, and toxicity of aqueous synthesized cadmium-containing quantum dots. *Biomaterials*, 2011, 32(25): 5855–62.

36. E.B. Souto, et al., Feasibility of lipid nanoparticles for ocular delivery of anti-inflammatory drugs. *Current Eye Research*, 2010, 35(7): 537–52.

37. R. Koelsch, et al., Incorporation of chemically modified proteins into liposomes. *Acta Biologica et Medica Germanica*, 1981, 40(3): 331–35.

38. R.C. Nagarwal, et al., Polymeric nanoparticulate system: A potential approach for ocular drug delivery. *Journal of Controlled Release: Official Journal of the Controlled Release Society*, 2009, 136(1): 2–13.

39. V.P. Torchilin, PEG-based micelles as carriers of contrast agents for different imaging modalities. *Advanced Drug Delivery Reviews*, 2002, 54(2): 235–52.

40. W. Drexler, and J.G. Fujimoto, State-of-the-art retinal optical coherence tomography. *Progress in Retinal and Eye Research*, 2008, 27(1): 45–88.

41. J.G. Fujimoto, Optical coherence tomography for ultrahigh resolution *in vivo* imaging. *Nature Biotechnology*, 2003, 21(11): 1361–67.

42. J.A. Izatt, et al., Micrometer-scale resolution imaging of the anterior eye *in vivo* with optical coherence tomography. *Archives of Ophthalmology*, 1994, 112(12): 1584–89.

43. U. Morgner, et al., Spectroscopic optical coherence tomography. *Optics Letters*, 2000, 25(2): 111–13.

44. B. Hermann, et al., Precision of extracting absorption profiles from weakly scattering media with spectroscopic time-domain optical coherence tomography. *Optics Express*, 2004, 12(8): 1677–88.

45. S.A. Boppart, Advances in contrast enhancement for optical coherence tomography. *Proceedings of the Annual International Conference of the IEEE Engineering in Medicine and Biology Society. IEEE Engineering in Medicine and Biology Society*, 2006, 1: 121–24.

46. J.S. Bredfeldt, et al., Molecularly sensitive optical coherence tomography. *Optics Letters*, 2005, 30(5): 495–97.

47. T.M. Lee, et al., Engineered microsphere contrast agents for optical coherence tomography. *Optics Letters*, 2003, 28(17): 1546–48.

48. D.A. Giljohann, et al., Gold nanoparticles for biology and medicine. *Angewandte Chemie*, 2010, 49(19): 3280–94.

49. S.A. Boppart, et al., Optical probes and techniques for molecular contrast enhancement in coherence imaging. *Journal of Biomedical Optics*, 2005, 10(4): 41208.

50. L.R. Hirsch, et al., Metal nanoshells. *Annals of Biomedical Engineering*, 2006, 34(1): 15–22.

51. A. Jayagopal, et al., Hairpin DNA-functionalized gold colloids for the imaging of mRNA in live cells. *Journal of the American Chemical Society*, 2010, 132(28): 9789–96.

52. C.J. Murphy, et al., Gold nanoparticles in biology: Beyond toxicity to cellular imaging. *Accounts of Chemical Research*, 2008, 41(12): 1721–30.

53. R. John, and S.A. Boppart, Magnetomotive molecular nanoprobes. *Current Medicinal Chemistry*, 2011, 18(14): 2103–14.

54. R. John, et al., *In vivo* magnetomotive optical molecular imaging using targeted magnetic nanoprobes. *Proceedings of the National Academy of Sciences of the United States of America*, 2010, 107(18): 8085–90.

55. B.A. Berkowitz, and R. Roberts, Prognostic MRI biomarkers of treatment efficacy for retinopathy. *NMR in Biomedicine*, 2008, 21(9): 957–67.

56. F. Herranz, et al., The application of nanoparticles in gene therapy and magnetic resonance imaging. *Microscopy Research and Technique*, 2011, 74(7): 577–91.

57. R.D. Engberink, et al., Magnetic resonance imaging of monocytes labeled with ultrasmall superparamagnetic particles of iron oxide using magnetoelectroporation in an animal model of multiple sclerosis. *Molecular Imaging*, 2010, 9(5): 268–77.

58. J.W. Bulte, *In vivo* MRI cell tracking: Clinical studies. AJR. *American Journal of Roentgenology*, 2009, 193(2): 314–25.

59. H.B. Na, et al., Development of a T1 contrast agent for magnetic resonance imaging using MnO nanoparticles. *Angewandte Chemie*, 2007, 46(28): 5397–401.

60. C.E. Gavin, et al., Manganese and calcium transport in mitochondria: Implications for manganese toxicity. *Neurotoxicology*, 1999, 20(2–3): 445–53.

61. R. Koole, et al., Magnetic quantum dots for multimodal imaging. *Wiley Interdisciplinary Reviews. Nanomedicine and Nanobiotechnology*, 2009, 1(5): 475–91.

62. P.A. Jarzyna, et al., Multifunctional imaging nanoprobes. *Wiley Interdisciplinary Reviews. Nanomedicine and Nanobiotechnology*, 2010, 2(2): 138–50.

63. W.J. Mulder, et al., Quantum dots for multimodal molecular imaging of angiogenesis. *Angiogenesis*, 2010, 13(2): 131–34.

64. Y. Wang, et al., Preparation of nanobubbles for ultrasound imaging and intracelluar drug delivery. *International Journal of Pharmaceutics*, 2010, 384(1–2): 148–53.

65. C. Kim, et al., Multifunctional microbubbles and nanobubbles for photoacoustic and ultrasound imaging. *Journal of Biomedical Optics*, 2010, 15(1): 010510.

66. Early Treatment Diabetic Retinopathy Study Research Group. Early photocoagulation for diabetic retinopathy. ETDRS report number 9. *Ophthalmology*, 1991, 98(5 Suppl): 766–85.

67. R. Bernardes, et al., Noninvasive evaluation of retinal leakage using optical coherence tomography. *Ophthalmologica. Journal International D'ophtalmologie. International Journal of Ophthalmology. Zeitschrift fur Augenheilkunde*, 2011, 226(2): 29–36.

68. J. Cunha-Vaz, et al., Blood-retinal barrier. *European Journal of Ophthalmology*, 2010, 21(S6): 3–9.

69. J. Cunha-Vaz, and G. Coscas, Diagnosis of macular edema. *Ophthalmologica. Journal International D'ophtalmologie. International Journal of Ophthalmology. Zeitschrift fur Augenheilkunde*, 2010, 224 (Suppl 1): 2–7.

70. J.G. Cunha-Vaz, and M. Shakib, Ultrastructural mechanisms of breakdown of the blood-retina barrier. *The Journal of Pathology and Bacteriology*, 1967, 93(2): 645–52.

71. P.K. Russ, et al., Retinal vascular permeability determined by dual-tracer fluorescence angiography. *Annals of Biomedical Engineering*, 2001, 29(8): 638–47.

72. J.H. Kim, et al., Intravenously administered gold nanoparticles pass through the blood-retinal barrier depending on the particle size, and induce no retinal toxicity. *Nanotechnology*, 2009, 20(50): 505101.

73. C.W. Jones, et al., Vitreous fluorophotometry in the alloxan- and streptozocin-treated rat. *Archives of Ophthalmology*, 1982, 100(7): 1141–45.

74. D.C. Metrikin, et al., Measurement of blood-retinal barrier breakdown in endotoxin-induced endophthalmitis. *Investigative Ophthalmology & Visual Science*, 1995, 36(7): 1361–70.

75. J. Tang, and T.S. Kern, Inflammation in diabetic retinopathy. *Progress in Retinal and Eye Research*, 2011, 30(5): 343–58.

76. H. Chibli, et al., Cytotoxicity of InP/ZnS quantum dots related to reactive oxygen species generation. *Nanoscale*, 2011, 3(6): 2552–59.

77. A.P. Adamis, and A.J. Berman, Immunological mechanisms in the pathogenesis of diabetic retinopathy. *Seminars in Immunopathology*, 2008, 30(2): 65–84.

78. S. Miyahara, et al., *In vivo* three-dimensional evaluation of leukocyte behavior in retinal microcirculation of mice. *Investigative Ophthalmology & Visual Science*, 2004, 45(11): 4197–201.

79. H. Tamura, et al., *In vivo* evaluation of ocular inflammatory responses in experimental diabetes. *The British Journal of Ophthalmology*, 2005, 89(8): 1052–57.

80. G. Zhang, et al., Multi-emissive difluoroboron dibenzoylmethane polylactide exhibiting intense fluorescence and oxygen-sensitive room-temperature phosphorescence. *Journal of the American Chemical Society*, 2007, 129(29): 8942–43.

81. A. Pfister, et al., Boron polylactide nanoparticles exhibiting fluorescence and phosphorescence in aqueous medium. *ACS Nano*, 2008, 2(6): 1252–58.

82. G. Zhang, et al., A dual-emissive-materials design concept enables tumour hypoxia imaging. *Nature Materials*, 2009, 8(9): 747–51.
83. G.M. Palmer, et al., Optical imaging of tumor hypoxia dynamics. *Journal of Biomedical Optics*, 2010, 15(6): 066021.
84. S.K. Chang, et al., *In vivo* optical molecular imaging of vascular endothelial growth factor for monitoring cancer treatment. *Clinical Cancer Research: An Official Journal of the American Association for Cancer Research*, 2008, 14(13): 4146–53.
85. A.M. Joussen, et al., Retinal vascular endothelial growth factor induces intercellular adhesion molecule-1 and endothelial nitric oxide synthase expression and initiates early diabetic retinal leukocyte adhesion *in vivo*. *The American Journal of Pathology*, 2002, 160(2): 501–09.
86. A.P. Adamis, and D.T. Shima, The role of vascular endothelial growth factor in ocular health and disease. *Retina*, 2005, 25(2): 111–18.
87. G.J. Strijkers, et al., Paramagnetic and fluorescent liposomes for target-specific imaging and therapy of tumor angiogenesis. *Angiogenesis*, 2010, 13(2): 161–73.
88. W.J. Mulder, and A.W. Griffioen, Imaging of angiogenesis. *Angiogenesis*, 2010, 13(2): 71–74.
89. E. Kluza, et al., Synergistic targeting of alphavbeta3 integrin and galectin-1 with hetero-multivalent paramagnetic liposomes for combined MR imaging and treatment of angiogenesis. *Nano Letters*, 2010, 10(1): 52–58.
90. H.M. Sanders, et al., Morphology, binding behavior and MR-properties of paramagnetic collagen-binding liposomes. *Contrast Media & Molecular Imaging*, 2009, 4(2): 81–88.
91. S.W. Cousins, et al., Monocyte activation in patients with age-related macular degeneration: A biomarker of risk for choroidal neovascularization?. *Archives of Ophthalmology*, 2004, 122(7): 1013–18.
92. D.G. Espinosa-Heidmann, et al., Macrophage depletion diminishes lesion size and severity in experimental choroidal neovascularization. *Investigative Ophthalmology & Visual Science*, 2003, 44(8): 3586–92.
93. E. Sakurai, et al., Macrophage depletion inhibits experimental choroidal neovascularization. *Investigative Ophthalmology & Visual Science*, 2003, 44(8): 3578–85.
94. M.M. Sadeghi, et al., Imaging atherosclerosis and vulnerable plaque. *Journal of Nuclear Medicine: Official Publication, Society of Nuclear Medicine*, 2010, 51 (Suppl 1): 51S–65S.
95. S.A. Wickline, et al., Molecular imaging and therapy of atherosclerosis with targeted nanoparticles. *Journal of Magnetic Resonance Imaging: JMRI*, 2007, 25(4): 667–80.
96. R.F. Spaide, and C.A. Curcio, Drusen characterization with multimodal imaging. *Retina*, 2010, 30(9): 1441–54.
97. R.J. Haynes, et al., Imaging of optic nerve head drusen with the scanning laser ophthalmoscope. *The British Journal of Ophthalmology*, 1997, 81(8): 654–57.
98. J.W. Crabb, et al., Drusen proteome analysis: An approach to the etiology of age-related macular degeneration. *Proceedings of the National Academy of Sciences of the United States of America*, 2002, 99(23): 14682–87.
99. L. Wang, et al., Abundant lipid and protein components of drusen. *PloS One*, 2010, 5(4): e10329.

20 Photoresponsive Polymers for Ocular Drug Delivery

Laura A. Wells and Heather Sheardown

CONTENTS

20.1 INTRODUCTION

Light in the form of lamps and more typically lasers has been successfully implemented in a variety of ophthalmic disease treatments and surgeries. The potential of lasers and light has yet to be realized with myriad research and recent success stories focused on the synthesis and implementation of photoresponsive (light-responsive) polymer platforms for the on-demand delivery of drugs in a smart and tuned fashion. The transparency of the eye and advanced technological developments in ophthalmic lasers allow for strict focal point control and high-specificity delivery of monochromatic ultraviolet (UV), visible, and infrared (IR) wavelengths (Gibson and Kernohan 1995; Qiu and Park 2001; Alvarez-Lorenzo et al. 2009). Light provides quick outcomes with low lag times since its absorption by molecules is instantaneous and does not rely on the relatively slow diffusion typical of pH- or temperature-responsive materials (Schmaljohann 2006). Light and lasers can provide safe, noninvasive stimuli for smart drug delivery systems and are expected to be well tolerated by patients who are already familiar with laser and light use as surgical and diagnostic tools in the clinic, particularly relevant for the eye.

Light and lasers can be used in several ways as photosensitive drug delivery systems for the eye. Drugs can be photoactivated (Misiuk-hojlo et al. 2006) or drug delivery vehicles photolysed (Zhang and Smith 1999) *in situ* at the site of delivery in the anterior or posterior segment of the eye to provide specific delivery to the site of interest. Reversible polymer delivery systems have the potential to provide on-demand delivery, which can be turned on or off and can be tunable to change the delivery rate (Wells et al. 2011b), with light and lasers being used to adjust delivery according to the progression of the disease. Light can also be used to solidify injectable systems (Baroli 2006) to provide noninvasive methods of introducing systems to poorly accessible areas such as the lens capsule or posterior segment. In addition, light can be used to degrade polymer-based delivery systems after their delivery life. Taken together, there is enormous potential for lasers and light to be used as a tool to optimize drug delivery through the creation of photoresponsive polymer-based drug delivery platforms.

20.2 LIGHT TRANSMISSION THROUGH THE EYE

UV, visible, and IR light are the components of the electromagnetic spectrum that can penetrate the eye with the potential to provide noninvasive stimuli for ophthalmic drug delivery. Table 20.1 describes the various wavelengths associated with UV, visible, and IR light and their general absorption trends in tissue.

Laser, light, and all forms of electromagnetic radiation can pass through, be reflected, be scattered, or be absorbed when exposed to a medium. Therefore, a defining

TABLE 20.1
Description of the Absorption of Various UV and IR Wavelengths

Wavelength	Designation	Characteristics (tissue)
100–280 nm	UV-C	Superficial absorption
280–315 nm	UV-B	Penetration
315–400 nm	UV-A	Deeper penetration
380–780 nm	Visible light	Day and night vision
780–1400 nm	IR-A	Deep penetration
1400–3000 nm	IR-B	Slight penetration (water absorbs)
3–1000 μm	IR-C	Very superficial absorption

Source: Blume, Y., D. J. Durzan, and P. Smertenko. 2006. *Cell biology and instrumentation: UV radiation, nitric oxide and cell death in plants.* The Netherlands: IOS Press.

parameter in photoresponsive ophthalmic drug delivery is the wavelengths that are able to penetrate the various tissues of the eye. In a study by Boettner and Wolter (1962), fresh human eyes from various-aged humans revealed the transmittance of different tissues, as illustrated in Figure 20.1 and summarized in Table 20.2. Transmittance is the fraction of incident light that passes through a sample at defined wavelengths.

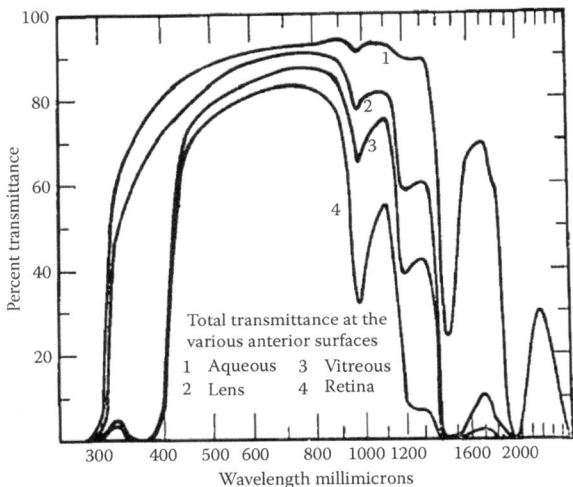

FIGURE 20.1   The spectrum of wavelengths that can transmit through sections of the eyes. (Reproduced with permission of Investigative Ophthalmology & Visual Science by the Association for Research in Vision and Ophthalmology, Transmission of the ocular media by Boettner, E. A., and J. R. Wolter, *Investigative Ophthalmology*, 1, 776–783, 1962. Copyright permission conveyed through Copyright Clearance Center, Inc.)

TABLE 20.2

Transmission of UV, Visual, and IR Wavelengths through Different Ocular Tissues

Tissue	Transmittance (nm)	Additional Information
Cornea	300–2500	Bands in IR region No age dependence
Aqueous humor	220–2400 **Not at 265	Absorption band at 265 nm due to proteins
Lens	320, then 390–1900	Absorption at 360 nm Strong age dependence, only children transmit 320
Vitreous humor	300–1400	
Whole eye	390–1400	Indirect transmittance maximum 83.5% Direct transmittance maximum 55%

Source: Boettner, E. A. and Wolter, J. R., *Investigative Ophthalmology*, 1, 776–83, 1962.

** Because of the proteins in the aqueous humor, there is an absorbance maximum at 265 nm. However, the fluid is otherwise transparent.

It is clear from Figure 20.1 and Table 20.2 that different wavelengths of light are able to penetrate the anterior and posterior segments. It may seem that UV light has difficulty penetrating the cornea and lens to the posterior segment due to the low transmittance of these tissues. However, the use of specialized lasers that can perform two-photon absorption (TPA) will allow for an equivalent delivery of UV light energy to target tissues and drug delivery platforms as will be discussed in Section 20.3. With technology, most wavelengths of light, or equivalents from TPA, allow for the use of UV, visible, and IR wavelengths of light as stimuli for polymer-based drug delivery systems.

20.3 LIGHT AND LASER TECHNOLOGY

With photoresponsive drug delivery, light and lasers are directed at a polymeric device to affect changes in the device that lead to changes in the rate of drug delivery but not to changes in the tissues within the eye. Lessons learned with photodynamic therapy (PDT) implementation, including the development of high-power diode lasers, suggest that lasers and light can easily be directed at photoresponsive drug delivery devices located in the anterior or posterior segments of the eye. To obtain precise and highly controllable exposures to devices placed within the eye, lasers are commonly employed because they allow for high-intensity monochromatic (single wavelength) light to be directed at precise spots up to 8000 μm (Schmidt-Erfurth and Hasan 2000). Therefore, light technologies in drug delivery that are currently under development will likely eventually be developed to use lasers for the light stimuli. In addition to ophthalmology, there is potential for systems to be developed in gynecology, dermatology, otolaryngology, gastroenterology, and physiotherapy, as these are also amenable to light- and laser-responsive devices and are medical areas that are already well versed in the application of lasers (Gibson and Kernohan 1995).

Lasers have been used in the eye since the 1960s for surgery and disease treatment. Ruby lasers (red wavelengths) were first used in the 1960s to reattach retinas; however, they cannot photocoagulate (cannot close blood vessels), somewhat limiting their use. Argon ion lasers (green and blue wavelengths) were then used as an improved option to ruby red for retinas since their wavelengths are absorbable by proteins and converted to heat to allow for coagulation (Wright et al. 2007) when melanin is present to allow for tissue absorption (Misiuk-hojlo et al. 2006). The Krypton ion laser (red and yellow wavelengths, 647 and 568 nm) delivers two wavelengths; the red is absorbed by choroid and the yellow by the pigment epithelium and the xanthophylls in the macula. Nd:YAG (neodymium:yttrium-aluminum-garnet) and ND:YLF are pulsed laser beams that are used to cut or break down lenses, respectively (Wright et al. 2007).

High-wavelength light, such as IR, is known to better penetrate the eye. Despite possible interactions low-wavelength light faces when penetrating the eye, in many instances, losses are likely negligible when irradiation is on the order of 600 mW/cm^2 (Schmidt-Erfurth and Hasan 2000). TPA offers an opportunity to deliver energy equivalent to low-wavelength UV light to the back of the eye. Ocular tissues to the front of the eye may absorb UV, leading to deleterious effects (Thompson et al. 1992). TPA, however, uses wavelengths in the visible range to allow penetration through the cornea and lens with high three-dimensional (3D) spatial selectivity

(Bhawalkar et al. 1996). With TPA, two separate photons are absorbed almost simultaneously to deliver the energy of one lower-wavelength/higher-energy photon and has been demonstrated to effectively deliver UV equivalent light (<280 nm) using lasers in the visible region (532 nm) to chromophores such as coumarin (Hartner et al. 2007). Overall, TPA has the potential to be an effective and safe method to allow high-energy equivalents of UVA and UVB light to affect molecules while safely penetrating ocular tissues.

20.4 CURRENT LIGHT AND LASER USE IN OPHTHALMOLOGY

20.4.1 CURRENT APPLICATIONS

Light and laser treatments are common in ophthalmology, with several well-established treatments used in both the anterior and posterior segments. In retinal treatment, lasers are used to stop detachment or coagulate hemorrhages and bleeding to treat retinal holes, retinal detachment, diabetic retinopathy, central vein occlusion, and senile macular degeneration (Wright et al. 2007). Lasers that induce thermal changes may be used in the vitreous humor for the treatment of membranes and neovascularizations extending from the retina into the vitreous humor. Lasers are commonly used for lens fragmentation to allow for lens removal during cataract treatment with intraocular lenses (Wright et al. 2007) and to remove secondary cataracts (Thompson et al. 1992). Nd:YAG laser pulses are used to create cuts in the lens capsule and Nd:YLF is used to fragment the lens (Wright et al. 2007). To treat closed-angle glaucoma, laser iridotomies, with argon ion or Nd lasers, are used to create perforations in the iris to allow drainage from the aqueous humor to the trabecula (Wright et al. 2007). To treat open-angle glaucoma, trabeculoplasty, with argon ion lasers, is used to shrink the trabecular meshwork to widen the canal to allow for drainage of the aqueous humor (Wright et al. 2007). Other treatments of open-angle glaucoma aim to create channels with sclerostomies. Visible light with thermal CO_2 lasers provides an external method, and a combination of this with India ink–dyed tissue or the use of Nd:YAG lasers offers an internal method (Wright et al. 2007). Noninvasive, painless corneal surgeries to remove pathological conditions and for refractive surgery can be performed with ArF excimer lasers and clean corneal excisions can be performed with Nd lasers. Laser *in situ* karatomileusis (LASIK) uses femtosecond lasers (an alternative to surgical knives) to remove a flap and ArF laser or femtosecond laser (Nd) to remove intrastromal tissue to cause changes in refraction with high precision (Wright et al. 2007). The use of lasers in drug delivery began with PDT to light-activated drugs for the treatment of macular edema.

20.4.2 PHOTODYNAMIC THERAPY: DRUG DELIVERY AND ACTIVATION

The first clinical example of photoresponsive drugs in ophthalmology is the use of PDT for the treatment of choroidal neovascularization secondary to age-related macular degeneration. Photocoagulation alone is not particularly effective in the majority of cases where lesions are not classic, small, and with well-defined boundaries. In addition, it is nonselective and may cause necrosis to healthy retinal tissue, leading to irreversible vision loss (Schmidt-Erfurth and Hasan 2000). With PDT, a photosensitizer drug such as Verteporfin is administered intravenously and then activated locally after 15 minutes with 689 nm laser light in the retina. Activation of the photosensitizer drug produces singlet oxygen molecules that induce coagulation and lower vessel growth to reduce pathological fluid leakage (Misiuk-hojlo et al. 2006). It is both effective on undefined boundaries and selective to specific tissue areas since Verteporfin will be present where it is exposed by edema in the areas of choroidal neovascularization (Schmidt-Erfurth and Hasan 2000). Unfortunately, since Verteporfin is systemically delivered, it and other similar prodrugs/photodynamic therapies have potential systemic side effects with the potential for activation in the nontarget tissue. After having undergone Verteporfin PDT, patients must avoid external light sources for 5 days to allow the prodrug to clear their system. Otherwise, there is potential for damage to, for example, the eyes or the skin (QLT 2005). Regardless, PDT illustrates the feasibility of using light and lasers as stimuli for drug delivery to the eye. While prodrugs are a valuable therapy, the use of localized photoresponsive polymer delivery systems would eliminate the risk of photosensitization of other tissues.

20.5   PHOTORESPONSIVE POLYMERS

Photoresponsive polymers are polymers grafted with photoresponsive molecules which, when designed and synthesized effectively, can introduce photoswitchable or photoactivatable attributes into the polymer system. Due to the length of polymers, photosensitive molecules may be bound along the polymer chain, within the chain, or at the ends of the chain (see Figure 20.2). Polymers may also be cross-linked (bound) to form 3D networks, and photosensitive molecules can also potentially be designed to form these cross-links. The changes that occur after the photosensitive molecules (chromophores) absorb light directly reflect the changes that occur within a polymer material that is bound with those molecules.

When a molecule absorbs light, it goes from a ground state to an excited state. After absorbing light in the form of energy, molecules may undergo fluorescence, nonradiative decay (heat), phosphorescence, luminescence, and a variety of other inter- and intramolecular processes (Wayne and Wayne 1996).

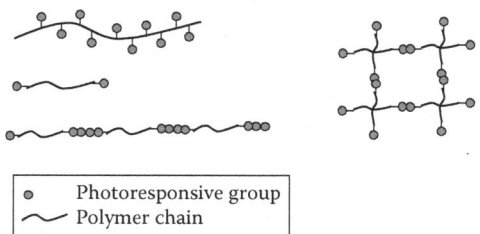

FIGURE 20.2   Popular methods to create photoresponsive polymers are to graft photosensitive groups along the polymer chain backbone (a) or at the ends of a polymer (b), to polymerize photoresponsive molecules within a polymer chain (c), or to create polymer networks cross-linked with photosensitive groups (d).

The molecules associated with photoresponsive drug delivery systems generally undergo inter- or intramolecular changes. Intramolecular changes occur within certain molecules and may alter their solubility or length, which could also impact polymer solubility or polymer/cross-link length. Intermolecular changes can result in the binding or dissociation of separate molecules, to alter binding within a polymer system, or result in the activation of a photosensitive molecule to induce permanent changes to a polymer system, such as cross-linking or polymer degradation. These are but a few examples of how photosensitive molecules may affect polymer systems. Researchers have effectively introduced molecules and created a variety of interesting irreversible and reversible polymer-based drug delivery systems.

20.6   EMERGING PHOTORESPONSIVE POLYMER TECHNOLOGIES

Reversible and irreversible light-responsive polymers can create drug delivery platforms with on-demand delivery or alterable drug release rates. Many of these technologies are in their infant stages but hold promise in the ability to optimize drug delivery in the near future.

20.6.1   IRREVERSIBLE POLYMER SYSTEMS

Molecules that undergo an irreversible transition with light and laser stimuli can create polymer systems that are photodegradable or phototriggered to delivery or alter the delivery rate of drugs from polymer matrices.

20.6.1.1   *In Situ* Photo-Cross-Linking and Photopolymerization

Injectable polymer delivery systems that can be cross-linked/polymerized *in vivo* are desirable in order to reduce the invasiveness of delivery to hard-to-reach areas (Baroli 2006) such as the posterior segment or lens capsule. Low-viscosity polymers with the drug in solution may be injected; light activation of the photoinitiator could then be used as a mechanism to cross-link the polymers (Poshusta and Anseth 2001). This entraps drugs, which can then slowly diffuse from the new polymer material and be delivered to the site. Unfortunately, in most currently reported systems, the photoinitiators remain as an impurity that could leach out and potentially cause side effects. A variety of injectable photo-cross-linkable systems are under development, which involve acrylic groups combined with polyethylene glycol (PEG) and a variety of hydrogels and other synthetic polymers (Baroli 2006). The focus of *in situ* photopolymerization has typically been on transdermal and cartilage tissue applications (Nguyen and West 2002); however, the eye will likely be a focus as the technology develops and moves toward requiring no additional photoinitiators, as discussed with reversible systems in Section 20.6.2.

20.6.1.2   Photoresponsive Degradation

A challenge for internal ocular delivery systems is their removal after release is complete. Light and lasers can also be used as a trigger for polymer degradation with light-labile polymers. Photoinduced degradation could also be used to increase drug release or to aid in the removal of drug release carriers after their release life.

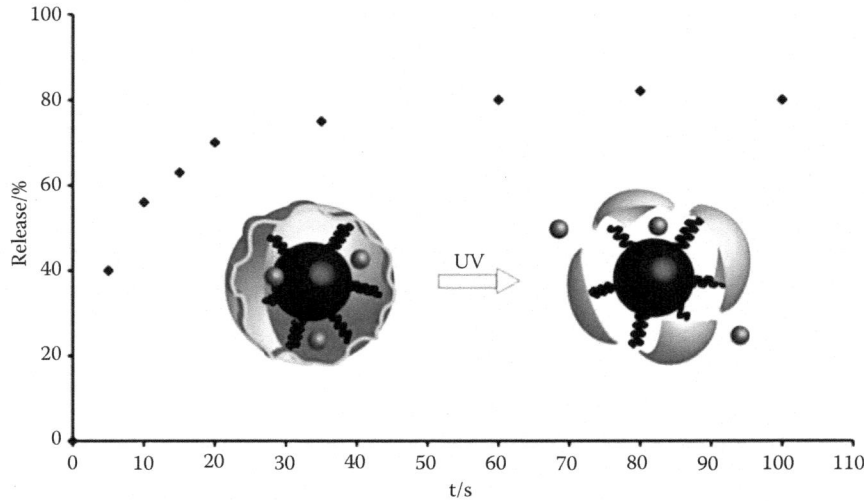

FIGURE 20.3 Release of Rose Bengal from 350 nm treated hyperbranched polyglycerols nanocarriers containing photodegradable shells. (Burakowska, E. et al.: Photoresponsive crosslinked hyperbranched polyglycerols as smart nanocarriers for guest binding and controlled release. *Small*, 2009, 5, 2199–204. Copyright Wiley-VCH Verlag GmbH & Co. KGaA, Weinheim. Reprinted with permission.)

Modification of polymer chains with photolabile groups allows for their degradation or bulk modification *in situ*. Photolabile groups, such as nitrobenzyloxycarbonyl, are photocleavable and may cause polymer degradation when activated with UV light (Johnson et al. 2007). In addition, PEG hydrogel scaffolds that contain similar nitrobenzyl ether groups use light, to release bound peptides (Kloxin et al. 2009). The creation of micelles and nanoparticles with photocleavable groups that cause their dissolution and release of their inner contents has been a focus for site-specific drug delivery. Nanocarriers based on hyperbranched polyglycerols with photodegradable *o*-nitrobenzyl-containing shells have shown increased delivery of Rose Bengal with exposure to 350 nm monochromatic light, as illustrated in Figure 20.3 (Burakowska et al. 2009). Alternatively, liposomes made with photolabile lipids can degrade or destabilize after UV light exposure, leading to the release of entrapped dye or drug molecules (Zhang and Smith 1999). Liposomes created with the photoreactive lipid 1,2-bis[10-(2′,4′-hexadienolyoxy-decanonyl]-*sn*-glycero-3-phosphocholine have UV light-induced permeability with the potential to alter drug release (Bondurant et al. 2001; Spratt et al. 2003). All of these mechanisms of photodegradation are tools both to deliver drugs and to remove the drug delivery platforms in a noninvasive fashion.

20.6.1.3 Phototriggered Systems

A novel approach is to use light to phototrigger drug release or drug activation *in situ*. PDT is a phototriggered system not involving polymers. As noted earlier, prodrugs are injected and delivered via the blood stream and are then activated in the posterior segment with appropriate lasers. However, in addition to the activation of prodrugs, light and lasers may also be applied to polymers to induce the release and/or activation of drugs to the eye.

One approach has been to create light-sensitive conjugates of current drugs so that when released from scaffolds, UV exposure would cleave them into an active form. Such systems with acetyl salicylic acid, ibuprofen, and ketoprofen delivery and activation from methacrylate-based scaffolds have been successfully created (McCoy et al. 2007). This method holds promise for the creation of similar prodrugs for ophthalmic applications. Alternatively, the photoactivation of a molecule may trigger secondary reactions in surrounding polymers to induce a response to aid in drug delivery. More specifically, the photocleavage of a molecule into charged groups can alter the charge density and osmotic pressure of a matrix to alter diffusion and swelling properties. Triphenylmethane leuco derivatives cleave to introduce ionic groups when incorporated into hydrogels (Irie and Kunwatchakun 1986; Mamada et al. 1990; Misu et al. 2009). This causes an increase in the osmotic pressure of the gel, potentially altering drug delivery properties. Recent work has also looked at controlling the binding and delivery of nanoparticles to cells by triggering their ability to bind. YIGSR ligands bound on the outside of particles to direct cell binding are "photocaged" with 4,5-dimethoxy-2-nitrobenzyl. When the "cages" are cleaved with light, the ligands become active and the particles can bind to cells (Dvir et al. 2010).

20.6.2 Photoreversible Polymer Systems

Photoinduced, reversible changes in classes of molecules that isomerize and dimerize may be used to switch the properties of polymer chains and matrices to create photoreversible drug delivery systems suitable for anterior and posterior segment drug delivery. Although most are in their infancy, there are many systems under development, which may well suit long-term ocular drug delivery needs.

20.6.2.1 Photoisomer-Based Systems

The grafting of polymers with molecules that have photoinduced reversible isomers with differing physical properties can lead to the creation of polymers with alterable chain solubilities and alterable cross-link mesh sizes, leading to control over drug delivery rates. Two common photoisomerizing molecules that have been grafted to multiple polymer platforms are azobenzene (Kumar and Neckers 1989) and spirobenzopyran (Minkin 2004). Transitions tend to occur with visible and UV light making polymer systems that contain these molecules be potentially affected by ambient visible light levels in the eye. Azobenzene becomes a *cis* isomer when exposed to UV light and a *trans* isomer when exposed to visible blue light, as shown in Figure 20.4. Since its *cis* form is hydrophilic and its *trans* form is hydrophobic, the exposure of polymers bound with azobenzene to different light treatments can alter their solubility in different environments and to different drugs.

Polymer-based nanoparticle suspensions are injectable delivery formulations for hard-to-reach tissues such as those in the posterior segment of the eye. *Trans/cis* forms of azobenzene can act as a switch between hydrophilic and hydrophobic solubility in polymers. Grafting of azobenzene groups to a hydrophilic polymer chain can alter solubility via light-induced *cis/trans* isomerization and has been used to create micro- and nanocarrier-based drug delivery systems (Zhao 2007). Amphiphilic block copolymer micelles are microparticulates that have cores and shells of alternate solubilities. By binding azobenzene to one end of a hydrophilic polymer, the *trans* form will cause that end to be hydrophobic and can be used to drive the formation of micelles. Exposure to light or lasers of UV wavelengths will change the azobenzene to a *cis* form that can be used to drive destabilization of the micelles, resulting in their dissolution and release of their contents. As shown in Figure 20.5, this can

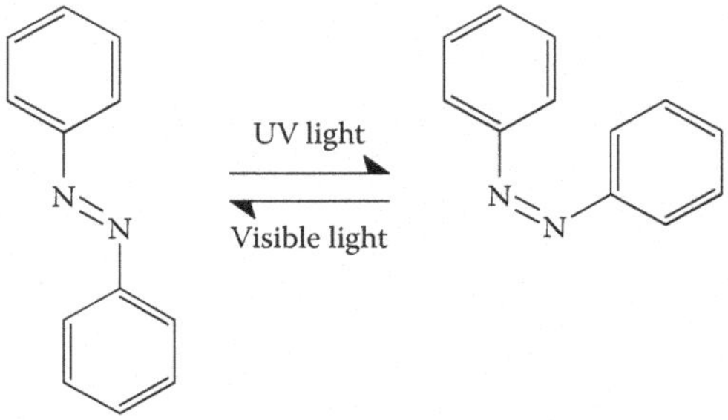

FIGURE 20.4   The reversible *trans* to *cis* isomerization of azobenzene that occurs with UV and visible light.

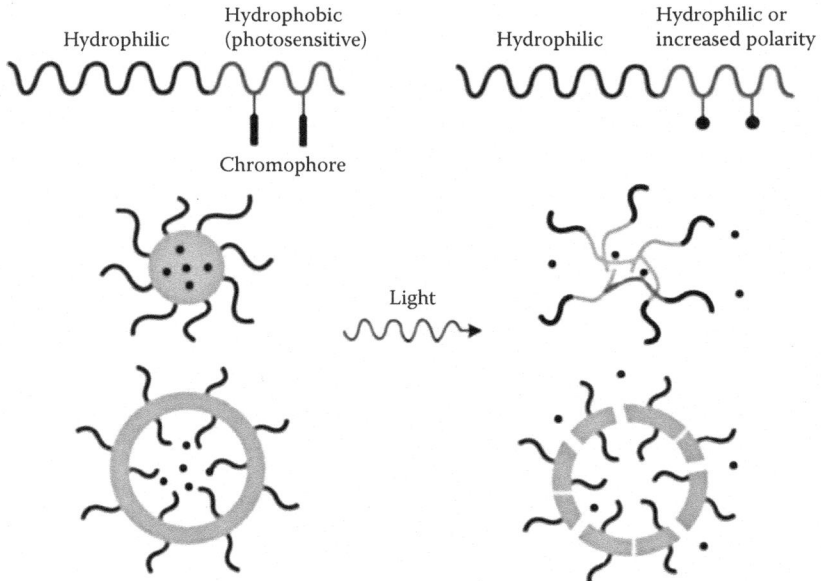

FIGURE 20.5   An overview of the design core-cross-linked micelles and shell-cross-linked vesicles synthesized with azobenzene-based chromophores. (Zhao, Y., *The Chemical Record,* 7, 286–94, 2007. Copyright The Japan Chemical Journal Forum and Wiley Periodicals, Inc. Reproduced with permission.)

be used to create light-labile micelles that can carry a drug and then deliver it upon rupture from light/laser exposure at the target delivery site (Zhao 2007). There are several example systems that use this technology. For example, hydrophilic polymers of poly(tert-butyl acrylate-co-acrylic acid) with azobenzene-containing polymethacrylic acid groups create micelles that undergo reversible formation with UV and visible light exposure (Wang et al. 2004; Kuiper and Engberts 2004). Similarly, liposomes, which are similar to micelles but formed by bilayers, can have alterable permeability by *cis/trans* alterations of azobenzene-bound lipids. The *cis*-isomer causes disruptions in the lipid membrane resulting in the creation of pores that allow the diffusion and release of drugs from the core. This concept has been demonstrated with doxorubicin release in response to UV light treatment of liposomes containing the photochromic lipid 1,2-(4' *n*-butylphenyl)azo-4'(γ-phenylbutyroyl)-glycero-3-phosphocholine (Bisby et al. 2000a,b).

During its isomer transition, azobenzene also undergoes a length change from 9.0 to 5.5 Å between its *trans* and *cis* isomer forms (El Halabieh et al. 2004). Drug depot systems with alterable cross-linking densities provide an opportunity to alter the diffusion and release of drugs. When azobenzene is grafted to a polymer at both of its ends as a crosslinker, light can be used to cause a switch between the *trans/cis* isomer forms, leading to changes in the length of the grafted azobenzene. This can cause slight alterations in the mesh size of the network to create larger or smaller pore sizes and an increase or decrease in the diffusion and release of drugs (Tomer and Florence 1993). The ability of visible light to decrease and UV light to increase the release of caffeine from azobenzene-cross-linked *N*-isopropylacrylamide-based hydrogels has been demonstrated to occur, albeit slightly, and was attributed to the change in size associated with the different light treatments (Tomer and Florence 1993).

20.6.2.2 Photodimer-Based Systems

Molecules that bind to like molecules with one specific wavelength of light and that then unbind with a different, separate, specific wavelength of light can act as a reversible switch to create unique drug delivery systems when incorporated into polymer materials. Reversible isomerization through $[2\pi + 2\pi]$ photoaddition can cause coumarin (Trenor et al. 2004a), nitrocinnamate (Zheng et al. 2001), and cinnamylidene acetate (Eiselt et al. 1999) to reversibly bind and form cyclobutane rings (Smets 1975; Ishigama et al. 1976; Ziffer et al. 1988; Trenor et al. 2004b) and through $[4\pi + 4\pi]$ photocycloaddition can cause anthracene to reversibly bind to other anthracene molecules (Greene et al. 1955; Bouas-Laurent et al. 2000, 2001; Zheng et al. 2002). As illustrated in Figure 20.6, anthracene and nitrocinnamate dimerization/binding occurs with wavelengths of light over 300 nm, and dissociation/debinding occurs with wavelengths of light under 300 nm. Direct modification of polymers with dimerizing molecules can lead to the creation of gels, micelles, and other drug delivery devices that rarely require external photosensitizers and have minimal side products (Zheng et al. 2002).

PEG materials modified with photodimerizing groups have been a focus because PEG has been shown to be compatible with biological systems and chemical moieties with varying hydrophilicities (Peppas et al. 1999). Reversible swelling ratios and alterations in the flux of proteins across cinnamylidene acetate end-capped star-PEG

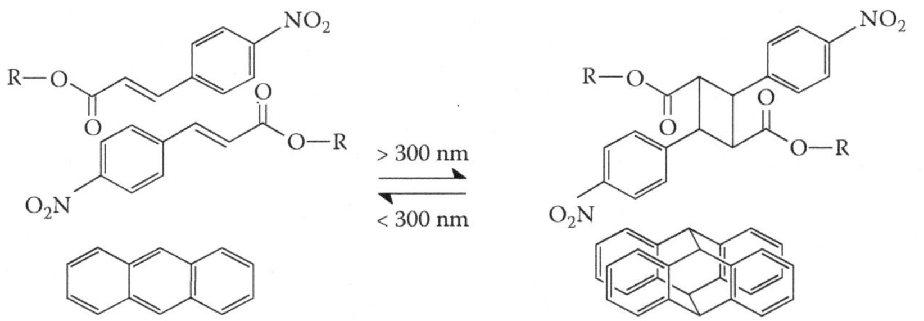

FIGURE 20.6 Examples of photodimerization. Nitrocinnamate (top) dimerizes via $[2\pi + 2\pi]$ photoaddition and anthracene (bottom) dimerizes via $[4\pi + 4\pi]$ photocycloaddition.

gels or cinnamylidene acetate-modified organophosphorus hydrolase upon light exposure helped to establish photodimerizing groups as a potential method for altering the diffusion and release of proteins (Andreopoulos et al. 1996, 1999). Work by Andreopoulos and Persaud (2006) showed that gels made of nitrocinnamate-capped eight-arm PEG were able to slow the release of fibroblast growth factor-2 with a 20% decrease in release over a 5-day period with UV treatment, demonstrating the potential of photodimerizing groups as reversible light-controlled drug delivery systems.

Polymers modified by grafting of photodimerizing groups create highly photosensitive systems whose bulk properties can be tailored by the main polymer chains. Incorporation of nitrocinnamate onto the backbone of gelatine results in a system with alterable cross-linking and topography with light treatment. An azobenzene-modified DNA-based photo-cross-linker that can bind to hydrogels and then associate with complimentary DNA to cause cross-linking also shows promise for the delivery of nanoparticles and drugs up to 90 minutes (Kang et al. 2011). Wells et al. (2011a,b) have demonstrated that the grafting of anthracene via a short PEG chain (PEG-anthracene) onto alginate and hyaluronic acid has highly sensitive photoresponsive properties that alter the release of small model drugs and proteins. Specifically, these materials have shown altered release rates and the capability to turn drug release off and on with these systems, with delivery lasting for months (Wells and Sheardown 2011). The concept and release of small dye molecules is shown in Figure 20.7.

Dedimerization has been used as a method to release photoimmobilized macro-molecules from photoresponsive materials. For example, coumarin has been used as a vehicle to release immobilized coumarin-modified heparin from polymer matrices (Nakayama and Matsuda 1993). Since heparin associates with many proteins, its release may be valuable in applications with heparin-associated proteins.

The core or shell of micelles made with polymers modified with photodimerizing groups can be created or destroyed with UV light to store or release drugs. For example, diblock polymers of PEG and poly [2-(2-methoxyethoxy)ethyl methacrylate-co-4-methyl-[7-(methacryloyl)oxyethyloxy]coumarin form micelles that can be reversibly core-cross-linked to potentially serve as micellar carriers (Jiang et al. 2007; He et al. 2009). These can be cross-linked with >300 nm light to encapsulate drugs that can subsequently, after delivery to a target site, be destroyed with light or lasers <260 nm in order to deliver the drugs. Similarly, nitrocinnamate-bound polymers have been used

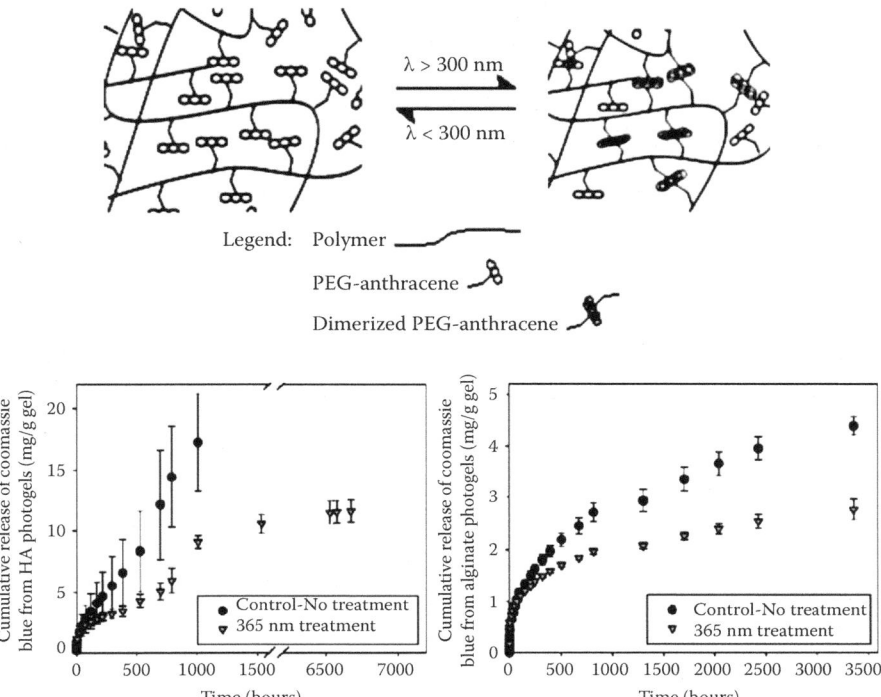

FIGURE 20.7 The concept of PEG-anthracene-modified polymers and release profiles of Coomassie blue dye from Alginate and HA anthracene-modified photogels. (Wells, L. A. et al. Generic, anthracene-based hydrogel crosslinkers for photo-controllable drug delivery. *Macromolecular Biosciences*, 2011a, 11, 988–98. Copyright Wiley-VCH Verlag GmbH & Co. KGaA, Weinheim. Reproduced with permission.)

to create photoreversible shells that may be used as microcontainers for drug delivery (Yuan et al. 2005). Nanoparticles can be reinforced through the photo-cross-linking of polymers modified with photodimerizing groups and then be dissolved with light. Alternatively, when assembled into a film, nanoparticles can be used to create photo reversible pore sizes to control the flux of drugs (Zhao et al. 2009).

20.6.3 PHOTORESPONSIVE CONTROL OF THERMORESPONSIVE POLYMERS

Photoresponsive groups can be used to alter the response profiles of thermoresponsive drug delivery systems. In their simplest form, molecules that release heat after the absorption of light can create a localized change in temperature, which can act as a stimuli for photoresponsive polymers. One of the most common thermoresponsive polymers is poly(N-isopropylacrylamide) (polyNIPAAm), which has a lower critical solution temperature (LCST) that is typically 32°C in aqueous solution but can be varied by polymerization with different hydrophilic and hydrophobic groups (Klouda and Mikos 2008; He et al. 2008). Below the LCST, polyNIPAAm is hydrophilic and will dissolve in aqueous solutions. Above the LCST, polyNIPAAm becomes more hydrophobic and collapses into a gel (Klouda and Mikos 2008). The LCST may be used to create *in situ* cross-linking polymer systems and may be used to cause localized controlled release of drugs (He et al. 2008). Localized heating by activated photoresponsive groups can cause polyNIPAAm gelation. For example, trisodium salt of copper chlorophyllin that releases heat with exposure to visible light and gold nanorods that release heat upon exposure to IR light have demonstrated the ability to cause phase changes in polyNIPAAm-based polymer systems (Suzuki and Tanaka 1990; Gorelikov et al. 2004; Budhlall et al. 2008).

Since the LCST of polyNIPAAm is driven by its hydrophilicity/hydrophobicity, molecules such as azobenzene that undergo changes in hydrophilicity with light due to *cis–trans* photoisomerization can lead to the creation of polyNIPAAm systems with light-alterable LCSTs. PolyNIPAAm containing azobenzene-functionalized acrylamides has a lower LCST upon exposure to visible light since azobenzene is in its *trans* form, and has a higher LCST upon exposure to UV light since azobenzene is then in its *cis* form (Akiyama and Tamaoki 2004). Changes can be fine-tuned, as shown with azobenzene-functionalized polyNIPAAm, which has incremental alterations in its LCST dependent upon the ratio of *cis* to *trans* azobenzene groups (Akiyama and Tamaoki 2007).

20.6.4 RELATED, PHOTORESPONSIVE POLYMER SYSTEMS

Photodimerizing and photoisomerizing polymers have been exploited to create photosensitive surface patterning and shape-memory materials that may be of use one day in ophthalmic systems. Shape-memory polymers are of interest in the bio-medical community to control the orientation of polymer materials *in vivo*. Films containing azobenzene have been created that bend upon UV light exposure and flatten upon visible light exposures to create sheets with alterable orientations (Yu et al. 2003; Barrett et al. 2007). Similarly, coumarin-containing polymer macromolecules that have photoinduced dimerization-driven cross-linking on one side of a film also bend and flatten with appropriate light exposures (He et al. 2009). Perhaps the most extreme photoresponsive shape-memory polymers are the polymer coils of Lendlein et al. (2005). Cinnamylidene-containing poly(hydroxyethylmethacrylate) coils relax and uncoil with exposure to light with a wavelength less than 260 nm due to de-crosslinking from the dedimerization of cinnamylidene groups and reform coils upon light exposures of above 360 nm due to cross-linking from the dimerization of the cinnamylidene groups (Lendlein et al. 2005).

Surface modification with light could be an effective method to perform controlled alteration of biomaterials *in vitro* and *in vivo*. The hydrophilicity of azobenzene-modified surfaces can be controlled by light exposure (Yuan et al. 2006), which could ultimately alter its interactions with cells, drugs, and other biomolecules. Materials that can undergo photodegradation could be used to pattern surfaces. For example, nitroveratryloxycarbonyl-bound molecules can be removed from surfaces with light exposure (Sieczkowska et al. 2007) and polymer-brush-modified surfaces that are end-capped with photolabile groups can have the brushes cleaved by light to remove the brush and expose the underlying surface (Brown et al. 2009). Other interesting applications have been demonstrated with spirobenzene-containing systems to create micropatterns that control cell adhesion to surfaces (Edahiro et al. 2005) and to create photosensitive microvalves for on-chip fluid control (Suguira et al. 2009). Photoresponsive polymer applications in drug delivery and biomedical applications relevant to ophthalmology will only continue to grow in the future.

20.7 CONCLUSIONS

Photoresponsive polymers have enormous potential to improve ophthalmic drug delivery technology. The potential future ability of clinicians to externally control drug delivery by altering the released dose of drugs or turning delivery on or off in a noninvasive fashion will optimize ocular disease treatments and increase patient compliance. Both reversible and irreversible light-responsive systems present potential noninvasive smart technologies to aid the delivery of drugs and the insertion and removal of polymer drug delivery devices in both the anterior and the posterior of the eye. Through discoveries in polymer science and parallel laser development, photoresponsive drug delivery will continue to develop from the bench to the clinic. Past historical use of light and lasers by patients in the clinic make photoresponsive drug delivery systems a likely well-tolerated form of treatment, making photoresponsive polymers particularly useful for ocular drug delivery.

REFERENCES

Akiyama, H., and N. Tamaoki. 2004. Polymers derived from N-isopropylacrylamide and azobenzene-containing acrylamides: Photoresponsive affinity to water. *Journal of Polymer Science Part A: Polymer Chemistry* 42: 5200–14.

Akiyama, H., and N. Tamaoki. 2007. Synthesis and photoinduced phase transitions of poly(N-isopropylacrylamide) derivative functionalized with terminal azobenzene units. *Macromolecules* 40: 5129–32.

Alvarez-Lorenzo, C., L. Bromberg, and A. Concheiro. 2009. Light-sensitive intelligent drug delivery systems. *Photochemistry and Photobiology* 85: 848–60.

Andreopoulos, F. M., C. R. Deible, M. T. Stauffer, S. G. Weber, W. R. Wagner, E. J. Backman, and A. J. Russell. 1996. Photoscissable hydrogel synthesis via rapid photopolymerization of novel PEG-based polymers in the absence of photoinitiator. *Journal of the American Chemical Society* 118: 6235–40.

Andreopoulos, F. M., and I. Persaud. 2006. Delivery of basic fibroblast growth factor (bFGF) from photoresponsive hydrogel scaffolds. *Biomaterials* 27: 2468–76.

Andreopoulos, F. M., M. J. Roberts, M. D. Bentley, J. M. Harris, E. J. Beckman, and A. J. Russell. 1999. Photoimmobilization of organophosphorus hydrolase within a PEG-based hydrogel. *Biotechnology & Bioengineering* 65: 579–88.

Baroli, B. 2006. Photopolymerization of biomaterials: Issues and potentialities of drug delivery, tissue engineering, and cell encapsulation applications. *Journal of Chemical Technology and Biotechnology* 81: 491–49.

Barrett, C. J., J. Mamiya, K. G. Yager, and T. Ikeda. 2007. Photo-mechanical effects in azobenzene-containing soft materials. *Soft Matter* 3: 1249–61.

Bhawalkar, J. D., G. S. He, and P. N. Prasad. 1996. Nonlinear multiphoton processes in organic and polymeric materials. *Reports on Progress in Physics* 59: 1041–70.

Bisby, R. H., C. Mead, and C. G. Morgan. 2000a. Active uptake of drugs into photosensitive liposomes and rapid release on UV photolysis. *Photochemistry and Photobiology* 72: 57–61.

Bisby, R. H., C. Mead, and C. G. Morgan. 2000b. Wavelength-programmed soluted release from photosensitivie liposomes. *Biochemical and Biophysical Research Communications* 276: 169–73.

Blume, Y., D. J. Durzan, and P. Smertenko. 2006. *Cell biology and instrumentation: UV radiation, nitric oxide and cell death in plants*. The Netherlands: IOS Press.

Boettner, E. A., and J. R. Wolter. 1962. Transmission of the ocular media. *Investigative Ophthalmology* 1: 776–83.

Bondurant, B., A. Mueller, and D. F. O'Brien. 2001. Photoinitiated destabilization of sterically stabilized liposomes. *Biochimica et Biophysica Acta* 1511: 113–22.

Bouas-Laurent, H., A. Castellan, J. P. Desvergne, and R. Lapouyade. 2000. Photodimerization of anthracene in fluid solution: Structural aspects. *Chemical Society Reviews* 29: 43–55.

Bouas-Laurent, H., A. Castellan, J. P. Desvergne, and R. Lapouyade. 2001. Photodimerization of anthracenes in fluid solutions: (Part 2) mechanistic aspects of the photocycloaddition and of the photochemical and thermal cleavage. *Chemical Society Reviews* 30: 248–63.

Brown, A. A., O. Azzaroni, and W. T. S. Huck. 2009. Photoresponsive polymer brushes for hydrophilic patterning. *Langmuir* 25: 1744–49.

Budhlall, B. M., M. Marquez, and O. D. Velev. 2008. Microwave, photo- and thermo responsive PNIPAm-gold nanparticle microgels. *Langmuir* 24: 11959–66.

Burakowska, E., S. C. Zimmerman, and R. Haag. 2009. Photoresponsive crosslinked hyperbranched polyglycerols as smart nanocarriers for guest binding and controlled release. *Small* 5: 2199–204.

Dvir, T., M. R. Banghart, B. P. Timko, R. Langer, and D. S. Kohane. 2010. Photo-targeted nanoparticles. *Nano Letters* 10: 250–54.

Edahiro, J., K. Sumaru, Y. Tada, K. Ohi, T. Takagi, M. Kameda, T. Shinbo, T. Kanamori, and Y. Yoshimi. 2005. In situ control of cell adhesion using photoresponsive culture surface. *Biomacromolecules* 6: 970–74.

Eiselt, P., K. Y. Lee, and D. J. Mooney. 1999. Rigidity of two-component hydrogels prepared from alginate and poly(ethylene glycol)-diamines. *Macromolecules* 32: 5561–66.

El Halabieh, R. H., O. Mermut, and C. J. Barrett. 2004. Using light to control physical properties of polymers and surfaces with azobenzene chromophores. *Pure and Applied Chemistry* 76: 1445–65.

Gibson, K. F., and W. G. Kernohan. 1995. Lasers in medicine—a review. *Journal of Medical Engineering and Technology* 17: 51–57.

Gorelikov, I., L. M. Field, and E. Kumacheva. 2004. Hybrid microgels photoresponsive in the near-infrared spectral range. *Journal of the American Chemical Society* 126: 15938–39.

Greene, F. D., S. L. Misrock, and J. R. Jr Wolfe. 1955. The structure of anthracene photodimers. *Journal of the American Chemiical Society* 77: 3852–55.

Hartner, S., H. C. Kim, and N. Hampp. 2007. Phototriggered release of photolabile drugs via two-photon absorption-induced cleavage of polymer-bound dicoumarin. *Journal of Polymer Science, Part A: Polymer Chemistry* 45: 2443–52.

He, C., S. Kim, and D. S. Lee. 2008. In situ gelling stimuli-sensitive block copolymer hydrogels for drug delivery. *Journal of Controlled Release* 127: 189–207.

He, J., X. Tong, and Y. Zhao. 2009. Photoresponsive nanogels based on photocontrollable cross-links. *Macromolecules* 42: 4845–52.

He, J., Y. Zhao, and Y. Zhao. 2009. Photoinduced bending of a coumarin-containing supramolecular polymer. *Soft Matter* 5: 308–10.

Irie, M., and D. Kunwatchakun. 1986. Photoresponsive polymers. 8. Reversible photostimulated dilation of polyacrylamide gels having triphenylmethane leuco derivatives. *Macromolecules* 19: 2476–80.

Ishigama, T., T. Murata, and T. Endo. 1976. The solution photodimerization of (E)-*p*-nitrocinnamates. *Bulletin of the Chemistry Society of Japan* 49: 3578–83.

Jiang, J., B. Qi, M. Lepage, and Y. Zhao. 2007. Polymer micelle stabilization on demand through reversible photo-crosslinking. *Macromolecules* 40: 790–92.

Johnson, J. A., M. G. Finn, J. T. Koberstein, and N. J. Turro. 2007. Synthesis of photocleavable linear macromonomers by ATRP and star macromonomers by a tandem ATRP-click reaction: Precursors to photodegradable model networks. *Macromolecules* 40: 3589–98.

Kang, H., H. Liu, X. Zhang, J. Yan, Z. Zhu, L. Peng, H. Yang, Y. Kim and W. Tan. 2011. Photoresponsive DNA-cross-linked hydrogels for controllable release and cancer therapy. *Langmuir* 27: 399–408.

Klouda, L., and A. Mikos. 2008. Thermoresponsive hydrogels in biomedical applications. *European Journal of Pharmaceutics and Biopharmaceutics* 65: 34–45.

Kloxin, A. M., A. M. Kasko, C. N. Salinas, and K. S. Anseth. 2009. Photodegradable hydrogels for dynamic tuning of physical and chemical properties. *Science* 324: 59–63.

Kuiper, J. M., and J. B. F. N. Engberts. 2004. H-aggregation of azobenzene-substituted amphiphiles in vesicular membranes. *Langmuir* 20: 1152–60.

Kumar, G. S., and D. C. Neckers. 1989. Photochemistry of azobenzene-containing polymers. *Chemistry Reviews* 89: 1915–25.

Lendlein, A., H. Jiang, O. Junger, and R. Langer. 2005. Light-induced shape-memory polymers. *Nature* 434: 879–82.

Mamada, A., T. Tanaka, D. Kungwachakun, and M. Irie. 1990. Photo-induced phase transition of gels. *Macromolecules* 23: 1517–19.

McCoy, C. P., C. Rooney, C. R. Edwards, D. S. Jones, and S. P. Gorman. 2007. Light-triggered molecule-scale drug dosing devices. *Journal of the American Chemical Society* 129: 9572–73.

Minkin, V. I. 2004. Photo-, thermo-, solvato-, and electrochromic spiroheterocyclic compounds. *Chemistry Reviews* 104: 2751–76.

Misiuk-hojlo, M., P. Krzyzanowska, and A. Hill-Bator. 2006. Therapeutic application of lasers in ophthalmology. Paper presented at the Symposium on Photonics Technologies for 7th Framework Program. Wrocław, Poland.

Misu, M., H. Furukawa, H. J. Kwon, K. Shikinaka, A. Kakugo, T. Satoh, Y. Osada, and J. P. Gong. 2009. Photoinduced in situ formation of various F-actin assemblies with a photoresponsive polycation. *Journal of Biomedical Material Research* 89A: 424–31.

Nakayama, Y., and T. Matsuda. 1993. Novel surface fixation technology of hydrogel based on photochemical method: Heparin-immobilized hydrogelated surface. *Journal of Polymer Science Part A: Polymer Chemistry* 31: 977–82.

Nguyen, K. T., and J. L. West. 2002. Photopolymerizable hydrogels for tissue engineering applications. *Biomaterials* 23: 4307–14.

Peppas, N. A., K. B. Keys, M. Torres-Lugo, and A. M. Lowman. 1999. Poly(ethylene glycol)-containing hydrogels in drug delivery. *Journal of Controlled Release* 62: 81–87.

Poshusta, A. K., and K. S. Anseth. 2001. Photopolymerized biomaterials for application in the temporomandibular joint. *CTO* 169: 272–78.

Qiu, Y., and K. Park. 2001. Environment-sensitive hydrogels for drug delivery. *Advanced Drug Delivery Reviews* 53: 321–39.

QLT. 2005. Visudyne (verteporfin for injection). Novartis product insert.

Schmaljohann, D. 2006. Thermo- and pH-responsive polymers in drug delivery. *Advanced Drug Delivery Reviews* 58: 1655–70.

Schmidt-Erfurth, U., and T. Hasan. 2000. Mechanisms of action of photodynamic therapy with verteporfin for the treatment of age-related macular degeneration. *Survey of Ophthalmology* 45: 195–213.

Sieczkowska, B., M. Millaruelo, M. Masserschmidt, and B. Voit. 2007. New photolabile functional polymers for patterning onto gold obtained by clock chemistry. *Macromolecules* 40: 2361–70.

Smets, G. 1975. Photochemical reactions in polymeric systems. *Pure and Applied Chemistry* 42: 509–26.

Spratt, T., B. Bondurant, and D. F. O'Brien. 2003. Rapid release of liposomal contents upon photoinitiated destabilization with UV exposure. *Biochimica et Biophysica Acta* 1611: 35–43.

Suguira, S., A. Szilagyi, K. Sumaru, K. Hattori, T. Takagi, G. Filipesei, M. Zrinyi, and T. Kanamori. 2009. On-demand microfluidic control by micropatterned light irradiation of a photoresponsive hydrogel sheet. *Lab on a Chip* 9: 196–98.

Suzuki, A., and T. Tanaka. 1990. Phase transition in polymer gels induced by visible light. *Nature* 346: 345–47.

Thompson, K. P., Q. S. Ren, and J. M. Parel. 1992. Therapeutic and diagnostic application of lasers in ophthalmology. *Proceedings of the IEEE* 80: 838–60.

Tomer, R., and A. T. Florence. 1993. Photo-responsive hdyrogels for potential responsive release applications. *International Journal of Pharmaceutics* 99: R5–R8.

Trenor, S. R., T. E. Long, and B. J. Love. 2004a. Photoreversible chain extension of poly(ethylene glycol). *Macromolecular Chemistry and Physics* 205: 715–23.

Trenor, S. R., A. R. Shultz, B. J. Love, and T. E. Long. 2004b. Coumarins in polymers: From light harvesting to photo-cross-linkable tissue scaffolds. *Chemistry Reviews* 104: 3059–77.

Wang, G., X. Tong, and Y. Zhao. 2004. Preparation of azobenzene-containing amphiphilic diblock copolymers for light-responsive micellar aggregates. *Macromolecules* 37: 8911–17.

Wayne, C. E., and R. P. Wayne. 1996. *Photochemistry*. New York: Oxford University Press.

Wells, L. A., M. A. Brook, and H. Sheardown. 2011a. Generic, anthracene-based hydrogel cross-linkers for photo-controllable drug delivery. *Macromolecular Biosciences* 11: 988–98.

Wells, L. A., S. Furukawa, and H. Sheardown. 2011b. Photoresponsive PEG-anthracene grafted hyaluronan as a controlled-delivery biomaterial. *Biomacromolecules* 12: 923–32.

Wells, L. A., and H. Sheardown. 2011. Photosensitive controlled release with polyethylene glycol-anthracene modified alginate. *European Journal of Pharmaceutics and Biopharmaceutics*. In press. doi:10.1016/j.ejpb.2011.03.023.

Wright, C. H. G., S. F. Barrett, and A. J. Welch. 2007. Laser-tissue interactions. In *Medical Applications of Lasers*. M. H. Niemz, 21–56. Berlin, Germany: Springer-Verlag.

Yu, Y., M. Nakano, and T. Ikeda. 2003. Photomechanics: Directed bending of a polymer film by light. *Nature* 425: 145.

Yuan, X., K. Fischer, and W. Schartl. 2005. Photocleavable microcapsules built from photoreactive nanospheres. *Langmuir* 21: 9374–80.

Yuan, W., G. Jiang, J. Wang, G. Wang, Y. Song, and L. Jiang. 2006. Temperature/light dual-responsive surface with tunable wettability created by modification with an azobenzene-containing copolymer. *Macromolecules* 39: 1300–03.

Zhang, Z. Y., and B. D. Smith. 1999. Synthesis and characterization of NVOC-DOPE, a caged photoactivatable derivative of dioleoylphosphatidylethanolamine. *Bioconjugate Chemistry* 10: 1150–52.

Zhao, Y. 2007. Rational design of light-controllable polymer micelles. *The Chemical Record* 7: 286–94.

Zhao, Y., J. Bertrand, X. Tong, and Y. Zhao. 2009. Photo-cross-linkable polymer micelles in hydrogen-bonding-built layer-by-layer films. *Langmuir* 25: 13151–57.

Zheng, Y., F. M. Andreopoulos, M. Micic, Q. Huo, S. M. Pham, and R. M. Leblanc. 2001. A novel photoscissile poly(ethylene glycol)-based hydrogel. *Advanced Functional Materials* 11: 37–40.

Zheng, Y., M. Micic, S. V. Mello, M. Mabrouki, F. M. Andreopoulos, V. Konka, S. M. Pham, and R. M. Leblanc. 2002. PEG-based hydrogel synthesis via the photodimerization of anthracene groups. *Macromolecules* 35: 5228–34.

Ziffer, H., A. Bax, R. J. Highet, and B. Green. 1988. Investigation by two-dimensional NMR of the structure and stereochemistry of a methyl *p*-nitrocinnamate photodimer. *Journal of Organic Chemistry* 5: 895–96.

21 Nanosuspensions in Ocular Drug Delivery Systems

Himanshu Bhattacharjee and Sonia Bedi

CONTENTS

21.1　INTRODUCTION

According to accepted conventions, chemical constructs or molecules are considered drug-like if they adhere to specific conditions that impart drug-like physicochemical characteristics. Some of these conditions are exemplified by extensively studied drug substances and marketed products. Lipinski's Rule of Five is a rule of thumb used in drug discovery that evaluates drug-like entities and indicates if a chemical compound with certain physicochemical properties and/or biological activity has attributes that would make it a likely drug candidate in humans. The rule was formulated by Christopher A. Lipinski in 2001, based on the observation that most medication drugs are relatively small, low-molecular-weight compounds with acceptable lipophilicity (Lipinski et al. 2001). However, these rules are pertinent to orally active drugs and may not be indicative for drugs used through other routes of administration. Moreover, in the past decade, a number of molecules that do not adhere to this rule have been approved and evaluated. Although these molecules may or may not be orally active, they have been found to be effective therapeutically when administered by other routes such as parenteral, topical, ocular, and otic routes. Many of these molecules have excellent biological activity but show modest to almost no solubility in aqueous mediums, thus making them poorly bioavailable. To harness their therapeutic activity, a number of approaches have been evaluated and have led to novel drug delivery systems. Nanosuspension is one such system that helps overcome the limitations of poor bioavailability and pharmacokinetic profiles.

The most pristine definition of a *nanosuspension* relates to pure drug particles in submicron colloidal dispersions stabilized by surfactants. Not all therapeutically relevant drugs can be formulated into pure nanosuspensions, thus necessitating the use of a carrier system composed of biodegradable polymeric matrices. A more recent concept of nanosuspension is the incorporation of drugs in biodegradable and biocompatible polymer, which are formulated as submicron colloidal dispersion. These systems are exemplified by polymeric colloidal carriers of drugs such as nanospheres, nanocapsules, nanoreservoirs, and solid lipid nanoparticles. This chapter, however, considers the former definition of nanosuspensions that are currently being used or developed as ophthalmic drug delivery systems. More specifically, the chapter discusses important concepts of nanosuspensions with respect to their compendial requirements, physicochemical characteristics, biopharmaceutical considerations, preparation methods, excipients used, characterization, and application in ocular drug delivery. A brief discussion as to how the newer polymeric ocular drug delivery systems are being evaluated and shaping the field of ocular drug delivery is also included in the chapter.

21.2 OPHTHALMIC SUSPENSIONS

Ophthalmic preparations are sterile liquid preparations that are essentially free from foreign particles and are suitably formulated to be instilled in the eye. However, according to the United States Pharmacopeia (USP), ophthalmic suspensions are considered to be sterile products containing solid particles dispersed in a liquid vehicle intended for application to the eye. It is important that such suspensions contain particles in a certain size range to prevent local irritation and/or abrasion on the surface of the cornea. Ophthalmic suspensions are generally considered to contain micronized drug particles that are relatively insoluble in the aqueous vehicle containing a suitable suspending and dispersing agent. The vehicle can also be considered as a saturated solution of the drug. In ophthalmic nanosuspensions, the drug particles are further finely divided into submicron size particles in the colloidal ranges. The U.S. pharmacopeia does not specifically address the requirement for emerging nanoparticulate polymeric ocular systems, and it would be prudent to follow the present compendial requirements for the existing particulate preparations.

21.3 COMPENDIAL REQUIREMENTS

According to the USP 33 (NF 29; 2010), nanosuspensions are not exclusively defined under ophthalmic products. However, ophthalmic suspensions have been discussed. According to Chapter <1151>, ophthalmic suspensions are "sterile liquid preparations [that] contain solid particles dispersed in a liquid vehicle intended for application to the eye." It also states that such systems should contain micronized drug particles to prevent irritation to the cornea. Additionally, it states that such systems should not be administered if evidence of agglomeration or caking exists. Although nanosuspensions are not discussed, the USP provides adequate information for other dispersed systems and colloidal systems. Since nanoparticulate colloidal systems adhere to the basic fundamentals of dispersed systems comprising specific physicochemical considerations, the requirements of dispersed systems can act as a roadmap for nanoparticulate dosage forms. Additionally, many of the limitations of suspensions, such as sedimentation and caking, can be overcome by the sheer nature of nanosuspensions. This is due to the existence of the particles' nanosize range (Brownian motion vs. effect of gravity; the effect due to Brownian motion is greater than the effect of gravity), charge characteristics of the particles, and relatively dilute dispersion. Other considerations include maintaining isotonicity with the lacrimal fluids, use of appropriate buffers, maintaining sterility of formulation, viscosity considerations, and use of preservatives, among others.

21.4 PHYSICOCHEMICAL PROPERTIES

Because nanosuspensions or suspensions in general are heterogeneous systems, their stability and therapeutic efficacy is significantly dependent on the physicochemical characteristics of the system. Within the confines of these features, the physicochemical properties of the dispersed phase (suspended solids) are even more critical in determining the stability of nanoparticulate systems. These features include, but are not limited to, solubility, particle size, crystal habits of the suspended solids, and the interactions resulting from some of these factors. As nanosuspension systems are essentially colloidal in nature, a number of features defining colloidal systems can be employed to elucidate the physicochemical characteristics of nanosuspensions. Some of the significant physicochemical aspects are discussed in this section.

21.4.1 Particle Size

Optimum particle size is an important factor for optimum therapeutic efficacy. Since a suspension will tend to accumulate in the cul-de-sac, the contact time of the particles is considerably greater than comparable ophthalmic solutions. This increased contact time is related to a longer duration of action and also affords a sustained depot effect that is dissolution controlled. These aspects are dependent on the characteristics of the drug (solubility) and are significantly affected by the particle size. Smaller particle size generally translates into a larger surface area, which ultimately affects dissolution and bioavailability. As the main focus of any ocular drug delivery system is to increase bioavailability, the size of the suspended particles becomes an important factor in determining their therapeutic efficacy. This concept is related to the degree and the velocity of drug dissolution (drug solubility) into its surrounding environment and the mean particle size of the suspended particles. Ocular suspensions have to be within an acceptable range of particle size (less than 10 mm for ophthalmic suspensions) to prevent excessive irritation of the corneal surface. Nanosuspensions are relatively free from these limitations as far as the particle size is concerned as they exist in submicron ranges (generally less than 1 mm). However, nanosuspensions are not completely devoid of irritation potential due to the use of certain surfactants and excipients employed to formulate these systems. A detailed description of the acceptable excipients will be discussed later in Section 21.6. Thus, one important consideration of nanosuspension dosage form design is to minimize irritation as irritation of the corneal surface is detrimental to the dosage efficacy, generally leading to excessive tear formation and drainage.

The stability of ophthalmic suspensions and nanosuspensions is an important factor to be considered. Unstable dispersed systems can lead to particle size increase, thereby causing sedimentation and affecting its dispersion characteristics. A stable suspension should be devoid of agglomeration of dispersed particles, exhibit high sedimentation volume, be devoid of caking effects, and be easily redispersed upon agitation. Lack of uniform dispersion is usually associated with lack of dose uniformity. Other stability factors include polymorphic changes, storage conditions, and microbial contamination.

21.4.2 Solubility

Ocular suspensions offer advantages such as prolonged residence time in a cul-de-sac and prevention of hypertonicity resulting from solubilization of water-soluble drugs. However, their therapeutic potential is dependent on the intrinsic solubility of the drug in lacrimal fluids. Thus, the intrinsic dissolution rate of the drug in lacrimal fluid is a rate-limiting step that dictates its release profile and ocular bioavailability.

More specifically, intrinsic solubility determines the extent of a drug available in a solution form, which can be absorbed upon administration or instillation. A drug with higher intrinsic solubility will lead to higher concentrations of the drug in the solution form and thus more of the drug will be available for absorption.

Instillations containing drug particles suspended in a dispersion medium upon administration in the eye initiate the solubilization of the drug particles in the lacrimal fluid. The dissolved drug in the outer layer of the suspended particle is absorbed, thus creating a drug-depleted region between the solid drug particle and the corneal absorption surface. This region creates a concentration gradient that initiates further dissolution of the drug particle, thus making more drugs available in the solution for it to be absorbed (Figure 21.1). This is true for drugs in suspension for both micronized and nanosuspension systems.

It is also necessary to consider the rate of dissolution (a factor partly dependent on the intrinsic solubility) and residence time in the eye for particulate systems. In other words, the dissolution must be optimal to maintain therapeutic levels within a specified residence time. If the drug dissolves at a slower rate, it would take longer to reach the steady state that could be greater than the residence time. This can result in low or no significant therapeutic effect due to drug administration and in many instances can be compared to effects observed with dilute solution instillation. Additionally, the particle size distribution (PSD) of the suspension plays an important role in this phenomenon as a smaller particle results in larger surface area available for dissolution.

The relationship between the dissolution rate (D_τ) and the particle size can be explained by the Noyes–Whitney equation:

$$D_\tau = \frac{D.A(C_s - C)}{hV} \tag{21.1}$$

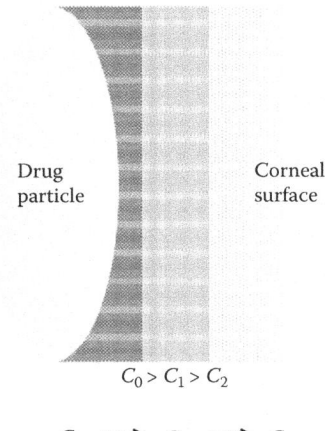

Drug particle

Corneal surface

$C_0 > C_1 > C_2$

$C_0 \rightleftarrows C_1 \rightleftarrows C_2$

FIGURE 21.1 Schematic representation of the effect of intrinsic solubility of a drug on absorption.

where D and A are the diffusion coefficients and the solute surface area respectively, C_s and C are concentrations of the solute at the particle surface and the bulk medium, respectively, V is the volume of the dissolution medium, and h is a constant that depends on the viscosity and the degree of agitation of the medium. C_s can be considered as a rate-limiting variable that depends on the nature of the solid dispersed material, the dispersion medium, and the temperature. This is particularly true for micronized suspensions containing drug particles. However, below a certain size range (1–2 µm), C_s depends on the particle size, and this relationship is greater below 1000 nm. Thus, the smaller the particle, greater is the surface area available for dissolution, resulting in greater solubilization.

21.4.3 Charge Characteristics

One of the most critical factors influencing the stability of colloidal nanosuspensions is the charge carried by the dispersed particles. The charge characteristics of the particles in nanosuspension can be determined in terms of an electrokinetic property called "zeta potential." According to the DLVO theory (Derjaguin, Landau, Verwey, and Overbeek), the stability of the colloidal particles is a function of their total potential energy function (V_T), which is a collective term used for the combination of three different forces acting among particles in a suspension system. These three forces are the force of attraction (represented by attraction potential, V_A), the force of repulsion (represented by repulsion potential, V_R), and the potential energy of solvent (V_S):

$$V_T = V_A + V_R + V_S \tag{21.2}$$

The last term, that is, the potential energy of solvent, is quite negligible compared to the first two terms, which play a significant role in determining the overall stability of a colloidal dispersion. The colloidal system would be very stable if V_R is greater than V_A, that is, the particles can resist contact with each other and would stay separate, thereby maintaining the dispersion stability. The repulsive potential, which is a bigger determinant of overall suspension stability, is given by the following expression (Equation 21.3):

$$V_R = 2\pi \xi A^2 \exp^{(-kD)} \tag{21.3}$$

where A is the particle radius, π is the solvent permeability, k is the function of ionic composition, and ξ is the zeta potential. As indicated by Equation 12.3, zeta potential is one of the direct measures of the net repulsive potential between two particles and therefore can be employed to derive practical information on suspension stability.

DLVO theory suggests that the net suspension stability is determined by the combination of both attraction and repulsion forces and that the particles under constant Brownian motion are approaching each other randomly due to force of attraction. However, they are prevented from making contact due to the force of repulsion in operation. This results in a net energy barrier between the particles that has to be overcome for the particles to contact each other. The repulsive forces prevent this from happening, thereby resulting in a stable dispersion. Once,

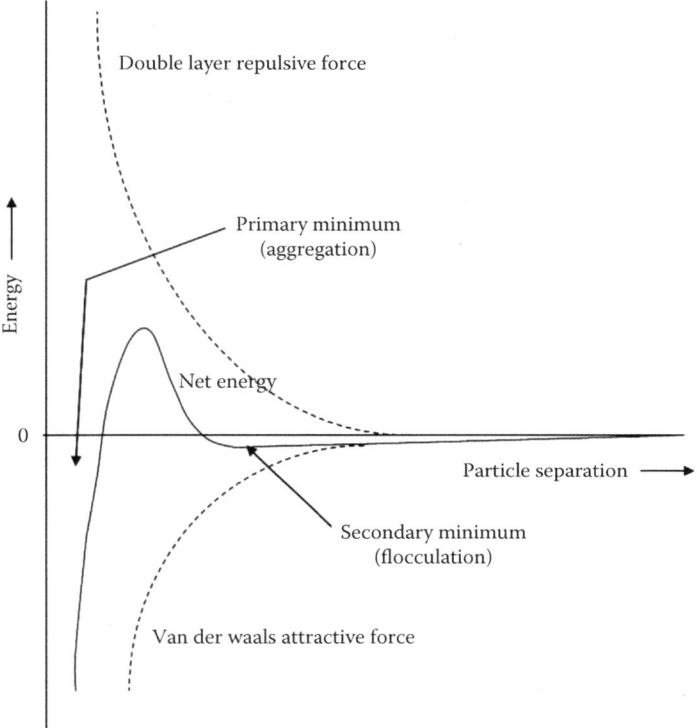

FIGURE 21.2 Schematic diagram of free energy change as a function of particle separation.

however, the energy barrier is overcome, the forces of attraction take over and it is difficult for the particles to go back into the random Brownian motion, eventually resulting in flocculation and aggregation. A schematic diagram of free energy change based on particle separation is shown in Figure 21.2. Sometimes, in the presence of high salt concentration, there exists a secondary minimum in free energy, which reflects the formation of weak flocs due to particle contact. Since the energy dip in the secondary minimum is smaller compared to the primary minimum, such flocs can be broken and dispersed relatively easily into individual particles by agitation.

These forces of attraction and repulsion in essence arise from the charge distribution on particles. The particles in the suspension system attain a certain amount of charge due to either ionization of some groups on the particle surface or by adsorption of other charged species. The surface charge further results in attraction of oppositely charged ions from the dispersion medium, leading to formation of a stern layer right adjacent to the particle surface and consisting of tightly packed oppositely charged ions. The potential at the outermost surface of this stern layer is known as *Nernst potential* or *thermodynamic potential*. Following the stern layer is another layer of loosely held oppositely charged ions known as the diffuse layer, which is loosely

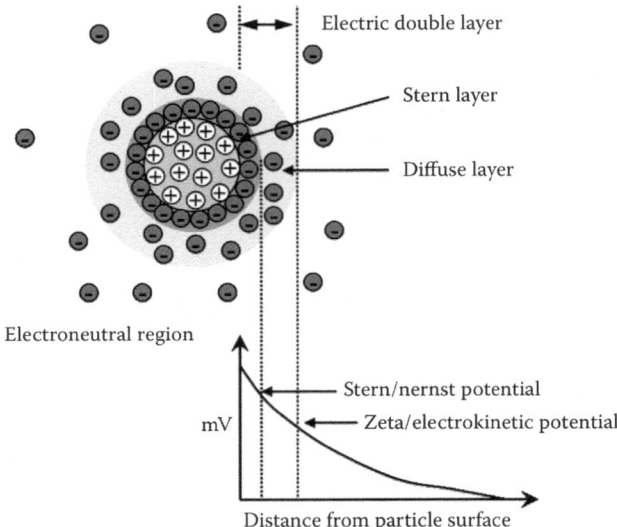

FIGURE 21.3 Schematic representation of the electric double layer.

held due to a larger separation distance between the surface charge on the particle and the diffuse layer (Figure 21.3). The potential on the outer surface of this diffuse layer is known as the zeta potential or the electrokinetic potential. The two charged layers (Stern layer and diffuse layer) together form an electric double layer. The extent of the double layer is a function of the valency of the ions forming the two layers. Multivalent ions are capable of forming a much stronger double layer and therefore resist flocculation much more efficiently than univalent ions.

Zeta potential is more meaningful compared to *Nernst* potential for determining colloid stability as it relates to the kinetic motion or the actual movement of particles under the influence of an electric field, and gives a better estimate of particle aggregation. A high value of zeta potential means more repulsive forces and hence greater stability of the suspension system. Usually, systems with higher values, around −30 or +30 mV, are considered to be stable. Electrokinetic measurements can provide fundamental insights on the effects of various excipients used in pharmaceutical ophthalmic nanosuspensions that could affect the suspension stability. Changes in the net particle charge and therefore the suspension stability can also be determined with respect to the pH of the formulation.

Zeta potential measurement is conducted based on the principles of electrokinetic effects, one of which is electrophoresis. During the process of electrophoresis, particles in the medium move toward the oppositely charged electrode under the influence of an applied electric field. This motion is opposed by the viscosity of the dispersing medium, thereby slowing down the particle velocity. Eventually, equilibrium is attained between the two forces, resulting in the movement of particles with a constant velocity, which is a function of the strength of the electric filed, the dielectric constant and viscosity of the medium, and the zeta potential. This velocity is known as the *electrophoretic mobility*, which is what is actually measured by zeta potential–measuring instruments. The value of electrophoretic mobility thus obtained is then substituted into *Henry's equation* to derive the value of zeta potential (Equation 21.4):

$$U_E = \frac{2\varepsilon\xi f(k_a)}{3} \tag{21.4}$$

where U_E is the electrophoretic mobility, ξ is the zeta potential, ε is the dielectric constant, η is the viscosity, and $f(k_a)$ is Henry's function.

21.4.4 Sedimentation

Sedimentation is an effect observed in coarse suspensions and is a feature that is often considered to be detrimental to the stability of the formulation. This effect, also known as the settling rate, is represented by the Stokes equation, which is expressed as the velocity of sedimentation v and is given by the following equation (Equation 21.5):

$$v = \frac{d^2(\rho_S - \rho_0)g}{18\eta_0} \tag{21.5}$$

where v is the velocity in cm/s, d is the diameter of the particle in centimeters, ρ_S and ρ_0 are the densities of the dispersed phase and dispersed medium, respectively, g is the acceleration due to gravity, and η_0 is the viscosity of the dispersion medium in poise. The critical size for settling can be calculated from the criterion of Overbeek, which states that according to Stokes' law, colloidal particles that settle at a rate of only 1 mm in 24 hours will never settle in actual practice because of Brownian motion. For a density of particles of 1.15 and $\eta = 1$, the critical particle diameter is calculated to be 300 nm. This can alter the stabilization strategy for suspensions that have a particle size constrained to this limit.

21.5 PHARMACOKINETICS OF OPHTHALMIC SUSPENSIONS

21.5.1 Challenges in Corneal Drug Bioavailability

Ophthalmic suspensions are one of the available dosage forms for treatment of ocular diseases or infections. As far as intraocular drug delivery is concerned, ophthalmic formulations can be made to deliver drugs in the aqueous humor compartment or the vitreous humor compartment, depending on the need for therapeutic treatment. For example, drug delivery to the aqueous humor is mostly to treat ocular infections, including corneal infections or even for immunosuppressive agents after corneal transplantation (Yuan et al. 2009). Drug delivery to the vitreous humor, on the other hand, is mainly for the treatment of diseases such as retinitis pigmentosa (degenerative disorder), diabetes retinopathy (vascular disorder), proliferative vitroretinopathy (proliferative disorder), and other disorders such as glaucoma and optic neuritis.

Ocular solution–based formulations are commonly available to treat most of the aforementioned disorders related to both the aqueous and vitreous humor

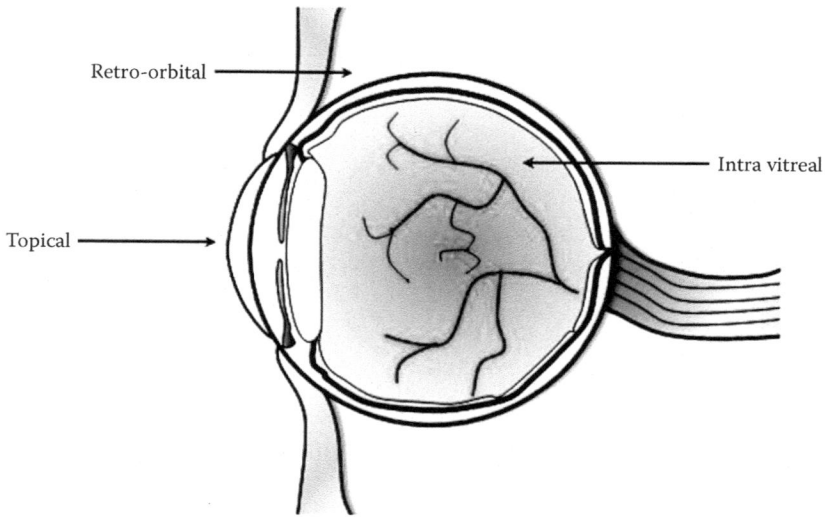

FIGURE 21.4 Routes for ocular drug delivery for nanosuspensions.

·compartments. Nonetheless, suspension dosage forms have found applications for topical ocular delivery and as injectable systems. Various routes of ocular drug delivery that could be employed for drug delivery by solution- and suspension-based formulations are depicted in Figure 21.4.

Before we discuss the essential pharmacokinetic differences between solution- and suspension-based ophthalmic dosage forms, it would be more meaningful to understand the pharmacokinetic challenges that a drug molecule has to face before it reaches the ocular site of action. The challenges for the overall bioavailability of drugs through the ocular route can be mainly categorized based on drug dilution and drainage, protein binding, and drug solubility issues, each of which is discussed in detail in the following sections (Sections 21.5.1.1 to 21.5.2.5).

21.5.1.1 Dilution of Administered Dose

Dilution effect and drainage issue: The eye (cul-de-sac) on an average can hold about 30 µL of fluid at one time (Mishima et al. 1966). The total amount of fluid in the cul-de-sac under normal conditions is around 7 µL, which is maintained as a result of two processes occurring simultaneously, that is, lacrimation (tear production) and drainage through the nasolacrimal canal. Of the two processes, the lacrimation rate has a dominant effect in the sense that drainage through the nasolacrimal canal generally compensates for the rate of lacrimation. The question is therefore, what happens upon instillation of eyedrops into the cul-de-sac? Shell has reported that the additional fluid from the eyedrops in the cul-de-sac is drained quite rapidly through the nasolacrimal route, thereby almost diminishing the major fluid volume from the drops. The nasolacrimal route therefore responds to the fluid volume and leads to drainage of excess fluid accordingly (Shell 1982).

Such nasolacrimal drainage often results in systemic drug absorption that might exacerbate the systemic side effects of the drug. The side effects of most drugs are more pronounced when administered by the topical ocular route than by oral administration. The overall effect is greater for drugs undergoing first-pass metabolism in the intestinal mucosa or liver. Drugs administered orally undergo a first-pass effect whereby a certain percentage of the drug is metabolized, thereby limiting the intensity of the side effects. This is not the case with topical ocular administration, as there is no first pass and therefore the drug is almost completely available to the systemic circulation. In this sense, ocular administration is closer to intravenous drug delivery. The overall intensity of the side effects is therefore much higher with drugs administered by the ocular route.

21.5.1.2 Tonicity of the Ophthalmic Dosage Form

The tonicity of the formulation is another factor affecting drainage. Those formulations for topical ocular administration with high tonicity end up drawing more water from the ocular tissues, thereby resulting in increased fluid volume in the cul-de-sac. This increased fluid volume eventually causes more nasolacrimal drainage, resulting in loss of formulation. Studies have also shown that the nasolacrimal drainage is directly proportional to the volume of fluid in the cul-de-sac. The overall loss of formulation is therefore reduced when low volumes are instilled in the eye compared to larger volumes (Chrai et al. 1974). This would essentially mean that a concentrated drug solution is bound to show less formulation loss due to drainage and thus more drugs available for corneal absorption. The preparation and stabilization of concentrated drug solutions is, however, another cause of concern. Drug suspensions, on the other hand, can show significant potential in this aspect.

21.5.1.3 Protein Binding and Other Issues Related to Drug Loss

The protein content of tears in rabbits is around 0.5% and around 0.7% in humans, mainly consisting of albumin, globulin, and lysozyme. Once the drug in the cul-de-sac is bound to tear proteins, it is no longer available for corneal absorption, thereby causing profound effects on drug bioavailability. Mikkelson et al. (1973) has shown reduced ocular bioavailability of pilocarpine in rabbits upon addition of serum albumin in the formulation. Apart from protein binding, other factors that might result in overall low ocular bioavailability of drugs are conjunctival absorption and precorneal metabolism (Shell 1982).

21.5.1.4 Drug Solubility and Permeability Issues

Most drugs are either weak acids or weak bases, which make their solubility dependent on the pH of the external fluid. The property of the drug molecules that determines their solubility relative to the pH of external fluid is the pK_a, also known as the negative log of dissociation constant. The relationship between the pH of the external medium and the pK_a of a drug is captured in the *Henderson–Hasselbalch equation*, which for a weak base is represented as follows (Equations 21.5 and 21.7):

$$pH = pK_a + \log\frac{[A^-]}{[HA]} \tag{21.6}$$

or

$$pH = pK_a - \log \frac{[HA]}{[A^-]} \qquad (21.7)$$

Here, [HA] is the concentration of the unionized form of the drug and [A⁻] is the concentration of ionized form of the drug. It is clearly understood from the equation that the concentration of the ionized drug is equal to that of the unionized form when pH is equal to pK_a of the drug molecule. It is desirable to achieve maximum drug concentrations in topical ocular solutions, which for most drugs that show pH-dependent solubility can be obtained by modulating the pH of the formulation relative to the pK_a of the drug.

This results in attaining the ionizable form of the drug in solution, which oftentimes does not have the required lipophilicity to permeate through the cornea. This phenomenon is dependent on the corneal structure that is divided into a multilayer epithelium, stroma, and a single-layer endothelium (Figure 21.5). Drugs can be absorbed either by partitioning through the cells (intracellular) or by passing between the cells (paracellular) (Ghate and Edelhauser 2006). The corneal epithelium is quite lipophilic in nature, thereby allowing only the drugs with sufficient lipophilicity to be taken up into the epithelium. Once the drug is taken up into the epithelium, further diffusion into the stroma is mainly dependent on its water solubility (Shell 1982). To permeate across the cornea into the aqueous humor, a drug molecule needs to have a balance between the hydrophilic and lipophilic natures. Once into the stroma, the corneal endothelium presents less of a challenge to the drug absorption into the aqueous humor. Drugs that are mainly hydrophilic with limited lipophilicity are taken up through the paracellular route.

21.5.1.5 Dosage Form Effects on Overall Pharmacokinetic Behavior

A drug molecule has to face all the above-mentioned challenges before it can be absorbed through the cornea and become bioavailable. One of the biggest concerns, especially for ocular preparations meant for topical applications, is the overall effective

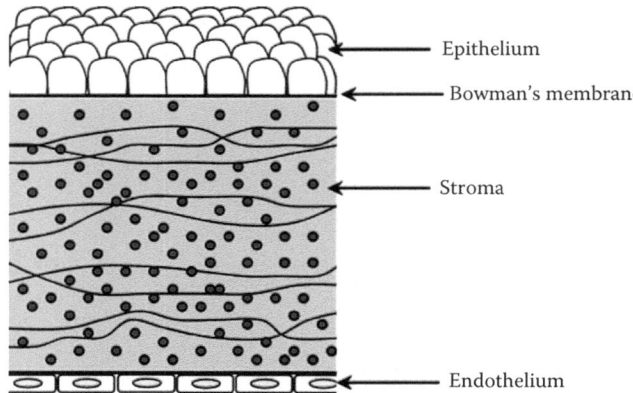

FIGURE 21.5 Anatomical structure of the cornea.

contact time of the drug with the corneal membrane in the cul-de-sac. This contact time, due to some of the aforementioned reasons (including drainage and protein binding), is limited for drug solutions. Suspensions, on the other hand, have a greater residence time in the cul-de-sac, so the drug is available for a longer time in the corneal vicinity to be absorbed. The total residence time of suspended drug particles in the cul-de-sac depends to some extent on the particle size, with the residence time decreasing with reduction in particle size (Schoenwald and Stewart 1980). A particle size <10 μm is recommended for topical ocular applications to minimize irritation effects. The particle size is also an important parameter for suspension-based formulations, as it determines the rate of dissolution and replenishment of free drug in solution available to be absorbed. Since suspensions can be retained in the cul-de-sac for a longer time, they tend to show greater overall bioavailability compared to solution formulations. Yuan et al. have demonstrated that highest levels of tacrolimus concentrations could be achieved in the conjunctiva, cornea, and sclera within 60 minutes of topical instillation with detectable drug concentrations in the aqueous humor maintained for about 500 minutes. The drug release from suspension formulations follows a pseudo zero-order behavior until the drug particles are completely dissolved or disappear. This results in a constant supply of the drug available for absorption through the cornea until the time the drug particles are available in the cul-de-sac, which is another major advantage of suspensions over solutions. The two factors together, that is, greater residence time and constant replenishment of free drug in the cul-de-sac, are functional for increased drug corneal bioavailability from suspensions.

Ophthalmic suspensions can prolong not only the drug absorption and increased bioavailability upon topical instillation but also upon injection into the posterior compartment of eye. Studies have indicated suspension of drug particles forming an aggregate or precipitate, giving the depot effect of slow and steady drug release within the vitreous compartment. Durairaj et al. have shown that concentrations of diclofenac in the vitreous humor could be maintained for almost eight times longer with the suspension system than with solution formulation form. Although the C_{max} of drug was lower with suspension formulation, a higher $AUC_{0\text{-inf}}$ could be achieved compared to the solution system. Other suspension-based dosage forms that could be given intravitreally are microspheres and nanospheres, where the drug is encapsulated in a biocompatible and biodegradable polymer. Such systems can provide sustained drug delivery for weeks or even months and can increase patient compliance.

Since the duration of the total pharmacologic effect is a direct function of the concentration of drug administered, it is logical to maintain high concentrations of the drug in the cul-de-sac for longer durations for the drug to be efficiently absorbed across the cornea and not get lost due to drainage.

$$T = \frac{\log\left(C/C_1\right)}{a} \tag{21.8}$$

where T is the duration of the drug effect, C is the concentration of the administered drug, C_1 is the least-effective concentration, and a is the rate constant of effect disappearance.

Such high drug concentrations for longer durations can be maintained much more efficiently with suspension formulations than with solution-based systems.

21.5.2 BIOPHARMACEUTICAL CONSIDERATIONS

21.5.2.1 Release Kinetics

One of the significant advantages of nanosuspensions over simple drug solutions is not only their higher saturation solubility as mentioned earlier but also an increased dissolution velocity. The increase in dissolution velocity is tangible because of a significant increase in the specific surface area of the drug particles when reduced from the micron range to the nanosized range. This is very well indicated by the *Nernst–Brunner* and *Levich* modification of *the Noyes–Whitney* dissolution model equation represented as Equation 21.9:

$$\frac{\mathrm{d}x}{\mathrm{d}t} = \left[\left(\frac{D \cdot A}{h} \right) \left(\frac{C_s - X}{V} \right) \right] \tag{21.9}$$

where $\mathrm{d}x/\mathrm{d}t$ is the dissolution velocity, D is the diffusion coefficient, A is the surface area of the particle, h is the diffusional distance or the thickness of hydrodynamic boundary layer, C_s is the saturation solubility of the drug, X is the concentration of the drug in the surrounding or bulk liquid, and V is the volume of the dissolution medium (Patravale and Ambarkhane 2003). This equation demonstrates the direct relationship between dissolution velocity and the surface area of the drug particle. The equation, however, makes a significant assumption that the surface area of the drug particle remains unchanged during the dissolution process. This is, however, extremely unlikely, as the drug molecules are in constant flux and diffuse out into the bulk medium, resulting in an overall dynamic reduction of the mass and volume of the particle. A modification has therefore been introduced into this equation by *Hixson* and *Crowell*, where the dissolution rate has been normalized for the decrease in surface area. The equation, known as the Hixson–Crowell cube root law (Equation 21.10), is represented as follows:

$$M^{\frac{1}{3}} = M_0^{\frac{1}{3}} - kt \tag{21.10}$$

where M_0 is the mass of drug particles before dissolution, M is the mass of drug particles undissolved at any time t, and k is the cube root dissolution rate constant (Judefeind and de villers 2009). Although this equation is capable of explaining the contribution of changing surface area, it still does not account for the shape factor of the particles as far as release kinetics are concerned. In fact, the Hixson–Crowell equation assumes a constant shape of the drug particle during dissolution.

The overall process of drug dissolution as described by the *Noyes–Whitney* equation and *Hixson–Crowell* equation mainly focuses on the drug diffusion phenomenon. For more practical considerations, the dissolution process must be considered to essentially consist of two steps: (1) the dissociation of solute molecules from the solid surface, which is termed *solvation* (a first-order process) and (2) the diffusion of solvated drug molecules from the surface to the bulk of the solution. Solvation involves breaking of the intermolecular bonds in the solute, the separation of the molecules of the solvent to provide space for the solute, and interaction between solute and solvent (Figures 21.6 and 21.7).

This two-step dissolution model, consisting of both the solvation and drug molecule diffusion steps, is depicted in Figure 21.6 (modified from Crisp 2007). Here,

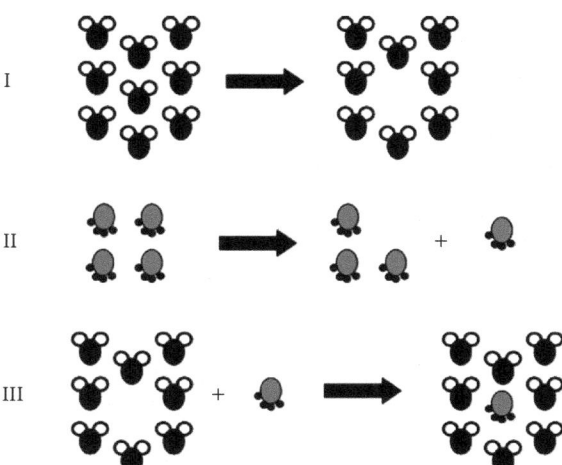

FIGURE 21.6 Schematic representation of drug solubilization and the release mechanism from nanosuspensions. Step I indicates creation of free space among solvent molecules, step II indicates dissociation of solute molecules from drug nanoparticles, and step III represents solvation of drug molecules by solvent molecules.

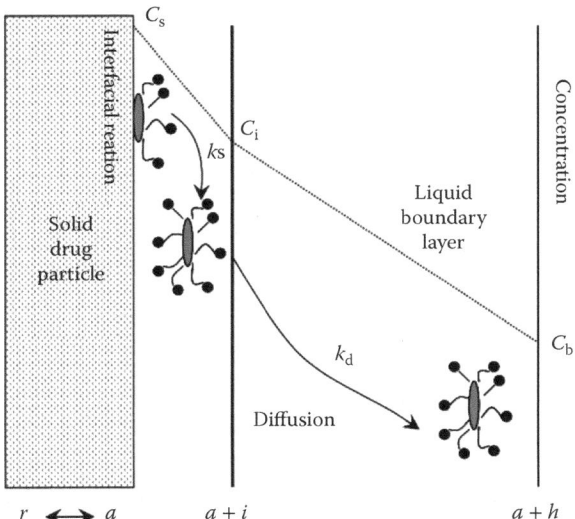

FIGURE 21.7 Schematic representation of the two-step drug dissolution process from nanosuspensions (C_i is the concentration of the drug at the interface $a + i$).

C_b is the bulk concentration, $r = a + h$ is the boundary layer, $r = a$ is the solid–liquid interface, and $r = a + i$ is the distance where interfacial reaction occurs.

The overall dissolution rate is therefore composed of two individual rate constants k_s and k_d representing the solvation and diffusion processes, respectively (Equation 21.11):

$$\frac{1}{k} = \frac{1}{k_s} + \frac{1}{k_d} \tag{21.11}$$

The diffusion rate constant is a function of the diffusion coefficient (D) of the drug and the thickness of boundary layer (h) (Equation 21.12):

$$k_d = \frac{D}{h} \tag{21.12}$$

Generally, the diffusion layer around the drug particles is thick enough to limit the overall rate of diffusion and drug dissolution. This phenomenon is mostly observed in larger particles with thick boundary layers. The reduction in the particle size, however, decreases the diffusion distance due to increasing curvature of the nano-sized or even microsized particles, such that the particle does not appear as a flat surface to the flow of liquid.

The boundary layer thickness is described by the *Prandtl* equation (Equation 21.13) as follows:

$$h_H = k \left\{ \frac{L^{\frac{1}{2}}}{V^{\frac{1}{2}}} \right\} \tag{21.13}$$

where L is the length of the surface in the direction of flow, k is a constant, V is the relative velocity of the flowing liquid against a flat surface, and h_H is the hydrodynamic boundary layer thickness. The high curvature of these nanosized particles results in exposure of only a small surface in the direction of fluid flow, thereby resulting in a small boundary layer thickness. Furthermore, the smaller thickness of the boundary layer remains significantly undisturbed in small-size particles, as reported by Bisrat and Nyström (1988), where they demonstrated no change in the boundary layer thickness at sizes <5 μm, but a significant change in the thickness at larger particle sizes. Thus, considering that the thickness of the boundary layer for small-size particles is significantly reduced, the question arises whether the diffusion of drug molecules into the bulk medium would still be the rate-limiting step. In a study conducted by Shekunov et al. (2006), it was found that the overall dissolution rate is controlled by solvation as the rate-limiting step for particles in size range of 100 nm to 1 μm, which was further confirmed by Crisp et al. (2007).

The shape of particles also has a significant impact on the dissolution rate as demonstrated by Mosharraf and Nyström (1995), where they used Heywood's shape factor to show the effect of particle size and shape on dissolution rate. They found that small spherical particles with a shape factor close to 6 dissolve faster than large irregular particles with a shape factor >6.

Nanosuspensions not only consist of solid drug particles suspended in an aqueous vehicle for ocular drug delivery purposes but also include drug molecules trapped in polymeric carriers. Drug polymer systems formed as nanoparticles demonstrate characteristic drug release kinetics. Depending on the spatial distribution of the drug inside the polymer, these systems are classified as *matrix* or *reservoir* systems, and the drug release kinetics are also described accordingly. Drug release from reservoir systems is controlled by the size of the nanocapsules or nanoreservoir, thickness of the polymer coating, if any, and diffusion properties of the core. The drug is released depending on the partition coefficient of the drug from the core to the membrane and to the bulk medium. The process comprises the phenomena in which the drug permeates through the intact membrane that is related to the diffusion coefficient. The drug molecules simply diffuse out through the preexisting pores in the membrane, while the overall drug concentration in the core acts as the driving force for drug release. Polymeric systems exhibiting such release characteristics result in either a *lag time* in drug release or a *burst effect* that can ultimately lead to the failure of the delivery system to deliver the required dose. The former effect is observed due to the time it takes for the drug to diffuse to the surface, while the latter effect can be explained by the adsorbed drug at the surface of the nanoreservoir during the preparation process. The failure of such polymeric systems might result in dose dumping issues, which in the case of poorly soluble drugs might lead to crystallization.

Drug release from reservoir to nanoreservoir is mainly described by Fick's first law (Equation 21.14):

$$\frac{\mathrm{d}M}{\mathrm{d}t} = \frac{D.K.A.C_\mathrm{d}}{h} \tag{21.14}$$

where D is the diffusion coefficient of drug, A is the surface area, K is the partition coefficient of drug, C_d is the drug concentration in the core of nanocapsules, and h is the thickness of the capsule shell. Marchal-Heussler et al. (1993) made carteolol-loaded poly(epsilon-caprolactone) nanocapsules where the drug was entrapped in an oily core inside the nanocapsules. Results indicated that the drug was readily available to the eye as it was predispersed in an oily core. Furthermore, the cardiovascular side effects of the drug were minimized due to reduced undesired noncorneal absorption.

The release of a drug from a nanoreservoir can be complicating, but can be adequately explained by a number of kinetic release models. Such carriers can be likened to a drug homogeneously dispersed throughout the reservoir matrix. The drug dissolves in the polymer matrix and diffuses out from the system, resulting in a constant increase (see the depletion zone in Figure 21.8) in the distance for diffusion. This phenomenon can be represented by the Higuchi equation (Equation 21.15) as follows:

$$F = k_\mathrm{H}\sqrt{t} \tag{21.15}$$

where F represents the cumulative drug release at time t, K_h is the Higuchi constant, and t is the time in hours. This equation explains the diffusion component of drug release alone such that the plot of the square root of time versus the cumulative drug release yields a straight line. Although a number of mathematical expressions

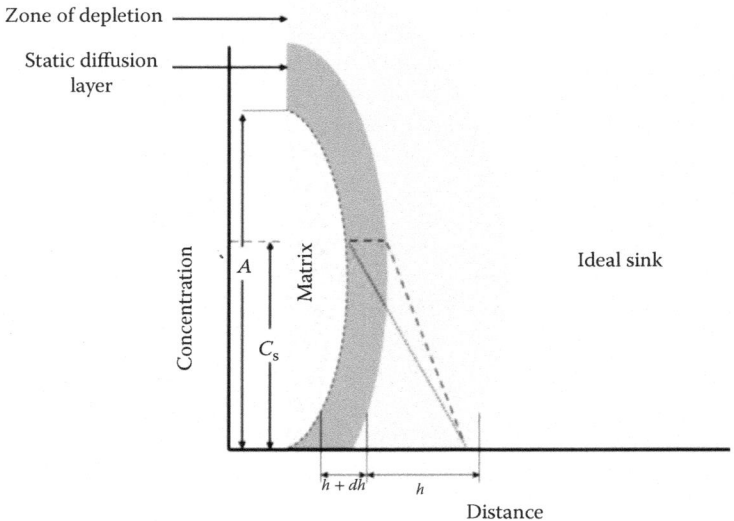

FIGURE 21.8   Release kinetics of drugs from a nanoreservoir system.

have been employed to decipher the release characteristics of these systems, no single equation or expression can be considered to be optimum. A more in-depth review of the kinetic analysis of drug release from nanoparticles and their impact on drug disposition have been extensively discussed in a review (Barzegar-Jalali et al. 2008).

21.6 PREPARATION METHODS

Nanosuspensions can be prepared by using one of two formulation approaches: (1) nanoprecipitation of dissolved solute by harnessing the physicochemical characteristics of the solute in its molecular state and (2) comminution of larger micron-sized particles into nanosized particles. In each of the methods employed in manufacturing these systems, new particles are generated, thereby forming a new surface area. This process is an energy-expensive process, uses a free energy, ΔG, and is directly proportional to the new surface area generated, ΔA, and the interfacial tension $\gamma_{s/l}$ (Rabinow 2004):

$$\Delta G = \Delta A \cdot \gamma_{s/l} \tag{21.16}$$

Thus, to decrease the free energy of the newly formed system, a reduction in the increased surface area occurs. Such a reduction in surface area results in solvation of the incipient crystalline nuclei or by agglomeration of smaller particles. Agglomeration of particles can be manipulated by reducing the $\gamma_{s/l}$ or in other words reducing the free energy of the system. Surface-active agents or surfactants have been employed to achieve this reduction of the free energy and have been shown to confer protection only during the crystal formation process. Both ionic and nonionic surfactants have been employed to stabilize these particles. The ionic surfactants prevent agglomeration by altering the electrostatic interactions, while the nonionic surfactants prevent agglomeration by steric effects. In such systems, interactions between the particles are minimized such that particles in proximity fail to agglomerate due to a high energy barrier formed by the surfactant at the solid–liquid interface (Figure 21.9). When nonionic surfactants are employed to stabilize these particles, the hydrophobic part of the surfactant molecule is oriented toward the solid surface and the hydrophilic tail is projected toward the aqueous environment. This orientation is responsible for a loss of entropy due to interactions between these polymer coatings at the solid–liquid interfaces of the suspended particles and is not considered favorable. On the other hand, when such particles approach each other, they tend to repel and prevent agglomeration. The presence of the surfactant at the solid–liquid interface additionally inhibits crystal growth and also has been shown to reduce particle size.

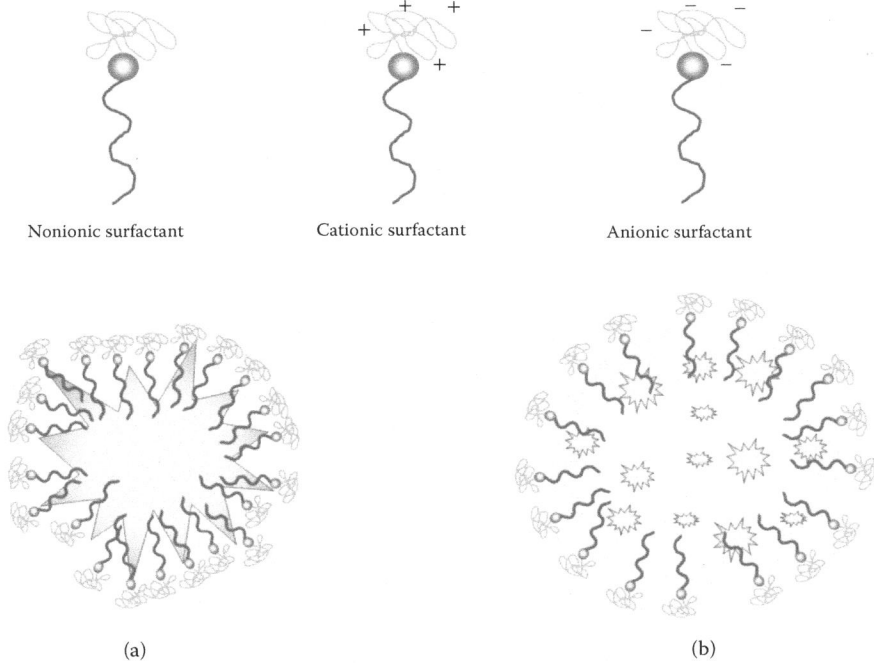

Nonionic surfactant   Cationic surfactant   Anionic surfactant

(a)   (b)

FIGURE 21.9 Schematic representation of nanosuspension systems. (a) The figure represents a pure drug nanosuspension system wherein the surfactant is associated at the solid–liquid interface of the dispersion system. (b) The figure represents a nanoreservoir system wherein the surfactant is associated at the solid–liquid interface of the dispersed particle. The surfactant can be uncharged (nonionic) or can be charged (cationic-anionic) or a combination of both can be utilized to impart suspension stability.

Although particle agglomeration can be prevented by steric interaction between the surfactants alone, it is prudent to employ a combination of both steric and electrostatic interactions. This is attainable by employing a combination of both ionic and anionic surfactants that are complementary to each other. As a rule of thumb, dispersed particles in the colloidal range are predominantly stabilized by surface charges on the particles. In other words, the presence of surface potential gives rise to a charge, positive or negative, that depends on the nature of the particulate, and this surface charge is responsible for the repulsive forces between the particles. Thus, it is prudent to harness the advantages of both steric and ionic contributions by employing nonionic surfactants in combination with ionic surfactants. The selection of a surfactant additionally depends on its ability to interact with the core of the crystalline drug in the nanosuspension or the characteristics of the polymeric carrier in which the drug is embedded. This interaction is more pronounced in the nanoreservoir-type systems in which the matrix is composed of polymeric substrates and can be formulated either as a hydrophobic core or a hydrophilic core.

Nanosuspensions can be manufactured by employing a number of processes. These processes when evaluated collectively can be broadly classified into two classes: (1) bottom-up process and (2) top-down process. In the first case, particles are generated by controlling crystal growth of the drug substances or drug polymeric complexes, whereas in the second process larger particles are broken down into smaller particles by a comminutive or attrition process (Figure 21.10). In both instances, additional excipients are introduced to stabilize the formed particles by preventing agglomeration. This attempt to stabilize agglomeration for nanosuspensions is considerably different from approaches employed to formulate oral suspensions, such that weak attractive forces are harnessed to form floccules that help in redispersion of the sediment upon agitation.

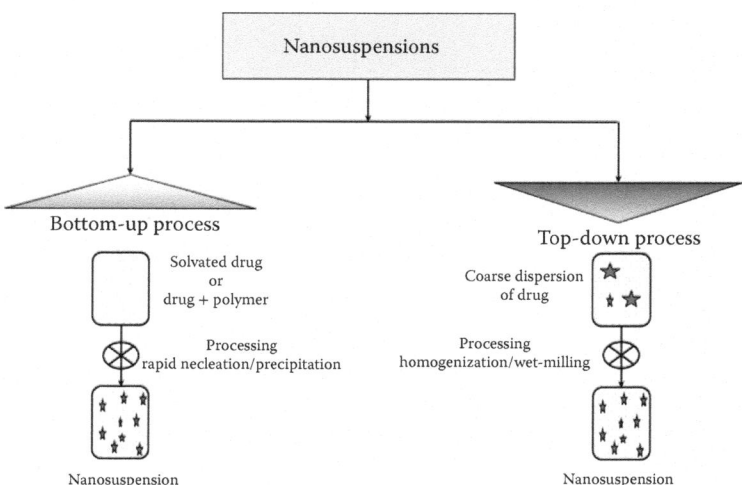

FIGURE 21.10 Concepts employed for manufacturing of nanosuspensions.

21.6.1 PRECIPITATION

Precipitation of solvated drug and controlling the growth of crystals is an example of a bottom-up process. The size of the crystals in such a manufacturing technique is controlled by altering the precipitation conditions. To control this step, it is very important to understand the process of drug crystal formation from a solvated state. The process occurs in two overlapping stages: (a) formation of a drug crystal nuclei and (b) progression of crystal growth. Both the processes are dependent on the temperature and the supersaturated state of the solvated drug. Nucleation of the solvated drug can be selectively achieved by controlling the temperature alone. This is possible because the energy required for crystal nuclei formation may be lower than that required for crystal growth. On the other hand, rapid nucleation can be achieved by the addition of a solution of the drug in a water-soluble solvent into a polar environment under vigorous agitation. This initiates a rapid dilution of the solvent leads to higher supersaturation environment that ultimately causes spontaneous nucleation. However, the dilution of the solvent also results in a reduction of supersaturation conditions near the nucleating crystals and thus retards crystal growth. The overall concept of this processing technique is to form a stable nanosuspension with the smallest particle size attained by rapid nucleation but slower crystal growth rate. Another technique is a modification of this bottom-up technique. Here, a drug dissolved in a volatile organic solvent is sprayed into a hot aqueous dispersed phase. The high temperature results in the evaporation of the solvent, leading to drug precipitation. The process can be carried out with a surfactant dissolved in the hot aqueous phase, resulting in the deposition of the surfactant at the solid–liquid interface (Sarkari et al. 2002).

A similar concept is employed for the manufacture of novel colloidal systems such as nanoreservoir systems. The process is termed *nanoprecipitation* and represents a simple instantaneous process that results in the formation of a colloidal nanosuspension. The nanoprecipitation technique (or solvent displacement method) for nanoparticle manufacture was first developed and patented by Fessi and coworkers (Alvarez-Roman et al. 2004; Galindo-Rodriguez et al. 2004). In this process, both the polymer and the drug are solvated in a nonpolar solvent and the nanosuspension is obtained by rapid mixing of this phase with a second solvent in which the former solvent is soluble. As the polymer-containing solvent diffuses into the dispersion media, the polymer precipitates, leading to immediate drug entrapment. The method has been shown to produce particles in the range of 100–300 nm that exhibit unimodal distributions. The technique has been employed for encapsulating a number of drugs into various biodegradable polymers such as poly (dl-lactide-coglycolide) (PLGA), cellulose derivatives, or polycaprolactones. The method has been shown to be gentle enough to incorporate proteins without any adverse effects on the structure of the macromolecule. Although surfactants can be employed to stabilize these systems, it is not necessary to have one at the solid–liquid interface.

21.6.2 QUASI-EMULSION SOLVENT DIFFUSION

The quasi-emulsion solvent diffusion (QESD) method is another technique that has been employed in a number of instances to yield nanosuspensions. This technique is generally applied to the nanoreservoir-type nanosupension systems. Briefly, in this technique, a solution of the drug and the polymer is made in a solvent (prefer-ably organic) and is introduced under rigorous agitation into a low-polarity solvent (aqueous). The aqueous phase may contain surfactants to stabilize the system. The difference in the polarity of the two liquids results in the formation of an unstable biphasic system (quasi-emulsion). The diffusion of the former solvent into the aque-ous dispersed phase results in the precipitation of the polymer that entraps the drug, yielding a nanosuspension. The choice of solvents used in this system is a crucial factor as the properties of the resulting nanosuspension are significantly affected by the nature of the solvents. In practice, the solvent selected for solvating the drug should also have acceptable solubility for the polymer to give a homogeneous phase. The density of the solvents should not be considerably different, while they can exhibit miscibility in each other. An excellent example of this concept is reported by Pignatello et al. (2006), wherein the researchers have successfully incorporated the drug cloricromene in Eudragit®-containing nanoparticles for ocular delivery.

21.6.3 INOTROPIC GELATION AND COACERVATION

Ionotropic gelation is a novel approach that is based on the ability of polyelectrolytes to cross-link in the presence of counterions to form hydrogels. This technique is a method that has been employed to manufacture nanosuspensions containing hydro-gel systems. The core concept behind this technique is the cross-linking of poly-electrolytic polymers with polyvalent cations to yield a matrix system in which the drug can be incorporated. These systems can not only help in increasing the load-ing of drugs into the polymeric core but can additionally impart controlled release characteristics to the formulation. Natural polyelectrolytes such as alginates, gel-lan gum, chitosan, and cellulose contain certain anions on their chemical structure. These anions form a meshwork structure by combining with the polyvalent cations and induce gelation by binding mainly to the anion blocks. The hydrogel beads are produced by dropping a drug-loaded polymeric solution into the aqueous solution of polyvalent cations. The cations diffuse into the drug-loaded polymeric drops, form-ing a three-dimensional lattice of ionically cross-linked moiety. A number of prom-ising nanosuspension ocular dosage forms have been prepared by this technique and are reported in Table 21.1. Coacervation is another process during which a homog-enous solution of charged macromolecules undergoes liquid–liquid phase separation producing a phase-separated system; one phase comprises polymer-rich particles.

21.6.4 SUPERCRITICAL FLUID TECHNOLOGY

Supercritical or compressed fluids have been utilized as an alternative way to pre-pare biodegradable nanoparticles (Thote and Gupta 2005). This new technique is advantageous over the other techniques as it does not use the toxic organic solvents that are associated with other conventional methods. Two techniques are most com-monly used for preparing nanoparticles using this concept: (1) supercritical antisol-vent (SAS) technique and (2) rapid expansion of critical solution (RESS) technique. In the SAS technique, solutes are dissolved in methanol, which in turn is completely miscible with supercritical fluids. The extraction of methanol by the supercritical

TABLE 21.1

Compilation of Various Drugs and Preparation Methods Used for Formulation of Ocular Drug Delivery Systems

Drug	Ocular Dosage Form	Preparation Method	References
Hydrocortisone	Drug nanosuspension	Microfluidic nanoprecipitation/milling	Ali et al. (2011)
Ibuprofen	Eudragit RS100 nanosuspension	Quasi-emulsion solvent diffusion	Pignatello et al. (2002), Kawashima et al. (1989)
Flurbiprofen	Eudragit RS100R and RL100R nanosuspension	Quasi-emulsion solvent diffusion	Rosario Pignatelloa (2002)
Acyclovir	Eudragit RS100 nanosuspension	Quasi-emulsion solvent evaporation	Dandagi et al. (2009)
Hydrocortisone, dexamethasone	Drug nanosuspension	High-pressure homogenization	Kassem et al. (2007)
5-Fluorouracil	Chitosan nanosuspension	Ionotropic gelation	Nagarwal et al. (2011)
Acyclovir	Chitosan nanosuspension	Ionotropic gelation	Rajendran et al. (2010)
Cyclosporin A	Chitosan nanosuspension	Ionotropic gelation	de Campos et al. (2004)
Sparfloxacin	PLGA nanosuspension	Nanoprecipitation	Gupta et al. (2010)
Flurbiprofen	PLGA nanosuspension	Nanoprecipitation	Vega et al. (2006)
Amphotericin B	Eudragit RL100 nanosuspension	Nanoprecipitation	Das et al. (2010)
Corticosteroids	Drug nanosuspension	High-pressure homogenization	Kassem et al. (2007)
Cyclosporin A	PLGA and Eudragit RL nanosuspension	Emulsion and solvent evaporation	Aksungur et al. (2011)
Mycophenolate mofetil	Drug nanosuspension and chitosan-modified drug nanosuspension	High-pressure homogenization	Wu et al. (2010)
Gatifloxacin/prednisolone bitherapy	Eudragit RS100 and RL100 nanosuspension	Emulsion and solvent diffusion	Ibrahim et al. (2010)

Note: The list contains drug nanosuspension systems as well as nanocarrier systems.

fluids leads to an instantaneous precipitation of the nanoparticles (Thote and Gupta 2005). Dexamethasone phosphate nanoparticles were prepared by this method. In the RESS method, solutes are dissolved in the supercritical fluid and the solution is expanded through a small nozzle into a region of lower pressure. The solutes eventually precipitate as nanoparticles.

21.6.5 HIGH-PRESSURE HOMOGENIZATION

Another approach used in the manufacturing of nanosuspensions is high-pressure homogenization. This technique involves the attrition of a coarse suspension under pressure through a narrow aperture. This is an example of a top-down process. This process is dependent on Bernoulli's law, wherein the passage of a suspension through a narrow aperture leads to an increase in the flow velocity. This increase in velocity is compensated by a reduction in the static pressure, resulting in the formation of bubbles of water vapor that cavitate and collapse as they leave the aperture. The increase in energy due to cavitation is responsible for breaking the larger particles of the coarse suspension. This process is dependent on the tensile strength of crystals, which can be significantly different for different drugs. This is especially true for pure crystals; however, pure crystals are seldom used in manufacturing processes. An advantage of this process is that it can be used in conjunction with rapid nucleation (a bottom-up process). The rapid nucleation process forms crystals with lower tensile strength and deformities that can be further broken down into smaller particles.

21.6.6 WET-MILLING

Another manufacturing technique for preparing nanosuspensions is wet-milling, which is also an example of a top-down process. In this process, the active agent, in the presence of a surfactant, is comminuted by milling media. Particle size here is determined by stress intensity, a function of the kinetic energy of the grinding beads, and the number of contact points. The number of contact points can be increased by utilizing smaller grinding media. A drive shaft, attached to rotating disks, provides the energy to a charge/load of cross-linked polystyrene beads to comminute the drug crystals by a compression–shear action.

21.7 CHARACTERIZATION

Characterization of nanoparticulate systems, such as nanosuspensions, is a field of its own. The science behind the elucidation of key physicochemical characteristics of these systems is a broad field of knowledge and requires a thorough appreciation of the physical and chemical attributes of these systems. Section 21.7.1 will not try to elaborate on all the working principles of the characterization technique but will simply touch on the scope and application of the commonly used methods.

Nanosuspensions are colloidal systems, and hence the properties of such dispersed systems are closer to colloidal dispersion. Properties such as particle size and distribution as well as particle charge are of utmost importance. Section 21.7.1 is devoted to particle size characterization and surface charge determination of these systems, as inadequacy in these areas generally results in an unstable product.

21.7.1 PARTICLE SIZE

21.7.1.1 Laser Light Diffraction

Laser light scattering is the alteration of the direction and intensity of a laser light beam that strikes an object. This is caused by the combined effects of reflection, refraction, and diffraction. Two different laser light scattering techniques are used to determine the PSD of solid liquid dispersions: static laser light scattering and dynamic light scattering (DLS).

21.7.1.1.1 Static Laser Light Diffraction

The SLLD technique is also known as laser diffraction, Rayleigh scattering, low-angle laser light scattering (LALLS), and Fraunhofer diffraction. The basic hypothesis of laser diffraction, when determining particle size, is based on the fact that a particle passing through a laser beam will scatter light at an angle based on its size and the refractive index of the material under study. The intensity of the scattered light can be affected by particle size, refractive index, number of particles, angle of observation, and the wavelength of light. The higher the refractive index of the particles relative to the medium, the lighter is the scattering. Scattering intensity will be high for larger particles at narrow angles, whereas intensity will be low and isotropic for smaller particles at wider angles. The instrument reports the volume fraction of a given size range rather than the number of particles since the scattering intensity is proportional to the cross-section of the particles. Two different optical models are used to calculate the PSD for laser light scattering experiments: the Fraunhofer approximation (Hecht 1987) and Mie theory (van de Hulst 1981). The Fraunhofer approximation assumes the particles are opaque, two-dimensional, and large circular discs and describes light scattering from the edges of an object. The Mie theory is a powerful tool to calculate size distribution based on refractive index differences between the particles and the dispersion medium. The Mie theory holds true for spherical, isotropic particles illuminated by monochromatic light.

The advantages for this technique are its flexibility to different sample types, wide size range, rapid processing, and high precision. There are few limitations with these instruments. At a given particle concentration, multiple scattering can affect the results when using a single scattering model. Additionally, nonspherical particles show a widening of PSD due to differences in the cross-sections of a single particle.

21.7.1.1.2 Dynamic Light Scattering

DLS, also known as photon correlation spectroscopy (PCS), quasi-elastic light scattering (Pecora 1972), and diffusing wave spectroscopy (Pine et al. 1988), is a technique used to determine the particle size and size distribution. This technique can also measure the polydispersity index (PDI) of various types of samples including nanoparticles, colloids, gels, emulsions, pigments, liquid crystals, DNA, polymers, and proteins (Pecora 1972). It can measure submicron particles and shows sensitivity between the ranges of 0.6 nm and 6 μm.

The DLS technique gives particle size as a hydrodynamic diameter, while PSD is based on scattering intensities. The scattered light undergoes either construction or destructive interference by the surrounding particles. This fluctuation in the intensity of scattered light at a given scattering angle (usually 90°) is due to particle movement rising from the Brownian motion. Assuming particles are spherical without interparticulate interactions, the *Stokes–Einstein equation* is used to convert the diffusion coefficients into particle sizes. The lower size limit of detection with this instrument depends on differences in the refractive index between the particles and the medium and on experimental noise (arising from electronic noise as well as temperature fluctuations and environmental disturbances). Medium viscosity and the density of the materials determine the upper size limit of detection.

The semiclassical light scattering theory is based on assumptions that the particles are in random Brownian motion and that all the particles are spherical in diameter. Assuming the spherical shape of the particle with no interparticulate interaction, the hydrodynamic radio of a particle is calculated using the Stokes–Einstein equation based on the translational diffusion coefficient. The equation is given as follows:

$$D = \frac{kT}{3\pi\eta d} \tag{21.17}$$

where D is the diffusion coefficient, k is the Boltzmann constant, T is the temperature in degrees Kelvin, η is the viscosity of the medium, and d is the diameter of the particle.

In PCS, the temporal fluctuation in the intensity of scattered laser light (in microseconds or milliseconds) by the particles in suspension is processed with the autocorrelation function. This provides information about the PSD, particle motion in the medium, and the dynamics of the dispersed particles. This autocorrelation function corresponds to the decay constants and diffusion coefficients of the particles. DLS provides three types of diameters of samples: Z-average diameter, number mean diameter, and volume mean diameter. Z-average diameter is the mean diameter based on laser scattering intensity. It is obtained from an exponential fit. The PDI can also be measured from DLS instruments. PDI is an index of width or spread or variation within the PSD. Monodisperse samples have a lower PDI value, whereas a higher value of PDI indicates a wider PSD and the polydisperse nature of the sample. PDI can be calculated by the following equation:

$$PDI = \frac{\Delta d}{d_{avg}} \tag{21.18}$$

where Δd is the standard deviation of particle size and d_{avg} is the average particle size. The usual range of PDI values is 0–0.05 (monodisperse standard), 0.05–0.08

(nearly monodisperse), 0.08–0.7 (mid-range polydispersity), and >0.7 (very polydisperse).

21.7.1.2 Surface Charge

Laser Doppler electrophoresis (LDE) combines the principles of laser Doppler anemometry (LDA) and electrophoresis (Desai and Armstrong 2003). LDA is the technique employed to measure the speed of moving particles with respect to the Doppler shift of scattered light by particles in motion. It is the most widely used technique to determine electrophoretic mobility, particle surface charge, or zeta potential for colloidal dispersions.

Electrophoretic mobility is measured in a cell with electrodes at either end to which a potential is applied. While the particles are moving to the electrode with opposite charge, their velocity is measured by laser Doppler velocimetry. The light scattered by moving particles is measured at a certain angle to the incident beam, and the resulting fluctuating intensity signal is measured with time (Figure 21.11).

The co-relation between the zeta potential on a colloidal particle and its stability has been extensively studied. Zeta potential values between ±0 and ±5 mV are indicative of rapid coagulation or flocculation; values between ±10 and ±30 mV indicate stability issues; and values between ±30 and ±40 mV indicate moderate stability of the colloidal dispersion. Charges between ±40 and ±60 mV indicate good stability and charges over ±60 mV indicate excellent stability characteristics. It is important to note that although these values represent the colloidal stability of the formulation, they may not be representative of its *in vivo* stability.

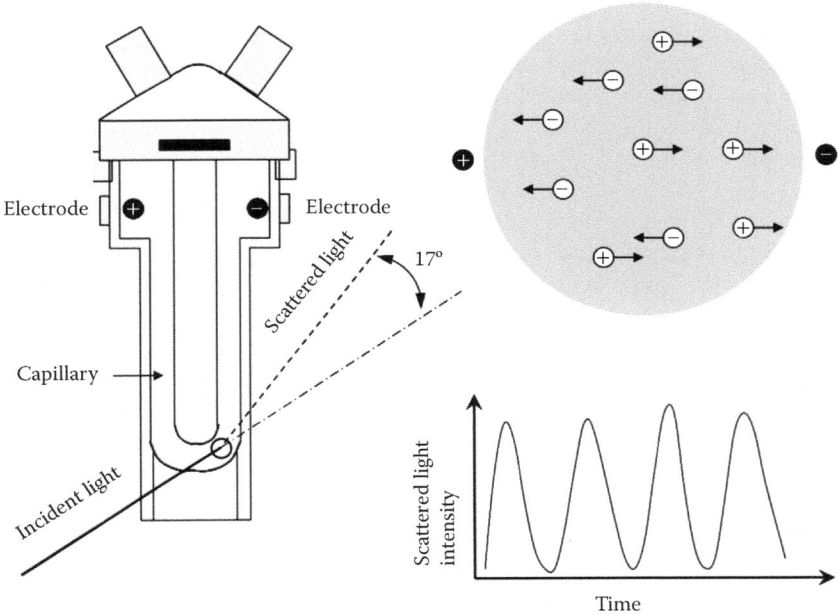

FIGURE 21.11 Schematics of electrophoretic mobility measurement.

21.7.1.3 Crystallinity and Solid State Characterization

True drug nanosuspensions can be compared to nanocrystals that may or may not have different crystal morphology from the parent drug. It is a known fact that different crystal polymorphs can lead to different physicochemical properties and can have a profound effect on solubility. The assessment of the crystalline state and particle morphology together helps in understanding the polymorphic or morphological changes that a drug might undergo when subjected to manufacturing. Additionally, when nanosuspensions are prepared, drug particles in an amorphous state are likely to be generated. Hence, it is essential to investigate the extent of amorphous drug nanoparticles generated during the production of nanosuspensions. The changes in the physical state of the drug particles as well as the extent of the amorphous fraction can be determined by x-ray diffraction analysis (Muller and Bohm 1998) and can be corroborated by differential scanning calorimetry and scanning electron microscopy is preferred.

21.8 NANOCARRIER-BASED OPHTHALMIC SUSPENSION SYSTEMS

A thorough literature review indicates a number of novel ophthalmic suspension systems. Most of the research being conducted based on nanoparticluate ocular drug delivery systems is directed to nanocarrier systems or nanoreservoir systems (Bhattacharjee et al. 2010). These systems include polymer drug-loaded nanosuspensions such as chitosan-loaded nanosuspension, PLGA (polylactic acid poly glycolic acid)-loaded drug suspensions, and Eudragit drug-loaded nanosuspensions, among others. A more synoptic view of these applications is indicated in Table 21.1.

Use of biocompatible as well as biodegradable polymer to incorporate drugs into nanocarrier systems has been shown to be beneficial in attaining better therapeutic outcomes in the treatment of ocular diseases in experimental settings. In a recent report, researchers have loaded ibuprofen into Eudragit RS 100, as a model nonsteroidal anti-inflammatory drug for the treatment used to contrast the miosis induced by surgical traumas, such as cataract extraction. The nanosuspensions were prepared by a modification of the QESD technique using variable formulation parameters and yielded particles with a mean size around 100 nm and a positive charge (zeta potential of +40 to +60 mV), making this method a potential candidate for ophthalmic applications. The preparation was shown to be stable at 4°C for 2 years and was evaluated in an ocular trauma model on the rabbit eye. Inhibition of the miotic response to the surgical trauma was achieved, comparable to a control aqueous eyedrop formulation, even though a lower concentration of free drug in the conjunctival sac was reached by the nanoparticle system. Drug levels in the aqueous humor were additionally higher after application of the nanosuspensions, and the drug-loaded nanosuspensions did not show toxicity in ocular tissues (Pignatello et al. 2002).

In another study, Gupta et al. have developed and evaluated a new colloidal system, composed of poly (dl-lactide-co-glycolide) (PLGA)-based nanoparticles and loaded sparfloxacin as an ophthalmic nanosuspension, to improve precorneal residence time and ocular penetration. Nanosuspensions were prepared by the

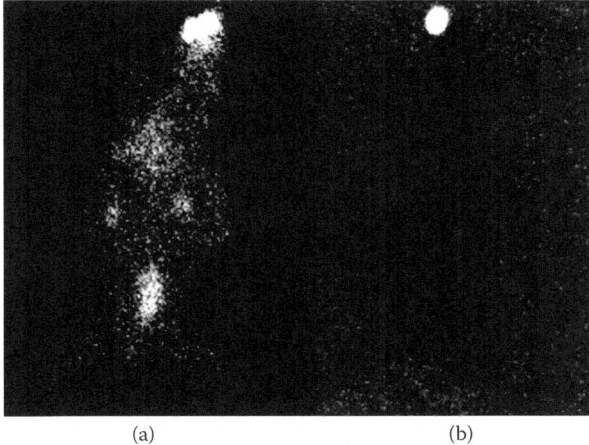

(a)   (b)

FIGURE 21.12   **(See color insert.)** Gamma scintigraphic static whole-body images after 6 hours of administration: (a) marketed formulation and (b) sparfloxacin-PLGA nanosuspension.

nanoprecipitation technique and characterized for various properties such as particle size, zeta potential, *in vitro* drug release, statistical model fitting, and stability. Precorneal residence time was studied in albino rabbits by gamma scintigraphy after radiolabeling of sparfloxacin by ^{99m}Tc, and ocular tolerance of the developed nanosuspension was also studied by the Hen Egg Test-Chorioallantoic Membrane (HET-CAM) method. The developed nanosuspension showed a mean particle size in the range of 180–190 nm, suitable for ophthalmic application with a zeta potential of –22 mV. *In vitro* release from the developed nanosuspension showed an extended-release profile of sparfloxacin. Additionally, acquired gamma camera images showed good retention over the entire precorneal area for the developed nanosuspension compared with that of a marketed formulation (Figure 21.12). The developed nanosuspension was cleared at a very slow rate and remained at the corneal surface for a longer duration, and no radioactivity was observed in the systemic circulation compared to significant radioactivity recorded in kidney and bladder after 6 hours of ocular administration of the solution formulation. HET-CAM assay with 0 score in 8 hours indicated the nonirritant property of the developed nanosuspension.

21.9   SUMMARY

The primary goal to formulate drugs into drug nanosuspension systems is to alleviate the limitation of poor solubility shown by some drugs. However, continued research has yielded a lot of understanding of the fundamental concepts with respect to their physicochemical characteristics, their therapeutic advantages, and also their limitations. This has driven the field of knowledge into new areas of not only drug nanosuspensions but also drug-polymer nanosuspensions.

A number of drugs that were once considered to be not suitable for development for ocular drug delivery can now be evaluated by employing some of the concepts of nanosuspension drug delivery systems. These systems have additionally exhibited superior drug product characteristics with respect to therapeutic as well as product stability parameters. Although scale-up and commercialization of these products is a challenge, newer approaches and new understanding of the drug delivery system have opened novel avenues for manufacturing these systems. The system that was once considered a delivery strategy in its infancy can now be viewed as an approach to delivery of drugs in a therapeutically effective alternative.

REFERENCES

Aksungur, P., Demirbilek, M., Denkbas, E. B., Vandervoort, J., Ludwig, A. & Unlu, N. 2011. Development and characterization of cyclosporine A loaded nanoparticles for ocular drug delivery: Cellular toxicity, uptake, and kinetic studies. *J Control Release*, 151, 286–94.

Ali, H. S., York, P., Ali, A. M. & Blagden, N. 2011. Hydrocortisone nanosuspensions for ophthalmic delivery: A comparative study between microfluidic nanoprecipitation and wet milling. *J Control Release*, 149, 175–81.

Alvarez-Roman, R., Naik, A., Kalia, Y. N., Guy, R. H. & Fessi, H. 2004. Enhancement of topical delivery from biodegradable nanoparticles. *Pharm Res*, 21, 1818–25.

Barzegar-Jalali, M., Adibkia, K., Valizadeh, H., Shadbad, M. R., Nokhodchi, A., Omidi, Y., Mohammadi, G., Nezhadi, S. H. & Hasan, M. 2008. Kinetic analysis of drug release from nanoparticles. *J Pharm Pharm Sci*, 11, 167–77.

Bhattacharjee, H., Balabathula, P. & Wood, G. C. 2010. Targeted nanoparticulate drug-delivery systems for treatment of solid tumors: A review. *Ther Deliv*, 1, 713–35.

Bisrat, M. & Nyström, C. 1988. Physicochemical aspects of drug release. VIII. The relation between particle size and surface specific dissolution rate in agitated suspensions. *Int J Pharm*, 47, 223–31.

Chrai, S. S., Makoid, M. C., Eriksen, S. P. & Robinson, J. R. 1974. Drop size and initial dosing frequency problems of topically applied ophthalmic drugs. *J Pharm Sci*, 63, 333–8.

Crisp, M. T., Tucker, C. J., Rogers, T. L., Williams, R. O. 3rd & Johnston, K. P. 2007. Turbidimetric measurement and prediction of dissolution rates of poorly soluble drug nanocrystals. *J Control Release*, 117, 351–9.

Dandagi, P., Kerur, S., Mastiholimath, V., Gadad, A. P. & Kulkarni, A. 2009. Polymeric ocular nanosuspension for controlled release of acyclovir: *In vitro* release and ocular distribution. *Ira J Pharm Res*, 8, 79–86.

Das, S., Suresh, P. K. & Desmukh, R. 2010. Design of Eudragit RL 100 nanoparticles by nanoprecipitation method for ocular drug delivery. *Nanomedicine*, 6, 318–23.

De Campos, A. M., Diebold, Y., Carvalho, E. L., Sanchez, A. & Alonso, M. J. 2004. Chitosan nanoparticles as new ocular drug delivery systems: *In vitro* stability, *in vivo* fate, and cellular toxicity. *Pharm Res*, 21, 803–10.

Desai, M. J. & Armstrong, D. W. 2003. Separation, identification, and characterization of microorganisms by capillary electrophoresis. *Microbiol Mol Biol Rev*, 67, 38–51, table of contents.

Galindo-Rodriguez, S., Allemann, E., Fessi, H. & Doelker, E. 2004. Physicochemical parameters associated with nanoparticle formation in the salting-out, emulsification-diffusion, and nanoprecipitation methods. *Pharm Res*, 21, 1428–39.

Ghate, D. & Edelhauser, H. F. 2006. Ocular drug delivery. *Expert Opin Drug Deliv*, 3, 275–87.

Gupta, H., Aqil, M., Khar, R. K., Ali, A., Bhatnagar, A. & Mittal, G. 2010. Sparfloxacin-loaded PLGA nanoparticles for sustained ocular drug delivery. *Nanomedicine*, 6, 324–33.

Hecht, E. 1987. *Optics.* Boston, MA: Addison-Wesley.

Ibrahim, H. K., El-Leithy, I. S. & Makky, A. A. 2010. Mucoadhesive nanoparticles as carrier systems for prolonged ocular delivery of gatifloxacin/prednisolone bitherapy. *Mol Pharm*, 7, 576–85.

Judefeind, A. & de Villers, M. M. 2009. *Drug loading and in vitro release from nanosized drug delivery systems.* New York: Springer.

Kassem, M. A., Abdel Rahman, A. A., Ghorab, M. M., Ahmed, M. B. & Khalil, R. M. 2007. Nanosuspension as an ophthalmic delivery system for certain glucocorticoid drugs. *Int J Pharm*, 1, 126–33.

Kawashima, Y., Niwa, T., Handa, T., Takeuchi, H., Iwamoto, T. & Itoh, K. 1989. Preparation of controlled-release microspheres of ibuprofen with acrylic polymers by a novel quasi-emulsion solvent diffusion method. *J Pharm Sci*, 78, 68–72.

Lipinski, C. A., Lombardo, F., Dominy, B. W. & Feeney, P. J. 2001. Experimental and computational approaches to estimate solubility and permeability in drug discovery and development settings. *Adv Drug Deliv Rev*, 46, 3–26.

Marchal-Heussler, L., Sirbat, D., Hoffman, M. & Maincent, P. 1993. Poly(epsilon-caprolactone) nanocapsules in carteolol ophthalmic delivery. *Pharm Res*, 10, 386–90.

Mikkelson, T. J., Chrai, S. S. & Robinson, J. R. 1973. Altered bioavailability of drugs in the eye due to drug-protein interaction. *J Pharm Sci*, 62, 1648–53.

Mishima, S., Gasset, A., Klyce, S. D. Jr. & Baum, J. L. 1966. Determination of tear volume and tear flow. *Invest Ophthalmol*, 5, 264–76.

Mosharraf, M. & Nyström, C. 1995. The effect of particle size and shape on the surface specific dissolution rate of microsized practically insoluble drugs. *Int J Pharm*, 122, 35–47.

Muller, R. H. & Bohm, B. S. 1998. *Nanosuspension.* Stuttgart: Medpharm Sceintific Publication.

Nagarwal, R. C., Singh, P. N., Kant, S., Maiti, P. & Pandit, J. K. 2011. Chitosan nanoparticles of 5-fluorouracil for ophthalmic delivery: Characterization, *in-vitro* and *in-vivo* study. *Chem Pharm Bull (Tokyo)*, 59, 272–8.

Patravale, V. B. & Ambarkhane, A. V. 2003. Study of solid lipid nanoparticles with respect to particle size distribution and drug loading. *Pharmazie*, 58, 392–5.

Pecora, R. 1972. Quasi-elastic light scattering from macromolecules. *Annu Rev Biophys Bioeng*, 1, 257–76.

Pignatello, R., Bucolo, C., Ferrara, P., Maltese, A., Puleo, A. & Puglisi, G. 2002. Eudragit RS100 nanosuspensions for the ophthalmic controlled delivery of ibuprofen. *Eur J Pharm Sci*, 16, 53–61.

Pignatello, R., Ricupero, N., Bucolo, C., Maugeri, F., Maltese, A. & Puglisi, G. 2006. Preparation and characterization of eudragit retard nanosuspensions for the ocular delivery of cloricromene. *AAPS PharmSciTech*, 7, E27.

Pine, D. J., Weitz, D. A., Chaikin, P. M. & Herbolzheimer, E. 1988. Diffusing wave spectroscopy. *Phys Rev Lett*, 60, 1134–7.

Rabinow, B. E. 2004. Nanosuspensions in drug delivery. *Nat Rev Drug Discov*, 3, 785–96.

Rajendran, N. N., Natrajan, R., Kumar, R. S. & Selvaraj, S. 2010. Acyclovir-loaded chitosan nanoparticles for ocular delivery. *Asian J Pharm*, 4, 220–4.

Sarkari, M., Brown, J., Chen, X., Swinnea, S., Williams, R. O. 3rd & Johnston, K. P. 2002. Enhanced drug dissolution using evaporative precipitation into aqueous solution. *Int J Pharm*, 243, 17–31.

Schoenwald, R. D. & Stewart, P. 1980. Effect of particle size on ophthalmic bioavailability of dexamethasone suspensions in rabbits. *J Pharm Sci*, 69, 391–4.

Shekunov, B. Y., Chattopadhyay, P., Seitzinger, J. & Huff, R. 2006. Nanoparticles of poorly water-soluble drugs prepared by supercritical fluid extraction of emulsions. *Pharm Res*, 23, 196–204.

Shell, J. W. 1982. Pharmacokinetics of topically applied ophthalmic drugs. *Surv Ophthalmol*, 26, 207–18.

Thote, A. J. & Gupta, R. B. 2005. Formation of nanoparticles of a hydrophilic drug using supercritical carbon dioxide and microencapsulation for sustained release. *Nanomedicine*, 1, 85–90.

Van De Hulst, H. C. 1981 *Light scattering by small particles*. New York: John Wiley & Sons.

Vega, E., Egea, M. A., Valls, O., Espina, M. & Garcia, M. L. 2006. Flurbiprofen loaded biodegradable nanoparticles for ophtalmic administration. *J Pharm Sci*, 95, 2393–405.

Wu, X. G., Xin, M., Yang, L. N. & Shi, W. Y. 2010. The biological characteristics and pharmacodynamics of a mycophenolate mofetil nanosuspension ophthalmic delivery system in rabbits. *J Pharm Sci*, Epub, 2010 Oct 22.

Yuan, J., Zhai, J. J., Chen, J. Q., Ye, C. T. & Zhou, S. Y. 2009. Preparation of 0.05% FK506 suspension eyedrops and its pharmacokinetics after topical ocular administration. *J Ocul Pharmacol Ther*, 25, 345–50.

Index